Second Course in Fundamentals of Mathematics

Allyn and Bacon, Inc.
Boston Rockleigh, N.J. Atlanta Dallas Belmont, Calif.

Second Course in Fundamentals of
MATHEMATICS

Edwin I. Stein

© Copyright 1978, by Allyn and Bacon, Inc.

All rights reserved. No part of the material protected by this copyright notice may be reproduced or utilized in any form or by any means, electronic or mechanical, including photocopying and recording, or by any information storage and retrieval system, without written permission from the copyright owner.

Printed in the United States of America.

Library of Congress Card No. 77-18315

3 4 5 6 7 8 9 86 85 84 83 82 81 80 79

Preface

Second Course in Fundamentals of Mathematics is the second textbook of a comprehensive general mathematics program for junior and senior high schools.

Like the *First Course* of the series, this edition of the *Second Course* contains a wealth of material in all the basic topics of arithmetic, providing computational practice for the development and maintenance of skills. The opening inventory test of arithmetic skills, previously studied, is keyed to related sections of the text where explanatory materials and practice exercises are found.

Second Course in Fundamentals of Mathematics includes a review of number, a comprehensive study of the metric system, informal geometry, and positive and negative numbers. Each of these subject areas has a keyed inventory test so that a student may locate specific deficiencies and may be directed to individualized help. This makes the book adaptable to the specific needs of each individual student.

Second Course in Fundamentals of Mathematics also features introductory algebra, probability, statistics, and an abundance of consumer applications. Enrichment materials such as flow charts and the study of logic are also included.

The algebra sections now include a simplified study of the number plane along with all other introductory basic topics such as the language of algebra, algebraic operations, solutions of equations and inequalities, and graphing on the number line.

The study of probability includes sample spaces, odds, complementary events, mutually exclusive events, and independent events.

The chapter on statistics reviews the bar, line, and circle graphs and develops the concepts of frequency distribution, measures of central tendency, percentiles and quartiles, and measures of dispersion.

The comprehensive chapter on informal geometry contains geometric measurement, constructions, relationships, and facts using both metric and customary units.

The wide range of practical applications includes essential consumer activities such as finding income, take-home pay, income tax, installment buying, annual percentage rate, life and fire insurance, property tax, sales and excise taxes, buying and renting a house, banking, simple and compound interest, borrowing money, financing a car, amortizing a mortgage, stocks and bonds, and other business, industrial, and aviation applications.

The practice material is varied and usually graded by difficulty. Keyed Chapter Reviews and Achievement Tests are provided throughout the textbook. Estimating Answers and Check by Calculator are timely additions to this edition. Important explanatory material and sample problems are emphasized by color screens and boxes.

Second Course in Fundamentals of Mathematics meets the many and varied requirements of today's programs in mathematics for all students.

EDWIN I. STEIN

Contents

1 REVIEW OF NUMBER 1

Introduction 2
Arithmetic Inventory Test 3

DECIMAL SYSTEM OF NUMERATION 5

WHOLE NUMBERS—COMPUTATION 6
1–1 Rounding Whole Numbers 6
1–2 Addition of Whole Numbers 7
1–3 Subtraction of Whole Numbers 10
1–4 Multiplication of Whole Numbers 12
1–5 Division of Whole Numbers 15

RATIONAL NUMBERS 19

CHANGING FORM OF A FRACTION OR OF A MIXED NUMBER 19
1–6 Expressing Fractions in Lowest Terms 20
1–7 Expressing Fractions in Higher Terms 21
1–8 Expressing Improper Fractions as Mixed or Whole Numbers 23
1–9 Expressing Mixed and Whole Numbers as Improper Fractions 23
1–10 Expressing Fractions as Equivalent Fractions With a Common Denominator 25
1–11 Comparing Fractions; Equivalent Fractions 27

COMMON FRACTIONS AND MIXED NUMBERS—COMPUTATION 29
1–12 Addition of Fractions and Mixed Numbers 29
1–13 Subtraction of Fractions and Mixed Numbers 32
1–14 Multiplication of Fractions and Mixed Numbers 34
1–15 Division of Fractions and Mixed Numbers 36

DECIMAL FRACTIONS—COMPUTATION 38
1–16 Rounding Decimals 39
1–17 Comparing Decimals 41
1–18 Addition of Decimals 42
1–19 Subtraction of Decimals 45
1–20 Multiplication of Decimals 47
1–21 Multiplying by Powers of Ten 49
1–22 Division of Decimals 50
1–23 Dividing by Powers of Ten 53
1–24 Shortened Names for Large Whole Numbers 54
1–25 Scientific Notation 55
1–26 Expressing Common Fractions as Decimal Fractions 57
1–27 Expressing Decimal Fractions as Common Fractions 59

PERCENT—COMPUTATION 61
1–28 Changing Percents to Decimal Fractions and Common Fractions 61

1−29	Changing Decimal Fractions and Common Fractions to Percents	63
1−30	Finding a Percent of a Number	65
1−31	Finding What Percent One Number is of Another	68
1−32	Finding the Number When a Percent of it is Known	70
1−33	Percent—Solving by Proportion	72
1−34	Common Fraction, Decimal Fraction, and Percent Number Relationships— Solving by Equation	73

SQUARES; SQUARE ROOTS; IRRATIONAL NUMBERS 76

1−35	Square of a Number; Exponents	76
1−36	Square Root	79

Achievement Test 83
Estimating Answers 85
Check by Calculator 86

NUMBER THEORY 88

1−37	Even and Odd Numbers	88
1−38	Factors	89
1−39	Prime and Composite Numbers	91
1−40	Number Multiples	92
1−41	Tests For Divisibility	93
1−42	Checking by Casting Out Nines	96

OPERATIONS AND PROPERTIES 98

1−43	Binary Operations; Inverse Operations	98
1−44	Properties	99

Chapter Review 104

2 UNITS OF MEASURE AND MEASUREMENT 107

Inventory Test 108

UNITS OF MEASURE 109

Changing to a Smaller Unit of Measure 109
Changing to a Larger Unit of Measure 111

METRIC SYSTEM 112

2−1	Measure of Length—Metric	114
2−2	Measure of Mass or Weight—Metric	118
2−3	Measure of Capacity—Metric	121
2−4	Measure of Area—Metric	123
2−5	Measure of Volume—Metric	125

CUSTOMARY SYSTEM 128

Fundamental Operations with Denominate Numbers 129

2−6	Measure of Length—Customary	130
2−7	Measure of Weight—Customary	131
2−8	Liquid Measure—Customary	132
2−9	Dry Measure—Customary	134
2−10	Measure of Area—Customary	135
2−11	Measure of Volume—Customary	135
2−12	Volume, Capacity, and Weight Relationships—Customary	136
2−13	Measure of Time	137
2−14	Measure of Angles and Arcs	138
2−15	Rates of Speed	139
2−16	Temperature	140
2−17	Twenty-Four Hour Clock	142
2−18	Time Zones	143

MEASUREMENT	144	2–21 Approximate Numbers	149
2–19 Precision and Accuracy in Measurement	144	Estimating Answers	151
2–20 Significant Digits	147	Chapter Review	153
		Achievement Test	155

3 POSITIVE AND NEGATIVE NUMBERS 157

Inventory Test	158	3–10 Addition of Integers; Properties	181
EXTENSION OF THE NUMBER SYSTEM	161	3–11 Addition of Rational Numbers; Properties	186
3–1 Integers	161	3–12 Subtraction on the Number Line	193
3–2 Rational Numbers; Real Numbers	163	3–13 Subtraction of Integers; Properties	194
3–3 Absolute Value	166	3–14 Subtraction of Rational Numbers; Properties	198
3–4 The Real Number Line; Graph of a Group of Numbers	167	3–15 Multiplication of Integers; Properties	201
3–5 Comparing Integers	170	3–16 Multiplication of Rational Numbers; Properties	209
3–6 Opposite Meanings	171		
3–7 Opposite Directions; Vectors	172	3–17 Division of Integers; Properties	215
3–8 Additive Inverse of a Number; More About Opposites	175	3–18 Division of Rational Numbers; Properties	220
		3–19 Computation—Using Numerals with Centered Signs	224
COMPUTATION WITH POSITIVE AND NEGATIVE NUMBERS	177	Chapter Review	226
3–9 Addition on the Number Line	177	Achievement Test	229

4 ALGEBRA 233

LANGUAGE OF ALGEBRA	234	**EVALUATION**	244
4–1 Symbols—Review	234	4–5 Algebraic Expressions	244
4–2 Mathematical Phrases or Mathematical Expressions	236	4–6 Formulas	246
4–3 Mathematical Sentences—Open Sentences; Equations; Inequalities	239	**OPERATIONS**	247
		4–7 Addition	248
		4–8 Subtraction	250
		4–9 Multiplication	252
4–4 Formulas	241	4–10 Division	254

4–11	Zero Exponents; Negative Exponents	256

OPEN SENTENCES IN ONE VARIABLE 260

4–12	Introduction to the Solution of Equations in One Variable	260
4–13	Inverse Operations; Properties of Equality	263
4–14	Solving Equations in One Variable	266
4–15	Solving Equations When the Replacements for the Variable are Restricted	273
4–16	Equations Involving Absolute Values	275
4–17	Evaluating Formulas	276
4–18	Algebraic Representation	278
4–19	Problems	280
4–20	Graphing an Equation in One Variable on the Number Line	282
4–21	Ratio	284
4–22	Proportions	287
4–23	Solving Inequalities in One Variable	289
4–24	Graphing on the Number Line an Inequality in One Variable in Simplest Form	296
4–25	Graphing on the Number Line an Inequality in One Variable Not in Simplest Form	299
4–26	Graphing an Inequality on the Number Line When the Replacements for the Variable are Restricted	300

POINTS ON A NUMBER PLANE 301

4–27	Introduction	301
4–28	Locating Points on a Number Plane	303
4–29	Plotting Points on a Number Plane	306
4–30	Variation	308
4–31	Transforming Formulas	310
4–32	Quadratic Equations	311
	Chapter Review	313
	Achievement Test	318

5 GRAPHS AND STATISTICS 323

5–1	Graphs	324
5–2	Frequency Distribution	326
5–3	Averages—Measures of Central Tendency	328
5–4	Percentiles and Quartiles	332
5–5	Measures of Dispersion	334
	Chapter Review	336

6 PROBABILITY 337

6–1	Probability; Sample Spaces; Complementary Events; Odds	338
6–2	Probability of A or B; Probability of A and B	346
	Chapter Review	350

7 FLOW CHARTS AND LOGIC 351

7–1	Flow Charts	352
7–2	Logic	353
	Chapter Review	356

8 GEOMETRY 357

Inventory Test 358

POINTS, LINES, PLANES, AND SPACE 360
- 8–1 Naming Points, Lines, and Planes 361
- 8–2 Kinds of Lines 364
- 8–3 Position of Lines 365
- 8–4 Intersecting, Parallel, and Perpendicular Lines 365
- 8–5 More Facts About Points, Lines, and Planes 367

ANGLES 369
- 8–6 Naming Angles 369
- 8–7 Kinds of Angles 371
- 8–8 Geometric Figures 373

MEASUREMENT AND CONSTRUCTIONS 378
- 8–9 Drawing and Measuring Line Segments 378
- 8–10 Copying a Line Segment 379
- 8–11 Scale 380
- 8–12 Measuring Angles 385
- 8–13 Drawing Angles 386
- 8–14 Angles in Navigation 388
- 8–15 Constructing Triangles 390
- 8–16 Constructing a Perpendicular to a Given Line at or through a Given Point on the Given Line 394
- 8–17 Constructing a Perpendicular to a Given Line from or through a Given Point Not on the Given Line 395
- 8–18 Bisecting a Line Segment 396
- 8–19 Copying a Given Angle 397
- 8–20 Bisecting an Angle 398
- 8–21 Constructing a Line Parallel to a Given Line through a Given Point Not on the Given Line 400
- 8–22 Constructing Regular Polygons and Other Polygons 401
- 8–23 Sum of Angles of Polygons 403
- 8–24 Pairs of Angles 404
- 8–25 Parallel Lines and Angle Relationships 407
- 8–26 Congruent Triangles 409
- 8–27 Similar Triangles 410

INDIRECT MEASUREMENT 412
- 8–28 By Rule of Pythagoras 412
- 8–29 By Similar Triangles 414
- 8–30 By Scale Drawing 415
- 8–31 Numerical Trigonometry 416

PERIMETER AND CIRCUMFERENCE 421
- 8–32 Perimeter of a Rectangle 421
- 8–33 Perimeter of a Square 422
- 8–34 Perimeter of a Triangle 423
- 8–35 Circumference of a Circle 423

AREA 426
- 8–36 Area of a Rectangle 426
- 8–37 Area of a Square 428
- 8–38 Area of a Parallelogram 429
- 8–39 Area of a Triangle 430
- 8–40 Area of a Trapezoid 431
- 8–41 Area of a Circle 433
- 8–42 Total Area of a Rectangular Solid and a Cube 435
- 8–43 Lateral Area and Total Area of a Right Circular Cylinder 436

VOLUME—MEASURE OF SPACE 438
- 8–44 Volume of a Rectangular Solid 438
- 8–45 Volume of a Cube 439
- 8–46 Volume of a Right Circular Cylinder 441

8–47	Volume of a Sphere, Right Circular Cone, and Pyramid	442		Chapter Review Achievement Test		444 450

9 APPLICATIONS OF MATHEMATICS 457

INCOME; TAKE-HOME PAY; INCOME TAX   458
9–1 Computing Income 458
9–2 Withholding Tax 461
9–3 Social Security 464
9–4 State and City Income Tax—Wage Tax 464
9–5 Take-Home Pay 465
9–6 Income Tax 467

CONSUMER SPENDING   470
9–7 Installment Buying; Annual Percentage Rate; Deferred Payment Purchase Plans 470
9–8 Fire Insurance 475
9–9 Life Insurance 477
9–10 Sales Tax 479
9–11 Excise Tax 480
9–12 Property Tax 480
9–13 Buying and Renting a House 482

MANAGING MONEY   484
9–14 Simple Interest 484
9–15 Banking 488
9–16 Borrowing Money 490
9–17 Compound Interest 496
9–18 Stocks, Bonds, Mutual Funds, Savings Certificates 500

APPLICATIONS IN BUSINESS   507
9–19 Discount 507
9–20 Promissory Notes—Bank Discount 512
9–21 Commission 513
9–22 Profit and Loss 517
9–23 Balance Sheet—Profit and Loss Statement 523
9–24 Payrolls 524
9–25 Invoices 526
9–26 Inventories and Sales Summaries 528
9–27 Parcel Post 530

APPLICATIONS IN INDUSTRY   532
9–28 Lumber 532
9–29 Excavation, Brickwork, and Concrete Mixtures 533
9–30 Lathing, Plastering, Paperhanging, Painting, and Roofing 533
9–31 Rim Speeds 535
9–32 Gears and Pulleys 535
9–33 Electricity 536
9–34 Levers 537
9–35 Reading a Micrometer 538
9–36 Tolerance 539
9–37 Finding Missing Dimensions 540
9–38 Areas of Irregular Figures 541

AVIATION   542
9–39 The Magnetic Compass 542
9–40 Aviation—Some Interesting Facts 543

Table of Squares and Square Roots 547
Tables of Measure 547

Index 550

Mathematics is the science dealing with quantities and forms. There is nothing in the world around us to which it does not somehow apply. Look at the illustrations that begin each chapter. Here are examples of many ways mathematics is used in careers and in our daily lives.

1

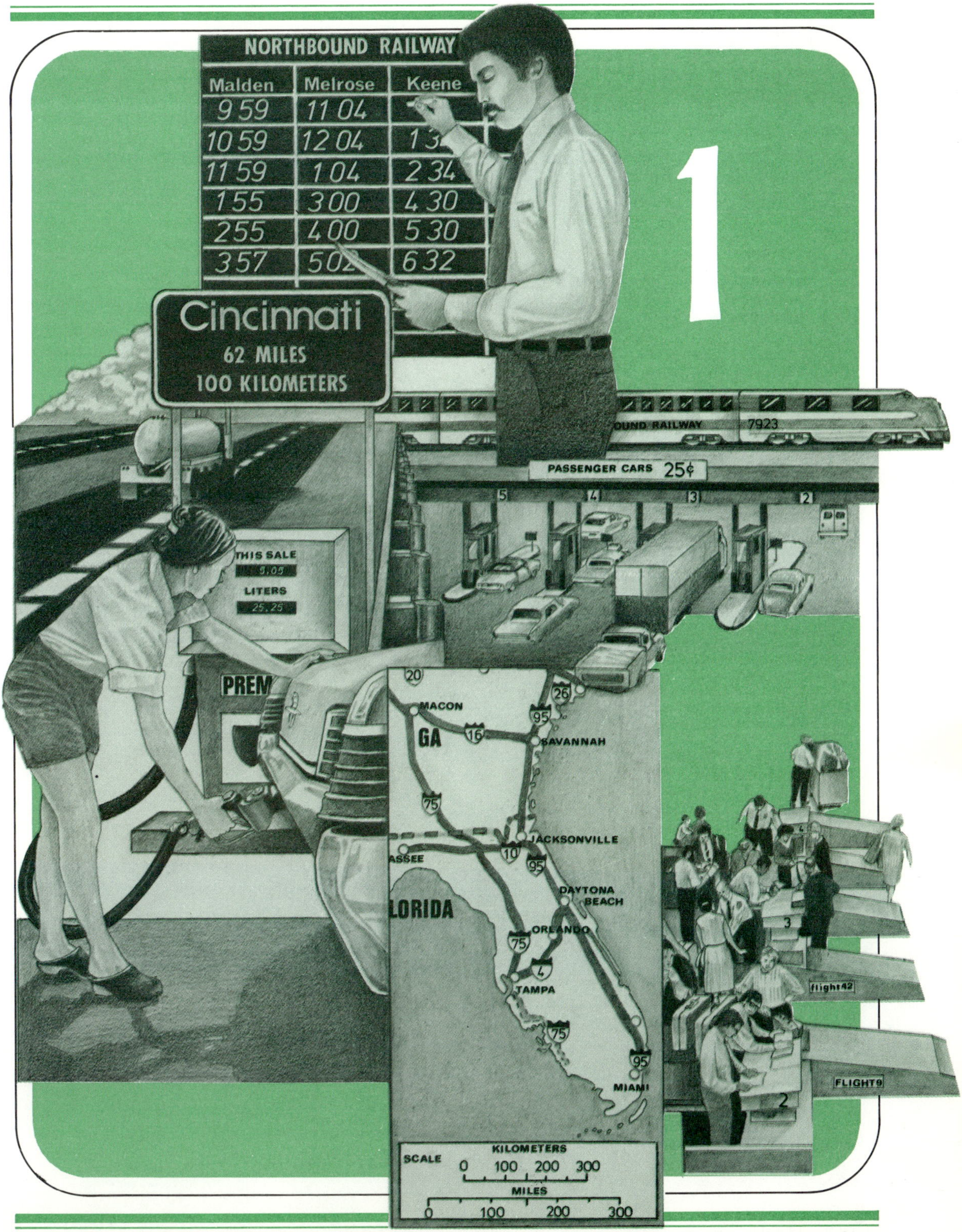
NORTHBOUND RAILWAY
Malden
Melrose
Keene
959 1104
1059 1204
1159 104 234
155 300 430
255 400 530
357 632
Cincinnati
62 MILES
100 KILOMETERS
PASSENGER CARS 25¢
THIS SALE
LITERS
PREM
MACON
GA
SAVANNAH
JACKSONVILLE
DAYTONA BEACH
ORLANDO
TAMPA
MIAMI
LORIDA
flight42
FLIGHT9
SCALE
KILOMETERS
0 100 200 300
MILES
0 100 200 300

CHAPTER 1
Review of Number

Introduction

The number symbols we write and see are not numbers but names for numbers. Although 25 is commonly called a *number,* it is a *numeral* or a group of number symbols which represent the number named *twenty-five*.

However, we generally use number symbols to denote both numbers and numerals. We read and write numerals but we compute with numbers, as: "Write the numeral 57" and "Add the numbers 24 and 83."

A number may have many names; it may be represented by symbols in many ways.

The number named *twenty-five* may be represented by any one of the following:

25   XXV   19 + 6   37 − 12   25 × 1   100 ÷ 4

There are many other names for this number *twenty-five* but 25 is considered to be the simplest numeral.

Although the same number symbols are generally used throughout the world, word names for numbers differ because of the differences in language. For example, in Spanish *two* is *dos;* in French, *deux;* in Italian, *due;* in German, *zwei*.

ARITHMETIC INVENTORY TEST

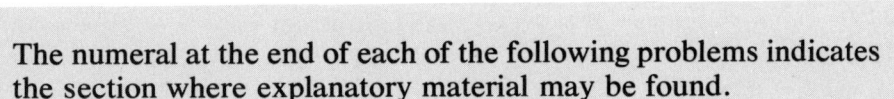

The numeral at the end of each of the following problems indicates the section where explanatory material may be found.

1. Round 96,854,703 to the nearest million. (1-1)

2. Add:
   84,627
   59,998
   47,596
   28,698
   37,989   (1-2)

3. Subtract:
   5,040,061
   4,932,972   (1-3)

4. Multiply:
   6,089
     857   (1-4)

5. Divide:
   869)605,693   (1-5)

6. a. Express $\frac{63}{84}$ in lowest terms. (1-6)
   b. Raise to higher terms: $\frac{3}{5} = \frac{?}{100}$. (1-7)
   c. Express $\frac{50}{16}$ as a mixed number. (1-8)
   d. Change $2\frac{3}{4}$ to an improper fraction. (1-9)

7. Add:
   $3\frac{7}{8}$
   $1\frac{1}{3}$ (1-12)

8. Subtract:
   $8\frac{1}{2}$
   $7\frac{2}{3}$ (1-13)

9. Multiply:
   $3\frac{1}{7} \times 5\frac{3}{8}$ (1-14)

10. Divide:
    $4 \div 6\frac{2}{5}$ (1-15)

11. Which of the following statements are true? (1-11)
    a. $\frac{12}{16}$ and $\frac{54}{72}$ are equivalent fractions.
    b. $\frac{2}{3} > \frac{3}{4}$
    c. .0020 < .020

12. Round 7.5046 to the nearest thousandth. (1-16)

Review of Number

13. **a.** Write 42.9 billion as a complete numeral.
    b. Write the shortened name for 5,170,000 in millions. (1–24)

14. Add:
    1.48 + .327 + 28.5 (1–18)

15. Subtract:
    94 − 2.6 (1–19)

16. Multiply:
    .005 × .042 (1–20)

17. Divide:
    1.5 ÷ .02 (1–22)

18. Multiply 342.4982 by 1,000 (1–21)

19. Divide 195.63 by 100 (1–23)

20. **a.** 78 is what fractional part of 91? (1–34)
    b. 25 is what decimal part of 40? (1–34)

21. **a.** $\frac{5}{8}$ of what number is 70? (1–34)
    b. .6 of what number is 36? (1–34)

22. Write in scientific notation: (1–25)
    a. 63,500,000,000
    b. .0000027

23. **a.** Express $\frac{4}{5}$ as a decimal numeral. (1–26)
    b. Express 129% as a decimal numeral. (1–28)
    c. Express $\frac{3}{7}$ as a numeral naming a repeating decimal. (1–26)

24. **a.** Express .85 as a numeral naming a common fraction. (1–27)
    b. Express $16\frac{2}{3}$% as a numeral naming a common fraction. (1–28)

25. **a.** Express .328 as a percent. (1–29)
    b. Express $\frac{13}{25}$ as a percent. (1–29)

26. Find $8\frac{1}{2}$% of $34. (1–30, 1–33, 1–34)

27. What percent of 72 is 27? (1–31, 1–33, 1–34)

28. 3 is 2% of what number? (1–32, 1–33, 1–34)

29. Find the square of $2\frac{5}{8}$. (1–35)

30. **a.** Find the square root of 19 by estimation, division, and average. (1–36)
    b. Find the square root of 24,068,836. (1–36)

DECIMAL SYSTEM OF NUMERATION

A *system of numeration* is a method of naming numbers by writing numerals. In the *decimal system of numeration* the number symbols 0, 1, 2, 3, 4, 5, 6, 7, 8, 9 are used to write numerals. Each of these is called a *digit* or figure. The numeral 936 has three digits: 9, 3, and 6. The numerals representing numbers greater than nine are formed by writing two or more of these basic number symbols next to each other in different positions or places.

Our decimal system of numeration is built on the base ten. It takes 10 in any one place to make 1 in the next higher place. The *base* of a system of numeration is the number it takes in any one place to make 1 in the next higher place.

Our decimal system of numeration is a positional system. It uses *place value* to represent each power of ten (10; 100; 1,000; etc.). The value of each place in the decimal number scale is ten times the value of the next place to the right. Thus in a numeral the value of each symbol depends not only on what the symbol is but also on its position in the numeral. The numeral 6 in the decimal numeral 63 means *6 tens* but in the decimal numeral 683 means *6 hundreds*.

Trillions	Billions	Millions	Thousands	Ones
quadrillions	hundred trillions, ten trillions, trillions	hundred billions, ten billions, billions	hundred millions, ten millions, millions	hundred thousands, ten thousands, thousands	hundreds, tens, ones
4,	4 4 4,	4 4 4,	4 4 4,	4 4 4,	4 4 4

In the number scale the ones (sometimes called units) are located on the right. One place to the left of the ones position is the tens place; one place to left of the tens position is the hundreds place. These three places form a group or *period*.

The scale illustrates that the periods in increasing order are: *ones, thousands, millions, billions, trillions, and quadrillions.* The names of other periods in increasing order are *quintillion, sextillion, septillion, octillion, nonillion, and decillion.* A decillion is written as:

1,000,000,000,000,000,000,000,000,000,000,000

Review of Number

WHOLE NUMBERS—COMPUTATION

The numbers beginning with zero named by the numerals 0, 1, 2, 3, 4, 5, 6, 7, 8, 9, 10, 11, 12, 13, 14, . . . are called *whole numbers*.

The whole numbers beginning with 1 that are used in counting are called *counting* or *natural numbers*.

Whole numbers have order. Every whole number is followed by another whole number which is one (1) greater. There is no greatest whole number and no last whole number.

1–1 ROUNDING WHOLE NUMBERS

To round a whole number, we find the place in the numeral to which the number is to be rounded (the nearest ten, hundred, thousand, etc.). We then rewrite the given digits to the left of the required place and we write a zero in place of each digit to the right of the required place. If the first digit dropped is 5 or more, we increase the given digit in the required place by 1; otherwise we write the same digit as given. However, as many digits to the left are changed as necessary when the given digit in the required place is 9 and one (1) is added.

> 7,684,295 rounded to the nearest:
> hundred is   7,684,300
> thousand is   7,684,000
> million is   8,000,000

EXERCISES

Round each of the following numbers to the nearest:

1. Ten:

| 53 | 85 | 31 | 498 | 6,176 |

2. Hundred:

| 260 | 427 | 8,052 | 3,109 | 56,983 |

3. Thousand:

| 7,690 | 2,449 | 19,604 | 88,525 | 90,358 |

4. Ten thousand:
   38,000   73,914   61,221   507,499   496,520

5. Hundred thousand:
   640,000   191,285   983,054
   4,263,971   39,972,109

6. Million:
   6,854,000   8,200,000   17,375,129
   2,645,722,008   549,534,072

7. Billion:
   1,613,000,000   2,090,700,000   7,498,260,000
   99,503,004,121   480,750,000,000

8. Trillion:
   4,900,000,000,000   7,050,000,000,000   29,630,000,000,000
   52,427,840,293,524   389,725,000,000,000

1–2 ADDITION OF WHOLE NUMBERS

The terms generally used in addition are: *addends,* the numbers that we add; and *sum,* the answer in addition. The symbol used to indicate addition is the *plus* sign (+).

To add whole numbers, we add each column, beginning with the ones column. If the sum of any column is ten or more, we write the last digit of the sum in the answer and carry the other digits to the next column to the left. We check by adding the columns in the opposite direction.

```
468  ←
 54  ←  Addends
379  ←
───
901  ←  Sum

Answer, 901
```

EXERCISES

1. Add and check:

   a. 0   1   5   3   2   1   4   2   8   7   9
       2   6   4   3   7   4   9   3   0   1   5

   b. 1   3   4   2   1   4   0   9   3   2   6
       7   5   0   6   1   3   4   6   4   2   4

Review of Number   7

c. 5   4   1   7   0   1   3   0   3   2   9
    2   4   8   3   9   5   1   0   6   4   9
    —   —   —   —   —   —   —   —   —   —   —

d. 4   8   9   5   6   5   9   8   0   4   7
    2   6   0   5   9   7   3   9   6   5   8
    —   —   —   —   —   —   —   —   —   —   —

e. 7   5   6   3   4   9   6   7   0   6   9
    6   0   8   2   6   4   5   4   8   1   7
    —   —   —   —   —   —   —   —   —   —   —

f. 6   4   2   3   7   5   7   0   3   1   9
    3   8   1   0   7   3   5   1   8   2   2
    —   —   —   —   —   —   —   —   —   —   —

g. 8   0   6   9   2   5   2   6   8   2   1
    3   5   7   2   0   1   8   6   1   5   9
    —   —   —   —   —   —   —   —   —   —   —

h. 0   9   4   1   8   6   8   6   5   7   3
    3   1   7   3   4   2   7   0   9   2   9
    —   —   —   —   —   —   —   —   —   —   —

i. 3   5   9   8   7   8   1   2   5   7   8
    7   6   8   5   0   8   0   9   8   9   2
    —   —   —   —   —   —   —   —   —   —   —

2. Add and check:

   a.   3   13   23   63   83   c.   8   18   58   68   88
        5   5   5   5   5   6   6   6   6   6
        —   —   —   —   —   —   —   —   —   —

   b. 42   31   4   2   60   d. 27   39   5   6   78
        7   6   25   72   9   9   8   87   54   9
        —   —   —   —   —   —   —   —   —   —

3. Add and check:

   a.   16   47   325   7,385   9,756   63,416   32,745
        23   55   442   2,415   8,648   29,974   99,258
        —   —   —   —   —   —   —

   b. 41,329   68,318   97,967   563,849   785,499
        7,093   35,276   56,498   274,679   986,978
        ——   ——   ——   ——   ——

   c.   3   5   22   68   225   412   583
        4   9   45   95   193   87   699
        7   3   32   44   377   9   586
        —   —   —   —   —   —   —

   d.   39   8,215   4,096   15,249   578,655
       478   1,904   8,327   31,754   298,579
     9,614   6,737   3,849   58,648   764,083
     ——   ——   ——   ——   ——

8   CHAPTER 1

e.
5	7	26	55	413	968	637
2	9	38	60	804	469	596
2	7	18	79	129	753	978
8	9	29	43	734	145	964

f.
4,386	3,427	2,097	38,486	975,269
597	9,748	1,135	90,145	286,879
88	8,977	4,526	29,076	826,957
9	5,589	3,242	42,293	657,838

g.
1	9	27	86	372	964	4,175
2	5	29	99	864	879	8,089
4	7	18	47	495	558	5,148
2	6	9	56	398	688	9,267
9	8	14	78	286	997	8,057

h.
5,862	6,589	82,637	45,584	485,876
9,104	7,896	58	96,137	576,988
7,467	5,945	4,397	30,598	859,649
5,399	3,287	6	58,943	360,795
3,875	6,759	845	76,456	698,489

i.
8	27	364	507	8,682	4,265
5	76	58	389	39	9,508
9	95	67	846	4	5,786
7	87	899	353	487	8,679
4	95	637	795	4,928	9,188
9	38	89	468	75	6,899

j.
3,146	68,215	43	614,859
65,298	59,836	82,514	928,678
402	34,477	567,827	348,399
72,298	83,568	9,549	567,285
87	59,343	699	905,671
4,969	27,259	491,875	786,987

k.
6	34	29	891	6	7,217
7	28	593	475	58	5,456
5	19	8	306	8,493	6,085
2	63	46	699	94	1,729
9	79	357	840	9,508	8,354
9	54	82	167	767	9,279
8	97	776	684	325	3,528

Review of Number

l.	818	39,206	459,317	6,915,825
	19,624	44,798	925,884	8,740,643
	4,075	29,613	709,603	2,325,791
	39	53,075	895,792	7,839,287
	8,596	19,680	600,599	4,056,130
	79,278	56,447	182,943	8,479,829
	903	84,238	750,856	3,984,568

4. Add horizontally:

    a. 36 + 58 + 62   **d.** 55 + 74 + 8 + 96
    b. 507 + 96 + 849   **e.** 384 + 69 + 5 + 678
    c. 697 + 9 + 4,528   **f.** 8,079 + 1,982 + 5,376 + 6,939

5. Arrange in columns, then find the sum of each of the following:

    a. 35; 58; 94   **d.** 788; 26; 8; 659
    b. 496; 948; 659   **e.** 9,000; 70; 800; 30
    c. 6,084; 7,925; 5,867   **f.** 687; 48; 7,856; 67

1–3 SUBTRACTION OF WHOLE NUMBERS

The terms generally used in subtraction are: *minuend*, the number from which we subtract; *subtrahend*, the number we subtract; and *remainder* or *difference*, the answer in subtraction. The symbol used to indicate subtraction is the *minus* sign (−).

To subtract whole numbers, we subtract the digits in the subtrahend from the corresponding digits in the minuend, starting from the ones place. When any digit in the subtrahend is greater than the corresponding digit in the minuend, we increase this digit in the minuend by 10 by taking 1 from the preceding digit in the next higher place and changing it to 10 of the next lower place. This is called *regrouping*. We check by adding the remainder to the subtrahend. Their sum should equal the minuend.

```
 946  ← Minuend
−372  ← Subtrahend
 574  ← Remainder or
         Difference
```
Answer, 574

EXERCISES

1. Subtract and check:

   a. | 7 | 6 | 3 | 2 | 9 | 10 | 8 | 4 | 6 | 1 | 10 |
      |---|---|---|---|---|----|---|---|---|---|----|
      | 6 | 4 | 2 | 0 | 5 | 1 | 3 | 4 | 3 | 1 | 9 |

   b. | 8 | 9 | 10 | 11 | 0 | 7 | 12 | 8 | 14 | 4 | 13 |
      |---|---|----|----|---|---|----|---|----|---|----|
      | 8 | 2 | 4 | 8 | 0 | 4 | 5 | 6 | 7 | 1 | 6 |

   c. | 4 | 6 | 9 | 7 | 16 | 8 | 11 | 5 | 12 | 10 | 11 |
      |---|---|---|---|----|---|----|---|----|----|----|
      | 1 | 5 | 6 | 3 | 8 | 1 | 2 | 4 | 9 | 8 | 6 |

   d. | 9 | 7 | 11 | 10 | 16 | 8 | 11 | 15 | 12 | 10 | 11 |
      |---|---|----|----|----|---|----|----|----|----|----|
      | 8 | 7 | 9 | 6 | 7 | 2 | 9 | 6 | 3 | 7 | 2 |

   e. | 10 | 6 | 12 | 11 | 18 | 10 | 7 | 8 | 12 | 13 | 16 |
      |----|---|----|----|----|----|---|---|----|----|----|
      | 7 | 1 | 8 | 4 | 9 | 3 | 2 | 5 | 7 | 5 | 9 |

   f. | 5 | 9 | 6 | 15 | 10 | 7 | 11 | 13 | 17 | 11 | 6 |
      |---|---|---|----|----|---|----|----|----|----|---|
      | 3 | 4 | 6 | 7 | 5 | 1 | 3 | 4 | 9 | 7 | 2 |

   g. | 8 | 9 | 13 | 14 | 5 | 8 | 15 | 7 | 12 | 17 | 12 |
      |---|---|----|----|---|---|----|---|----|----|----|
      | 2 | 1 | 8 | 5 | 0 | 7 | 6 | 5 | 3 | 8 | 6 |

   h. | 9 | 13 | 15 | 5 | 14 | 15 | 9 | 8 | 11 | 12 | 13 |
      |---|----|----|---|----|----|---|---|----|----|----|
      | 3 | 7 | 9 | 2 | 6 | 8 | 9 | 4 | 5 | 4 | 9 |

2. Subtract and check:

   a. | 89 | 48 | 65 | 76 | 90 | 61 | 73 | 85 |
      |----|----|----|----|----|----|----|----|
      | 4 | 9 | 25 | 72 | 47 | 56 | 38 | 69 |

   b. | 568 | 736 | 842 | 465 | 677 | 305 | 423 | 861 |
      |-----|-----|-----|-----|-----|-----|-----|-----|
      | 25 | 124 | 829 | 238 | 392 | 248 | 95 | 353 |

   c. | 6,875 | 2,754 | 5,129 | 9,046 | 7,365 | 8,004 | 7,948 | 9,507 |
      |-------|-------|-------|-------|-------|-------|-------|-------|
      | 2,336 | 1,619 | 2,496 | 4,527 | 6,788 | 7 | 7,898 | 699 |

   d. | 86,479 | 61,758 | 50,000 | 40,844 | 82,465 | 32,600 |
      |--------|--------|--------|--------|--------|--------|
      | 32,146 | 52,693 | 49,081 | 978 | 5,067 | 26,075 |

   e. | 579,625 | 952,548 | 147,212 | 401,002 | 800,000 |
      |---------|---------|---------|---------|---------|
      | 143,502 | 294,756 | 9,625 | 197,064 | 452,188 |

Review of Number

f. 8,962,744   6,348,290   7,040,603   6,004,908   8,569,142
    2,611,432   5,029,365   6,989,594   4,093,868   3,789,679

g. 39,842,567   91,532,408   63,000,500   42,753,962   70,000,000
    16,230,437   35,150,939   53,783,496   42,749,189   67,080,192

3. Subtract horizontally:

- **a.** 75 − 63
- **b.** 35 − 9
- **c.** 86 − 59
- **d.** 100 − 78
- **e.** 300 − 198
- **f.** 424 − 6
- **g.** 2,635 − 1,845
- **h.** 7,901 − 94
- **i.** 98,450 − 39,667
- **j.** 50,000 − 3
- **k.** 42,173 − 9,081
- **l.** 605,000 − 587,752

4. a. From 400 subtract 289

   b. Subtract 826 from 902

   c. Take 5,081 from 6,000

   d. From 6,146 take 4,097

   e. Subtract 72,987 from 69,053

   f. From 31,004 subtract 19,898

5. Find the difference between:

   a. 2,500 and 750

   b. 30,000 and 300

1–4 MULTIPLICATION OF WHOLE NUMBERS

The terms generally used in multiplication are: *multiplicand,* the number that we multiply; *multiplier,* the number by which we multiply; *product,* the answer in multiplication; *partial product,* the product obtained by multiplying the multiplicand by any figure in the multiplier containing two or more figures; and *factor,* any one of the numbers used in multiplication to form a product. The *times sign* (×) or the *raised dot* (·) are the symbols used to indicate multiplication.

One-Digit Multipliers

To multiply by a one-digit multiplier, we multiply each digit in the multiplicand by the one-digit multiplier, starting from the right. If the product or total for any place is ten or more, we write the last digit in the answer and add the other digit to the next product.

> 325 ← Multiplicand
> ×3 ← Multiplier
> 975 ← Product
> *Answer,* 975

Multipliers of Two or More Digits

To multiply by a multiplier containing two or more digits, we multiply all the digits in the multiplicand by each digit in the multiplier, starting from the right, to find the partial products. The partial products are written under each other with the numerals placed so that the right-hand digit of each partial product is directly under its corresponding digit in the multiplier. Then we add the partial products. We check by interchanging the multiplier and multiplicand and multiplying again or by division if it has been studied.

```
    24
   ×76
   ───
   144  ←┐
   168  ←┘  Partial Products
  ─────
  1,824
```
Answer, 1,824

EXERCISES

1. Multiply and check:

a.	5 3	7 0	1 2	2 8	6 1	4 2	8 5	7 2	3 1	2 6	8 7
b.	2 4	8 1	5 9	9 6	4 0	1 5	7 3	6 4	9 2	6 7	8 6
c.	3 6	2 5	8 4	4 3	7 6	0 0	4 7	1 6	8 8	1 3	3 5
d.	9 3	3 4	7 7	5 1	6 3	4 8	7 5	8 2	5 6	2 9	7 1
e.	4 5	5 7	3 0	8 3	1 1	7 9	9 1	5 8	4 9	2 2	9 9
f.	7 4	6 2	1 7	0 6	2 3	9 8	6 9	3 7	1 4	6 6	9 5

Review of Number

g. | 3 | 4 | 9 | 2 | 3 | 1 | 0 | 5 | 3 | 8 | 4
    | 2 | 6 | 4 | 7 | 8 | 9 | 8 | 2 | 3 | 9 | 4

h. | 1 | 0 | 5 | 9 | 7 | 0 | 3 | 6 | 5 | 4 | 6
    | 8 | 9 | 5 | 7 | 8 | 5 | 9 | 8 | 4 | 1 | 5

2. Multiply and check:

   a. | 23 | 71 | 48 | 99 | 642 | 905 | 700 | 869
       | 3 | 9 | 5 | 8 | 4 | 6 | 5 | 9

   b. | 9,152 | 7,589 | 6,080 | 8,954 | 34,285 | 69,524 | 58,697 | 97,006
       | 4 | 7 | 9 | 5 | 8 | 9 | 6 | 8

3. Multiply and check:

   a. | 23 | 17 | 49 | 83 | 58 | 80 | 94 | 60 | 9 | 82
       | 12 | 54 | 26 | 68 | 97 | 75 | 50 | 60 | 36 | 10

   b. | 487 | 790 | 566 | 603 | 75 | 978 | 70 | 542 | 876 | 659
       | 18 | 59 | 87 | 40 | 289 | 68 | 300 | 95 | 49 | 38

   c. | 6,583 | 2,240 | 8,007 | 54 | 24,315 | 69,206 | 83,259 | 90,803
       | 74 | 25 | 86 | 7,030 | 39 | 75 | 42 | 67

   d. | 345 | 898 | 702 | 469 | 900 | 124 | 935 | 788
       | 286 | 697 | 845 | 500 | 300 | 875 | 144 | 869

   e. | 2,243 | 5,614 | 6,895 | 5,280 | 14,324 | 60,000 | 73,586 | 34,652
       | 321 | 628 | 769 | 450 | 212 | 807 | 569 | 973

   f. | 6,529 | 5,100 | 6,080 | 8,799 | 5,601 | 6,259 | 4,006 | 8,973
       | 1,728 | 2,454 | 8,167 | 9,376 | 2,240 | 3,894 | 7,086 | 9,845

   g. | 49,486 | 65,002 | 35,827 | 72,149 | 14,231 | 67,005 | 89,614 | 92,983
       | 8,395 | 8,030 | 6,768 | 9,336 | 32,123 | 20,094 | 56,742 | 87,369

   h. | 9 | 6 | 7 | 40 | 95 | 8 | 720 | 61 | 303 | 4,509
       | 36 | 475 | 890 | 800 | 6,257 | 5,000 | 2,240 | 96,542 | 40,907 | 83,274

4. Do as directed in each of the following:

   a. 24 times 70   288 times 35   709 times 809
   b. 12 × 87   450 × 300   598 × 962
   c. 36 × 18 × 9   53 × 61 × 47   9,000 × 50 × 80
   d. Multiply 45 by 58.   Multiply 190 by 72.   Multiply 948 by 781.
   e. Find the product of   Find the product of   Find the product of
      60 and 50.   893 and 86.   6,080 and 124.

1–5 DIVISION OF WHOLE NUMBERS

The terms generally used in division are: *dividend,* the number that we divide; *divisor,* the number by which we divide; *quotient,* the answer in division; *remainder,* the number left over when the division is not exact; *partial dividend,* the first part of the given dividend or digits of the dividend annexed to the remainder. The symbols ÷ and ⟌ and the fraction bar (—) are used to indicate division.

To divide a whole number by a whole number:

(1) We first find the *quotient digit* by dividing the one-digit divisor or the trial divisor, when the divisor contains more than one digit, into the first digit of the dividend. When the divisor contains two or more digits, we use as the trial divisor the first digit of the divisor if the next digit on the right is 0, 1, 2, 3, 4, or 5 and increase the first digit of the divisor by one (1) if the next digit on the right is 6, 7, 8, or 9. The greatest digit that can be used in the quotient at any time is 9.

> (1) Divide:
>
> $$\text{Divisor} \rightarrow 26 \overline{)884} \begin{array}{l} \leftarrow \text{Quotient (3)} \\ \leftarrow \text{Dividend} \end{array}$$

(2) We multiply the divisor by the quotient digit and write this product under the corresponding digits in the dividend. This product must be the same or less than the partial dividend. If it is greater, then we use as the trial quotient digit one (1) less than the digit first tried.

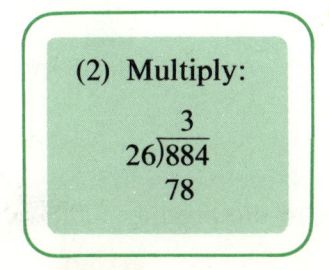

> (2) Multiply:
>
> $$\begin{array}{r} 3 \\ 26\overline{)884} \\ 78 \end{array}$$

Review of Number

(3) We subtract this product from the corresponding numbers in the dividend. The remainder must be less than the whole divisor. If it is not, then we use as the trial quotient digit one (1) more than the digit tried.

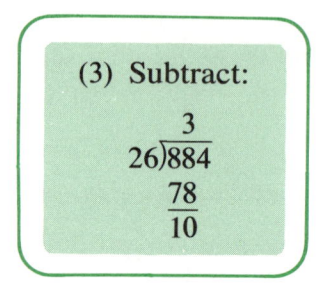

> (3) Subtract:
>
> $$\begin{array}{r} 3 \\ 26\overline{)884} \\ 78 \\ \hline 10 \end{array}$$

(4) We bring down the next digit of the dividend and annex it to the remainder, if any.

> (4) Bring down digit:
>
> $$\begin{array}{r} 3 \\ 26\overline{)884} \\ 78 \\ \hline 104 \end{array} \leftarrow \text{Partial Dividend}$$

(5) (6) (7) Then using the remainder and annexed numbers as partial dividends, we repeat the above steps for each partial dividend.

> (5) Divide:
>
> $$\begin{array}{r} 34 \\ 26\overline{)884} \\ 78 \\ \hline 104 \end{array}$$
>
> *Answer, 34*
>
> (6) Multiply:
>
> $$\begin{array}{r} 34 \\ 26\overline{)884} \\ 78 \\ \hline 104 \\ 104 \end{array}$$
>
> (7) Subtract:
>
> $$\begin{array}{r} 34 \\ 26\overline{)884} \\ 78 \\ \hline 104 \\ 104 \\ \hline 0 \end{array}$$

We check by multiplying the quotient by the divisor and adding the remainder, if any, to the product. The result should equal the dividend.

When a division is not exact, the remainder may be written in the answer using the letter *R* to indicate it (like 478 R5) or it may be written as a fraction in lowest terms or it may be indicated (when the divisor is a large number) by a + sign in the answer.

EXERCISES

1. Divide and check:

   a. 1)7   3)3   2)6   8)16   6)48   9)54   5)15   4)20   3)18
   b. 5)35   1)4   7)28   3)6   8)40   2)10   9)18   3)21   4)36
   c. 7)21   3)24   1)2   6)30   2)12   8)72   4)4   3)15   9)27
   d. 5)45   4)32   3)9   2)4   1)9   7)42   6)24   9)0   8)64
   e. 2)14   5)40   4)28   8)8   2)16   1)1   7)63   6)54   9)81
   f. 6)18   3)27   5)10   7)35   2)2   5)20   9)36   7)7   8)56
   g. 4)16   5)5   7)49   8)32   4)12   1)6   9)45   6)42   2)18
   h. 3)12   2)8   1)5   7)14   5)30   8)48   6)36   9)9   4)24
   i. 9)72   6)12   4)8   1)3   5)25   7)56   8)24   6)6   9)63

2. Divide and check:

   3)63   5)75   6)810   7)903   8)9,712   6)8,526   8)99,984   4)91,648

3. Divide each of the following. Be careful where you place the first quotient figure.

   8)568   3)279   9)774   6)534   4)3,284   7)1,855   8)59,152   5)46,430

4. Watch the zeros in each of the following division problems:

   2)80   7)420   4)812   6)9,600   5)5,300   3)60,903   8)56,200   5)90,010

5. Divide each of the following:

   a. By 2:   86   738   1,402   93,574   81,250
   b. By 5:   60   975   5,700   37,825   63,065
   c. By 3:   72   561   7,125   62,982   22,017
   d. By 4:   40   648   3,496   25,700   75,376
   e. By 7:   91   686   4,963   36,148   63,483
   f. By 6:   78   828   6,324   51,570   70,962
   g. By 8:   96   392   5,000   83,656   47,984
   h. By 9:   54   117   7,434   49,068   86,571

Review of Number   17

6. Find the quotient and the remainder in each of the following:

    6)47   3)34   7)968   6)145   4)3,851   9)5,329   8)95,739   5)83,472

7. Divide. Write the remainder as a fraction in lowest terms:

    8)52   9)96   4)890   9)641   6)3,669   5)9,354   4)61,958   8)34,468

8. Do as indicated:

    a. 8,472 ÷ 6
    b. 2,931 ÷ 3
    c. 5,036 ÷ 4
    d. 3,984 ÷ 8
    e. 91,278 ÷ 2
    f. 34,860 ÷ 5
    g. 78,525 ÷ 9
    h. 40,173 ÷ 7

9. Divide and check:

    a. 10)70   23)69   16)96   36)288   90)900   49)882   27)972   74)666
    b. 12)8,364   24)2,136   27)9,423   96)5,472   72)3,096
    c. 28)91,252   75)49,425   19)98,306   58)54,984   90)80,010
    d. 50)352,500   94)928,626   86)803,928   79)608,142   99)300,960

10. Find the quotient and the remainder in each of the following:

    43)135   32)1,747   17)5,963   91)67,482   59)43,900

11. Divide. Write the remainder as a fraction in lowest terms:

    60)400   36)1,932   75)4,250   40)24,330   28)17,514

12. Divide and check:

    a. 279)837   517)4,136   825)53,625   642)371,718   886)792,084
    b. 700)371,000   804)389,136   630)427,770   506)256,036   868)555,520

13. Find the quotient and the remainder in each of the following:

    246)7,013   695)5,184   125)12,300   538)91,575   752)696,935

14. Divide and check:

    5,625)208,125   9,248)813,824   6,080)1,854,400   3,691)25,191,075

15. Find the quotient and the remainder in each of the following:

    2,240)64,972   3,600)315,000   6,437)6,482,060   4,856)4,382,699

16. Do as indicated:

    a. 4,602 ÷ 78
    b. 39,711 ÷ 93
    c. 28,391 ÷ 319
    d. 387,871 ÷ 407
    e. 4,738,602 ÷ 6,527
    f. 26,183,632 ÷ 9,004

17. Divide and check:

    43)4,128   74)28,786   362)19,186   906)792,750   6,851)34,926,398

RATIONAL NUMBERS

A *rational number* is a number that can be expressed as a quotient of two whole numbers, with division by zero excluded.

Common fractions, decimal fractions, and percents name rational numbers.

When a unit or thing or group of things is divided into equal parts, the numeral that expresses the relation of one or more of the equal parts to the total number of equal parts is called a *fraction*.

Although "fraction" is generally used to mean both the fractional number and the numeral written for the number, some mathematicians use the term "fraction" to mean the symbol and "rational number" to mean the fractional number.

Fractions that are expressed by a symbol consisting of a pair of numerals, one written above the other with a horizontal bar between them, such as $\frac{2}{3}$, are called *common fractions*. The number above the bar is called the *numerator* and the number below the bar is called the *denominator*. The denominator cannot be zero.

In the fraction $\frac{2}{3}$, the numbers 2 and 3 are the *terms* of the fraction. The fraction $\frac{2}{3}$ means 2 parts of 3 equal parts.

A fraction may be used to indicate that one number is divided by another number. The fraction $\frac{2}{3}$ means 2 divided by 3 and represents $2 \div 3$ or $3\overline{)2}$. Thus 10 divided by 2 may be written as $\frac{10}{2}$ and read "ten over two." The numerator is the number that is divided and the denominator is the number by which we divide. Thus whole numbers may be expressed in fraction form, 2 as $\frac{2}{1}$, 3 as $\frac{3}{1}$, 4 as $\frac{4}{1}$, 5 as $\frac{5}{1}$, etc., and are therefore rational numbers.

Changing Form of a Fraction or of a Mixed Number

A fraction whose numerator is smaller than its denominator, such as $\frac{2}{3}$, is called a *proper fraction*. The value of a proper fraction is always less than one.

A fraction whose numerator is equal to or greater than its denominator, such as $\frac{4}{4}$ or $\frac{9}{5}$, is called an *improper fraction*. The value of an improper fraction is one or more than one.

A *mixed number* consists of a whole number and a fraction, such as $2\frac{5}{6}$.

Fractions that name the same number are called *equivalent fractions*. The fractions $\frac{1}{2}$, $\frac{2}{4}$, $\frac{3}{6}$, and $\frac{4}{8}$ are equivalent fractions. They all name the same number which in simplest form is $\frac{1}{2}$.

Review of Number

1-6 EXPRESSING FRACTIONS IN LOWEST TERMS

A fraction is in its *lowest terms* or in *simplest form* when its numerator and denominator cannot be divided exactly by the same number, except by 1.

To express a fraction in lowest terms:

(1) Either we divide both its numerator and denominator by the greatest whole number that can be divided exactly into both;

(2) Or we factor the numerator and denominator, using as one of the factors the greatest factor common to both. (Use whole numbers other than 1 or 0.) Then we divide both the numerator and the denominator by this common factor or we use the multiplicative identity one (1) as shown below.

Express $\frac{18}{24}$ in lowest terms.

$$\frac{18}{24} = \frac{18 \div 6}{24 \div 6} = \frac{3}{4} \quad \text{or} \quad \frac{18}{24} = \frac{3 \cdot 6}{4 \cdot 6} = \frac{3}{4}$$

$$\text{or} \quad \frac{18}{24} = \frac{3 \cdot 6}{4 \cdot 6} = \frac{3}{4} \times \frac{6}{6} = \frac{3}{4} \times 1 = \frac{3}{4}$$

Answer, $\frac{3}{4}$

EXERCISES

1. Express each of the following in lowest terms:

a. $\frac{3}{6}$	$\frac{4}{32}$	$\frac{4}{64}$	$\frac{2}{10}$	$\frac{6}{72}$
b. $\frac{3}{66}$	$\frac{5}{30}$	$\frac{8}{64}$	$\frac{7}{49}$	$\frac{9}{36}$
c. $\frac{7}{14}$	$\frac{6}{24}$	$\frac{5}{40}$	$\frac{4}{20}$	$\frac{11}{33}$
d. $\frac{13}{39}$	$\frac{21}{84}$	$\frac{27}{81}$	$\frac{14}{70}$	$\frac{17}{85}$
e. $\frac{6}{27}$	$\frac{21}{24}$	$\frac{16}{20}$	$\frac{10}{16}$	$\frac{39}{52}$
f. $\frac{32}{48}$	$\frac{84}{96}$	$\frac{27}{36}$	$\frac{21}{56}$	$\frac{72}{108}$
g. $\frac{10}{40}$	$\frac{20}{60}$	$\frac{75}{100}$	$\frac{24}{800}$	$\frac{500}{1200}$

h. $\frac{3500}{6000}$ $\frac{3000}{9000}$ $\frac{480}{6400}$ $\frac{2400}{3600}$ $\frac{300}{4800}$

i. $\frac{104}{288}$ $\frac{48}{80}$ $\frac{40}{56}$ $\frac{210}{280}$ $\frac{504}{648}$

j. $\frac{450}{800}$ $\frac{180}{216}$ $\frac{147}{168}$ $\frac{102}{114}$ $\frac{546}{630}$

2. Which fractions are not expressed in lowest terms?

   a. $\frac{3}{5}$ $\frac{7}{9}$ $\frac{3}{9}$ $\frac{2}{8}$ d. $\frac{1}{5}$ $\frac{4}{8}$ $\frac{3}{8}$ $\frac{3}{4}$

   b. $\frac{8}{10}$ $\frac{9}{10}$ $\frac{3}{10}$ $\frac{5}{10}$ e. $\frac{6}{17}$ $\frac{13}{45}$ $\frac{7}{35}$ $\frac{13}{18}$

   c. $\frac{13}{17}$ $\frac{5}{8}$ $\frac{27}{36}$ $\frac{19}{30}$ f. $\frac{21}{28}$ $\frac{11}{16}$ $\frac{25}{37}$ $\frac{14}{19}$

3. Find which fractions are equivalent in each group.

   a. $\frac{6}{9}$ $\frac{21}{28}$ $\frac{28}{35}$ $\frac{8}{12}$ d. $\frac{20}{64}$ $\frac{6}{16}$ $\frac{18}{48}$ $\frac{12}{32}$

   b. $\frac{14}{28}$ $\frac{3}{6}$ $\frac{28}{56}$ $\frac{12}{24}$ e. $\frac{20}{24}$ $\frac{7}{14}$ $\frac{15}{18}$ $\frac{10}{12}$

   c. $\frac{4}{12}$ $\frac{18}{30}$ $\frac{5}{15}$ $\frac{18}{27}$ f. $\frac{30}{45}$ $\frac{24}{30}$ $\frac{28}{30}$ $\frac{12}{15}$

1–7 EXPRESSING FRACTIONS IN HIGHER TERMS

To express a fraction in higher terms when the higher denominator is known, we divide the higher denominator by the denominator of the given fraction. Then we multiply both the numerator and denominator of the given fraction by this quotient. This is the same as multiplying the fraction by the multiplicative identity one (1).

Express $\frac{7}{12}$ as 60ths   or $\frac{7}{12} = \frac{?}{60}$

60 divided by 12 equals 5.

$\frac{7}{12} = \frac{7 \times 5}{12 \times 5} = \frac{35}{60}$ or $\frac{7}{12} = \frac{7}{12} \times 1 = \frac{7}{12} \times \frac{5}{5} = \frac{35}{60}$

Answer, $\frac{35}{60}$

Review of Number

START

EXERCISES

1. Express as fractions in the indicated higher terms:
   a. $\frac{1}{4} = \frac{?7}{28}$ $\frac{1}{3} = \frac{?9}{27}$ $\frac{1}{2} = \frac{?8}{16}$ $\frac{1}{5} = \frac{?6}{30}$
   $\frac{1}{8} = \frac{?4}{32}$ $\frac{1}{6} = \frac{?3}{18}$ $\frac{1}{12} = \frac{?5}{60}$ $\frac{1}{16} = \frac{?3}{48}$
   b. $\frac{3}{4} = \frac{?27}{36}$ $\frac{2}{3} = \frac{?8}{24}$ $\frac{17}{32} = \frac{34}{64}$ $\frac{3}{5} = \frac{?12}{20}$
   $\frac{5}{8} = \frac{?40}{64}$ $\frac{5}{6} = \frac{?40}{48}$ $\frac{17}{24} = \frac{?68}{96}$ $\frac{11}{16} = \frac{55}{80}$
   c. $\frac{21}{50} = \frac{?42}{100}$ $\frac{19}{25} = \frac{?76}{100}$ $\frac{13}{20} = \frac{?65}{100}$ $\frac{2}{5} = \frac{40}{100}$
   → $\frac{3}{4} = \frac{?75}{100}$ $\frac{7}{10} = \frac{?70}{100}$ $\frac{1}{4} = \frac{25}{100}$ $\frac{1}{2} = \frac{50}{100}$

2. Express:
   a. $\frac{1}{4}$ as 32nds 8
   b. $\frac{1}{2}$ as 8ths 4/8
   c. $\frac{2}{3}$ as 72nds 48
   d. $\frac{5}{6}$ as 24ths 20
   e. $\frac{21}{25}$ as 100ths 84
   f. $\frac{3}{5}$ as 20ths 12
   g. $\frac{9}{12}$ as 72nds 54
   h. $\frac{4}{10}$ as 60ths 24
   i. $\frac{9}{32}$ as 64ths 18
   j. $\frac{3}{4}$ as 32nds 24

3. Equivalent fractions may be developed from the same fractional number in simplest form by expressing it in successively higher terms.
   Examples are $\frac{1}{2}, \frac{2}{4}, \frac{3}{6}, \frac{4}{8}, \ldots$ and $\frac{2}{5}, \frac{4}{10}, \frac{6}{15}, \frac{8}{20}, \ldots$
   Write the group of fractions equivalent to:
   a. $\frac{1}{6}$ 2/12 3/18   e. $\frac{1}{5}$ 2/10   i. $\frac{3}{4}$ 4/8   m. $\frac{1}{15}$ 2/30   q. $\frac{17}{25}$
   b. $\frac{7}{9}$ 8/18   f. $\frac{3}{16}$ 4/32   j. $\frac{5}{12}$ 6/24   n. $\frac{7}{13}$ 8/32   r. $\frac{3}{11}$
   c. $\frac{2}{7}$ 3/14   g. $\frac{2}{7}$ 3/14   k. $\frac{5}{6}$ 6/12   o. $\frac{3}{4}$ 4/8   s. $\frac{7}{30}$
   d. $\frac{7}{10}$ 8/20   h. $\frac{1}{8}$ 2/16   l. $\frac{2}{3}$ 4/6   p. $\frac{23}{50}$ 24/100   t. $\frac{3}{20}$

4. For each of the following fractions, write the simplest form and three other equivalent fractions:
   a. $\frac{12}{18}$ 4/6 2/3   c. $\frac{56}{64}$ 7/8   e. $\frac{40}{96}$ 5/12   g. $\frac{24}{84}$ 4/14   i. $\frac{143}{208}$
   b. $\frac{27}{36}$ 3/4   d. $\frac{12}{44}$ 3/11   f. $\frac{30}{48}$ 5/8   h. $\frac{42}{189}$ j. $\frac{30}{100}$ 3/10

1-8 EXPRESSING IMPROPER FRACTIONS AS MIXED OR WHOLE NUMBERS

To express an improper fraction as a mixed or whole number, we divide the numerator by the denominator and write the remainder, if any, as a fraction expressed in lowest terms.

> Express $\frac{8}{4}$ as a whole number.
>
> $$\frac{8}{4} = 2$$
>
> Answer, 2
>
> Express $\frac{9}{5}$ as a mixed number.
>
> $$\frac{9}{5} = 1\frac{4}{5}$$
>
> Answer, $1\frac{4}{5}$

EXERCISES

Express each of the following as a numeral for a whole or a mixed number:

1. $\frac{4}{4}$ $\frac{9}{9}$ $\frac{12}{12}$ $\frac{20}{20}$ $\frac{18}{3}$ $\frac{14}{7}$ $\frac{54}{6}$ $\frac{35}{5}$ $\frac{40}{10}$ $\frac{96}{16}$
2. $\frac{6}{5}$ $\frac{13}{8}$ $\frac{11}{6}$ $\frac{7}{3}$ $\frac{9}{2}$ $\frac{22}{9}$ $\frac{31}{4}$ $\frac{43}{10}$ $\frac{37}{12}$ $\frac{56}{25}$
3. $\frac{14}{8}$ $\frac{15}{6}$ $\frac{38}{4}$ $\frac{45}{10}$ $\frac{30}{18}$ $\frac{42}{24}$ $\frac{34}{20}$ $\frac{54}{12}$ $\frac{69}{15}$ $\frac{100}{16}$

1-9 EXPRESSING MIXED AND WHOLE NUMBERS AS IMPROPER FRACTIONS

To express a mixed number as an improper fraction, we multiply the whole number by the denominator of the fraction and add the numerator of the fraction to this product. Then we write this sum over the denominator of the fraction.

> Express $3\frac{2}{5}$ as an improper fraction.
>
> Briefly, $3\frac{2}{5} = \frac{15 + 2}{5} = \frac{17}{5}$
>
> or $3\frac{2}{5} = 3 + \frac{2}{5} = \frac{15}{5} + \frac{2}{5} = \frac{17}{5}$
>
> Answer, $\frac{17}{5}$

Review of Number

To express a whole number as a fraction with a specified denominator, we multiply the given whole number by the fraction equivalent to one (1) that has the specified denominator as both the numerator and denominator.

> Change 4 to tenths.
> $$\frac{4}{1} \times 1 = \frac{4}{1} \times \frac{10}{10} = \frac{40}{10}$$
> Answer, $\frac{40}{10}$

EXERCISES

1. Express each of the following as an improper fraction:
   a. $1\frac{1}{4}$ $1\frac{1}{3}$ $1\frac{1}{9}$ $1\frac{1}{5}$ $1\frac{1}{12}$ $1\frac{1}{10}$ $1\frac{1}{8}$ $1\frac{1}{16}$ $1\frac{1}{60}$ $1\frac{1}{25}$
   b. $1\frac{2}{3}$ $1\frac{7}{8}$ $1\frac{3}{5}$ $1\frac{9}{10}$ $1\frac{13}{16}$ $1\frac{5}{12}$ $1\frac{17}{20}$ $1\frac{19}{24}$ $1\frac{11}{15}$ $1\frac{67}{100}$
   c. $9\frac{1}{2}$ $2\frac{1}{3}$ $10\frac{1}{5}$ $7\frac{1}{4}$ $3\frac{1}{7}$ $4\frac{1}{10}$ $2\frac{1}{9}$ $8\frac{1}{6}$ $3\frac{1}{12}$ $5\frac{1}{16}$
   d. $2\frac{5}{6}$ $6\frac{4}{5}$ $9\frac{7}{8}$ $3\frac{4}{7}$ $6\frac{9}{10}$ $5\frac{13}{20}$ $18\frac{2}{3}$ $4\frac{12}{25}$ $11\frac{15}{16}$ $4\frac{53}{100}$

2. Express 1 as:
   a. tenths
   b. eighteenths
   c. fiftieths
   d. thirty-sixths

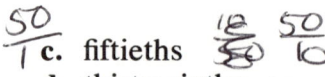

3. How many fourths are in sixteen?

4. How many twelfths are in 7?

5. How many:
   a. quarter inches are in 5 inches?
   b. half inches are in 3 inches?
   c. sixteenth inches are in 7 inches?
   d. eighth inches are in 14 inches?

6. Find the missing numbers:
   a. $1 = \frac{?}{8}$
   b. $5 = \frac{?}{7}$
   c. $3 = \frac{?}{10}$
   d. $8 = \frac{?}{12}$
   e. $6 = \frac{?}{100}$
   f. $10 = \frac{?}{5}$

1–10 EXPRESSING FRACTIONS AS EQUIVALENT FRACTIONS WITH A COMMON DENOMINATOR

A *common denominator* is a number that can be divided exactly by the denominators of all the given fractions. It is a multiple of the denominators of the given fractions.

The *lowest common denominator* (L.C.D.) is the smallest whole number, with zero excluded, that can be divided exactly by the denominators of all the given fractions. It is the least common multiple of the given denominators.

The L.C.D. may be the denominator of one of its fractions.

> The L.C.D. of $\frac{1}{4}$ and $\frac{1}{2}$ is 4.

The L.C.D. may be the product of the denominators of the given fractions.

> The L.C.D. of $\frac{1}{3}$ and $\frac{1}{2}$ is 6.

When the L.C.D. is greater than any given denominator but smaller than the product of all the denominators, to find this L.C.D. we may follow the method used in finding the least common multiple (p. 93.)

> The L.C.D. of $\frac{5}{6}$ and $\frac{3}{4}$ is 12.

To express two or more fractions each to its equivalent fraction having the lowest common denominator as its denominator, we first find the L.C.D., then we change each given fraction to higher terms as an equivalent fraction with the L.C.D. as its denominator.

Review of Number

> Change $\frac{5}{6}$ and $\frac{3}{4}$ each to an equivalent fraction having the L.C.D. as its denominator.
>
> $\frac{5}{6}$ and $\frac{3}{4}$ L.C.D. = 12
>
> $\frac{5}{6} = \frac{5 \times 2}{6 \times 2} = \frac{10}{12}$ $\frac{3}{4} = \frac{3 \times 3}{4 \times 3} = \frac{9}{12}$
>
> *Answer*, $\frac{10}{12}$ and $\frac{9}{12}$

EXERCISES

Find the lowest common denominator (L.C.D.) of the given fractions in each of the following; then change each fraction to an equivalent fraction with the L.C.D. as its denominator:

1. a. $\frac{1}{3}$ and $\frac{1}{6}$
   b. $\frac{1}{4}$ and $\frac{11}{12}$
   c. $\frac{5}{8}$ and $\frac{9}{16}$
   d. $\frac{9}{10}$ and $\frac{1}{2}$
   e. $\frac{1}{12}$ and $\frac{2}{3}$
   f. $\frac{11}{32}$ and $\frac{3}{4}$
   g. $\frac{13}{18}$ and $\frac{1}{2}$
   h. $\frac{17}{36}$ and $\frac{7}{12}$
   i. $\frac{7}{8}$ and $\frac{21}{32}$
   j. $\frac{3}{10}$ and $\frac{79}{100}$
   k. $\frac{29}{36}$ and $\frac{5}{6}$
   l. $\frac{4}{5}$ and $\frac{17}{60}$

2. a. $\frac{1}{2}$ and $\frac{1}{5}$
   b. $\frac{1}{3}$ and $\frac{1}{4}$
   c. $\frac{2}{3}$ and $\frac{6}{7}$
   d. $\frac{5}{6}$ and $\frac{3}{5}$
   e. $\frac{1}{7}$ and $\frac{1}{9}$
   f. $\frac{9}{10}$ and $\frac{1}{3}$
   g. $\frac{2}{5}$ and $\frac{7}{8}$
   h. $\frac{1}{4}$ and $\frac{13}{15}$
   i. $\frac{11}{12}$ and $\frac{4}{5}$
   j. $\frac{21}{25}$ and $\frac{3}{4}$
   k. $\frac{11}{20}$ and $\frac{2}{3}$
   l. $\frac{5}{9}$ and $\frac{7}{16}$

3. a. $\frac{7}{8}$ and $\frac{1}{6}$
   b. $\frac{3}{4}$ and $\frac{7}{10}$
   c. $\frac{11}{12}$ and $\frac{3}{8}$
   d. $\frac{13}{16}$ and $\frac{5}{12}$
   e. $\frac{5}{6}$ and $\frac{3}{10}$
   f. $\frac{18}{25}$ and $\frac{14}{15}$
   g. $\frac{19}{24}$ and $\frac{11}{16}$
   h. $\frac{7}{12}$ and $\frac{23}{32}$
   i. $\frac{17}{20}$ and $\frac{9}{16}$
   j. $\frac{5}{18}$ and $\frac{13}{24}$
   k. $\frac{13}{48}$ and $\frac{39}{64}$
   l. $\frac{67}{100}$ and $\frac{31}{75}$

4. a. $\frac{1}{5}, \frac{1}{10}$, and $\frac{1}{20}$
   b. $\frac{3}{4}, \frac{1}{2}$, and $\frac{7}{16}$
   c. $\frac{53}{100}, \frac{7}{10}$, and $\frac{4}{5}$
   d. $\frac{11}{24}, \frac{5}{12}$, and $\frac{1}{6}$
   e. $\frac{1}{4}, \frac{1}{5}$, and $\frac{1}{3}$
   f. $\frac{2}{3}, \frac{5}{8}$, and $\frac{1}{2}$
   g. $\frac{5}{6}, \frac{4}{5}$, and $\frac{3}{8}$
   h. $\frac{1}{2}, \frac{2}{3}$, and $\frac{3}{4}$
   i. $\frac{7}{8}, \frac{1}{4}$, and $\frac{1}{6}$
   j. $\frac{5}{6}, \frac{2}{3}$, and $\frac{3}{5}$
   k. $\frac{11}{16}, \frac{5}{12}$, and $\frac{7}{8}$
   l. $\frac{9}{10}, \frac{13}{18}$, and $\frac{1}{4}$
   m. $\frac{3}{8}, \frac{7}{12}$, and $\frac{3}{10}$
   n. $\frac{9}{16}, \frac{5}{8}$, and $\frac{17}{24}$
   o. $\frac{2}{3}, \frac{1}{6}$, and $\frac{3}{10}$
   p. $\frac{11}{12}, \frac{17}{20}$, and $\frac{15}{16}$

1–11 COMPARING FRACTIONS; EQUIVALENT FRACTIONS

To determine whether two fractions are equivalent, we first express each given fraction in lowest terms. If the resulting fractions are the same, then the given fractions are equivalent.

> The symbol $>$ means "is greater than."
> The symbol $<$ means "is less than."

To compare two fractions with the same (like) *denominator,* we select the fraction with the greater numerator as the greater fraction.

> $\frac{9}{10}$ is greater than $\frac{3}{10}$ or $\frac{9}{10} > \frac{3}{10}$.

To compare two fractions with different (unlike) *denominators,* we first express the given fractions as fractions with a common denominator and then select the given fraction that is equivalent to the fraction having the greater numerator and common denominator as the greater fraction.

> Which is smaller: $\frac{4}{5}$ or $\frac{5}{6}$?
> Since $\frac{4}{5} = \frac{24}{30}$ and $\frac{5}{6} = \frac{25}{30}$,
> then $\frac{4}{5}$ is less than $\frac{5}{6}$ or $\frac{4}{5} < \frac{5}{6}$.

To compare two fractions by the cross product method, if a represents the numerator of the first fraction; b, the denominator of the first fraction; c, the numerator of the second fraction; and d, the denominator of the second fraction, then

(1) $\frac{a}{b} = \frac{c}{d}$ if $a \times d = c \times b$ The two fractions are equivalent.

(2) $\frac{a}{b} > \frac{c}{d}$ if $a \times d > c \times b$ The first fraction is greater.

(3) $\frac{a}{b} < \frac{c}{d}$ if $a \times d < c \times b$ The first fraction is smaller.

Review of Number

EXERCISES

Test whether each of the following pairs of fractions are equivalent by using the method of:

1. Lowest terms:
   - a. $\frac{2}{8}$ and $\frac{3}{12}$
   - b. $\frac{20}{25}$ and $\frac{24}{40}$
   - c. $\frac{5}{10}$ and $\frac{18}{24}$
   - d. $\frac{6}{21}$ and $\frac{14}{49}$
   - e. $\frac{13}{39}$ and $\frac{15}{60}$
   - f. $\frac{35}{45}$ and $\frac{28}{36}$
   - g. $\frac{15}{24}$ and $\frac{40}{64}$
   - h. $\frac{9}{21}$ and $\frac{12}{32}$
   - i. $\frac{30}{36}$ and $\frac{45}{54}$
   - j. $\frac{28}{49}$ and $\frac{36}{63}$

2. Equal cross products:
   - a. $\frac{2}{3}$ and $\frac{12}{15}$
   - b. $\frac{15}{27}$ and $\frac{10}{18}$
   - c. $\frac{12}{14}$ and $\frac{15}{20}$
   - d. $\frac{16}{28}$ and $\frac{18}{33}$
   - e. $\frac{18}{45}$ and $\frac{16}{40}$
   - f. $\frac{21}{30}$ and $\frac{14}{20}$
   - g. $\frac{20}{45}$ and $\frac{8}{18}$
   - h. $\frac{5}{8}$ and $\frac{20}{28}$
   - i. $\frac{21}{24}$ and $\frac{3}{4}$
   - j. $\frac{14}{22}$ and $\frac{35}{55}$

3. Either method:
   - a. $\frac{6}{42}$ and $\frac{8}{56}$
   - b. $\frac{8}{10}$ and $\frac{10}{15}$
   - c. $\frac{32}{48}$ and $\frac{28}{42}$
   - d. $\frac{16}{20}$ and $\frac{24}{42}$
   - e. $\frac{34}{41}$ and $\frac{14}{21}$
   - f. $\frac{6}{10}$ and $\frac{5}{12}$
   - g. $\frac{39}{52}$ and $\frac{30}{40}$
   - h. $\frac{27}{33}$ and $\frac{54}{66}$
   - i. $\frac{24}{64}$ and $\frac{18}{48}$
   - j. $\frac{36}{44}$ and $\frac{24}{33}$

4. Use the common denominator method to determine which of the following statements are true:
   - a. $\frac{1}{6} > \frac{1}{8}$
   - b. $\frac{1}{10} > \frac{1}{5}$
   - c. $\frac{2}{3} > \frac{5}{6}$
   - d. $\frac{6}{7} > \frac{10}{14}$
   - e. $\frac{5}{6} > \frac{13}{18}$
   - f. $\frac{3}{8} > \frac{5}{12}$
   - g. $\frac{3}{10} < \frac{4}{15}$
   - h. $\frac{2}{5} < \frac{3}{10}$
   - i. $\frac{3}{4} < \frac{4}{7}$
   - j. $\frac{7}{10} < \frac{5}{8}$
   - k. $\frac{4}{5} < \frac{7}{9}$
   - l. $\frac{1}{3} < \frac{3}{8}$
   - m. $\frac{11}{18} > \frac{16}{27}$
   - n. $\frac{3}{5} < \frac{7}{13}$
   - o. $\frac{19}{30} > \frac{53}{90}$
   - p. $\frac{27}{28} < \frac{20}{21}$
   - q. $\frac{9}{10} > \frac{7}{8}$
   - r. $\frac{24}{36} < \frac{18}{24}$
   - s. $\frac{9}{16} > \frac{5}{9}$
   - t. $\frac{5}{18} < \frac{4}{15}$

5. Use the cross product test to determine which of the following statements are true:
   - a. $\frac{1}{6} > \frac{1}{8}$
   - b. $\frac{1}{10} > \frac{1}{3}$
   - c. $\frac{3}{4} > \frac{5}{6}$
   - d. $\frac{2}{5} > \frac{1}{3}$
   - e. $\frac{5}{6} > \frac{9}{10}$
   - f. $\frac{3}{4} > \frac{13}{18}$
   - g. $\frac{6}{7} < \frac{7}{8}$
   - h. $\frac{11}{24} < \frac{12}{25}$
   - i. $\frac{2}{3} < \frac{7}{11}$
   - j. $\frac{8}{15} < \frac{3}{5}$
   - k. $\frac{9}{10} < \frac{10}{11}$
   - l. $\frac{4}{9} < \frac{7}{16}$
   - m. $\frac{29}{36} > \frac{4}{5}$
   - n. $\frac{11}{20} > \frac{17}{30}$
   - o. $\frac{5}{16} < \frac{7}{24}$
   - p. $\frac{7}{12} < \frac{13}{20}$
   - q. $\frac{9}{16} > \frac{3}{5}$
   - r. $\frac{28}{35} < \frac{24}{42}$
   - s. $\frac{1}{2} > \frac{51}{100}$
   - t. $\frac{7}{15} < \frac{19}{45}$

6. Arrange each of the following groups of fractions according to size, greatest first:
   - a. $\frac{13}{25}, \frac{7}{25},$ and $\frac{19}{25}$
   - b. $\frac{3}{8}, \frac{3}{10},$ and $\frac{3}{5}$
   - c. $\frac{5}{6}, \frac{17}{24},$ and $\frac{3}{4}$
   - d. $\frac{13}{24}, \frac{9}{16},$ and $\frac{7}{12}$
   - e. $\frac{3}{4}, \frac{6}{7},$ and $\frac{2}{3}$
   - f. $\frac{7}{30}, \frac{2}{15},$ and $\frac{1}{5}$

7. Arrange each of the following groups of fractions according to size, smallest first:
   - a. $\frac{3}{7}, \frac{2}{7},$ and $\frac{5}{7}$
   - b. $\frac{5}{6}, \frac{5}{12},$ and $\frac{5}{16}$
   - c. $\frac{7}{10}, \frac{3}{5},$ and $\frac{1}{2}$
   - d. $\frac{5}{8}, \frac{2}{3},$ and $\frac{3}{5}$
   - e. $\frac{16}{25}, \frac{13}{20},$ and $\frac{63}{100}$
   - f. $\frac{9}{16}, \frac{3}{5},$ and $\frac{11}{20}$

COMMON FRACTIONS AND MIXED NUMBERS—COMPUTATION

1–12 ADDITION OF FRACTIONS AND MIXED NUMBERS

(1) *To add fractions with like denominators*, we add the numerators and write the sum over the common denominator. We express the answer in simplest form.

> (1) Vertically:
> $$\frac{3}{10}$$
> $$+\frac{5}{10}$$
> $$\overline{\frac{8}{10}} = \frac{4}{5}$$
>
> or
>
> Horizontally:
> $$\frac{3}{10} + \frac{5}{10} = \frac{3+5}{10}$$
> $$= \frac{8}{10} = \frac{4}{5}$$
>
> Answer, $\frac{4}{5}$

(2) *To add fractions with unlike denominators*, we express the given fractions as equivalent fractions having a common denominator and add as we do with like denominators.

> (2) Add:
> $$\frac{2}{3} = \frac{8}{12}$$
> $$\frac{3}{4} = \frac{9}{12}$$
> $$\overline{\frac{17}{12}} = 1\frac{5}{12}$$
>
> Answer, $1\frac{5}{12}$

(3) *To add a fraction and a whole number*, we annex the fraction to the whole number.

> (3) $9 + \frac{3}{8} = 9\frac{3}{8}$
>
> Answer, $9\frac{3}{8}$

(4) *To add a mixed number and a whole number*, we add the whole numbers and annex the fraction to this sum.

> (4) $4\frac{9}{16} + 7 = 11\frac{9}{16}$
>
> Answer, $11\frac{9}{16}$

Review of Number

(5) *To add a mixed number and a fraction*, we add the fractions and annex their sum to the whole number.

(5) Add:
$$3\tfrac{1}{2} = 3\tfrac{5}{10}$$
$$\tfrac{2}{5} = \tfrac{4}{10}$$
$$\overline{3\tfrac{9}{10}}$$

Answer, $3\tfrac{9}{10}$

(6) *To add mixed numbers*, we first add the fractions and then add this sum to the sum of the whole numbers. We express answers in simplest form.

(6) Add:
$$2\tfrac{7}{8} = 2\tfrac{7}{8}$$
$$5\tfrac{3}{4} = 5\tfrac{6}{8}$$
$$\overline{7\tfrac{13}{8}} = 8\tfrac{5}{8}$$

Answer, $8\tfrac{5}{8}$

EXERCISES

Add:

1. $\tfrac{1}{5}$ $\quad$ $\tfrac{3}{7}$ $\quad$ $\tfrac{3}{4}$ $\quad$ $\tfrac{7}{16}$ $\quad$ $\tfrac{7}{18}$ $\quad$ $\tfrac{4}{16}$ $\quad$ $\tfrac{1}{4}$ $\quad$ $\tfrac{3}{8}$ $\quad$ $\tfrac{5}{12}$
   $\tfrac{3}{5}$ $\quad$ $\tfrac{2}{7}$ $\quad$ $\tfrac{1}{4}$ $\quad$ $\tfrac{9}{16}$ $\quad$ $\tfrac{11}{18}$ $\quad$ $\tfrac{9}{16}$ $\quad$ $\tfrac{1}{4}$ $\quad$ $\tfrac{1}{8}$ $\quad$ $\tfrac{4}{12}$

2. $\tfrac{2}{5}$ $\quad$ $\tfrac{3}{4}$ $\quad$ $\tfrac{5}{9}$ $\quad$ $\tfrac{5}{12}$ $\quad$ $\tfrac{5}{6}$ $\quad$ $\tfrac{6}{16}$ $\quad$ $\tfrac{7}{16}$ $\quad$ $\tfrac{2}{3}$ $\quad$ $\tfrac{7}{10}$
   $\tfrac{4}{5}$ $\quad$ $\tfrac{3}{4}$ $\quad$ $\tfrac{5}{9}$ $\quad$ $\tfrac{10}{12}$ $\quad$ $\tfrac{3}{6}$ $\quad$ $\tfrac{13}{16}$ $\quad$ $\tfrac{13}{16}$ $\quad$ $\tfrac{2}{3}$ $\quad$ $\tfrac{9}{10}$

3. $\tfrac{3}{4}$ $\quad$ $\tfrac{3}{5}$ $\quad$ $\tfrac{9}{16}$ $\quad$ $\tfrac{2}{3}$ $\quad$ $\tfrac{7}{12}$ $\quad$ $\tfrac{1}{6}$ $\quad$ $\tfrac{23}{60}$ $\quad$ $\tfrac{7}{20}$ $\quad$ $\tfrac{27}{32}$
   $\tfrac{1}{8}$ $\quad$ $\tfrac{3}{10}$ $\quad$ $\tfrac{1}{4}$ $\quad$ $\tfrac{1}{6}$ $\quad$ $\tfrac{7}{24}$ $\quad$ $\tfrac{1}{2}$ $\quad$ $\tfrac{8}{15}$ $\quad$ $\tfrac{21}{100}$ $\quad$ $\tfrac{3}{8}$

4. $\tfrac{1}{4}$ $\quad$ $\tfrac{3}{4}$ $\quad$ $\tfrac{2}{3}$ $\quad$ $\tfrac{1}{4}$ $\quad$ $\tfrac{3}{5}$ $\quad$ $\tfrac{1}{3}$ $\quad$ $\tfrac{1}{5}$ $\quad$ $\tfrac{3}{4}$ $\quad$ $\tfrac{5}{8}$
   $\tfrac{1}{3}$ $\quad$ $\tfrac{1}{6}$ $\quad$ $\tfrac{1}{4}$ $\quad$ $\tfrac{3}{5}$ $\quad$ $\tfrac{3}{8}$ $\quad$ $\tfrac{1}{10}$ $\quad$ $\tfrac{1}{2}$ $\quad$ $\tfrac{2}{3}$ $\quad$ $\tfrac{2}{3}$

5. $\tfrac{3}{4}$ $\quad$ $\tfrac{1}{4}$ $\quad$ $\tfrac{5}{8}$ $\quad$ $\tfrac{5}{16}$ $\quad$ $\tfrac{5}{8}$ $\quad$ $\tfrac{11}{12}$ $\quad$ $\tfrac{3}{8}$ $\quad$ $\tfrac{1}{4}$ $\quad$ $\tfrac{3}{10}$
   $\tfrac{1}{6}$ $\quad$ $\tfrac{3}{10}$ $\quad$ $\tfrac{1}{6}$ $\quad$ $\tfrac{7}{10}$ $\quad$ $\tfrac{5}{12}$ $\quad$ $\tfrac{13}{16}$ $\quad$ $\tfrac{9}{10}$ $\quad$ $\tfrac{5}{6}$ $\quad$ $\tfrac{9}{16}$

6.
$\frac{1}{5}$	$\frac{3}{4}$	$\frac{11}{16}$	$\frac{5}{6}$	$\frac{9}{24}$	$\frac{3}{8}$	$\frac{3}{4}$	$\frac{3}{10}$	$\frac{7}{8}$
$\frac{1}{5}$	$\frac{1}{4}$	$\frac{9}{16}$	$\frac{2}{6}$	$\frac{12}{24}$	$\frac{1}{2}$	$\frac{2}{3}$	$\frac{1}{5}$	$\frac{11}{12}$
$\frac{1}{5}$	$\frac{3}{4}$	$\frac{14}{16}$	$\frac{5}{6}$	$\frac{19}{24}$	$\frac{5}{16}$	$\frac{1}{2}$	$\frac{29}{100}$	$\frac{9}{16}$

7. a. $\frac{3}{10} + \frac{3}{10}$ $\frac{49}{100} + \frac{51}{100}$ $\frac{1}{2} + \frac{1}{5}$ $\frac{3}{10} + \frac{4}{5}$

   b. $\frac{13}{100} + \frac{9}{10}$ $\frac{3}{100} + \frac{9}{10} + \frac{1}{5}$ $\frac{7}{10} + \frac{11}{100} + \frac{1}{2}$

8.
3	$\frac{3}{8}$	$4\frac{3}{4}$	8	$3\frac{3}{5}$	$\frac{1}{3}$	$\frac{7}{12}$	$7\frac{1}{3}$
$\frac{3}{5}$	5	3	$7\frac{13}{16}$	$\frac{1}{5}$	$2\frac{1}{2}$	$4\frac{1}{4}$	$\frac{2}{3}$

9. $3 + \frac{2}{3}$ $3\frac{1}{4} + 4$ $\frac{5}{6} + 9$ $8 + 3\frac{11}{16}$ $\frac{5}{12} + 2\frac{7}{12}$ $\frac{9}{16} + 3 + 7\frac{1}{2}$

10.
$2\frac{1}{4}$	$5\frac{2}{9}$	$4\frac{1}{6}$	$4\frac{3}{10}$	$6\frac{2}{5}$	$15\frac{7}{12}$	$3\frac{3}{5}$	$6\frac{3}{4}$
$3\frac{1}{4}$	$3\frac{5}{9}$	$2\frac{1}{6}$	$5\frac{1}{10}$	$5\frac{3}{5}$	$13\frac{5}{12}$	$2\frac{4}{5}$	$2\frac{3}{4}$

11.
$5\frac{3}{5}$	$3\frac{3}{8}$	$8\frac{2}{3}$	$3\frac{2}{5}$	$2\frac{3}{4}$	$3\frac{5}{12}$	$3\frac{1}{3}$	$12\frac{5}{12}$
$8\frac{3}{10}$	$2\frac{11}{16}$	$5\frac{1}{4}$	$5\frac{1}{2}$	$3\frac{3}{10}$	$5\frac{3}{8}$	$4\frac{1}{6}$	$13\frac{1}{4}$

12.
$3\frac{11}{16}$	$3\frac{5}{8}$	$3\frac{9}{10}$	$11\frac{5}{6}$	$8\frac{3}{4}$	$5\frac{5}{6}$	$3\frac{1}{2}$	$9\frac{7}{10}$
$6\frac{10}{16}$	$2\frac{7}{8}$	$4\frac{7}{10}$	$12\frac{5}{6}$	$7\frac{2}{3}$	$3\frac{5}{8}$	$8\frac{5}{6}$	$12\frac{1}{2}$

13.
$2\frac{3}{8}$	$3\frac{2}{5}$	$5\frac{7}{10}$	$3\frac{3}{16}$	$3\frac{1}{2}$	$5\frac{5}{6}$
$3\frac{3}{8}$	$4\frac{4}{5}$	$3\frac{3}{10}$	$4\frac{1}{4}$	$6\frac{1}{5}$	$2\frac{3}{4}$
$1\frac{1}{8}$	$6\frac{4}{5}$	$1\frac{9}{10}$	$2\frac{1}{2}$	$4\frac{1}{10}$	$1\frac{2}{3}$

14.
$7\frac{7}{10}$	$3\frac{1}{2}$	$6\frac{5}{6}$	$6\frac{1}{4}$	$\frac{11}{12}$	$2\frac{2}{3}$
$6\frac{1}{6}$	$7\frac{4}{5}$	$3\frac{3}{4}$	$3\frac{5}{8}$	$1\frac{2}{3}$	$\frac{3}{4}$
$4\frac{7}{8}$	$5\frac{2}{3}$	$8\frac{7}{8}$	$1\frac{13}{16}$	$8\frac{5}{8}$	$6\frac{1}{2}$

15. $1\frac{3}{4} + 5\frac{19}{32} + 3\frac{11}{16}$ $3\frac{7}{8} + 2\frac{5}{16} + 4\frac{1}{6}$ $1\frac{5}{8} + 6\frac{1}{6} + 5\frac{3}{4}$ $3\frac{15}{16} + 7\frac{13}{24} + 2\frac{11}{12}$

16. Determine which of the following statements are true:

   a. $\frac{3}{8} + \frac{5}{8} = \frac{5}{8} + \frac{3}{8}$

   b. $4\frac{2}{3} + 2\frac{1}{2} = 2\frac{1}{2} + 4\frac{2}{3}$

   c. $\left(\frac{3}{16} + \frac{1}{16}\right) + \frac{5}{16} = \frac{3}{16} + \left(\frac{1}{16} + \frac{5}{16}\right)$

   d. $2\frac{1}{3} + \left(1\frac{1}{6} + 3\frac{1}{2}\right) = \left(2\frac{1}{3} + 1\frac{1}{6}\right) + 3\frac{1}{2}$

Review of Number

1–13 SUBTRACTION OF FRACTIONS AND MIXED NUMBERS

(1) *To subtract fractions with like denominators,* we subtract the numerators and write the difference over the common denominator. We express the answer in lowest terms.

> (1) Vertically:
> $$\begin{array}{r}\frac{9}{10}\\-\frac{3}{10}\\\hline \frac{6}{10}=\frac{3}{5}\end{array}$$
> or
> Horizontally:
> $$\frac{9}{10} - \frac{3}{10} = \frac{9-3}{10}$$
> $$= \frac{6}{10} = \frac{3}{5}$$
> *Answer,* $\frac{3}{5}$

(2) *To subtract fractions with unlike denominators,* we express the fractions as equivalent fractions having a common denominator and subtract as we do with like denominators.

> (2) Subtract:
> $$\frac{7}{12} = \frac{7}{12}$$
> $$\frac{1}{3} = \frac{4}{12}$$
> $$\overline{\frac{3}{12}} = \frac{1}{4}$$
> *Answer,* $\frac{1}{4}$

(3) *To subtract a fraction or mixed number from a whole number,* we regroup by taking one (1) from the whole number and express it as a fraction making the numerator and denominator the same. Then we subtract the fractions and we subtract the whole numbers.

> (3) Subtract:
> $$6 = 5\frac{8}{8}$$
> $$2\frac{3}{8} = 2\frac{3}{8}$$
> $$\overline{3\frac{5}{8}}$$
> *Answer,* $3\frac{5}{8}$

(4) *To subtract a whole number from a mixed number,* we find the difference between the whole numbers and annex the fraction to this difference.

> (4) Subtract:
> $$9\frac{3}{4} - 4 = 5\frac{3}{4}$$
> *Answer,* $5\frac{3}{4}$

(5) *To subtract a fraction or a mixed number from a mixed number,* first we subtract the fractions then the whole numbers. If the fraction in the subtrahend is greater than the fraction in the minuend, we regroup by taking one (1) from the whole number in the minuend to increase the fraction of the minuend, then we subtract. We express answers in simplest form.

> (5) Subtract:
> $9\frac{1}{2} = 9\frac{3}{6} = 8\frac{9}{6}$
> $1\frac{2}{3} = 1\frac{4}{6} = 1\frac{4}{6}$
> _____
> $7\frac{5}{6}$
>
> Answer, $7\frac{5}{6}$

EXERCISES

Subtract:

1. $\frac{5}{9}$; $\frac{3}{5}$; $\frac{14}{15}$; $\frac{7}{10}$; $\frac{11}{12}$; $\frac{1}{2}$; $\frac{23}{32}$; $\frac{5}{6}$; $\frac{7}{8}$
   $\frac{3}{9}$; $\frac{1}{5}$; $\frac{2}{15}$; $\frac{5}{10}$; $\frac{7}{12}$; $\frac{1}{2}$; $\frac{15}{32}$; $\frac{5}{6}$; $\frac{3}{8}$

2. $\frac{1}{4}$; $\frac{11}{12}$; $\frac{7}{10}$; $\frac{1}{3}$; $\frac{4}{5}$; $\frac{7}{8}$; $\frac{3}{4}$; $\frac{5}{8}$; $\frac{5}{6}$
   $\frac{1}{8}$; $\frac{3}{4}$; $\frac{3}{20}$; $\frac{3}{12}$; $\frac{1}{2}$; $\frac{11}{24}$; $\frac{1}{3}$; $\frac{5}{10}$; $\frac{3}{4}$

3. $5\frac{3}{10}$; $6\frac{2}{5}$; 7 ; $9\frac{9}{16}$; 1 ; 8 ; $9\frac{3}{4}$; 6 ; $12\frac{5}{24}$
   2 ; 4 ; $4\frac{1}{2}$; 8 ; $\frac{5}{6}$; $\frac{2}{3}$; 5 ; $3\frac{1}{3}$; 11

4. $4\frac{3}{5}$; $8\frac{4}{7}$; $5\frac{5}{6}$; $1\frac{9}{10}$; $2\frac{11}{32}$; $16\frac{7}{12}$; $5\frac{1}{2}$; $4\frac{5}{6}$; $7\frac{3}{4}$
   $\frac{1}{5}$; $3\frac{3}{7}$; $\frac{5}{6}$; $\frac{7}{10}$; $2\frac{11}{32}$; $16\frac{5}{12}$; $\frac{1}{8}$; $\frac{3}{4}$; $2\frac{1}{3}$

5. $7\frac{2}{5}$; $8\frac{1}{8}$; $6\frac{1}{7}$; $8\frac{1}{2}$; $4\frac{1}{6}$; $5\frac{3}{10}$; $1\frac{5}{16}$; $11\frac{6}{7}$; $4\frac{7}{12}$
   $4\frac{3}{5}$; $1\frac{5}{8}$; $2\frac{3}{7}$; $3\frac{7}{8}$; $3\frac{2}{5}$; $1\frac{5}{16}$; $\frac{7}{16}$; $9\frac{13}{14}$; $\frac{3}{4}$

6. $5\frac{7}{12}$; $12\frac{7}{10}$; $20\frac{2}{3}$; 18 ; 10 ; 15 ; $11\frac{5}{9}$; $8\frac{1}{5}$; $9\frac{5}{8}$
   $2\frac{1}{4}$; $9\frac{1}{3}$; $7\frac{1}{6}$; $6\frac{9}{32}$; $9\frac{4}{5}$; $10\frac{3}{4}$; $2\frac{4}{5}$; $7\frac{1}{4}$; $8\frac{15}{16}$

7. a. $\frac{3}{4} - \frac{2}{3}$ b. $5\frac{2}{5} - 5$ c. $8\frac{5}{6} - \frac{3}{8}$ d. $7 - 6\frac{2}{3}$ e. $17\frac{4}{5} - 8\frac{7}{8}$

Review of Number

1-14 MULTIPLICATION OF FRACTIONS AND MIXED NUMBERS

To multiply fractions and mixed numbers, we first express each mixed number, if any, as an improper fraction and each whole number, if any, as a fraction by writing the whole number over 1. Then where possible we divide any numerator and denominator by the greatest possible number that is exactly contained in both (greatest common factor). This reduction simplifies computation. Finally we multiply the resulting numerators and multiply the resulting denominators, expressing the answer in simplest form.

$$\frac{1}{3} \times 5 = \frac{1}{3} \times \frac{5}{1} = \frac{5}{3} = 1\frac{2}{3}$$

Answer, $1\frac{2}{3}$

$$6 \times 2\frac{1}{4} = \frac{\cancel{6}^{3}}{1} \times \frac{9}{\cancel{4}_{2}} = \frac{27}{2} = 13\frac{1}{2}$$

Answer, $13\frac{1}{2}$

$$9\frac{1}{3} \times 1\frac{7}{8} = \frac{\cancel{28}^{7}}{\cancel{3}_{1}} \times \frac{\cancel{15}^{5}}{\cancel{8}_{2}} = \frac{35}{2} = 17\frac{1}{2}$$

Answer, $17\frac{1}{2}$

When we multiply a mixed number and a whole number, the vertical form may also be used.

$$\begin{array}{r} 13 \\ \times 8\frac{2}{5} \\ \hline 104 \\ 5\frac{1}{5} \\ \hline 109\frac{1}{5} \end{array} \qquad \frac{2}{5} \times 13 = \frac{26}{5} = 5\frac{1}{5}$$

Answer, $109\frac{1}{5}$

EXERCISES

Multiply:
1. $\frac{1}{4} \times 4$ $\frac{4}{9} \times 9$ $\frac{1}{14} \times 56$ $\frac{5}{8} \times 80$ $\frac{1}{24} \times 2$

   $\frac{7}{12} \times 4$ $\frac{1}{18} \times 8$ $\frac{3}{4} \times 3$ $\frac{1}{6} \times 32$ $\frac{3}{8} \times 28$

2. $7 \times \frac{1}{7}$ $12 \times \frac{7}{12}$ $15 \times \frac{1}{5}$ $15 \times \frac{3}{5}$ $4 \times \frac{1}{12}$

   $3 \times \frac{7}{9}$ $12 \times \frac{1}{15}$ $5 \times \frac{7}{8}$ $45 \times \frac{1}{10}$ $93 \times \frac{1}{6}$

3. $\frac{1}{7} \times 3$ $\frac{5}{14} \times 1$ $\frac{1}{5} \times 9$ $\frac{2}{5} \times 3$ $\frac{15}{6} \times 5$

   $5 \times \frac{1}{6}$ $2 \times \frac{2}{3}$ $17 \times \frac{1}{5}$ $21 \times \frac{5}{8}$ $10 \times \frac{19}{12}$

4. $\frac{1}{3} \times \frac{1}{2}$ $\frac{5}{8} \times \frac{3}{4}$ $\frac{1}{4} \times \frac{4}{15}$ $\frac{2}{5} \times \frac{1}{8}$ $\frac{2}{3} \times \frac{5}{24}$

   $\frac{3}{4} \times \frac{8}{15}$ $\frac{21}{32} \times \frac{8}{9}$ $\frac{13}{16} \times \frac{6}{13}$ $\frac{25}{14} \times \frac{7}{10}$ $\frac{22}{9} \times \frac{45}{16}$

5. $5 \times 2\frac{3}{5}$ $24 \times 2\frac{11}{12}$ $5 \times 2\frac{7}{10}$ $12 \times 2\frac{15}{32}$ $10 \times 5\frac{2}{3}$

   $3\frac{1}{3} \times 3$ $1\frac{1}{2} \times 4$ $7\frac{3}{10} \times 5$ $1\frac{15}{16} \times 20$ $3\frac{3}{8} \times 5$

6. $\frac{5}{9} \times 3\frac{1}{3}$ $\frac{3}{10} \times 3\frac{1}{6}$ $\frac{9}{16} \times 2\frac{4}{5}$ $\frac{2}{5} \times 1\frac{9}{16}$ $\frac{5}{8} \times 2\frac{2}{5}$

   $3\frac{5}{8} \times \frac{5}{6}$ $1\frac{1}{3} \times \frac{1}{6}$ $1\frac{5}{16} \times \frac{2}{3}$ $\frac{3}{5} \times 5\frac{5}{8}$ $3\frac{3}{4} \times \frac{2}{15}$

7. $2\frac{1}{2} \times 3\frac{1}{2}$ $4\frac{7}{8} \times 1\frac{4}{5}$ $7\frac{1}{3} \times 4\frac{9}{10}$ $2\frac{5}{6} \times 1\frac{4}{9}$ $2\frac{7}{8} \times 5\frac{1}{3}$

   $1\frac{5}{9} \times 3\frac{3}{7}$ $2\frac{4}{5} \times 3\frac{3}{16}$ $3\frac{1}{3} \times 1\frac{1}{2}$ $10\frac{2}{3} \times 3\frac{3}{8}$ $2\frac{7}{10} \times 1\frac{1}{6}$

8. $\frac{22}{5} \times \frac{9}{14} \times \frac{5}{8}$ $8 \times 1\frac{1}{15} \times \frac{3}{8}$ $4\frac{2}{5} \times 3\frac{1}{2} \times 1\frac{1}{3}$

   $1\frac{5}{6} \times 3\frac{1}{2} \times 3\frac{1}{3}$ $1\frac{3}{8} \times 2\frac{2}{3} \times 4\frac{1}{2}$ $6\frac{2}{5} \times \frac{2}{3} \times 4\frac{3}{8}$

9. Multiply vertically:

   $19\frac{2}{3}$ $24\frac{5}{8}$ $3\frac{1}{2}$ $14\frac{5}{6}$ $37\frac{13}{16}$ 72 13 35

   $\underline{9}$ $\underline{32}$ $\underline{5}$ $\underline{9}$ $\underline{26}$ $\underline{21\frac{7}{8}}$ $\underline{9\frac{1}{4}}$ $\underline{79\frac{11}{12}}$

10. a. Which of these statements are true?

    (1) $\frac{2}{3} \times \frac{5}{8} = \frac{5}{8} \times \frac{2}{3}$ (3) $\frac{1}{3} \times \left(\frac{3}{5} \times \frac{1}{2}\right) = \left(\frac{1}{3} \times \frac{3}{5}\right) \times \frac{1}{2}$

    (2) $7\frac{2}{3} \times 4\frac{4}{5} = 4\frac{4}{5} \times 7\frac{2}{3}$ (4) $\left(\frac{7}{10} \times 2\frac{1}{2}\right) \times 2\frac{3}{4} = \frac{7}{10} \times \left(2\frac{1}{2} \times 2\frac{3}{4}\right)$

   b. Does the commutative property hold for the multiplication of fractions?

   c. Does the associative property hold for the multiplication of fractions?

   d. Is $\frac{2}{3} \times (7 + 10) = \left(\frac{2}{3} \times 7\right) + \left(\frac{2}{3} \times 10\right)$ true? Does the distributive property of multiplication over addition hold when fractions are used?

Review of Number   35

11. Square each of the following:

   a. $\frac{3}{4}$ c. $\frac{6}{7}$ e. $\frac{5}{12}$ g. $2\frac{7}{8}$ i. $4\frac{2}{5}$
   b. $\frac{1}{3}$ d. $\frac{5}{8}$ f. $1\frac{1}{3}$ h. $3\frac{1}{7}$ j. $3\frac{5}{16}$

1–15 DIVISION OF FRACTIONS AND MIXED NUMBERS

To divide fractions and mixed numbers, we first express each mixed number, if any, as an improper fraction and each whole number, if any, as a fraction by writing the whole number over 1. We then invert the divisor (number after the division sign ÷) and multiply as in the multiplication of fractions, using reduction where possible. This inverted form of a given fraction is called the *reciprocal* or *multiplicative inverse* of the given fraction. Since both division by a number and multiplication by the reciprocal of this number give the same result, to divide a number by another number, we may instead multiply the first number by the reciprocal of the divisor.

$$\frac{3}{4} \div \frac{4}{5} = \frac{3}{4} \times \frac{5}{4} = \frac{15}{16}$$

Answer, $\frac{15}{16}$

$$15 \div \frac{9}{10} = \frac{\cancel{15}^{5}}{1} \times \frac{10}{\cancel{9}_{3}} = \frac{50}{3} = 16\frac{2}{3}$$

Answer, $16\frac{2}{3}$

$$7\frac{1}{2} \div 3 = \frac{\cancel{15}^{5}}{2} \times \frac{1}{\cancel{3}_{1}} = \frac{5}{2} = 2\frac{1}{2}$$

Answer, $2\frac{1}{2}$

$$2\frac{1}{4} \div 3\frac{3}{8} = \frac{\cancel{9}^{1}}{\cancel{4}_{1}} \times \frac{\cancel{8}^{2}}{\cancel{27}_{3}} = \frac{2}{3}$$

Answer, $\frac{2}{3}$

CHAPTER 1

A *complex fraction* is a fraction in which the numerator or denominator or both have a fraction as a term. *To simplify a complex fraction,* we divide the fraction in the numerator by the fraction in the denominator.

$$\frac{\frac{3}{5}}{\frac{3}{8}} = \frac{3}{5} \div \frac{3}{8} = \frac{\cancel{3}}{5} \times \frac{8}{\cancel{3}} = \frac{8}{5} = 1\frac{3}{5}$$

Answer, $1\frac{3}{5}$

EXERCISES

Divide:

1. $\frac{2}{3} \div 2$ $\frac{15}{16} \div 5$ $\frac{24}{25} \div 4$ $\frac{1}{3} \div 3$ $\frac{1}{4} \div 6$
   $\frac{9}{10} \div 3$ $\frac{5}{6} \div 3$ $\frac{2}{3} \div 12$ $\frac{3}{5} \div 60$ $\frac{12}{25} \div 24$

2. $3\frac{1}{2} \div 7$ $6\frac{2}{5} \div 8$ $7\frac{1}{7} \div 5$ $12\frac{1}{4} \div 15$ $3\frac{7}{8} \div 15$
   $5\frac{1}{2} \div 3$ $2\frac{4}{5} \div 3$ $3\frac{4}{15} \div 7$ $4\frac{2}{3} \div 3$ $5\frac{5}{6} \div 3$

3. $\frac{4}{5} \div \frac{1}{5}$ $\frac{1}{4} \div \frac{3}{4}$ $\frac{7}{8} \div \frac{7}{8}$ $\frac{1}{8} \div \frac{1}{3}$ $\frac{2}{5} \div \frac{1}{3}$
   $\frac{4}{7} \div \frac{4}{21}$ $\frac{1}{2} \div \frac{7}{8}$ $\frac{14}{15} \div \frac{7}{10}$ $\frac{15}{32} \div \frac{15}{16}$ $\frac{2}{3} \div \frac{4}{9}$

4. $6 \div \frac{1}{3}$ $4 \div \frac{4}{5}$ $15 \div \frac{5}{8}$ $3 \div \frac{9}{10}$ $12 \div \frac{8}{9}$
   $5 \div \frac{3}{4}$ $1 \div \frac{1}{8}$ $7 \div \frac{17}{32}$ $1 \div \frac{5}{12}$ $6 \div \frac{18}{25}$

5. $5\frac{2}{3} \div \frac{1}{3}$ $5\frac{1}{4} \div \frac{3}{4}$ $3\frac{1}{3} \div \frac{7}{9}$ $4\frac{7}{8} \div \frac{1}{2}$ $1\frac{9}{16} \div \frac{5}{8}$
   $2\frac{14}{15} \div \frac{11}{12}$ $4\frac{1}{5} \div \frac{7}{10}$ $2\frac{3}{4} \div \frac{11}{16}$ $3\frac{1}{3} \div \frac{1}{2}$ $3\frac{1}{5} \div \frac{7}{12}$

6. $\frac{1}{3} \div 1\frac{1}{3}$ $\frac{1}{3} \div 2\frac{2}{3}$ $\frac{9}{16} \div 2\frac{3}{4}$ $\frac{11}{12} \div 2\frac{1}{2}$ $\frac{3}{8} \div 3\frac{3}{4}$
   $\frac{11}{25} \div 4\frac{2}{5}$ $\frac{3}{10} \div 1\frac{1}{5}$ $\frac{1}{5} \div 1\frac{5}{12}$ $\frac{3}{8} \div 3\frac{1}{3}$ $\frac{2}{3} \div 1\frac{2}{5}$

7. $15 \div 3\frac{3}{4}$ $46 \div 6\frac{4}{7}$ $5 \div 6\frac{2}{3}$ $5 \div 13\frac{1}{3}$ $3 \div 2\frac{1}{4}$
   $11 \div 3\frac{3}{10}$ $2 \div 4\frac{2}{3}$ $8 \div 1\frac{1}{2}$ $15 \div 15\frac{3}{4}$ $1 \div 2\frac{2}{3}$

8. $2\frac{1}{8} \div 4\frac{2}{5}$ $1\frac{5}{6} \div 3\frac{2}{3}$ $12\frac{1}{4} \div 1\frac{3}{4}$ $37\frac{1}{2} \div 3\frac{1}{8}$ $1\frac{3}{4} \div 2\frac{3}{8}$
   $7\frac{1}{2} \div 11\frac{1}{4}$ $3\frac{1}{3} \div 3\frac{1}{3}$ $3\frac{7}{12} \div 2\frac{1}{2}$ $24\frac{1}{2} \div 2\frac{5}{8}$ $9\frac{2}{3} \div 1\frac{5}{24}$

Review of Number

9. $(\frac{4}{5} \div \frac{9}{20}) \div \frac{2}{3}$ $(8\frac{1}{3} \div 2\frac{1}{2}) \div 1\frac{1}{2}$ $5\frac{1}{4} \div (3\frac{5}{6} \div 1\frac{1}{3})$ $1\frac{4}{15} \div (3\frac{4}{5} \div 3\frac{3}{4})$

10. a. $\dfrac{\frac{3}{8}}{\frac{3}{4}}$ $\dfrac{\frac{2}{3}}{\frac{4}{5}}$ $\dfrac{\frac{9}{100}}{\frac{9}{10}}$ $\dfrac{2\frac{5}{8}}{3\frac{3}{4}}$ $\dfrac{4\frac{1}{4}}{\frac{3}{10}}$

   b. $\dfrac{\frac{4}{5}}{2}$ $\dfrac{1\frac{1}{3}}{6}$ $\dfrac{62\frac{1}{2}}{100}$ $\dfrac{12}{\frac{2}{3}}$ $\dfrac{7}{3\frac{1}{2}}$

DECIMAL FRACTIONS—COMPUTATION

A *decimal fraction* is a fractional number whose denominator is some power of ten (10; 100; 1,000; etc.) and is named by a numeral, such as .29, in which the denominator is not written as it is in a common fraction but is expressed by place value. Only the numerators appear in decimal notation.

On the decimal number scale each place has one tenth the value of the next place to the left. By extending the scale to the right of the ones place, we express parts of one. To separate the whole number from the parts or fraction we use a dot which is called a *decimal point*.

millions	hundred thousands	ten thousands	thousands	hundreds	tens	ones	tenths	hundredths	thousandths	ten-thousandths	hundred-thousandths	millionths
3,	6	9	7,	4	0	3.	8	2	1	5	4	6

Each of the following pairs of numerals name the same number: .1 and $\frac{1}{10}$, *tenth;* .01 and $\frac{1}{100}$, *hundredth;* .001 and $\frac{1}{1000}$, *thousandth;* .0001 and $\frac{1}{10,000}$, *ten-thousandth;* .00001 and $\frac{1}{100,000}$, *hundred-thousandth;* .000001 and $\frac{1}{1,000,000}$, *millionth;* etc.

A *mixed decimal* is a number consisting of a whole number and a decimal fraction. Thus, 49.3 is a mixed decimal.

Decimals like .6$\overline{6}$ . . . (sometimes written as .$\overline{6}$) and .2727 . . . (sometimes written as .$\overline{27}$) which have a digit or group of digits repeating endlessly are called *repeating decimals*. The bar indicates the repeating sequence (period) and the three dots indicate that the sequence repeats endlessly.

The decimal form such as .25 is called a *terminating decimal*. However, since .25 may be written in the repeating form .25$\overline{0}$ . . . , we see that a terminating decimal may also be thought of as a repeating decimal.

1–16 ROUNDING DECIMALS

To round a decimal, we find the place to which the number is to be rounded, rewriting the given digits to the left of the required place and dropping all the digits to the right of the required place. If the first digit dropped is 5 or more, we increase the given digit in the required place by 1, otherwise we write the same digit as given. However, as many digits to the left are changed as necessary when the given digit in the required place is 9 and one (1) is added.

.4728 rounded to the nearest:

tenth is	.5
hundredth is	.47
thousandth is	.473

EXERCISES

1. Select the numbers that are nearer to:

a. .6 than .7	.67	.62	.69	.64	.68
b. .2 than .3	.23	.29	.27	.21	.26
c. .97 than .98	.973	.978	.971	.974	.979
d. .80 than .81	.806	.804	.803	.807	.801
e. .356 than .357	.3560	.3566	.3567	.3563	.3568
f. 7 than 8	7.643	7.52	7.49	7.51	7.2

Review of Number

2. Any number from:
   a. .25 to .34 rounded to the nearest tenth is what number?
   b. .725 to .734 rounded to the nearest hundredth is what number?
   c. .4375 to .4384 rounded to the nearest thousandth is what number?
   d. 36.5 to 37.4 rounded to the nearest whole number is what number?

3. Round each of the following numbers to the nearest:
   a. Tenth:
.67	1.54	5.82	3.08	34.35
.36	.68	.835	3.22	6.963

   b. Hundredth:
.529	7.644	13.328	32.7352	8.3709
.846	.593	.3825	8.495	13.0794

   c. Thousandth:
.2817	4.9372	8.1085	37.74945	16.0607
.0362	.0346	.28536	9.7284	37.29588

   d. Ten-thousandth:
.43625	.09253	7.45296	17.380506	6.37926

   e. Hundred-thousandth:
.000037	.000483	.06080	1.451365	3.0906096

   f. Millionth:
.0000073	.0093255	.0000029	1.0293563	2.1848327

   g. Whole number:
3.6	22.4	39.26	9.08	88.501
52.3	7.91	84.56	6.475	2.499

4. When required to find an amount correct to the nearest cent or nearest dollar, we follow the same procedure as with decimals. When the amount contains a fractional part of a cent, we drop the fraction but add one cent when the fraction is $\frac{1}{2}$ or more.

> $1.574 rounded to the nearest cent is $1.57
>
> $.68$\frac{3}{4}$ rounded to the nearest cent is $.69

Round each of the following amounts to the nearest:

  a. Cent:

$.493	$.079	$2.384	$32.708	$8.3974
$.37$\frac{1}{2}$	$.47$\frac{1}{4}$	7.28\frac{3}{4}$	6.37\frac{2}{5}$	28.93\frac{7}{8}$
$.543	$.067	3.65\frac{4}{5}$	63.76\frac{3}{10}$	$82.0829

  b. Dollar:

| $8.49 | $4.27 | $5.79 | $45.28 | $326.85 |
| $7.25 | $7.76 | $38.49 | $57.98 | $430.78 |

1–17 COMPARING DECIMALS

To compare decimals, we express, if necessary, the given numbers as numerals containing the same number of decimal places and take the greater number as the greater decimal.

Which is greater: .04 or .4?

$$.04 = .04$$
$$.4 = .40$$

Answer, .4 is greater.

Is .25 < .115 a true statement?
Since .25 = .250, the statement .250 < .115 is not true.
Answer, No.

EXERCISES

1. Which is greater:

  a. .6 or .59?   **d.** .387 or .39?   **g.** .347 or .3469?   **j.** 7.56 or .742?
  b. .81 or .674?   **e.** 5.4 or .54?   **h.** 2.85 or .456?   **k.** 3.009 or 3.0101?
  c. .3 or .051?   **f.** .673 or 6.45?   **i.** 5.834 or 5.8339?   **l.** .04 or .008?

Review of Number

2. Which is smaller:

   a. .6 or .61?   d. .5 or .499?   g. 7.5 or 7.49?   j. 3.7982 or 3.79821?
   b. .80 or .08? e. .0082 or .009? h. 4.99 or 5.2? k. 7.8 or .79?
   c. .37 or .4?   f. 4.83 or .483? i. 7.54 or 7.054? l. .789 or .7888?

Arrange the following numbers according to size, writing the numeral for:

3. Greatest number first:

   a. 7.92, 79.5, .796, .0799
   b. .03, .3, .0003, .003
   c. .0189, .03, 3.46, .1
   d. 2.6, .06, .023, .28
   e. 5.06, 4.987, 4.99, 5.6

4. Smallest number first:

   a. .35, .0035, 3.5, .035
   b. .49, .485, 4.32, .57
   c. 4.07, .46, .4069, 4.23
   d. .75, .751, .7503, .7499
   e. 2.08, 0.978, 0.9789, 1.3

5. Which of the following are true statements?

   a. .38 < .4   c. .3 = 0.3   e. .059 > .06   g. .699 < .7   i. .0008 > .076
   b. .50 > .5   d. 8.9 < .89   f. .678 < .6781 h. 57.3 = 5.73   j. .898 < 1

1–18 ADDITION OF DECIMALS

To add decimals:

When it is necessary to write the numerals in columns, we write each addend so that the decimal points are under each other. Zeros may be annexed to the numerals naming decimal fractions so that the addends have the same number of decimal places.

Then we add as in the addition of whole numbers, placing the decimal point in the sum directly under the decimal points in the addends.

When an answer ends in one or more zeros to the right of the decimal point, the zeros may be dropped unless it is necessary to show the exact degree of measurement. We check by adding the columns in the opposite direction.

Add: 5.9 + .38

   5.90
    .38

   6.28

Answer, 6.28

Add: .27
      .14
      .29

      .70 = .7

Answer, .7

EXERCISES

1. Use this section of the metric ruler to find the sum of each of the following:

   ```
   |mmmmmmmmmmmmmmmmmmmmmmmmmmmmmmmmm|
   |   1    2    3    4    5    6    7    8    9   |
   |cm                                              |
   ```

a. .4 cm	.2 cm	.3 cm	4.8 cm
.1 cm	.7 cm	.6 cm	1.5 cm
b. 1.3 cm	2.8 cm	.6 cm	2.5 cm
.7 cm	3.7 cm	6.7 cm	5.5 cm

2. In each of the following, first add the common fractions, or mixed numbers, then the decimals. Compare the answers.

   a. $\frac{1}{10}$ .1   b. $\frac{19}{100}$ .19   c. $3\frac{9}{10}$ 3.9
   $+\frac{2}{10}$ $+.2$ $+\frac{16}{100}$ $+.16$ $+4\frac{61}{100}$ $+4.61$

3. Add:

   a.
.3	0.7	.06	.16	.25	4.05	1.3	.41
.2	0.5	.04	.24	.50	3.09	.8	7.
.1	0.8	.07	.56	.53	1.01	.31	.06

   b.
.8	.27	.97	7.4	9.32	1.3	6.741	56.91
.7	.14	.33	6.5	7.17	.86	8.918	43.12
.9	.44	.46	4.3	6.91	2.75	1.234	59.64
.6	.08	.63	2.7	1.31	.7	8.179	30.13

   c.
.6	.07	7.6	4.51	.58	7.186	.4139	463.82
.7	.51	8.9	6.98	3.7	5.825	.8614	501.39
.5	.24	3.1	8.09	0.5	3.491	.9036	379.81
.1	.36	4.5	1.49	6.	1.201	.1234	231.06
.2	.12	5.8	2.81	2.49	5.003	.1698	976.47

   d.
0.6	.68	8.9	7.50	.3	67.08	.13892	719.63
0.4	.19	6.3	6.39	5.69	29.46	.41976	846.45
0.7	.24	7.2	4.87	3.	3.98	.17596	541.61
0.9	.71	5.8	1.99	2.61	37.87	.51234	481.76
0.5	.41	6.1	5.19	.504	26.66	.6781	502.09
0.1	.64	1.7	2.64	.33	.96	.24689	190.86

Review of Number

e. | $.86 | $2.98 | $4.76 | $47.19 | $116.87 | $1.68 | $39.04 | $748.71
    | .94 | 1.71 | 3.41 | 76.96 | 3.41 | 15.73 | 16.87 | 3.14
    | .06 | .69 | .67 | 8.24 | 16.23 | .49 | 348.17 | 86.58
    | .19 | 4.16 | .18 | .47 | 486.19 | 5.86 | 67.23 | 142.60
    | .91 | .47 | 7.86 | 62.73 | 1.98 | 486.17| 119.65 | 238.71

f. | $.64 | $7.16 | $6.81 | $68.71 | $336.14 | $716.84 | $668.77 | $486.17
    | .98 | .96 | 3.29 | 8.73 | 280.03 | 971.19 | 13.48 | 986.19
    | .07 | .65 | 2.47 | 5.40 | 68.70 | 49.12 | 271.80 | 31.26
    | .35 | 1.76 | 5.86 | 34.76 | 3.94 | 132.75 | 76.36 | 640.08
    | .49 | 2.98 | 4.53 | 2.98 | 406.88 | 8.63 | 391.14 | 716.65
    | .76 | 5.04 | 1.78 | 45.13 | 75.63 | 381.07 | 5.88 | 239.99

4. Add as indicated:

a. .3 + .1
    .06 + 2.3
    5.9 + 1.6
    14 + .25
    0.67 + 0.3

b. 2.63 + 9
    0.1 + .063
    6 + .07
    .956 + .31
    1.301 + .23

c. .35 + .61 + .87
    .61 + 3.26 + 47
    3.14 + 71.4 + .961
    .007 + .06 + 0.5
    .103 + .09 + .32

d. .47 + .3
    .04 + 2.6
    8.7 + .35
    0.6 + .05
    .14 + 3

e. 3.16 + .37
    1.6 + 4.86
    6.41 + .4
    71.8 + .597
    60.2 + 7.56

f. 3.61 + .50 + .217
    1.16 + .528 + 40.3
    65.5 + 9.14 + 303
    .657 + .28 + .5
    0.08 + 0.013 + 0.5

g. $.38 + $.51 + $.76
    $16.84 + $33.90 + $47.14
    $.49 + $5.73 + $22.80
    $8.88 + $2.54 + $11.07
    $18.90 + $10.29 + $35.96

h. $.65 + $3.28 + $.74
    $3.38 + $4.27 + $41.92
    $.49 + $12.82 + $6.98
    $3.69 + $24.87 + $.37
    $40.23 + $96.81 + $57.16

1–19 SUBTRACTION OF DECIMALS

To subtract decimals:

When it is necessary to write the numerals, we write the subtrahend under the minuend so that the decimal points are directly under each other. Zeros may be annexed to the decimal fraction or a decimal point and zeros to a whole number in the minuend so that the minuend and subtrahend will have the same number of decimal places.

We subtract as in the subtraction of whole numbers, placing the decimal point in the remainder (or difference) directly under the decimal points of the subtrahend and minuend. When an answer ends in one or more zeros to the right of the decimal point, the zeros may be dropped unless it is necessary to show the exact degree of measurement. We check by adding the remainder to the subtrahend. Their sum should equal the minuend.

Subtract:
.87
.34
.53
Answer, .53

Subtract:
.65
.58
.07
Answer, .07

Subtract: 12 − 5.461
12.000
5.461
6.539
Answer, 6.539

Subtract:
28.15
3.95
24.20 = 24.2
Answer, 24.2

EXERCISES

1. Subtract, using the section of the metric ruler on page 43.

a.   .7 cm   .8 cm   1.0 cm   3.7 cm
    −.3 cm   −.2 cm   −.7 cm   −3.2 cm

b.   7.2 cm   5.0 cm   6.4 cm   5.3 cm
    −4.8 cm   −3.4 cm   −4.6 cm   −.8 cm

Review of Number

2. In each of the following, first subtract the common fractions or mixed numbers, then the decimals. Compare the answers.

   a. $\frac{8}{10}$ .8   b. $5\frac{3}{10}$ 5.3   c. $5\frac{9}{10}$ 5.9
      $-\frac{4}{10}$ $-.4$ $-2\frac{7}{10}$ -2.7 $-2\frac{43}{100}$ -2.43

3. Subtract:

   a. .9   .38   0.73   .08   7.8   8.3   7.2   6.3   .85   .52   7.8   .25
       .6   .15   0.38   .02   1.3   2.9   5.9   2.7   .78   .32   2.8   .22

   b. .37   .46   8.3   0.543 6.27   .426   0.703 42.5   76.54 74.2
       .07   .39   7.3   0.368 2.79   .397   0.067 7.2   23.95 6.5

   c. 5.036   .8695 35.26   7.614 .8507 5.52   .49325 .7625
       2.658 .6959   16.98   5.772 .1499 4.66   .18963 .7618

   d. 386.27 72.974 .06792 715.41 8.4157 6.000   7.0000
       187.79 63.989 .06697 229.87 1.4148 4.727   5.0693

   e. 8.57   .4   7   3.2   .06   6.7   8.   1.043
       0.7   .245   2.59   .48   .0575   .603   5.003 0.29

   f. $7.45 $3.07 $40.00 $37.15 $358.21 $532.06 $600.00 $5,000.00
       4.78   .39   37.46   28.65   219.44   77.97   439.60   3,807.21

4. Subtract as indicated:
   a. .62 − .18   f. .73 − .7   k. 6.7 − .42
   b. .038 − .029   g. .745 − .74   l. 8 − .735
   c. 0.318 − 0.198   h. 0.308 − 0.28   m. 3 − 2.983
   d. .3200 − .2708   i. .57 − .387   n. 2 − 0.003
   e. 7.048 − 3.178   j. .004 − .0002   o. 7.62 − 1.7009

5. Subtract as indicated:
   a. $.60 − $.46   f. $2.83 − $.95   k. $4 − $.73
   b. $7.23 − $3.54   g. $6.50 − $.75   l. $10 − $.34
   c. $18.70 − $5.95   h. $8.79 − $3   m. $3 − $2.25
   d. $32.76 − $29.47   i. $207.32 − $89   n. $8 − $5.07
   e. $118.50 − $24.76 j. $7.05 − $.67   o. $6 − $4.92

6. a. Subtract .76 from .9   c. From .082 take 0.07   e. Subtract 3.135 from 4
   b. Take $.54 from $4   d. From $5.70 subtract $3   f. From $4.24 take $.87

1-20 MULTIPLICATION OF DECIMALS

To multiply decimals:

We write the given numerals and multiply as in the multiplication of whole numbers. The decimal point in one factor does not necessarily have to be directly under the decimal point in the second factor. We point off in the product, counting from right to left, as many decimal places as there are in *both* factors together. When the product contains fewer digits than the required number of decimal places, we prefix as many zeros as are necessary.

When a decimal answer ends in one or more zeros to the right of the decimal point, the zeros may be dropped unless it is necessary to show the exact degree of measurement. We check by interchanging the two factors and multiplying again.

Multiply:

.6
.9

.54

Answer, .54

Multiply:

.25
.2

.050 = .05

Answer, .05

Multiply:

.02
.03

.0006

Answer, .0006

Multiply:

.875
7.2

1750
6 125

6.3000 = 6.3

Answer, 6.3

EXERCISES

1. In each of the following, first multiply the common fractions or mixed numbers, then multiply the decimals. Compare answers.

   a. $\frac{3}{10} \times \frac{4}{10}$ b. $\frac{2}{10} \times \frac{5}{10}$ c. $\frac{7}{10} \times \frac{4}{100}$ d. $\frac{17}{100} \times \frac{56}{100}$ e. $2\frac{3}{10} \times 1\frac{9}{10}$

   .3   .2   .7   .17   2.3
   ×.4   ×.5   ×.04   ×.56   ×1.9

Review of Number

2. Multiply:

   a.   .3   2   6   .5   16   .7   8.6   100   .3   163
        4   .2   .7   6   .5   60   37   .9   25   .7

   b.   .15   .19   8   .05   3.14   .175   35   .0103 700   4.0083
        7   76   .01   16   4   810   .007   26   .0002 30

   c.   .5   .2   .4   .19   .12   .1   .871   .3034 7.26   8.3
        .3   .6   .5   .6   .5   .08   .7   .5   .9   2.9

   d.   2.78   3.81   .4   6.19   10.36 2.718 3.102 8.0413 4.719 1.0905
        .4   .5   6.7   3.2   4.5   .8   7.3   3.6   .7   2.3

   e.   .34   .33   .16   .01   .7   .039   .005   13.50 3.04   .007
        .52   .32   .08   .01   .53   .78   .02   .08   3.04   .09

   f.   2.6   4.3   .5   8.1   14.8   3.048 7.849 7.138 .009   2.8015
        .25   .24   2.49   .76   2.87   .16   .06   6.17   4.38   11.07

   g.   .513   .004   .115   .6   13.25 6.002 .0019 2.675
        .726   .012   .137   2.731 2.395 .1055 1.295 2.675

   h.   .3937   2.1614   1.0875   .7124 287.5 40.75 4.6561 8.542
        .8   5.8   30.6   .39   .4676 .8133 700.8 2.3064

3. a. $.78 $5.96 $14.73 $302.60 $658 $7,000 $525.30
       14   40   .05   .19   .728   .036   .006

   b. $30 $9,000 $6.78 $.85 $360 $76.39 $48.60
       .62½   .04½   .33⅓   .16⅔   .06¼   .02¾   .66⅔

4. a. .5 × .8   b. ½ × $.68   c. ¼ × $.39   d. .75 × $2
      6 × .004   ⅔ × $.45   ⅞ × $1.67   .4 × $.60
      15 × .02   ⅜ × $1.68   ⅗ × $26   .02 × $7
      .03 × 30   1¼ × $16.29   2¼ × $10.40   .62½ × $12
      .01 × .016   1⅞ × $3.04   3⅞ × $9.13   .03 × $6.50

5. Square each of the following:
   a. .5   d. .002   g. 3.14   j. 13.13
   b. .7   e. .38   h. .898   k. 29.67
   c. .03   f. 2.17   i. 37.4   l. 5.4062

1–21 MULTIPLYING BY POWERS OF TEN*

To multiply a whole number by 10; 100; 1,000; etc. quickly, we annex as many zeros to the right of the given numeral as there are zeros in the given multiplier.

To multiply a decimal by 10; 100; 1,000; etc., we first write the digits of the given numeral, then we move the decimal point as many places to the right of the original position as there are zeros in the given multiplier. Where the product requires them, we use zeros as placeholders.

$$10 \times 3.94 = 39.4$$
$$100 \times .09 = 9$$
$$1{,}000 \times 2.8 = 2{,}800$$
$$1{,}000{,}000 \times 4.756 = 4{,}756{,}000$$

EXERCISES

1. Multiply each of the following numbers by 10:
   .6;   .18;   .865;   .07;   4.3;   3.21;   .1;   .065;   .006;   5.50;   62.7;   8.024

2. Multiply each of the following numbers by 100:
   .68;   .05;   .8;   .587;   .006;   5.14;   1.5;   33.1;   8.337;   .012;   .9;   .19

3. Multiply each of the following numbers by 1,000:
   .468;   .52;   .4;   .8024;   3.571;   2.35;   .0008;   15.3;   39.37;   .0027;   .5;   .02

4. Multiply by short method:

   a. 100,000 × 6.8
   b. 8.6 × 1,000,000
   c. 10,000,000 × 18.3
   d. 29.94 × 1,000,000
   e. 100,000,000 × 6.519
   f. .0058 × 1,000,000,000
   g. 24.7 × 10,000,000,000
   h. 1,000,000,000 × 35.15

---

* *See* page 77.

Review of Number

1-22 DIVISION OF DECIMALS

To divide a decimal fraction or mixed decimal by a whole number, we divide as in the division of whole numbers, placing the decimal point in the quotient directly above the decimal point in the dividend.

To divide a whole number by a whole number where the division is not exact, this division may be carried out to as many decimal places as are required in the quotient by annexing a decimal point and the required zeros to the dividend.

$$6\overline{)7.2} = 1.2 \qquad 9\overline{)\,.45} = .05 \qquad 8\overline{)29} = 8\overline{)29.000} = 3.625$$

Answer, 1.2   Answer, .05   Answer, 3.625

To divide a decimal fraction or mixed decimal by a decimal fraction or mixed decimal, we make the divisor a whole number by multiplying it by the proper power of ten so that its decimal point is moved to the right of the last digit indicating its new position by a caret (∧). We multiply the dividend by the same power of ten so that its decimal point is moved to the right as many places as we moved the decimal point in the divisor and indicate its new position by a caret (∧). The reason we may do this is illustrated below.

$$.04\overline{)\,.896} = \frac{.896}{.04} = \frac{.896}{.04} \times 1 = \frac{.896 \times 100}{.04 \times 100}$$

$$= \frac{89.6}{4} = 4\overline{)89.6}$$

Thus $.04\overline{)\,.896}$ becomes $4\overline{)89.6}$

Instead of doing all the work shown above, we may move decimal points directly in the given problem and indicate the new positions by carets. Then we divide as in the division of whole numbers, placing the decimal point in the quotient directly above the caret (∧) in the dividend.

$$.04\overline{).896} \text{ becomes } .04_\wedge\overline{).89_\wedge 6}$$

Answer, 22.4

When the dividend contains fewer decimal places than required, we annex as many zeros as are necessary to a decimal fraction dividend, and a decimal point and the required zeros to a dividend containing a whole number.

Divide 376.6 by .007:
$$.007_\wedge\overline{)376.600_\wedge}\text{quotient } 53\,800.$$
Answer, 53,800

Divide \$22 by \$2.75:
$$\$2.75_\wedge\overline{)\$22.00_\wedge}\text{quotient } 8.$$
Answer, 8

We check by multiplying the quotient by the divisor and adding the remainder, if any, to the product. The result should be equal to the dividend.

To find the quotient correct to the nearest required decimal place:

We find the quotient to one more than the required number of decimal places, then round it off.

Find quotient to nearest hundredth:

$$6\overline{)9.430}\text{quotient } 1.571$$

1.571 rounded to the nearest hundredth is 1.57

Answer, 1.57

Review of Number

Or find the quotient to the required number of decimal places, adding 1 to the last digit of the quotient if the remainder is equal to one-half or more of the divisor.

> Find quotient to nearest tenth:
>
> $$\begin{array}{r} 3.4 \\ 18\overline{)62.5} \\ \underline{54} \\ 85 \\ \underline{72} \\ 13 \end{array}$$
>
> $\frac{1}{2}$ of the divisor 18 is 9. The remainder 13 is greater than 9. So the digit 4 of the quotient is increased to 5.
>
> *Answer,* 3.5

EXERCISES

1. In each of the following annex zeros to the dividend until the division is exact:

$24\overline{)18}$　　$40\overline{)26}$　　$25\overline{)36}$　　$56\overline{)35}$　　$250\overline{)1}$

2. Divide:

a. $6\overline{)9.6}$	$3\overline{).426}$	$4\overline{).0028}$	$76\overline{).532}$	$87\overline{)260.13}$
b. $6\overline{)30.42}$	$6\overline{)5.154}$	$25\overline{)232.5}$	$64\overline{)48.00}$	$16\overline{)1.0000}$
c. $7.5\overline{)352.5}$	$.9\overline{)51.21}$	$2.7\overline{).648}$	$4.8\overline{).0384}$	$1.5\overline{).003}$
d. $.3\overline{)6.18}$	$.4\overline{).002}$	$.7\overline{).0063}$	$2.5\overline{)21.125}$	$3.8\overline{)19.0}$
e. $.53\overline{).318}$	$.66\overline{)1.6764}$	$5.39\overline{)46.354}$	$.05\overline{).02}$	$.45\overline{)13.5}$
f. $.07\overline{).00028}$	$.48\overline{)12.00}$	$.87\overline{)2.001}$	$26.71\overline{)224.364}$	$2.14\overline{)13.91}$
g. $.003\overline{).693}$	$.416\overline{)0.3328}$	$1.025\overline{)4.1}$	$.138\overline{).69}$	$2.864\overline{)11.456}$
∗**h.** $.003\overline{)25.236}$	$.062\overline{)2.48}$	$.625\overline{)250.000}$	$1.375\overline{)20.625}$	$.093\overline{).01674}$
i. $.6\overline{)42}$	$8.4\overline{)21}$	$.06\overline{)9}$	$3.25\overline{)26}$	$4.038\overline{)2019}$

3. Find the quotient correct to the nearest tenth:
   17)13   23.4)36.94   .675).4   .82)7.9   2.3)30

4. Find the quotient correct to the nearest hundredth:
   8)7   .9)4   .825)0.72   .67)48.9   14.29)3.72

5. Find the quotient correct to the nearest thousandth:
   13)92   4.4)6,000   0.029)3.75   8.5).983   1.36)5.03

6. Find the answer correct to the nearest cent:
   7)$.56   144)$382.52   84)$69.25   62)$7.28   106)$4,926.13

7. Divide:
   $.08)$40 $.20)$8.60 $1.25)$36 $.49)$16.91 $2.37)$243.26

8. Divide as indicated:
   87.5 ÷ .25   .0159 ÷ 5.3 $24 ÷ $.32

1–23   DIVIDING BY POWERS OF TEN*

To divide a whole number or a decimal by 10; 100; 1,000; etc. quickly, we first write the digits of the given numeral, then we move the decimal point as many places to the left of its original position as there are zeros in the given divisor. In a numeral naming a whole number the decimal point is understood after the ones digit.

$$900 \div 10 = 90 \qquad 375.2 \div 100 = 3.752$$
$$4 \div 100 = .04 \qquad 826.1 \div 1{,}000 = .8261$$

EXERCISES

1. Divide each of the following numbers by 10:
   30;   76;   7;   0.4;   3.7;   469;   0.34;   0.475;   34.7;   600;   5.38;   89.45

2. Divide each of the following numbers by 100:
   700;   3,000;   315;   78;   3;   0.8;   0.728;   40.7;   932.87;   460;   8.73;   30

---

* See page 77.

Review of Number   53

3. Divide each of the following numbers by 1,000:

   4,000; 6,265; 648; 67; 0.361; 0.6; 636.8; 8; 350,000; 1,320; 0.03; 5,000.5

4. Divide by the short method:

   a. 570,000 ÷ 100,000

   b. 6,400,000 ÷ 10,000

   c. 16,400,000 ÷ 1,000,000

   d. 437,850,000 ÷ 1,000,000

   e. 3,600,000,000 ÷ 1,000,000,000

   f. 325,000,000,000 ÷ 100,000,000

1–24 SHORTENED NAMES FOR LARGE WHOLE NUMBERS

To write the complete numeral for the shortened name of a whole number, we multiply the number named by the given numeral prefix by the power of ten (10; 100; 1,000; etc.) that is equivalent to the value of the period name (or place value) used.

> Write the complete numeral for 53.8 million:
> 53.8 million = 53.8 × 1,000,000 = 53,800,000
> *Answer,* 53,800,000

To write the shortened name for a large whole number when its numeral ends in a series of zeros, we divide the given number by the power of ten (10; 100; 1,000; etc.) that is equivalent to the value of the period name (or place value) used.

> Write the shortened name in billions for 49,500,000,000:
> 49,500,000,000 = 49.5 billion, since
> 49,500,000,000 ÷ 1,000,000,000 = 49.5
> *Answer,* 49.5 billion

EXERCISES

1. Write the complete numeral for each of the following:

   a. 45.7 million;   8.76 million;   163.1 million;   4.895 million
   b. 1.6 thousand;   38.49 thousand;   7.629 thousand;   586.2 thousand
   c. 29.8 billion;   106.3 billion;   2.89 billion;   86.803 billion
   d. 7.8 hundred;   35.48 hundred;   875.1 hundred;   22.98 hundred
   e. 9.8 trillion;   14.6 trillion;   3.45 trillion;   367.174 trillion

2. Write the shortened name for each of the following:

   a. In billions:
       3,700,000,000;   24,800,000,000;   385,480,000,000;   7,976,000,000
   b. In millions:
       54,800,000;   3,604,000;   493,800,000;   28,279,000
   c. In hundreds:
       7,860;   1,374;   18,250;   79,347
   d. In thousands:
       40,300;   7,130;   882,600;   6,463,530
   e. In trillions:
       6,900,000,000,000;   94,500,000,000,000;   1,370,000,000,000;
       581,026,000,000,000;   903,400,000,000,000

1–25 SCIENTIFIC NOTATION

A brief form of writing numerals, usually for very large or very small numbers, is known as *scientific notation*. A number is expressed in scientific notation when its numeral names a number that is greater than 1 but less than 10 multiplied by some power of ten.

> 5,800,000 is 5.8 million, which may be expressed in scientific notation as 5.8×10^6

To write a numeral in scientific notation, we rewrite the significant digits:

(1) As a numeral for a whole number if there is only one significant digit, as in 5,000,000.

Review of Number

(2) As a numeral for a mixed decimal if there are two or more significant digits, as in 5,800,000 or 5,870,000, using the first digit as a numeral for a whole number and all other digits as the numeral for a decimal fraction.

Then we indicate that this numeral is multiplied by the required power of ten.

The required power of ten may be determined by dividing the whole number or mixed decimal into the given number and changing the quotient into a power of ten or by counting the number of places the decimal point is being moved.

When the given number is 10 or greater, a positive integer is used for the exponent.

> Write 490,000,000 in scientific notation:
> 490,000,000 = 4.9 × ?
> $$\phantom{\text{Divide }4.9\,\big)}\overline{100,000,000}$$
> Divide 4.9$_\wedge$)490,000,000.0$_\wedge$
> However, 100,000,000 = 10^8
> 490,000,000 = 4.9 × 10^8
> Answer, 4.9 × 10^8

Thus, 490,000,000 = 4.9 × 10^8. Note that the decimal point is moved 8 places to the left.

When the given number is between 0 and 1, a negative integer is used for the exponent.

> Write .00000516 in scientific notation:
> .00000516 = 5.16 × ?
> $$\phantom{\text{Divide: }5.16\,\big)}\overline{.000001}$$
> Divide: 5.16$_\wedge$).00$_\wedge$000516
> However .000001 = 10^{-6}
> .00000516 = 5.16 × 10^{-6}
> Answer, 5.16 × 10^{-6}

Thus, .00000516 = 5.16 × 10^{-6}. Note that the decimal point is moved 6 places to the right.

CHAPTER 1

EXERCISES

1. Express each of the following numbers in scientific notation:
   - **a.** 20   75   800   279   6,000   3,780   9,030
   - **b.** 20,000   98,000   380,000   65,000,000   2,347,000
   - **c.** 4,000,000,000   5,600,000   289,000,000,000   69,203,000,000,000
   - **d.** 680000000000   27100000000000   9000000000000000   18000000000000000

2. Express each of the following numbers in scientific notation:
   - **a.** 7.   .5   .3   .24   .08   .53   .92
   - **b.** .067   .003   .587   .0723   .2004   .0048   .9107
   - **c.** .00029   .00136   .04145   .000372   .0000095
   - **d.** .0000031   .00000357   .000000065   .000000203   .0000000187
   - **e.** .000000000032   .0000000000204   .00000000003   .000000000000516

1–26 EXPRESSING COMMON FRACTIONS AS DECIMAL FRACTIONS

To express a numeral naming a common fraction as a decimal numeral, we divide the numerator by the denominator.

> Express $\frac{2}{5}$ as a decimal:
>
> $$\frac{2}{5} = 2 \div 5 = 5\overline{)2.0}^{\,.4}$$
>
> Answer, .4

When the common fraction has as its denominator some power of ten (10; 100; 1,000; etc.) we drop the denominator and rewrite the numerator, placing the decimal point so that the name of the part and the place value of the last digit correspond.

> Express $\frac{563}{1000}$ as a decimal:
>
> $$\frac{563}{1000} = .563$$
>
> Answer, .563

Review of Number

To express a numeral naming a mixed number as a decimal numeral, we express the numeral for the common fraction as a numeral for the decimal fraction and annex it to the numeral for the whole number.

> Express $8\frac{1}{4}$ as a mixed decimal:
> $$8\frac{1}{4} = 8.25$$
> since $\frac{1}{4} = .25$
>
> *Answer,* 8.25

Sometimes when we divide the numerator by the denominator to express a common fraction as a decimal fraction, the division is not exact and a remainder other than 0 may keep repeating. Then the sequence of digits obtained in the quotient between occurrences of this remainder also will repeat endlessly and thus result in a *repeating decimal.*

$$\begin{array}{r} .6363 \\ 11\overline{)7.0000} \\ \underline{6\;6} \\ 40 \\ \underline{33} \\ 70 \\ \underline{66} \\ 40 \\ \underline{33} \\ 7 \end{array}$$

Answer, $.6363\ldots$ or $.6\overline{3}\ldots$ or $.\overline{63}$

Observe that a bar is used to indicate the repeating sequence, and dots may be used to indicate that the sequence repeats endlessly.

CHAPTER 1

EXERCISES

1. Express each of the following as a 1-digit numeral naming a decimal fraction:

   $\frac{3}{10}$, $\frac{7}{10}$, $\frac{2}{10}$, $\frac{9}{10}$, $\frac{1}{10}$, $\frac{9}{15}$, $\frac{3}{5}$, $\frac{1}{5}$, $\frac{1}{2}$, $\frac{2}{5}$

2. Express each of the following as a 2-digit numeral naming a decimal fraction:

   $\frac{31}{100}$, $\frac{9}{100}$, $\frac{49}{100}$, $\frac{1}{2}$, $\frac{3}{4}$, $\frac{7}{25}$, $\frac{14}{50}$, $\frac{13}{20}$, $\frac{24}{32}$, $\frac{13}{52}$

3. Express each of the following as a 3-digit numeral naming a decimal fraction:

   $\frac{621}{1000}$, $\frac{1}{8}$, $\frac{5}{8}$, $\frac{38}{1000}$, $\frac{7}{8}$, $\frac{3}{8}$, $\frac{9}{1000}$, $\frac{20}{32}$, $\frac{21}{24}$, $\frac{45}{120}$

4. Express each of the following as a 2-digit numeral naming a decimal fraction. Retain the remainder as a common fraction.

   $\frac{1}{7}$, $\frac{1}{3}$, $\frac{2}{9}$, $\frac{5}{6}$, $\frac{5}{12}$, $\frac{35}{42}$, $\frac{12}{28}$, $\frac{45}{48}$, $\frac{7}{42}$, $\frac{10}{35}$

5. Express each of the following as a numeral naming a mixed decimal:

   $1\frac{1}{4}$, $3\frac{2}{5}$, $6\frac{7}{10}$, $2\frac{1}{8}$, $8\frac{7}{8}$, $\frac{7}{4}$, $\frac{5}{2}$, $\frac{8}{5}$, $\frac{12}{5}$, $\frac{18}{10}$

6. Express each of the following as a numeral naming a repeating decimal. Indicate the repeating sequence by a horizontal bar.

   a. $\frac{1}{6}$ e. $\frac{6}{13}$ i. $\frac{5}{6}$ m. $\frac{26}{33}$ q. $\frac{7}{12}$ u. $\frac{6}{17}$
   b. $\frac{5}{7}$ f. $\frac{8}{15}$ j. $\frac{1}{3}$ n. $\frac{17}{24}$ r. $\frac{8}{13}$ v. $\frac{3}{7}$
   c. $\frac{11}{12}$ g. $\frac{11}{18}$ k. $\frac{2}{9}$ o. $\frac{14}{27}$ s. $\frac{10}{11}$ w. $\frac{15}{18}$
   d. $\frac{7}{9}$ h. $\frac{9}{11}$ l. $\frac{5}{12}$ p. $\frac{15}{19}$ t. $\frac{9}{23}$ x. $\frac{8}{21}$

1–27 EXPRESSING DECIMAL FRACTIONS AS COMMON FRACTIONS

To express a numeral naming a decimal fraction as a numeral naming a common fraction, we rewrite the digits of the decimal numeral as the numerator and use a power of ten (10; 100; 1,000; etc.) corresponding to the place value of the last digit of the decimal numeral as the denominator. We then express the fraction in lowest terms if possible.

Review of Number

> Express .65 as a common fraction:
> $$.65 = \frac{65}{100} = \frac{13}{20}$$
> Answer, $\frac{13}{20}$

To express a numeral naming a mixed decimal as a numeral naming a mixed number, we write the numeral naming the decimal fraction as a numeral naming a common fraction in lowest terms and annex it to the numeral for the whole number.

> Express 3.8 as a mixed number:
> $$3.8 = 3\frac{8}{10} = 3\frac{4}{5}$$
> Answer, $3\frac{4}{5}$

EXERCISES

1. Express each of the following as a numeral naming a common fraction:

   a. .2; .5; .7; .4; .6
   b. .48; .33; 0.75; 0.90; 0.63
   c. .40; .04; .90; .05; .70
   d. .147; .375; 0.875; 0.125; 0.625
   e. $.37\frac{1}{2}$; $.62\frac{1}{2}$; $.08\frac{1}{4}$; $.87\frac{1}{2}$; $.12\frac{1}{2}$
   f. $.16\frac{2}{3}$; $0.66\frac{2}{3}$; $0.83\frac{1}{3}$; $0.33\frac{1}{3}$; $0.14\frac{2}{7}$

2. Express each of the following as a numeral naming a mixed number:

   a. 2.2; 4.5; 3.25; 8.625; 6.04
   b. $1.37\frac{1}{2}$; $4.33\frac{1}{3}$; $8.87\frac{1}{2}$; $10.66\frac{2}{3}$; $5.12\frac{1}{2}$

PERCENT—COMPUTATION

Percent is a fraction in which the numeral preceding the percent symbol % represents the numerator and the percent symbol represents the denominator 100.

Percent means "hundredths." It is equivalent to a two-place decimal fraction or to a common fraction having 100 as its denominator.

$$73\% = .73 = \frac{73}{100}$$

1–28 CHANGING PERCENTS TO DECIMAL FRACTIONS AND COMMON FRACTIONS

Percents to Decimal Fractions

To express a percent as a numeral for a decimal fraction, we rewrite the digits of the given percent but drop the percent symbol. Then we move the decimal point two places to the left.

When a percent is a decimal percent such as 27.4%, the answer contains more than 2 decimal places.

When a percent is 100% or more, the answer is a numeral naming a whole number or a mixed decimal.

When the percent is a common fraction less than one percent such as $\frac{3}{4}$%, the answer may be written as a two-place decimal numeral.

When the percent is a mixed number percent such as $63\frac{1}{4}$%, the answer may be written as a two-place decimal numeral.

Express 52% as a decimal numeral:
$$52\% = .52$$
Answer, .52

$$27.4\% = .274$$

$$500\% = 5$$
$$149\% = 1.49$$

$$\tfrac{3}{4}\% = .00\tfrac{3}{4}$$
(or as .0075)

$$63\tfrac{1}{4}\% = .63\tfrac{1}{4}$$
(or as .6325)

Review of Number 61

Percents to Common Fractions

To express a percent as a numeral for a common fraction, we write the digits of the given percent as the numerator over 100 as the denominator. Then, if possible, we express the fraction in lowest terms.

> Express 25% as a numeral naming a common fraction:
> $$25\% = \frac{25}{100} = \frac{1}{4}$$
> Answer, $\frac{1}{4}$

When the percent is 100% or more, the answer is a numeral naming a whole number or a mixed number.

> $$180\% = \frac{180}{100} = 1\frac{4}{5}$$
> or $180\% = 100\% + 80\% = 1 + \frac{4}{5} = 1\frac{4}{5}$

When the percent is a mixed number percent, we express the percent as a decimal fraction, then as a common fraction in its lowest terms.

> $$37\frac{1}{2}\% = .37\frac{1}{2} = .375 = \frac{375}{1000} = \frac{3}{8}$$
> or $37\frac{1}{2}\% = \frac{37\frac{1}{2}}{100} = 37\frac{1}{2} \div 100 = \frac{\overset{3}{\cancel{75}}}{2} \times \frac{1}{\underset{4}{\cancel{100}}} = \frac{3}{8}$

EXERCISES

1. Express each of the following percents as a decimal numeral:
   - **a.** 52%; 58%; 85%; 23%; 68%
   - **b.** 6%; 4%; 2%; 7%; 9%
   - **c.** 40%; 20%; 90%; 60%; 50%
   - **d.** 172%; 135%; 104%; 287%; 136%
   - **e.** 170%; 110%; 120%; 180%; 230%
   - **f.** $37\frac{1}{2}\%$; $6\frac{1}{4}\%$; $3\frac{3}{4}\%$; $33\frac{1}{3}\%$; $\frac{3}{5}\%$
   - **g.** 32.7%; 9.4%; 157.31%; 0.7%; 0.625%
   - **h.** 71%; 5%; 80%; $112\frac{1}{2}\%$; 0.875%

2. Express each of the following percents as a numeral naming a whole number:

   700%; 500%; 100%; 300%; 200%

3. Express each of the following percents as a numeral naming a common fraction in lowest terms:

   a. 70%; 40%; 25%; 50%; 30%; 10%; 60%; 80%; 20%; 90%

   b. 8%; 37%; 54%; 26%; 96%; 89%; 33%; 85%; 68%; 15%

   c. $62\frac{1}{2}$%; $37\frac{1}{2}$%; $12\frac{1}{2}$%; $6\frac{1}{4}$%; $66\frac{2}{3}$%; $8\frac{1}{3}$%; $16\frac{2}{3}$%; $33\frac{1}{3}$%; $83\frac{1}{3}$%; $87\frac{1}{2}$%

4. Express each of the following percents as a numeral naming a mixed number:

   175%; 130%; 110%; 230%; 280%; $137\frac{1}{2}$%; $162\frac{1}{2}$%; $166\frac{2}{3}$%; $233\frac{1}{3}$%; $312\frac{1}{2}$%

1–29 CHANGING DECIMAL FRACTIONS AND COMMON FRACTIONS TO PERCENTS

Decimal Fractions to Percents

To express a decimal fraction as a percent, we write the digits of the numeral naming the given decimal. We move the decimal point two places to the right and write the percent symbol after the numeral.

Express .64 as a percent:

.64 = 64%

Answer, 64%

Express .726 as a percent:

.726 = 72.6%

Answer, 72.6%

Express .3 as a percent:

.3 = .30 = 30%

Answer, 30%

A mixed decimal is equal to more than 100%.

Express 1.87 as a percent:

1.87 = 187%

Answer, 187%

Review of Number 63

A whole number is equivalent to some multiple of 100%.

$$1 = 1.00 = 100\%; \quad 2 = 2.00 = 200\%$$
$$3 = 3.00 = 300\%; \quad 4 = 4.00 = 400\%; \text{ etc.}$$

Common Fractions to Percents

To express a common fraction as a percent, we divide the numerator by the denominator, finding the quotient to two decimal places. We rewrite the digits in the quotient, drop the decimal point, and write the percent symbol % after the numeral.

Express $\frac{5}{8}$ as a percent:

$$\frac{5}{8} = 5 \div 8 = 8\overline{)5.00}^{\,.62\frac{1}{2}} = 62\frac{1}{2}\%$$

Answer, $62\frac{1}{2}\%$

A mixed number is equal to more than 100%.

$$1\frac{1}{2} = 150\% \text{ since } 1 = 100\% \text{ and } \frac{1}{2} = 50\%$$

EXERCISES

1. Express each of the following decimals as a percent:

   a. .24; .47; .61; .36; .92

   b. .08; .06; .03; .07; .05

   c. .3; .7; .4; .1; .2

   d. 4.00; 2.00; 8; 5; 3

   e. 1.34; 2.10; 1.3; 2.45; 1.6

   f. $.37\frac{1}{2}$; $.03\frac{1}{4}$; $.06\frac{1}{2}$; $.80\frac{3}{8}$; $1.16\frac{2}{3}$

   g. .625; .375; 2.103; .005; .0004

   h. $.00\frac{1}{4}$; $.00\frac{3}{8}$; $.00\frac{1}{2}$; $.00\frac{3}{4}$; $.00\frac{3}{5}$

2. Express each of the following fractions as a percent:
   a. $\frac{7}{100}; \frac{24}{100}; \frac{58}{100}; \frac{80}{100}; \frac{163}{100}$
   b. $\frac{4}{5}; \frac{1}{4}; \frac{1}{2}; \frac{1}{5}; \frac{3}{4}$
   c. $\frac{9}{10}; \frac{3}{5}; \frac{3}{10}; \frac{2}{5}; \frac{1}{10}$
   d. $\frac{7}{20}; \frac{9}{25}; \frac{17}{20}; \frac{7}{25}; \frac{3}{50}$
   e. $\frac{30}{600}; \frac{14}{200}; \frac{40}{800}; \frac{3}{900}; \frac{45}{500}$
   f. $\frac{1}{8}; \frac{5}{6}; \frac{3}{8}; \frac{1}{3}; \frac{1}{6}$
   g. $\frac{5}{8}; \frac{3}{6}; \frac{2}{3}; \frac{5}{12}; \frac{15}{16}$
   h. $\frac{18}{24}; \frac{21}{30}; \frac{36}{72}; \frac{12}{48}; \frac{33}{60}$
   i. $\frac{28}{32}; \frac{40}{48}; \frac{35}{56}; \frac{48}{60}; \frac{42}{63}$
   j. $\frac{6}{7}; \frac{24}{56}; \frac{40}{72}; \frac{30}{66}; \frac{54}{100}$

3. Express each of the following fractions as a percent:
   a. To the nearest whole percent: $\frac{5}{6}; \frac{3}{7}; \frac{1}{3}; \frac{13}{18}; \frac{7}{9}$
   b. To the nearest tenth of a percent: $\frac{4}{7}; \frac{5}{6}; \frac{8}{15}; \frac{11}{12}; \frac{25}{45}$
   c. To the nearest hundredth of a percent: $\frac{2}{9}; \frac{13}{16}; \frac{8}{11}; \frac{17}{24}; \frac{44}{60}$

4. Express each of the following mixed numbers as a percent:
   $1\frac{1}{5}; 1\frac{3}{4}; 3\frac{1}{2}; 2\frac{3}{8}; 2\frac{2}{3}$

5. Express each of the following improper fractions as a percent:
   a. $\frac{43}{43}; \frac{14}{14}; \frac{211}{211}; \frac{18}{9}; \frac{35}{7}$
   b. $\frac{5}{4}; \frac{7}{5}; \frac{5}{2}; \frac{19}{4}; \frac{13}{5}$

1–30 FINDING A PERCENT OF A NUMBER

To find a percent of a number, we change the percent to its equivalent decimal fraction or common fraction. Then we multiply the given number by this fraction.

Find 23% of 89:

23% = .23

$$\begin{array}{r} 89 \\ \times .23 \\ \hline 267 \\ 178 \\ \hline 20.47 \end{array}$$

Answer, 20.47

Find 60% of 450:

60% = $\frac{3}{5}$

$$\frac{3}{\cancel{5}} \times \frac{\cancel{450}^{90}}{1} = 270$$

Answer, 270

Review of Number

65

EXERCISES

Find:

1.
   a. 36% of 68; 10% of 60; 18% of 88; 2% of 71; 75% of 6,000
   b. 17% of 130; 2% of 76; 85% of 540; 40% of 763; 87% of 4,632
   c. 3% of 70; 8% of 65; 11% of 70; 7% of 500; 5% of 1,776
   d. 6% of 94; 15% of 126; 1% of 248; 18% of 286; 30% of 1,000
   e. 5% of 40; 30% of 8; 53% of 480; 40% of 145; 1% of 7,530

2.
   a. 5% of 8.6; 16% of 4.7; 7% of 6.16; 63% of 5.24; 78% of 6.80
   b. 14% of 65.42; 2% of 51.03; 36% of 69.24; 77% of 923.84; 1% of 347.09

3.
   a. $23\frac{1}{2}$% of 68; $86\frac{1}{2}$% of 600; $34\frac{3}{4}$% of 72; $16\frac{1}{2}$% of 468; $56\frac{1}{4}$% of 4,000
   b. $4\frac{1}{2}$% of 12; $3\frac{1}{4}$% of 36; $7\frac{1}{4}$% of 276; $1\frac{1}{4}$% of 2,187; $3\frac{3}{4}$% of 5,197
   c. $18\frac{5}{8}$% of 60; $4\frac{3}{8}$% of 150; $2\frac{1}{3}$% of 696; $52\frac{5}{6}$% of 2,400; $7\frac{5}{8}$% of 1,000

4.
   a. 60.2% of 18; 19.7% of 560; 20.6% of 9; 23.8% of 5.25; 41.6% of 4,600
   b. 3.8% of 90; 5.2% of 6; 9.4% of 7.21; 1.4% of 83.67; 5.6% of 3,000
   c. 5.85% of 600; 70.08% of 3,000; 6.74% of 680; 47.61% of 3,540

5.
   a. 107% of 308; 200% of 54; 105% of 55; 310% of 16.5; 600% of 500
   b. 118% of 7; 168% of 75; 500% of 240; 375% of 4.98; 128% of 4,800
   c. 100% of 90; 101% of 2,000; 124% of 158; 200% of 2.75; 306% of 2,000
   d. 138.4% of 60; 130% of 6; 240% of 454; 105% of 8.6; 180% of 5,500
   e. 150% of 48; 98.6% of 1,000; 120.42% of 7; 123.5% of 16.25
   f. $112\frac{1}{2}$% of 4; $323\frac{1}{4}$% of 840; $100\frac{1}{2}$% of 8,000; $106\frac{1}{4}$% of 32.6

6.
   a. $\frac{1}{4}$% of 600; $\frac{1}{8}$% of 16; $\frac{2}{3}$% of 288; $\frac{3}{4}$% of 56.92; $\frac{1}{2}$% of 105.23
   b. $\frac{1}{2}$% of 650; $\frac{1}{3}$% of 4.2; $\frac{5}{8}$% of 4,000; $\frac{7}{8}$% of 106.48; $\frac{5}{6}$% of 12,000
   c. .06% of 90; 0.4% of 6,000; 0.6% of 14; 0.5% of 7; 0.3% of 6,000
   d. 0.15% of 6; 0.25% of 80; 0.45% of 65; 0.63% of 14.8; 0.95% of 5,000

7.
   a. 20% of 10; 80; 190; 450; 2,000
   b. 80% of 20; 110; 350; 680; 2,850
   c. 50% of 16; 40; 112; 280; 784
   d. 25% of 24; 72; 136; 248; 1,000
   e. 10% of 25; 45; 70; 240; 625
   f. 75% of 8; 40; 76; 68; 1,720
   g. $33\frac{1}{3}$% of 12; 60; 78; 135; 960
   h. $12\frac{1}{2}$% of 64; 48; 40; 128; 528
   i. $16\frac{2}{3}$% of 96; 60; 192; 282; 4,188
   j. $37\frac{1}{2}$% of 32; 112; 72; 480; 1,200

k. $66\frac{2}{3}$% of 12; 60; 75; 342; 1,614
l. $87\frac{1}{2}$% of 100; 86; 202; 600; 2,812
m. 60% of 90; 130; 360; 45; 4,450
n. 40% of 60; 75; 100; 650; 2,600
o. 70% of 10; 150; 225; 500; 2,800
p. $83\frac{1}{3}$% of 9; 36; 84; 432; 1,203

q. $62\frac{1}{2}$% of 40; 72; 200; 648; 3,600
r. 30% of 30; 80; 140; 600; 1,500
s. $8\frac{1}{3}$% of 24; 96; 144; 1,728; 4,500
t. 5% of 20; 50; 130; 250; 1,100
u. $6\frac{1}{4}$% of 32; 224; 384; 800; 1,960
v. 90% of 30; 100; 500; 1,920; 4,000

8. Use fractional equivalents for percents to find:

a. 30% of 60   20% of 55   50% of 86   70% of 300   10% of 680
b. 20% of 54   90% of 432   25% of 800   60% of 680   75% of 1,260
c. $12\frac{1}{2}$% of 864 $37\frac{1}{2}$% of 480 $6\frac{1}{4}$% of 400 $62\frac{1}{2}$% of 292 $87\frac{1}{2}$% of 36
d. $8\frac{1}{3}$% of 1,200 $16\frac{2}{3}$% of 600 $33\frac{1}{3}$% of 96 $66\frac{2}{3}$% of 240 $83\frac{1}{3}$% of 288
e. 10% of 182 $33\frac{1}{3}$% of 186   20% of 96 $16\frac{2}{3}$% of 64   50% of 55

9. Use mixed numbers for percents to find:

a. 110% of 36   160% of 248   225% of 320   150% of 500   180% of 1,000
b. $137\frac{1}{2}$% of 220 $166\frac{2}{3}$% of 690 $112\frac{1}{2}$% of 3,800 $266\frac{2}{3}$% of 420 $133\frac{1}{3}$% of 280

10. Find each of the following to the nearest cent:

a. 24% of $80   66% of $9   45% of $178   90% of $1,420   78% of $3,000
b. 3% of $969   1% of $32   5% of $9   6% of $254   4% of $7,500
c. 92% of $65.38   60% of $215.75   14% of $6.48   90% of $.60   43% of $.75
d. 6% of $51.08   9% of $.70   2% of $572.68   7% of $3.67   3% of $.54
e. $\frac{3}{4}$% of $800 $12\frac{1}{2}$% of $272 $\frac{1}{2}$% of $78.43 $66\frac{2}{3}$% of $.98 $1\frac{1}{2}$% of $67.46
f. 0.2% of $96.47   3.8% of $80   5.85% of $2,600   14.52% of $30.08
g. 160% of $78   138% of $2.30 $100\frac{3}{4}$% of $3,000 $362\frac{1}{2}$% of $360

11. Use fractional equivalents for percents to find each of the following to the nearest cent:

a. 60% of $324   30% of $700   75% of $3,000 $16\frac{2}{3}$% of $1,800 $87\frac{1}{2}$% of $790
b. 50% of $.90   10% of $6.50   80% of $17.90   25% of $6.38 $66\frac{2}{3}$% of $8.74
c. 20% of $.64   30% of $1.76 $33\frac{1}{3}$% of $8.75 $8\frac{1}{3}$% of $12   90% of $48.79

Review of Number

1-31 FINDING WHAT PERCENT ONE NUMBER IS OF ANOTHER

To find what percent one number is of another, we find what fractional part one number is of the other expressed in lowest terms. Then we change this fraction to a percent, using the percent equivalent if it is known. Otherwise we change the fraction first to a 2-place decimal by dividing the numerator by the denominator, then we change the decimal to a percent.

What percent of 75 is 27?

$$\frac{27}{75} = \frac{9}{25} = 25\overline{)9.00} \quad .36 = 36\%$$

$$\phantom{\frac{27}{75} = \frac{9}{25} = 25\overline{)}}\underline{7\,5}$$
$$\phantom{\frac{27}{75} = \frac{9}{25} = 25\overline{)}}1\,50$$
$$\phantom{\frac{27}{75} = \frac{9}{25} = 25\overline{)}}\underline{1\,50}$$

Answer, 36%

Sometimes the fraction has a denominator that can be changed easily to the denominator 100.

$$\frac{9}{20} = \frac{9 \times 5}{20 \times 5} = \frac{45}{100} = 45\%$$

$$\frac{42}{700} = \frac{42 \div 7}{700 \div 7} = \frac{6}{100} = 6\%$$

An improper fraction is equal to 100% or more.

What percent of 4 is 7?

$$\frac{7}{4} = 1\frac{3}{4} = 175\%$$

Answer, 175%

EXERCISES

Find the answer to each of the following:

1. a. What percent of 50 is 19?
   b. 53 is what percent of 100?
   c. 9 is what percent of 25?
   d. What percent of 5 is 3?
   e. 1 is what percent of 2?
   f. What percent of 20 is 13?
   g. What percent of 4 is 3?
   h. 7 is what percent of 10?

2. a. What percent of 3 is 1?
   b. 5 is what percent of 8?
   c. What percent of 8 is 3?
   d. 5 is what percent of 6?

3. a. What percent of 7 is 3?
   b. What percent of 12 is 7?
   c. 17 is what percent of 22?
   d. 5 is what percent of 9?
   e. 7 is what percent of 18?
   f. What percent of 13 is 11?

4. a. What percent of 100 is 25?
   b. 30 is what percent of 45?
   c. What percent of 30 is 12?
   d. 35 is what percent of 70?
   e. What percent of 20 is 8?
   f. 12 is what percent of 75?

5. a. What percent of 100 is 20?
   b. 24 is what percent of 150?
   c. What percent of 400 is 28?
   d. 30 is what percent of 400?
   e. What percent of 200 is 12?
   f. 6 is what percent of 300?

6. a. What percent of 22 is 11?
   b. 14 is what percent of 35?
   c. What percent of 40 is 16?
   d. 6 is what percent of 60?
   e. 8 is what percent of 32?
   f. What percent of 30 is 21?
   g. 42 is what percent of 70?
   h. What percent of 72 is 54?

7. a. What percent of 27 is 9?
   b. 45 is what percent of 54?
   c. What percent of 21 is 14?
   d. 30 is what percent of 80?
   e. What percent of 56 is 35?
   f. 42 is what percent of 48?

8. a. What percent of 27 is 27?
   b. What percent of 100 is 300?
   c. 40 is what percent of 8?
   d. 210 is what percent of 210?
   e. 51 is what percent of 17?
   f. What percent of 400 is 1,600?

9. a. What percent of 5 is 9?
   b. 15 is what percent of 7?
   c. What percent of 14 is 21?
   d. 45 is what percent of 25?
   e. 7 is what percent of 2?
   f. What percent of 9 is 17?
   g. 81 is what percent of 30?
   h. What percent of 65 is 120?

10. Round to the nearest whole percent:
    a. What percent of 12 is 7?
    b. 56 is what percent of 64?
    c. What percent of 48 is 15?
    d. 5 is what percent of 26?
    e. What percent of 8 is 5?
    f. 40 is what percent of 28?

Review of Number

11. Round to the nearest tenth of a percent:

    a. 15 is what percent of 16?
    b. What percent of 33 is 15?
    c. 69 is what percent of 105?
    d. What percent of 18 is 11?
    e. 10 is what percent of 35?
    f. What percent of 24 is 40?

12. Round to the nearest hundredth of a percent:

    a. What percent of 11 is 3?
    b. 4 is what percent of 12?
    c. What percent of 317 is 100?
    d. 34 is what percent of 64?
    e. What percent of 18 is 3?
    f. 75 is what percent of 38?

13. a. What percent of 3.2 is 2.4?
    b. .6 is what percent of 2?
    c. What percent of 3.5 is 7?
    d. .02 is what percent of .1?
    e. What percent of 8 is .4?
    f. 6.16 is what percent of 30.8?

14. a. What percent of 10 is $2\frac{1}{2}$?
    b. $\frac{1}{5}$ is what percent of $\frac{3}{5}$?
    c. What percent of $4\frac{1}{3}$ is $2\frac{1}{6}$?
    d. $3\frac{1}{4}$ is what percent of $8\frac{1}{8}$?
    e. What percent of $\frac{1}{2}$ is $\frac{3}{4}$?
    f. 10 is what percent of $3\frac{1}{3}$?

15. a. What percent of $52 is $13?
    b. $.18 is what percent of $.30?
    c. What percent of $3 is $.75?
    d. $3.50 is what percent of $24.50?
    e. What percent of $.40 is $.92?
    f. $9.20 is what percent of $115?

1–32 FINDING THE NUMBER WHEN A PERCENT OF IT IS KNOWN

To find the number when a percent of it is known, first change the percent to its equivalent decimal fraction or common fraction. Then divide the given number representing the given percent of the unknown number by this decimal or common fraction.

75% of what number is 12?

75% = .75

$$.75 \overline{)12.00} = 16.$$

or

75% = $\frac{3}{4}$

$$12 \div \frac{3}{4} = \frac{\cancel{12}^{4}}{1} \times \frac{4}{\cancel{3}_{1}} = 16$$

Answer, 16   *Answer,* 16

EXERCISES

1. Use decimal equivalents of the percents to find the following missing numbers:

   a. 16% of what number is 8?
   b. 4% of what number is 10?
   c. 60 is 3% of what number?
   d. 30% of what number is 150?
   e. 25.8 is 43% of what number?
   f. 64 is 16% of what number?
   g. 6 is 8% of what number?
   h. 60% of what number is 48?
   i. 42 is 70% of what number?
   j. 9% of what number is 48.15?

2. Use fractional equivalents of the percents to find the following:

   a. 30% of what number is 42?
   b. 7 is 10% of what number?
   c. 80% of what number is 120?
   d. $62\frac{1}{2}$% of what number is 100?
   e. 346 is $33\frac{1}{3}$% of what number?
   f. 37 is 25% of what number?
   g. $66\frac{2}{3}$% of what number is 86?
   h. 700 is 70% of what number?
   i. 600 is $37\frac{1}{2}$% of what number?
   j. $83\frac{1}{3}$% of what number is 680?

Find the missing numbers:

3. a. 300% of what number is 24?
   b. 100% of what number is 19?
   c. 51 is 300% of what number?
   d. 171 is 100% of what number?
   e. 300 is 600% of what number?
   f. 200% of what number is 72?

4. a. 175% of what number is 105?
   b. 48 is $133\frac{1}{3}$% of what number?
   c. 109% of what number is 654?
   d. 54 is 150% of what number?
   e. $337\frac{1}{2}$% of what number is 810?
   f. $575 is 115% of what number?

5. a. 3.5% of what number is 7?
   b. 45 is 0.3% of what number?
   c. 101.1% of what number is 2,022?
   d. 549 is 36.6% of what number?
   e. 0.1% of what number is 30?
   f. 97.85% of what number is 39,140?

6. a. $4\frac{1}{2}$% of what number is 9?
   b. $\frac{1}{4}$% of what number is 4?
   c. 2,070 is $103\frac{1}{2}$% of what number?
   d. 4,386 is $87\frac{1}{2}$% of what number?
   e. 429 is $3\frac{3}{4}$% of what number?
   f. $\frac{1}{2}$% of what number is 70?

7. a. 6% of what amount is $300?
   b. $2.47 is $33\frac{1}{3}$% of what amount?
   c. 160% of what amount is $48?
   d. $\frac{1}{2}$% of what amount is $175?
   e. $35 is 14% of what amount?
   f. 80% of what amount is $.72?
   g. $6.72 is $5\frac{1}{4}$% of what amount?
   h. $7.26 is 0.3% of what amount?

Review of Number

1–33 PERCENT—SOLVING BY PROPORTION

The three basic types of percentage may be treated as one through the use of the proportion. See section 4–22.

(1) Find 75% of 24.

75% is the ratio of 75 to 100 or $\frac{75}{100}$

To find 75% of 24 means:
To determine the number (n) which compared to 24 is the same as 75 compared to 100.

The proportion $\frac{n}{24} = \frac{75}{100}$ is formed and solved.

(1)
$$\frac{n}{24} = \frac{75}{100}$$
$$100\,n = 1{,}800$$
$$n = 18$$
Answer, 18

(2) What percent of 24 is 18?
To find what percent of 24 is 18 means:
To find the number (n) per 100 or the ratio of a number to 100 which has the same ratio as 18 to 24.

The proportion $\frac{n}{100} = \frac{18}{24}$ is formed and solved.

(2)
$$\frac{n}{100} = \frac{18}{24}$$
$$24\,n = 1{,}800$$
$$n = 75$$
$$\frac{n}{100} = 75\%$$
Answer, 75%

(3) 75% of what number is 18?

75% is the ratio of 75 to 100 or $\frac{75}{100}$

To find the number of which 75% is 18 means:
To determine the number (n) such that 18 compared to this number (n) is the same as 75 compared to 100.

The proportion $\frac{18}{n} = \frac{75}{100}$ is formed and solved.

(3)
$$\frac{18}{n} = \frac{75}{100}$$
$$75\,n = 1{,}800$$
$$n = 24$$
Answer, 24

CHAPTER 1

EXERCISES

1. Write each of the following as a ratio:
   a. 6%   **c.** 20%   **e.** 4%   **g.** 300%   **i.** $\frac{1}{4}$%   **k.** $16\frac{2}{3}$%   **m.** 180%
   b. 19%   **d.** 75%   **f.** 125%   **h.** $87\frac{1}{2}$%   **j.** 5.4%   **l.** $2\frac{1}{2}$%   **n.** 0.5%

2. Find:
   a. 2% of 80
   b. 40% of 265
   c. 25% of 108
   d. $33\frac{1}{3}$% of 765
   e. 18% of 3,976
   f. 7% of 28.52
   g. 6% of $541
   h. 130% of 903
   i. $5\frac{3}{4}$% of $4,800

3. Find each of the following:
   a. What percent of 35 is 14?
   b. What percent of 54 is 18?
   c. 6 is what percent of 24?
   d. 95 is what percent of 76?
   e. What percent of $160 is $60?
   f. $1.40 is what percent of $2?
   g. 18 is what percent of 63?
   h. What percent of 117 is 65?

4. Find each of the following:
   a. 6% of what number is 30?
   b. 18% of what number is 45?
   c. 80% of what amount is $312?
   d. 120 is 5% of what number?
   e. 141 is 4.7% of what number?
   f. 57% of what number is 57?
   g. $.60 is $62\frac{1}{2}$% of what amount?
   h. 98 is 175% of what number?

5. Find each of the following:
   a. 27% of $930.
   b. What percent of 96 is 80?
   c. 75% of what number is 84?
   d. 150 is what percent of 225?
   e. 3.5% of 2,549.
   f. $2.40 is 8% of what amount?

1-34 COMMON FRACTION, DECIMAL FRACTION, AND PERCENT NUMBER RELATIONSHIPS— SOLVING BY EQUATION

The equation may be used to find:

(1) A fractional part or decimal part or percent of a number.
(2) What fractional part or decimal part or percent one number is of another.
(3) A number when a fractional part or decimal part or percent of it is known.

Review of Number

To use this method, we read each problem carefully to find the facts which are related to the missing number. We represent this unknown number by a letter. We form an equation by translating two equal facts, with at least one containing the unknown, into algebraic expressions and by writing one expression equal to the other. If necessary, we change the percent to a common fraction or to a decimal equivalent. Then we solve the equation and check.

(1)
Find $\frac{3}{4}$ of 24:

$\frac{3}{4} \times 24 = n$
$18 = n$
$n = 18$
Answer, 18

Find .75 of 24:

$.75 \times 24 = n$
$18 = n$
$n = 18$
Answer, 18

Find 75% of 24:

$75\% \times 24 = n$
$.75 \times 24 = n$
$18 = n$
$n = 18$
Answer, 18

(2)
What fractional part of 24 is 18?

$n \times 24 = 18$
$24n = 18$
$\frac{24n}{24} = \frac{18}{24}$
$n = \frac{3}{4}$
Answer, $\frac{3}{4}$

What decimal part of 24 is 18?

$n \times 24 = 18$
$24n = 18$
$\frac{24n}{24} = \frac{18}{24}$
$n = \frac{3}{4}$
$n = .75$
Answer, .75

What percent of 24 is 18?

$n\% \times 24 = 18$
$\frac{n}{100} \times 24 = 18$
$\frac{24n}{100} = 18$
$24n = 1,800$
$n = 75$
$n\% = 75\%$
Answer, 75%

(3)
$\frac{3}{4}$ of what number is 18?

$\frac{3}{4} \times n = 18$
$\frac{3}{4}n = 18$
$\frac{4}{3} \times \frac{3}{4}n = \frac{4}{3} \times 18$
$n = 24$
Answer, 24

.75 of what number is 18?

$.75 \times n = 18$
$.75n = 18$
$\frac{.75n}{.75} = \frac{18}{.75}$
$n = 24$
Answer, 24

75% of what number is 18?

$75\% \times n = 18$
$.75n = 18$
$\frac{.75n}{.75} = \frac{18}{.75}$
$n = 24$
Answer, 24

EXERCISES

1. Find:
   a. $\frac{1}{3}$ of 45
   b. $\frac{7}{12}$ of 108
   c. $\frac{3}{8}$ of 54
   d. .4 of 26
   e. .39 of 5
   f. .058 of 9.7
   g. 5% of 980
   h. 125% of 4,500
   i. $83\frac{1}{3}$% of $3.60
   j. .03 of 81
   k. $\frac{3}{4}$ of 15
   l. $2\frac{1}{4}$% of $6,000

2. Find each of the following:
   a. What fractional part of 6 is 5?
   b. 8 is what fractional part of 72?
   c. What fractional part of 25 is 15?
   d. 98 is what fractional part of 112?
   e. What decimal part of 15 is 6?
   f. 18 is what decimal part of 24?
   g. 49 is what decimal part of 70?
   h. What decimal part of 81 is 54?
   i. What percent of 25 is 19?
   j. What percent of 70 is 42?
   k. 36 is what percent of 54?
   l. $50 is what percent of $60?
   m. What decimal part of 72 is 45?
   n. What percent of $800 is $16?
   o. 77 is what fractional part of 132?
   p. 90 is what percent of 75?

3. Find each of the following:
   a. $\frac{1}{5}$ of what number is 52?
   b. $\frac{7}{8}$ of what number is 168?
   c. 412 is $\frac{2}{3}$ of what number?
   d. 85 is $\frac{5}{12}$ of what number?
   e. .3 of what number is 2.1?
   f. .18 is .02 of what number?
   g. .375 of what number is 600?
   h. 105 is .42 of what number?
   i. 15% of what number is 24?
   j. 4% of what amount is $2,500?
   k. 64 is 160% of what number?
   l. $1,000 is $62\frac{1}{2}$% of what amount?
   m. $\frac{5}{6}$ of what number is 360?
   n. 75% of what number is 96?
   o. 63 is .9 of what number?
   p. $.84 is 6% of what amount?

4. Find each of the following:
   a. $\frac{13}{24}$ of 9,264
   b. $16\frac{2}{3}$% of $.78
   c. .01 of 67.3
   d. What fractional part of 42 is 28?
   e. 51 is what decimal part of 68?
   f. What percent of 250 is 75?
   g. 30% of what amount is $12.60?
   h. $\frac{5}{9}$ of what number is 875?
   i. 36 is .045 of what number?
   j. 220 is $133\frac{1}{3}$ percent of what number?

Review of Number

SQUARES; SQUARE ROOTS; IRRATIONAL NUMBERS

The *square* of a rational number is the product obtained when the given number is multiplied by itself. The square of 6 is 36 because $6 \times 6 = 36$.

The *square root* of a number is that number which when multiplied by itself produces the given number. It is one of the two equal factors of a product. The square root of 36 is 6 because 6 is the number which when multiplied by itself equals 36.

An *irrational number* is a number that cannot be expressed as a quotient of two whole numbers (with division by zero excluded). A number that is both a non-terminating and non-repeating decimal like the square root of any arithmetic number other than perfect squares (numbers having an exact square root) is an irrational number.

$\sqrt{14} = 3.742 \ldots$ is an irrational number.

1–35 SQUARE OF A NUMBER; EXPONENTS

Square of a Number

To square a number, we multiply the given number by itself.

Square 5:
$5 \times 5 = 25$
Answer, 25

Square $\frac{2}{3}$:
$\frac{2}{3} \times \frac{2}{3} = \frac{4}{9}$
Answer, $\frac{4}{9}$

Square .7:
$.7 \times .7 = .49$
Answer, .49

Exponents

When two equal factors are used in multiplication, such as 5×5, they may be written in *exponential* form as 5^2. The small numeral written to the upper right of the repeated factor is called an *exponent*. It tells us how many times the factor is being used in multiplication. The factor that is being repeated is called the *base*.

> The numeral 5² uses 5 as the base and 2 as the exponent.

The numeral 5^2 is read "five squared" or "the square of five" or "five to the second power" or "the second power of five."

The product of equal factors is called a *power* of the factor. It may be expressed as an indicated product using an exponent.

> 25, the product of 5×5, or the indicated product 5^2 is the second power of 5.

Numbers like 4, 8, 16, and 32 which can be expressed as the indicated products 2^2, 2^3, 2^4, and 2^5 respectively are powers of 2. The number 4 (since $4 = 2 \times 2$) is the second power of 2, the number 8 (since $8 = 2 \times 2 \times 2$) is the third power of 2, the number 16 (since $16 = 2 \times 2 \times 2 \times 2$) is the fourth power of 2, etc.

EXERCISES

1. Read each of the following expressions:
   a. 9^2 b. 17^2 c. $\left(\frac{3}{4}\right)^2$ d. $(.8)^2$ e. 100^2 f. $\left(1\frac{2}{3}\right)^2$ g. $\left(\frac{8}{5}\right)^2$ h. $(.52)^2$

2. Write as a numeral:
   a. Fifteen squared   b. The square of twenty   c. Eleven to the second power

3. Express each of the following in exponential form:
   a. 4×4 b. $\frac{7}{12} \times \frac{7}{12}$ c. $.6 \times .6$ d. $2\frac{9}{16} \times 2\frac{9}{16}$

4. Square each of the following numbers:
   a. 5   c. 106   e. $\frac{1}{2}$ g. $1\frac{5}{8}$ i. .3   k. 4.5   m. 967   o. 8.25
   b. 31   d. 87   f. $\frac{11}{12}$ h. $\frac{15}{7}$ j. .06   l. 3.14   n. $4\frac{3}{5}$ p. 1,000

5. Find the value of each of the following:
   a. 8^2 b. 23^2 c. $\left(\frac{2}{5}\right)^2$ d. $(.1)^2$ e. $\left(2\frac{1}{2}\right)^2$ f. $(4.7)^2$ g. 685^2 h. $\left(\frac{12}{25}\right)^2$

Review of Number

6. Read, or write in words, each of the following:

   a. 6^3
   b. 3^5
   c. 8^7
   d. 5^6
   e. 12^4
   f. 10^{11}
   g. 4^1
   h. 1^{10}
   i. 25^2
   j. 100^{19}
   k. 7^{15}
   l. 9^{20}

7. Express each of the following as a numeral, using exponents:

   a. Eight to the fourth power
   b. Five cubed
   c. Ten to the ninth power
   d. Fourteen squared
   e. Thirty to the fifth power
   f. Seven to the twelfth power
   g. Nineteen to the tenth power
   h. Eleven to the sixteenth power

8. a. What is the exponent in each of the following?

   (1) 3^7 (2) 18^5 (3) 5^{10} (4) 10^{14} (5) 35^{21} (6) 24^{100}

   b. What is the base in each of the following?

   (1) 8^4 (2) 12^9 (3) 10^{15} (4) 43^{11} (5) 15^{30} (6) 69^8

9. How many times is the base being used as a factor in each of the following?

   a. 10^7 b. 3^5 c. 7^9 d. 6^{12} e. 18^5 f. 35^{17}

10. Use the exponential form to write:

    a. $6 \times 6 \times 6 \times 6$
    b. $14 \times 14 \times 14$
    c. $8 \times 8 \times 8 \times 8 \times 8$
    d. 47×47
    e. $2 \times 2 \times 2 \times 2 \times 2 \times 2 \times 2$
    f. $5 \times 5 \times 5 \times 5 \times 5 \times 5 \times 5 \times 5 \times 5$
    g. $3 \times 3 \times 3 \times 3 \times 3 \times 3 \times 3 \times 3 \times 3 \times 3 \times 3 \times 3$
    h. $1 \times 1 \times 1 \times 1 \times 1 \times 1 \times 1 \times 1 \times 1 \times 1 \times 1 \times 1 \times 1 \times 1$

11. Which power:

    a. Of 3 is 27?
    b. Of 10 is 1,000?
    c. Of 5 is 625?
    d. Of 4 is 256?
    e. Of 2 is 64?
    f. Of 10 is 1,000,000?

12. First write as an indicated product, using exponents, then find the product of each of the following:

    a. Fourth power of 7
    b. Third power of 12
    c. Eighth power of 10
    d. Sixth power of 3
    e. Fifth power of 4
    f. Twentieth power of 1

1-36 SQUARE ROOT

Square Root by Estimation, Division, and Average

We may find the *approximate* square root of a number by:
(1) Estimating the square root of the given number.
(2) Dividing the given number by the estimated square root.
(3) Finding the average of the resulting quotient and the estimated square root.
(4) Dividing the given number by the result of step (3).
(5) Finding the average of the divisor and quotient found in step (4).
(6) Continuing this process to obtain a greater degree of approximation as the divisor and quotient will eventually approximate each other.

To find the square root of 6 by this method, since the square root of 4 is 2 and the square root of 9 is 3, we estimate that the square root of 6 is between 2 and 3, perhaps 2.5.

Dividing 6 by 2.5 we get the quotient 2.4. Averaging 2.5 and 2.4, we get 2.45.

Dividing 6 by 2.45 we get the quotient 2.44. Averaging 2.45 and 2.44, we get 2.445.

Find the square root of 6:

```
        2.4              2.4
 2.5)6.0 0             +2.5
     5 0              2)4.9  = 2.45
     1 0 0
     1 0 0
```

```
         2.44            2.45
 2.45)6.00 00          +2.44
      4 90            2)4.89 = 2.445
      1 10 0
        98 0
        12 00
         9 80
         2 20
```

Answer, 2.445

EXERCISES

Find the square root of each of the following numbers by estimation, division, and average:

a. 5   **c.** 42   **e.** 85   **g.** 76   **i.** 26   **k.** 53   **m.** 18   **o.** 89
b. 12 **d.** 30   **f.** 67   **h.** 39   **j.** 94   **l.** 87   **n.** 46   **p.** 28

Square Root—Alternate Method

To find the square root of a number, we write its numeral under the square root symbol $\sqrt{}$, and separate the numeral into groups of two digits each to the left and then to the right of the decimal point. If a numeral naming a whole number contains an odd number of digits, there will be one group with one digit on the left. If a numeral naming a decimal fraction contains an odd number of digits, a zero is annexed so that each group contains two digits.

(1) We find the largest square (25) which can be subtracted from the first group at the left (33) and write it under this group.

(2) We write the square root (5) of this largest square as the first digit in the answer. Each digit in the answer is written directly over its corresponding group.

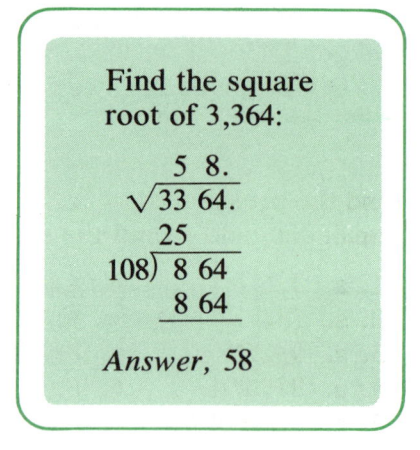

Find the square root of 3,364:

$$\begin{array}{r} 5\ 8. \\ \sqrt{33\ 64.} \\ 25 \\ \hline 108)\ 8\ 64 \\ 8\ 64 \\ \hline \end{array}$$

Answer, 58

(3) After we subtract the square, the next group (64) is annexed to the remainder (8).

(4) A trial divisor is formed by multiplying the root already found (5) by 2 and annexing a zero which is not written but used mentally.

(5) We divide the dividend (864) by this trial divisor (100) and annex the quotient (8) to the root already found. We also annex it to the trial divisor to form the complete divisor (8).

(6) We multiply the complete divisor by the new digit (8) of the root and subtract this product from the dividend.

(7) This process is continued until all groups are used or the required number of decimal places has been obtained. The decimal point in the answer is placed directly above the decimal point in the given numeral.

Find the square root of 57,121:

```
       2 3 9.
    √‾5 71 21.
      4
   43)1 71
      1 29
  469)  42 21
       42 21
```

Answer, 239

Find the square root of .04:

```
      .2
   √‾.04
```

Answer, .2

Find the square root of 3.4:

```
      1. 8
   √‾3.40
      1
   28)2 40
      2 24
        16
```

Answer, 1.8

EXERCISES

Find the square root of each of the following numbers. If there is a remainder, find the answer correct to the nearest hundredth:

1. a. 3,249 b. 1,936 c. 6,084 d. 3,600 e. 8,836
2. a. 784 b. 361 c. 900 d. 576 e. 729
3. a. 99,225 b. 24,336 c. 72,361 d. 52,900 e. 16,384
4. a. 876,096 b. 462,400 c. 299,209 d. 677,329 e. 474,721
5. a. 2,119,936 b. 4,695,889 c. 3,928,324 d. 7,070,281 e. 9,759,376
6. a. 69,639,025 b. 10,876,804 c. 35,868,121 d. 61,826,769 e. 99,361,024
7. a. 43,681 b. 93,025 c. 651,249 d. 820,836 e. 494,209
8. a. 16,072,081 b. 4,028,049 c. 50,126,400 d. 15,210,000 e. 81,144,064
9. a. .64 b. .0001 c. .0324 d. 31.36 e. 8.41
10. a. 29 b. 6.82 c. 914.5 d. 200 e. 4.9

Square Root by Use of Table

Using the table of squares and square roots (see page 547), we can find directly the square roots of whole numbers 1 to 99 inclusive and of the perfect squares (squares of whole numbers) given in the table.

Review of Number

To find the square root of any whole number from 1 to 99 inclusive, we first locate the given number in the "No." column and then move to the right to the corresponding "Square Root" column to obtain the required square root.

To find the square root of a perfect square given in the table, we first locate this number in the "Square" column and then move to the left to the corresponding "No." column to obtain the required square root.

EXERCISES

Find the square root of each of the following numbers:

1. a. 27   b. 19   c. 85   d. 51   e. 37   f. 63   g. 96   h. 44
2. a. 72   b. 33   c. 61   d. 42   e. 20   f. 80   g. 13   h. 99
3. a. 676   b. 1,156   c. 4,225   d. 2,704   e. 6,889   f. 9,409
4. a. 5,041   b. 9,604   c. 529   d. 1,681   e. 2,916   f. 3,364

Principal Square Root*

Since $(+5) \times (+5) = 25$ and $(-5) \times (-5) = 25$, it follows that 25 has two square roots: $+5$ and -5. Whole numbers, other than zero, each have two square roots, a positive square root and a negative square root. Zero has only one square root which is zero (0). The positive square root of a number is called the *principal square root*.

EXERCISES

1. Find the two square roots of each of the following numbers:
   a. 16   b. 100   c. 81   d. 144   e. 36   f. 121   g. 64   h. 1
2. What is the principal square root of each of the following numbers?
   a. 4   b. 9   c. 49   d. 100   e. 441   f. 225   g. 169   h. 324

---

* Take this section only if operations with positive and negative numbers have been studied.

ACHIEVEMENT TEST

The numeral at the end of each of the following problems indicates the section where explanatory material may be found.

1. Round 439,675,802 to nearest million. (1–1)

2. Add: 58,699
   48,985
   39,498
   47,626
   88,748 (1–2)

3. Subtract: 9,320,085
              8,925,479 (1–3)

4. Multiply: 5,608
             9,047 (1–4)

5. Divide: 783)545,751 (1–5)

6. a. Express $\frac{40}{56}$ in lowest terms. (1–6)
   b. Raise to higher terms: $\frac{7}{20} = \frac{?}{100}$ (1–7)
   c. Express $\frac{27}{12}$ as a mixed number. (1–8)
   d. Change $3\frac{5}{6}$ to an improper fraction. (1–9)

7. Add: $5\frac{7}{10}$
        $3\frac{4}{5}$ (1–12)

8. Subtract: $9\frac{5}{6}$
             $4\frac{7}{8}$ (1–13)

9. Multiply: $6\frac{2}{3} \times 3\frac{9}{10}$ (1–14)

10. Divide: $12 \div 1\frac{3}{5}$ (1–15)

11. Which of the following statements are true? (1–11)
    a. $\frac{15}{18}$ and $\frac{80}{96}$ are equivalent fractions.
    b. $\frac{3}{5} < \frac{7}{12}$
    c. .045 > .03

12. Round 10.5249 to the nearest hundredth. (1–16)

13. a. Write 86.9 million as a complete numeral. (1–24)
    b. Write the shortened name for 4,580,000,000 in billions. (1–24)

Review of Number

14. Add: 3.16 + .153 + 42.8 (1–18)
15. Subtract: 6.2 − .32 (1–19)
16. Multiply: .025 × .004 (1–20)
17. Divide: 2.7 ÷ .03 (1–22)
18. Multiply 86.531 by 100 (1–21)
19. Divide 95.07 by 1,000 (1–23)
20. a. 54 is what fractional part of 63? (1–34)
    b. 18 is what decimal part of 60? (1–34)
21. a. $\frac{3}{4}$ of what number is 81? (1–34)
    b. .9 of what number is 108? (1–34)
22. Write in scientific notation:
    a. 54,600,000,000 (1–25)
    b. .000092 (1–25)
23. a. Express $\frac{13}{25}$ as a decimal numeral. (1–26)
    b. Express 5.8% as a decimal numeral. (1–28)
    c. Express $\frac{11}{18}$ as a numeral naming a repeating decimal. (1–26)
24. a. Express .36 as a numeral naming a common fraction. (1–27)
    b. Express $66\frac{2}{3}$% as a numeral naming a common fraction. (1–28)
25. a. Express .045 as a percent. (1–29)
    b. Express $\frac{17}{20}$ as a percent. (1–29)
26. Find $8\frac{3}{4}$% of $6,400. (1–30, 1–33, 1–34)
27. What percent of 78 is 65? (1–31, 1–33, 1–34)
28. 7 is 28% of what number? (1–32, 1–33, 1–34)
29. Find the square of 1.05 (1–35)
30. a. Find the square root of 12 by estimation, division, and average. (1–36)
    b. Find the square root of 651,249. (1–36)

ESTIMATING ANSWERS

For each of the following select your nearest estimate:

1.	294 × 513 is approximately:	130,000	140,000	150,000
2.	$\frac{1}{2}$ of 2,589 is approximately:	1,200	1,300	1,400
3.	78,085 ÷ 97 is approximately:	850	800	750
4.	$6\frac{1}{2} \div 2\frac{1}{8}$ is approximately:	3	4	5
5.	594,362 + 405,473 is approximately:	900,000	950,000	1,000,000
6.	.31 ÷ .6 is approximately:	.05	.005	.5
7.	4.16 − .205 is approximately:	.2	.02	4
8.	3% of 197 is approximately:	600	6	60
9.	60 × $7.94 is approximately:	$420	$500	$480
10.	9,107 − 2,986 is approximately:	6,000	7,000	8,000
11.	$12\frac{1}{3} - 5\frac{7}{8}$ is approximately:	7	6	8
12.	11 is approximately what percent of 51?	40%	30%	20%
13.	$7.25 + $11.71 + $5.08 is approximately:	$23	$25	$24
14.	$10\frac{3}{4} + 5\frac{1}{6}$ is approximately:	14	15	16
15.	$7\frac{3}{4}$% of $6,000 is approximately:	$420	$470	$450
16.	$149.89 ÷ $.10 is approximately:	1,500	150	15
17.	$3\frac{7}{8} \div 1\frac{5}{6}$ is approximately:	4	3	2
18.	$83.09 − $31.98 is approximately:	$52	$5.25	$51
19.	.04 × .03 is approximately:	.12	.012	.0012
20.	73,500 − 4,675 is approximately:	6,900	69,000	72,000
21.	4,207 × 6,835 is approximately:	250,000	2,800,000	29,000,000
22.	$213 \div 4\frac{5}{6}$ is approximately:	50	1,000	40
23.	24 is approximately what percent of 74?	30%	40%	50%
24.	8,155 + 3,340 + 5,528 is approximately:	15,000	16,000	17,000
25.	$8\frac{4}{5} \times 3\frac{1}{10}$ is approximately:	24	27	30
26.	.83 + .69 + .19 is approximately:	.17	1.7	17
27.	357,888 ÷ 903 is approximately:	350	400	450

Review of Number

CHECK BY CALCULATOR

Use your calculator to check each of the following answers. Indicate which answers are incorrect. If a calculator is not available, check by computation.

1. Add:

    a. 9,589 + 423 + 6,594 + 2,778 + 696 = 21,080
    b. 8,296 + 5,948 + 9,397 + 4,859 + 3,775 = 32,275
    c. 2,878 + 71,509 + 8,458 + 89,326 + 7,698 = 179,869
    d. 49,603 + 82,959 + 57,416 + 69,576 + 38,869 = 297,423
    e. 687,948 + 57,839 + 382,094 + 47,548 + 589,967 = 1,765,396

2. Subtract:

    a. 9,436 − 5,395 = 4,041
    b. 67,307 − 9,208 = 58,199
    c. 78,051 − 68,786 = 10,265
    d. 874,005 − 385,172 = 488,833
    e. 3,961,084 − 1,493,596 = 2,467,488

3. Multiply:

    a. 926 × 473 = 437,998
    b. 285 × 857 = 244,245
    c. 3,962 × 708 = 2,815,096
    d. 9,428 × 5,984 = 56,417,152
    e. 7,080 × 8,907 = 63,061,560

4. Divide:

    a. 51,520 ÷ 64 = 805
    b. 26,691 ÷ 287 = 93
    c. 514,608 ÷ 906 = 568
    d. 683,886 ÷ 798 = 857
    e. 1,394,841 ÷ 1,569 = 899

5. Add:

    a. 2.79 + 8.36 + 3.57 = 14.72
    b. .741 + .689 + .276 = 1.706
    c. .684 + 4 = .688
    d. 5.6 + .738 = 6.338
    e. .063 + 2.31 + 75.4 = 77.773

6. Subtract:

   a. .91 − .85 = .6
   b. .5 − .29 = .21
   c. 8.57 − .2 = 8.55
   d. 20 − 3.67 = 16.33
   e. 7.865 − 4.64 = 3.225

7. Multiply:

   a. .3 × .47 = .141
   b. .007 × .064 = .448
   c. 8.6 × 9.2 = 79.12
   d. 3,000 × .0006 = 18
   e. 4.1317 × 16.09 = 66.479053

8. Divide:

   a. 5.04 ÷ 14 = 3.6
   b. 60.35 ÷ 8.5 = 7.1
   c. .04 ÷ .08 = .02
   d. 48 ÷ .6 = 80
   e. .00027 ÷ :03 = .009

9. Find:

   a. 4% of 78 = 312
   b. 65% of 859 = 558.35
   c. 8% of 30.46 = 243.68
   d. 135% of 20 = 27
   e. 5.9% of 16.7 = .9853

10. Find each of the following:

a. 63 is what percent of 84?	75%
b. What percent of 95 is 19?	25%
c. 168 is what percent of 350?	48%
d. 93 is what percent of 62?	150%
e. What percent of 900 is 27?	30%

11. Find each of the following:

a. 3 is 15% of what number?	5
b. 32% of what number is 80?	250
c. 125% of what number is 75?	40
d. 54 is 300% of what number?	18
e. 7.9% of what number is 237?	3,000

Review of Number

NUMBER THEORY

1-37 EVEN AND ODD NUMBERS

Whole numbers may be separated into even and odd numbers.

An *even number* is a whole number that can be divided exactly (is divisible) by two (2). Zero is considered an even whole number. Numbers whose numerals end with a 0, 2, 4, 6, or 8 are even numbers.

An *odd number* is a whole number that cannot be divided exactly by two (2). Numbers whose numerals end with a 1, 3, 5, 7, or 9 are odd numbers.

EXERCISES

1. Which of the following are odd numbers? Which are even numbers?

    a. 92   45   19   200   563   7,894

    b. 83   30   76   605   958   48,541

2. **a.** Write all the one-digit numerals that name even whole numbers.
    b. Write all the one-digit numerals that name odd natural numbers.

3. **a.** Write the numerals that name all the odd numbers greater than 58 and less than 71.
    b. Write the numerals that name all the even numbers less than 109 and greater than 97.

4. Is the sum an odd number or an even number when we add:

    a. Two odd numbers? Illustrate.
    b. Two even numbers? Illustrate.
    c. An odd and an even number? Illustrate.

5. Is the product an odd number or an even number when we multiply:

    a. Two odd numbers? Illustrate.
    b. Two even numbers? Illustrate.
    c. An odd number and an even number? Illustrate.

6. a. Is there an even whole number between every pair of odd whole numbers?
   b. Is there an odd whole number between every pair of even whole numbers?

7. a. Is one more than any whole number an even number?
   b. Is two more than any whole number an even number?
   c. Is two times any whole number an odd number or an even number?

8. a. Is one more than any odd number an odd number or an even number?
   b. Is one more than any even number an odd number or an even number?
   c. Is two more than any odd number an odd number or an even number?

9. a. Select any odd number and square it. Is the square of an odd number an odd number or an even number? Why?
   b. Select any even number and square it. Is the square of an even number an odd number or an even number? Why?

10. Using the odd numbers 1, 3, 5, 7, 9, 11, 13, and 15, find the sum of the first two odd numbers; first three odd numbers; first four odd numbers; first five odd numbers; first six odd numbers; first seven odd numbers; all eight odd numbers. Study the sums and the number of addends used in each case to find the relationship between the number of addends and their sums.

1–38 FACTORS

A *factor* is any one of the numbers used in multiplication to form the answer called the *product*. In $3 \times 7 = 21$, the numbers 3 and 7 are factors and the number 21 is called the product. To determine whether a particular number is a factor of a given whole number, we divide it into the given number. If the division is exact (with a zero remainder), the number is a factor.

Any number that is a factor of each of two or more given whole numbers is called a *common factor* of the numbers. Sometimes it is called the *common divisor* of the numbers. One (1) is a common factor of any set of numbers.

The *greatest common factor* of two or more whole numbers is the greatest whole number that will divide all the given numbers exactly.

> The factors of 8 are 1, 2, 4, and 8.
>
> The factors of 12 are 1, 2, 3, 4, 6, and 12.
>
> The common factors of 8 and 12 are 1, 2, and 4.
>
> The greatest common factor of 8 and 12 is 4.

Euclid, a Greek mathematician, devised a method of finding the greatest common factor (or greatest common divisor) of two given numbers. Euclid's method:

Divide the larger given number by the smaller number. Then divide the divisor by the remainder, then divide the next divisor by the next remainder, continuing in this way until the remainder is zero. The last non-zero remainder is the greatest common factor (G.C.F.).

> Find the G.C.F. of 48 and 18:
>
> $$18 \overline{)48} \; (2 \; R12$$
> $$12 \overline{)18} \; (1 \; R6$$
> $$6 \overline{)12} \; (2 \; R0$$
>
> *Answer,* G.C.F. = 6

EXERCISES

1. Write all the factors of each of the following numbers:

   a. 11 **d.** 48 **g.** 200
   b. 34 **e.** 72 **h.** 450
   c. 18 **f.** 108 **i.** 500

2. For each of the following groups of numbers write all the common factors:

   a. 8 and 10 **c.** 54 and 72 **e.** 36, 84, and 132
   b. 6 and 5 **d.** 18, 24, and 42 **f.** 200, 150, and 500

3. Find the greatest common factor of each of the following groups of numbers:

   a. 12 and 20
   b. 3 and 4
   c. 108 and 144
   d. 39, 65, and 91
   e. 72, 120, and 168
   f. 119, 51, and 85

4. Use Euclid's method to find the greatest common factor of:

   a. 48 and 56
   b. 35 and 63
   c. 120 and 216
   d. 80 and 224
   e. 250 and 625
   f. 144 and 240

1–39 PRIME AND COMPOSITE NUMBERS

Whole numbers other than 0 and 1 may be classified as either prime or composite numbers.

A *prime number* is a whole number other than 0 and 1 which is divisible (can be divided exactly) only by itself and by 1 and by no other whole number.

> 17 is a prime number;
> it can be divided exactly only by 17 and by 1.

A *composite number* is a whole number other than 0 and 1 which is not a prime number. It can be expressed as a product of two or more smaller whole numbers.

> 10 is a composite number; it can be divided exactly not only by 10 and by 1 but also by 2 and by 5.

A composite number can be expressed as a product of prime numbers. Each composite number has only one group of prime factors, but the factors may be arranged in different orders. For example, 12 = 2 · 2 · 3 or 2 · 3 · 2 or 3 · 2 · 2.

Two prime numbers are called *twin primes* if one number is two more than the other number. 29 and 31 are a pair of twin primes.

Two numbers are said to be *relatively prime* to each other when they have no common factors other than 1. The numbers do not necessarily have to be prime numbers. The numbers 14 and 25 are relatively prime to each other because 1 is the only whole number that will divide into both 14 and 25 exactly.

Review of Number

EXERCISES

1. Which of the following are prime numbers? Are all the other numbers named composite numbers?

    57   61   89   46   63   29   87   1   97   51

2. Name three composite numbers greater than 69 and less than 75.

3. **a.** Name all the one-digit prime numbers.
   b. Name all the one-digit even prime numbers.

4. What twin primes are greater than 34 and less than 62?

5. Which of the following are relatively prime?

    a. 10 and 18   **c.** 33 and 75   **e.** 91 and 26
    b. 7 and 22   **d.** 24 and 49   **f.** 59 and 95

1–40 NUMBER MULTIPLES

A *multiple* of a given whole number is a product of the given number and another whole number factor. The multiples of 8 are: 0, 8, 16, 24, ... A multiple of a given number is divisible by the given number.

Any number which is a multiple of two or more whole numbers is called the *common multiple* of the numbers. It is a number that is divisible by all the given numbers. Numbers may have many common multiples.

The *least common multiple* (L.C.M.) of two or more numbers is the smallest natural number which is the multiple of all of them. It is the smallest possible natural number that can be divided exactly by all the given numbers. Zero (0) is excluded when determining the least common multiple, although it is a common multiple of any group of numbers.

> The multiples of 8 are 0, 8, 16, 24, 32, 40, 48, . . .
> The multiples of 12 are 0, 12, 24, 36, 48, 60, 72, . . .
> The common multiples of 8 and 12 are 0, 24, 48, . . .
> The least common multiple of 8 and 12 is 24 since 0 is excluded.

The L.C.M. may be found by factoring the given numbers as primes and forming a product of these primes using each the greatest number of times it appears in the factored form of any one number.

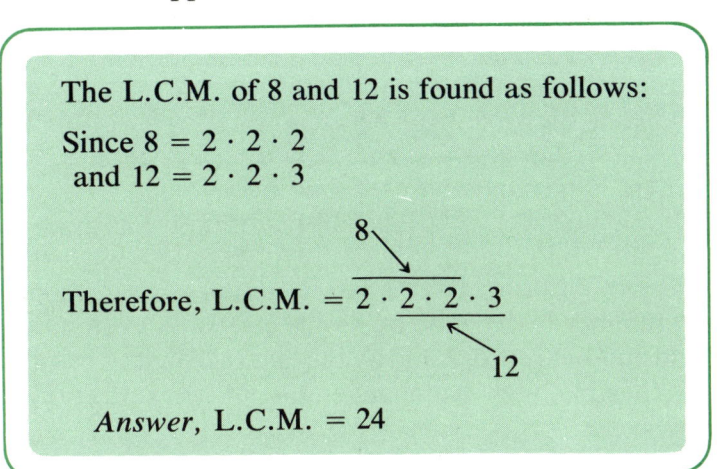

> The L.C.M. of 8 and 12 is found as follows:
>
> Since $8 = 2 \cdot 2 \cdot 2$
> and $12 = 2 \cdot 2 \cdot 3$
>
> Therefore, L.C.M. = $\underset{12}{\overset{8}{2 \cdot 2 \cdot 2 \cdot 3}}$
>
> *Answer*, L.C.M. = 24

EXERCISES

1. Write all the multiples of each of the following, listing the first five:

   a. 3 **b.** 7 **c.** 6 **d.** 10 **e.** 19 **f.** 25 **g.** 32 **h.** 60 **i.** 84 **j.** 275

2. Write all the common multiples of each of the following, listing the first four:

   a. 3 and 8   **c.** 8 and 18   **e.** 72 and 96   **g.** 125 and 200   **i.** 8, 10, and 12
   b. 10 and 4 **d.** 12 and 20 **f.** 56 and 42   **h.** 2, 4, and 5   **j.** 25, 30, and 75

3. Find the least common multiple for each of the following groups:

   a. 4 and 5   **c.** 10 and 16 **e.** 24 and 18   **g.** 144 and 108   **i.** 6, 10, and 16
   b. 8 and 6   **d.** 35 and 14 **f.** 60 and 84   **h.** 3, 4, and 5   **j.** 18, 45, and 54

1–41 TESTS FOR DIVISIBILITY

A number is said to be *divisible* by another number if it can be divided exactly by the second number with no remainder.

The following are some quick tests to determine whether a number is divisible by 2, 3, 4, 5, 6, 8, 9, 10.

Review of Number

(1) A number is divisible by 2 only if it ends in 0, 2, 4, 6, or 8. All even numbers are divisible by 2.
(2) A number is divisible by 3 only if the sum of its digits is divisible by 3.

> 6,528 is divisible by 3 because
> 6 + 5 + 2 + 8 = 21, and 21 is divisible by 3.
> 2,431 is not divisible by 3 because
> 2 + 4 + 3 + 1 = 10, and 10 is not divisible by 3.

(3) A number is divisible by 4 only if it is an even number and the number represented by the last two digits (tens and units) is divisible by 4. Numbers ending in two zeros are divisible by 4.

> 27,932 is divisible by 4 because 32 is divisible by 4.

(4) A number is divisible by 5 only if it ends in 5 or 0.

> 4,685 ends in 5, thus it is divisible by 5.
> 77,130 ends in 0, thus it is divisible by 5.

(5) A number is divisible by 6 only if it is an even number and the sum of its digits is divisible by 3.

> The even number 7,512 is divisible by 6 because
> 7 + 5 + 1 + 2 = 15 and 15 is divisible by 3.

(6) A number is divisible by 8 only if it is an even number and the number represented by the last three digits (hundreds, tens, and units digits) is divisible by 8. Numbers ending in three zeros are divisible by 8. Use this test only when the number is 1,000 or larger.

> The even number 19,304 is divisible by 8 because 304 is divisible by 8.

(7) A number is divisible by 9 only if the sum of its digits is divisible by 9.

> 83,115 is divisible by 9 because
> 8 + 3 + 1 + 1 + 5 = 18 and 18 is divisible by 9.

(8) A number is divisible by 10 only if it ends in 0.

> 6,370 is divisible by 10 because it ends in 0.

EXERCISES

1.
   a. Is 59,748 divisible by 4?
   b. Is 88,767 divisible by 9?
   c. Is 46,823 divisible by 2?
   d. Is 286,590 divisible by 5?
   e. Is 727,101 divisible by 3?
   f. Is 15,462 divisible by 6?
   g. Is 63,520 divisible by 8?
   h. Is 905,005 divisible by 10?
   i. Is 832,455 divisible by 9?
   j. Is 578,356 divisible by 4?

2.
   a. Is 94,584 divisible by 2? by 4? by 3? by 8? by 9?
   b. Is 807,810 divisible by 5? by 10? by 2? by 3? by 6?
   c. Is 56,000 divisible by 4? by 9? by 8? by 6? by 10?

3. Determine whether the following numbers are divisible:

   a. by 6:   752;   3,207;   4,068;   49,180;   585,774
   b. by 8:   4,694; 5,000;   9,959;   75,962;   654,392
   c. by 3:   551;   4,526;   1,782;   56,895;   569,994
   d. by 5:   306;   9,345;   1,830;   37,253;   344,900
   e. by 2:   978;   8,663;   9,082;   15,730;   862,114
   f. by 9:   243;   5,739;   35,496; 88,577;   796,698
   g. by 10: 670;   3,000;   83,405; 60,850;   196,230
   h. by 4:   512;   4,900;   7,674;   29,632;   852,175

Review of Number

1-42 CHECKING BY CASTING OUT NINES

The operations of addition, subtraction, multiplication, and division may be checked by a method called *casting out nines*. This method is not perfect since an incorrect answer sometimes will check.

In this checking process each given number is divided by nine, the nines (quotient) are then cast out (discarded) and only the remainder, called the *excess*, is used.

Suppose 4,251 is divided by 9. We find there are 472 nines and 3 ones. The remainder 3 indicates there are 3 ones in excess of an exact number of nines (472 nines).

$$\begin{array}{r} 472 \\ 9\overline{)4{,}251} \\ 3\ 6 \\ \hline 65 \\ 63 \\ \hline 21 \\ 18 \\ \hline 3 \end{array}$$

This excess in a number may be found by an easier method.

Suppose we add the digits in 4,251; we find the sum is 12. If we add the digits in 12, we find the sum is 3. This sum is the remainder we found when we divided.

$$4 + 2 + 5 + 1 = 12$$
$$1 + 2 = 3$$

Thus we may find the excess ones with respect to the number of groups of nine a number possesses by adding the digits in a given number, then adding the digits in the sum, continuing in this way until the final sum is a one-digit number less than 9. If the sum is 9, it is replaced by a zero since the original number is divisible by 9 and 0 is the remainder. This final sum is the excess.

To check addition

We find the excess in each addend and in the sum. The sum of the excesses in the addends should equal the excess in the sum of the addends.[1]

```
Addition:    392    (5)       (5) + (8) + (2)
             431    (8)       = (15) = (6)
            +380    (2)
            1,203   (6)
```

---

[1] If the sum or product of the excesses is a two-digit number, simplify and express as a single digit excess.

To check subtraction

We find the excess in the minuend, subtrahend, and answer. The sum of the excesses in the subtrahend and answer should equal the excess in the minuend. When the excess in the minuend is 0, use 9.

Subtraction: $\begin{array}{r}6{,}783 \\ -4{,}342 \\ \hline 2{,}441\end{array}$ (6)   (6) − (4) = (2)
(4)
(2)   or (4) + (2) = (6)

To check multiplication

We find the excess in the multiplicand, multiplier, and product. The product of the excesses in the factors should equal the excess in the product.[1]

Multiplication: $\begin{array}{r}65 \\ \times 31 \\ \hline 65 \\ 195 \\ \hline 2{,}015\end{array}$ (2)
(4)   (2) × (4) = (8)

(8)

To check division

We find the excess in the divisor, dividend, quotient, and remainder if any. The product of the excesses in the divisor and quotient increased by the excess in the remainder should equal the excess in the dividend.[2] When the excess in the divisor is 0, use 9.

Division: $\begin{array}{r}57 \\ 43\overline{)2{,}451} \\ 2\ 15 \\ \hline 301 \\ 301\end{array}$ (3)
(7))(3)

(7) × (3) = (21) = (3)

[1,2] If the sum or product of the excesses is a two-digit number, simplify and express as a single digit excess.

Review of Number

EXERCISES

Compute each of the following as directed and check by casting out nines.

1. Add:

7,156	4,468	84,526	75,213	865,523
483	9,325	35,145	8,856	914,643
97	7,842	63,368	42,101	887,574
4,588	6,587	71,456	5,639	348,206
625	3,958	92,061	67,482	768,516

2. Subtract:

| 8,424 | 42,103 | 85,976 | 523,418 | 4,562,177 |
| 3,842 | 7,346 | 43,819 | 178,993 | 3,259,689 |

3. Multiply:

| 416 | 587 | 7,524 | 8,747 | 25,633 |
| 231 | 428 | 967 | 6,852 | 58,246 |

4. Divide:

$84\overline{)5{,}628}$ $325\overline{)30{,}225}$ $214\overline{)157{,}290}$ $648\overline{)386{,}208}$ $907\overline{)805{,}416}$

OPERATIONS AND PROPERTIES

1–43 BINARY OPERATIONS; INVERSE OPERATIONS

An operation that is performed on only two numbers at a time is called a *binary operation*. The operations of addition, subtraction, multiplication, and division are binary operations.

Operations that undo each other are called *inverse operations*. Addition and subtraction are inverse operations; multiplication and division are inverse operations. They undo each other.

EXERCISES

Find the missing numbers or symbols of operation indicated by n, □, or ?:

1. a. $(15 - 7) + 7 = n$
    b. $(92 + 11) - 11 = $ □
    c. $(36 \div 4) \times 4 = ?$
    d. $(8 \times 9) \div 9 = n$

2. a. (40 × 3) ÷ ? = 40
   b. (18 + 12) − □ = 18
   c. (50 − 8) + n = 50
   d. (27 ÷ 9) × □ = 27

3. a. (75 + 25) ? 25 = 75
   b. (39 × 8) ? 8 = 39
   c. (84 ÷ 12) ? 12 = 84
   d. (60 − 17) ? 17 = 60

4. a. (32 ÷ 4) ? n = 32
   b. (48 − 21) ? n = 48
   c. (29 + 19) ? n = 29
   d. (33 × 41) ? n = 33

1–44 PROPERTIES

Operations have certain characteristics or *properties*. A property is not true unless it holds for all cases.

Commutative Property

Of Addition:

The *commutative property of addition* permits us to change the order of adding two numbers without affecting the sum.

$$7 + 5 = 5 + 7$$

Of Multiplication:

The *commutative property of multiplication* permits us to change the order of multiplying two numbers without affecting the product.

$$7 × 5 = 5 × 7$$

Associative Property

Of Addition:

The *associative property of addition* permits us to group or associate the first and second numbers and add their sum to the third number, or to group or associate the second and third numbers and add their sum to the first number. Either way we get the same final sum.

7 + 5 + 2 may be thought of as either:

(7 + 5) + 2 which is 12 + 2 = 14
or 7 + (5 + 2) which is 7 + 7 = 14

Thus, (7 + 5) + 2 = 7 + (5 + 2)

Review of Number

Of Multiplication:

The *associative property of multiplication* permits us to group or associate the first and second numbers and multiply their product by the third number, or to group or associate the second and third numbers and multiply their product by the first number. Either way we get the same final product.

> 7 × 5 × 2 may be thought of as either:
>
> (7 × 5) × 2 which is 35 × 2 = 70
>
> or 7 × (5 × 2) which is 7 × 10 = 70
>
> Thus, (7 × 5) × 2   =   7 × (5 × 2)

Distributive Property of Multiplication Over Addition

The *distributive property of multiplication over addition* tells us that when we multiply one number by the sum of a second and third number we get the same result as when we add the product of the first and second numbers to the product of the first and third numbers. Multiplication is being distributed over addition.

> To find the product of 3 × (6 + 2):
>
> 3 × (6 + 2) = 3 × 8 = 24
>
> or (3 × 6) + (3 × 2) = 18 + 6 = 24
>
> Thus, 3 × (6 + 2) = (3 × 6) + (3 × 2)

Closure

If, using all of a group of given numbers, we add any two numbers (or subtract or multiply or divide) and get as our answer in every case one of the given numbers, we say that the given group of numbers is closed under that operation. This property is called *closure*.

Identity Elements

Additive Identity:

A number which, when added to a given number, makes the given number the sum is called the *additive identity* (or *identity element for addition*). Thus, zero (0) is the additive identity.

$$0 + 3 = 3$$

Multiplicative Identity:

A number which, when multiplied by a given number, makes the given number the product, is called the *multiplicative identity* (or *identity element for multiplication*). Thus, one (1) is the multiplicative identity.

$$1 \times 4 = 4$$

Inverses

Additive Inverse:

If the sum of two numbers is zero (0), then each addend is said to be the *additive inverse* of the other. See section 3–8.

Multiplicative Inverse:

If the product of two numbers is one (1), then each factor is said to be the *multiplicative inverse* or *reciprocal* of the other.

The multiplicative inverse of 8 is $\frac{1}{8}$. Of $\frac{1}{8}$ it is 8.
The multiplicative inverse of $\frac{4}{3}$ is $\frac{3}{4}$. Of $\frac{3}{4}$ it is $\frac{4}{3}$.

Other Properties of Zero

A. Zero subtracted from any number is the number.
B. The difference between any number and itself is zero.
C. The product is zero:
   (1) When a non-zero number is multiplied by zero.
   (2) When zero is multiplied by a non-zero number.
   (3) When zero is multiplied by zero.

$$7 - 0 = 7$$
$$8 - 8 = 0$$
$$0 \times 6 = 0$$
$$5 \times 0 = 0$$
$$0 \times 0 = 0$$

Review of Number

D. Since the product of any number and zero is zero, it follows that, if the product of two numbers is zero, then one of the factors is zero or both factors are zero.

E. When zero is divided by any number other than zero, the quotient is zero.

$$0 \div 9 = 0$$

F. Division by zero is excluded.

> A statement like $4 \div 0 = ?$ is meaningless.
> A statement like $0 \div 0 = ?$ is indeterminate.

Other Properties of One

A. When a number is multiplied by one (1) or its equivalent, its value remains unchanged.

B. When any number, except zero, is divided by itself, the quotient is one (1).

$$1 = \frac{1}{1} = \frac{2}{2} = \frac{3}{3} = \frac{4}{4} = \frac{5}{5}, \text{ etc.}$$

C. One raised to any power is one.

$$1^7 = 1 \times 1 \times 1 \times 1 \times 1 \times 1 \times 1 = 1$$

EXERCISES

1. Find the numbers that will make each of the following a true statement:
   a. $18 + 33 = ? + 18$
   b. $24 \times ? = 15 \times 24$
   c. $(5 \times ?) \times 4 = 5 \times (7 \times 4)$
   d. $(9 + 6) + 11 = 9 + (6 + ?)$

2. Which of the following statements are true?
   a. $63 - 7 = 7 - 63$
   b. $63 \times 7 = 7 \times 63$
   c. $63 + 7 = 7 + 63$
   d. $63 \div 7 = 7 \div 63$
   e. $(36 + 12) + 3 = 36 + (12 + 3)$
   f. $(36 \times 12) \times 3 = 36 \times (12 \times 3)$
   g. $(36 \div 12) \div 3 = 36 \div (12 \div 3)$
   h. $(36 - 12) - 3 = 36 - (12 - 3)$

Which of the following statements are true?

3. a. $48 + (8 - 4) = (48 + 8) - (48 + 4)$
   b. $48 \times (8 + 4) = (48 \times 8) + (48 \times 4)$
   c. $48 + (8 \div 4) = (48 + 8) \div (48 + 4)$
   d. $48 \div (8 - 4) = (48 \div 8) - (48 \div 4)$
   e. $48 \times (8 \div 4) = (48 \times 8) \div (48 \times 4)$
   f. $48 \div (8 + 4) = (48 \div 8) + (48 \div 4)$
   g. $48 \times (8 - 4) = (48 \times 8) - (48 \times 4)$
   h. $48 - (8 \times 4) = (48 - 8) \times (48 - 4)$

4. a. $(40 + 10) \times 2 = (40 \times 2) + (10 \times 2)$
   b. $(40 + 10) \div 2 = (40 \div 2) + (10 \div 2)$
   c. $(40 - 10) \times 2 = (40 \times 2) - (10 \times 2)$
   d. $(40 - 10) \div 2 = (40 \div 2) - (10 \div 2)$
   e. $(40 - 10) - 2 = (40 - 2) - (10 - 2)$
   f. $(40 \times 10) + 2 = (40 + 2) \times (10 + 2)$
   g. $(40 \div 10) + 2 = (40 + 2) \div (10 + 2)$
   h. $(40 \times 10) - 2 = (40 - 2) \times (10 - 2)$

5. Which of the following groups of numbers are closed under the:
   a. Operation of addition?
      (1) 0, 1, 2, 3, 4, and 5
      (2) 0, 4, 8, 12, 16, . . .
      (3) 0, 7, 14, 21, . . . , 70
      (4) 1, 3, 5, 7, 9, . . .

   b. Operation of subtraction?
      (1) 2, 4, 6, 8, and 10
      (2) 0, 3, 6, 9, 12, . . .
      (3) 5, 10, 15, 20, . . . , 100
      (4) 0

   c. Operation of multiplication?
      (1) 0, 2, 4, 6, 8, . . . , 48
      (2) 5, 25, 125, 625, . . .
      (3) 1, 3, 5, 7, 9, . . .
      (4) 0 and 1

   d. Operation of division?
      (1) 1, 2, 10, and 20
      (2) 2, 4, 6, 8, . . . , 20
      (3) 1
      (4) 2, 4, 6, 8, 16, 32, . . .

Determine the value of each of the following:

6. a. $11 - 0$
   b. $49 - 49$
   c. $0 + 3$
   d. $53 + 0$
   e. 36×0
   f. $0 \div 12$
   g. 0×0
   h. $9 \times 2 \times 0 \times 4 \times 3$
   i. $\dfrac{10 + 15}{5 - 5}$

7. a. 1×93
   b. $\dfrac{8}{8}$
   c. 1^{12}
   d. 42×1
   e. $69 \times 1 \times 1 \times 1 \times 1 \times 1$
   f. 7×1^{15}
   g. $1^9 \times 1^{11}$
   h. $\dfrac{15 + 25}{32 + 8}$
   i. $1^{14} \times 3$

8. Which of the following name the number zero? The number one?
   a. $10 \div 0$
   b. 0×0
   c. $\dfrac{12}{12}$
   d. $12 - 12$
   e. 1^{16}
   f. 16^1

9. Write the multiplicative inverse of each of the following:
   a. 6
   b. $\dfrac{1}{4}$
   c. 0
   d. 1
   e. $\dfrac{7}{10}$
   f. 15
   g. $\dfrac{9}{5}$

Review of Number

CHAPTER REVIEW

1. Round 459,605,132 to the nearest million. (1–1)

2. Add:
   54,968
   28,437
   6,985
   98,564
   7,379   (1–2)

3. Subtract:
   8,219,056
   6,508,396   (1–3)

4. Multiply:
   5,964
   8,005   (1–4)

5. Divide:
   598)554,346   (1–5)

6. a. Express $\frac{32}{48}$ in lowest terms. (1–6)

   b. Raise to higher terms: $\frac{3}{10} = \frac{?}{100}$ (1–7)

   c. Express $\frac{29}{5}$ as a mixed number. (1–8)

   d. Change $3\frac{5}{6}$ to an improper fraction. (1–9)

7. Are $\frac{51}{68}$ and $\frac{39}{52}$ equivalent fractions? (1–11)

8. Which of the following statements are true? (1–11)
   a. $\frac{3}{5} > \frac{2}{3}$ b. $\frac{15}{16} < \frac{9}{10}$

9. Add:
   $2\frac{3}{5} + \frac{9}{10} + 3\frac{1}{2}$ (1–12)

10. Subtract:
    $17 - 6\frac{7}{8}$ (1–13)

11. Multiply:
    $\frac{5}{6} \times 4\frac{1}{2}$ (1–14)

12. Divide:
    $2\frac{1}{12} \div 3\frac{3}{4}$ (1–15)

13. Round 3.000478 to the nearest hundred-thousandth. (1–16)

14. Which of the following statements are true? (1–17)
    a. 7.05 < 1.064   b. .8 > .796

15. Add:
    81.4 + .427 + 3.59   (1–18)

16. Subtract:
    .628 − .18   (1–19)

17. Multiply:
    .35 × .002   (1–20)

18. Divide:
    .06)‾84   (1–22)

19. Multiply 94.71 by 1,000,000.   (1–21)

20. Divide 32,650 by 1,000.   (1–23)

21. a. Write 9.07 billion as a complete numeral.   (1–24)

    b. Write the shortened name for 62,300,000 in millions.

22. a. What fractional part of 72 is 30?   (1–34)

    b. 14 is what decimal part of 25?   (1–34)

23. a. 42 is $\frac{3}{4}$ of what number?   (1–34)

    b. .09 of what number is 630?   (1–34)

24. Write in scientific notation: (1–25)

    a. 407,000,000,000

    b. .000082

25. a. Express $\frac{30}{75}$ as a decimal numeral.   (1–26)

    b. Express $10\frac{1}{4}\%$ as a decimal numeral.   (1–28)

    c. Express $\frac{6}{11}$ as a numeral naming a repeating decimal.   (1–26)

26. a. Express .04 as a numeral naming a common fraction.   (1–27)

    b. Express $62\frac{1}{2}\%$ as a numeral naming a common fraction.   (1–28)

27. a. Express .7 as a percent.   (1–29)

    b. Express $\frac{13}{20}$ as a percent.   (1–29)

28. Find 9.6% of $2,500.   (1–30, 1–33, 1–34)

29. What percent of 40 is 15?   (1–31, 1–33, 1–34)

30. 3 is 12% of what number?   (1–32, 1–33, 1–34)

31. Square: a. 36   b. $4\frac{3}{5}$ c. .09   (1–35)

Review of Number

32. a. Find the square root of 32 by estimation, division and average. (1–36)

  b. Find the square root of 88,804.

33. a. Write the numerals that name all the even numbers greater than 27 and less than 36. (1–37)

  b. Write the numerals that name all the odd numbers less than 201 and greater than 189.

34. What is the greatest common factor of 54 and 81? (1–38)

35. Which of the following are prime numbers: 9, 19, 29, 39, 49, 59? (1–39)

36. What is the least common multiple of 12 and 18? (1–40)

37. Is 597,267 divisible by 3? by 2? by 6? by 9? (1–41)

38. What property is illustrated by each of the following? (1–44)

  a. $(61 + 37) + 94 = 61 + (37 + 94)$
  b. $25 \times 18 = 18 \times 25$
  c. $48 \times (21 + 39) = (48 \times 21) + (48 \times 39)$
  d. $(87 \times 12) \times 6 = 87 \times (12 \times 6)$
  e. $56 + 63 = 63 + 56$

39. Are the odd whole numbers closed under the operation of addition? Subtraction? Multiplication? Division? (1–44)

40. Which of the following name the number zero? The number one? (1–44)

  a. $16 \div 16$ **e.** 0×0
  b. $16 - 16$ **f.** $0 \div 5$
  c. 1^8 **g.** $5 \div 0$
  d. 8^1 **h.** 0×8

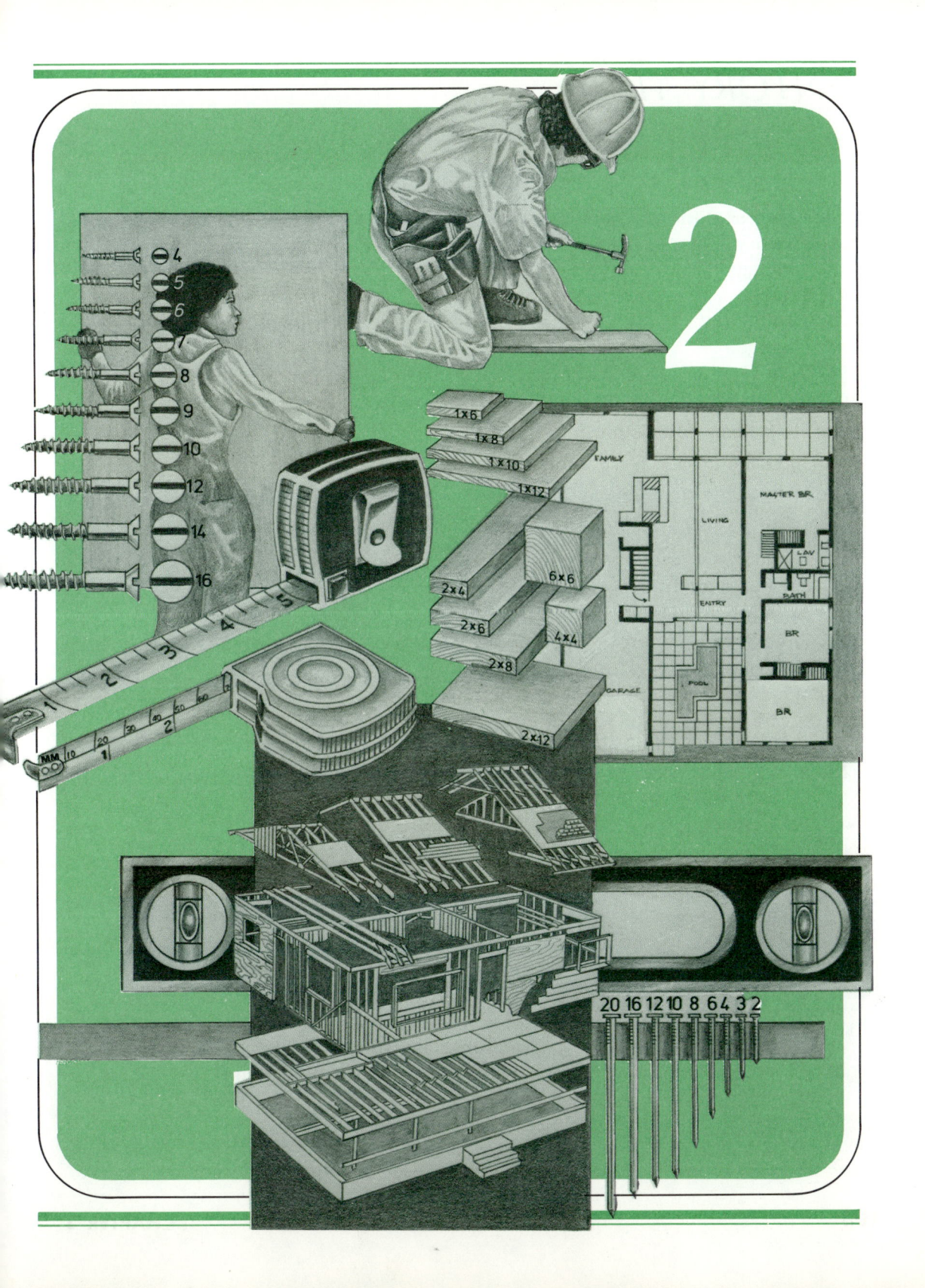

2

INVENTORY TEST

The numeral at the end of each problem indicates the section where explanatory material may be found.

1. Change: a. 5.9 km to m   b. 105 mm to cm   (2–1)
2. Change: a. 840 g to kg   b. 6.3 g to mg   (2–2)
3. How many: a. mL are in 8 L?   b. L are in 125 cL?   (2–3)
4. Change: a. 2.9 cm² to mm²   b. 17 km² to hectares   (2–4)
5. a. 9.75 L of water occupies ____ cm³ and weighs ____ kg.   (2–5)
    b. Change: 7,250 cm³ to dm³; 4.6 m³ to dm³   (2–5)
6. How many inches are in $8\frac{3}{4}$ yards?   (2–6)
7. Find the number of pounds in 176 ounces.   (2–7)
8. How many pints are in $6\frac{1}{2}$ gallons?   (2–8)
9. What part of a bushel is 3 pecks?   (2–9)
10. Change $10\frac{3}{4}$ sq. ft. to sq. in.   (2–10)
11. Change 243 cu. ft. to cu. yd.   (2–11)
12. Multiply 3 hr. 20 min. by 7. Simplify your answer.   (2–13)
13. Divide 19°32′45″ by 15.   (2–14)
14. Change 45 km/h to m/min.   (2–15)
15. Which is coldest, a temperature of 14°C, 280 K, or 50°F?   (2–16)
16. Express as A.M. or P.M. time: a. 2010;   b. 0050   (2–17)
17. If it is 11 A.M. in Chicago, what time is it in Miami?   (2–18)
18. a. Find the greatest possible error and the relative error in each of the following measurements: (2–19)

     37,000 km;   82.7 g;   90 cm; $3\frac{1}{8}$ in.; $6\frac{3}{4}$ lb.

    b. Which is more precise: 0.050 mm or 0.05 mm?   (2–19)
    c. Arrange the following measurements in order of precision, most precise first; and also in order of accuracy, most accurate first: 60 m   37 cm   920,000 km   8.1 mm   (2–19)
19. How many significant digits are in each of the following?   (2–20)
    a. 76,493,255   d. 7.86×10^9
    b. 8,040   e. 5.094
    c. 1,500,000   f. .000030
20. Compute the following approximate numbers as indicated: (2–21)
    a. 2.185 + 9.6 + 3.43   c. 8.07 × .368
    b. 65.72 − 8.934   d. $7.1\overline{)9.524}$

CHAPTER 2
Units of Measure and Measurement

Many years ago there were no standard units of measure. Early humans used their fingers, feet, and arms to measure length or distance. The width across the open hand at the base of the fingers, called the *palm,* the breadth of a finger, called a *digit,* the greatest stretch of the open hand, called the *span,* and the length of the forearm from the elbow to the end of the middle finger, called the *cubit,* were some of the units used. However, since these measurements varied depending upon the size of the person, they were unsatisfactory.

Today we use metric and customary units of measure which are standard. See pages 547–549 for the tables of measure. Probably in the near future we shall be using in the United States a modernized metric system, the International System of Units, generally known as SI. It should be noted that sometimes the metric unit meter is spelled as *metre* and liter as *litre*.

UNITS OF MEASURE

Changing to a Smaller Unit of Measure

To change a given number of units of one denomination to units of smaller denomination, we find the number of units of the smaller denomination that is equivalent to one unit of the larger denomination. This number is sometimes called the *conversion factor*. Then we multiply the given number of units of the larger denomination by this conversion factor.

Since each conversion factor in the metric system is some power of ten, short methods of computation may be used. See section 1-21.

On the next page are several examples of changing to a smaller unit of measure and references to sections where you will find additional practice exercises.

Units of Measure and Measurement

2–1 Metric—Length
37 centimeters to millimeters:
10 × 37 = 370
Answer, 370 millimeters

2–2 Metric—Weight
8 kilograms to grams:
1,000 × 8 = 8,000
Answer, 8,000 grams

2–3 Metric—Capacity
2 liters to centiliters:
100 × 2 = 200
Answer, 200 centiliters

2–4 Metric—Area
8 km² to m²:
1,000,000 × 8 = 8,000,000
Answer, 8,000,000 m²

2–5 Metric—Volume
3 dm³ to cm³:
1,000 × 3 = 3,000
Answer, 3,000 cm³

2–6 Customary—Length
4 feet to inches:
12 × 4 = 48
Answer, 48 inches

2–7 Customary—Weight
9 pounds to ounces:
16 × 9 = 144
Answer, 144 ounces

2–8 Customary—Liquid
22 gallons to quarts:
4 × 22 = 88
Answer, 88 quarts

2–9 Customary—Dry
7 bushels to pecks:
4 × 7 = 28
Answer, 28 pecks

2–10 Customary—Area
18 sq. yd. to sq. ft.:
9 × 18 = 162
Answer, 162 sq. ft.

2–11 Customary—Volume
4 cu. ft. to cu. in.:
1,728 × 4 = 6,912
Answer, 6,912 cu. in.

2–13 Time
16 hours to minutes:
60 × 16 = 960
Answer, 960 minutes

2–14 Angles and Arcs
32 degrees to minutes:
60 × 32 = 1,920
Answer, 1,920 minutes

Changing to a Larger Unit of Measure

To change a given number of units of one denomination to units of a larger denomination, we find the number of units of the smaller denomination that is equivalent to one of the larger denomination. Then we divide the given number of units of the smaller denomination by this conversion factor.

Since each conversion factor in the metric system is some power of ten, short methods of computation may be used. See section 1–23.

Below are several examples of changing to a larger unit of measure and references to sections where you will find additional practice exercises.

2–1 Metric—Length
5,000 meters to kilometers:

5,000 ÷ 1,000 = 5

Answer, 5 kilometers

2–2 Metric—Weight
43 milligrams to centigrams:

43 ÷ 10 = 4.3

Answer, 4.3 centigrams

2–3 Metric—Capacity
685 centiliters to liters:

685 ÷ 100 = 6.85

Answer, 6.85 liters

2–4 Metric—Area
730 mm² to cm²:

730 ÷ 100 = 7.3

Answer, 7.3 cm²

2–5 Metric—Volume
920,000 dm³ to m³:

920,000 ÷ 1,000 = 920

Answer, 920 m³

2–6 Customary—Length
27 feet to yards:

27 ÷ 3 = 9

Answer, 9 yards

2–7 Customary—Weight
18,000 lb. to short tons:

18,000 ÷ 2,000 = 9

Answer, 9 short tons

Units of Measure and Measurement

2–8 Customary—Liquid
16 pints to quarts:
16 ÷ 2 = 8
Answer, 8 quarts

2–9 Customary—Dry
20 pecks to bushels:
20 ÷ 4 = 5
Answer, 5 bushels

2–10 Customary—Area
432 sq. in. to sq. ft.:
432 ÷ 144 = 3
Answer, 3 sq. ft.

2–11 Customary—Volume
135 cu. ft. to cu. yd.:
135 ÷ 27 = 5
Answer, 5 cu. yd.

2–13 Time
96 months to years:
96 ÷ 12 = 8
Answer, 8 years

2–14 Angles and Arcs
360 minutes to degrees:
360 ÷ 60 = 6
Answer, 6 degrees

METRIC SYSTEM

Our monetary system is a decimal system in which the *dollar* is the basic unit. In the metric system, the *meter* (m) is the basic unit of length, the *gram* (g) is the basic unit of weight or mass, and the *liter* (L) is the basic unit of capacity (dry and liquid measures). Other metric units of length, weight, and capacity are named by adding the following prefixes to the basic unit of measure:

Prefix	Symbol		Value
kilo-	k	meaning	thousand (1,000)
hecto-	h	meaning	hundred (100)
deka-	da	meaning	ten (10)
deci-	d	meaning	one-tenth (.1 or $\frac{1}{10}$)
centi-	c	meaning	one-hundredth (.01 or $\frac{1}{100}$)
milli-	m	meaning	one-thousandth (.001 or $\frac{1}{1000}$)

The chart of units of measure below shows the relationship among units. Observe the similarity of the metric system of measures to our decimal numeration system and to our monetary system.

It should be noted that just as 4 dollars 7 dimes 6 cents may be written as a single numeral, 4.76 *dollars* or $4.76; so 4 meters 7 decimeters 6 centimeters may also be written as a single numeral, 4.76 *meters*.

Abbreviations or symbols are written without the period after the last letter. The same symbol is used for both one or more quantities. Thus, *cm* is the symbol for *centimeter* or *centimeters*.

UNITS OF MEASURE

	1,000	100	10	1	.1	.01	.001
Decimal Place Value	thousands	hundreds	tens	ones	tenths	hundredths	thousandths
United States Money	$1,000 bill	$100 bill	$10 bill	dollar	dime	cent	mill
Metric Length	kilometer km	hectometer hm	dekameter dam	meter m	decimeter dm	centimeter cm	millimeter mm
Metric Weight	kilogram kg	hectogram hg	dekagram dag	gram g	decigram dm	centigram cg	milligram mg
Metric Capacity	kiloliter kL	hectoliter hL	dekaliter daL	liter L	deciliter dL	centiliter cL	milliliter mL

The above chart reveals that in the metric system ten (10) of any unit of measure is equivalent to one (1) unit of the next larger size.

The prefix "micro" means one-millionth, and the prefix "mega" means one million. There is a *micrometer* (one-millionth of a meter), a *microgram* (one-millionth of a gram), and a *microliter* (one-millionth of a liter). A *megameter* is one million (1,000,000) meters, a *megagram* is one million grams, and a *megaliter* is one million liters.

Units of Measure and Measurement

2-1 MEASURE OF LENGTH—METRIC

Let us examine the following section of a metric ruler:

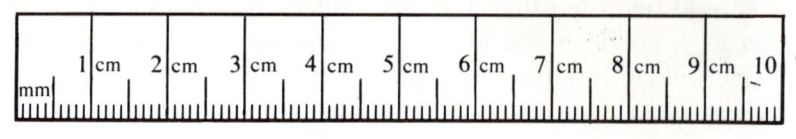

ONE DECIMETER

Each of the smallest subdivisions shown here indicates a measure of 1 *millimeter* (mm). Observe that ten (10) of these millimeter subdivisions form the next larger subdivision, called a *centimeter* (cm), and that ten (10) of the centimeter subdivisions form the next larger subdivision, called a *decimeter* (dm). The meter stick measuring one meter (m), the basic metric unit measuring length, has markings that show 10 decimeter divisions, 100 centimeter divisions, and 1,000 millimeter divisions.

To change a given number of units of one metric denomination to units of another metric denomination, we follow the procedures that are explained on pages 110–112. For short methods of computation, see sections 1–21 and 1–23.

EXERCISES

1. What metric unit of length does each of these symbols represent?

    a. mm   b. km   c. dm   d. m   e. dam   f. hm   g. cm

2. Use this section of the metric ruler to answer the following questions:

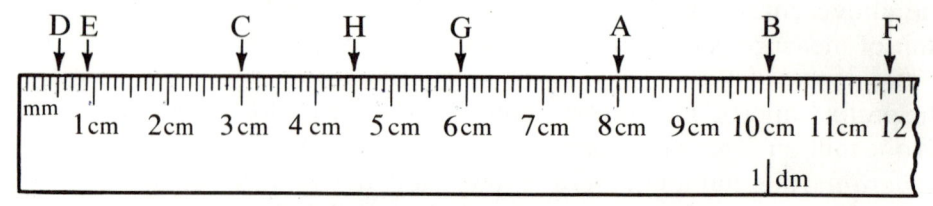

    a. What measurement is indicated by the point labeled:

       C?   A?   D?   H?   E?   G?   B?   F?

114   CHAPTER 2

b. How far from point A is point B? From point D is point H? From point C is point G? From point A is point E? From point F is point G?

c. How many centimeters is it from point C to point A? From point B to point C? From point E to point G?

d. How many millimeters is it from point D to point A? From point D to point G? From point E to point F?

e. How many millimeters are in the length from 0 to each of the following markings:

2 cm?   4 cm 6 mm?   8 cm 2 mm?   1 dm 1 cm 7 mm?

f. In each of the following find the sum of the measurements by locating the mark for the first measurement on the ruler and adding on to this the second measurement. Simplify each sum as indicated:

5 mm + 7 mm = _____ mm = _____ cm _____ mm

31 mm + 62 mm = _____ mm = _____ cm _____ mm

1 cm 6 mm + 5 cm 3 mm = _____ cm _____ mm

3 cm 7 mm + 4 cm 8 mm = _____ cm _____ mm = _____ cm _____ mm

2 cm 3 mm + 6 cm 7 mm = _____ cm _____ mm = _____ cm

1 dm 7 mm + 1 cm 2 mm = _____ dm _____ cm _____ mm

7 cm 4 mm + 2 cm 6 mm = _____ cm _____ mm = _____ dm

g. In each of the following locate the marking for the first measurement on the ruler and take away from this the second measurement to find your answer:

15 mm − 8 mm = _____ mm

7 cm 9 mm − 3 cm 4 mm = _____ cm _____ mm

8 cm − 3 mm = _____ cm _____ mm

3 cm 1 mm − 5 mm = _____ cm _____ mm

5 cm 2 mm − 4 cm 6 mm = _____ cm _____ mm

9 cm − 7 cm 2 mm = _____ cm _____ mm

1 dm − 6 cm 1 mm = _____ cm _____ mm

h. Find the 112 mm mark on the metric ruler. Subtract from it a measurement of 3.7 cm. What measurement does the mark you reach indicate?

Units of Measure and Measurement   115

3. Complete each of the following:
   a. _____ mm = 1 cm   e. _____ dam = 1 hm   i. _____ m = 1 dm
   b. _____ cm = 1 dm   f. _____ hm = 1 km   j. _____ dam = 1 m
   c. _____ dm = 1 m   g. _____ cm = 1 mm   k. _____ hm = 1 dam
   d. _____ m = 1 dam   h. _____ dm = 1 cm   l. _____ km = 1 hm

4. Complete each of the following:
   a. _____ mm = 1 m   h. _____ m = 1 cm
   b. _____ cm = 1 m   i. _____ m = 1 mm
   c. _____ dm = 1 m   j. _____ m = 1 dm
   d. _____ mm = 1 cm   k. 1 m = _____ dm = _____ cm = _____ mm
   e. _____ mm = 1 dm   l. 1 km = _____ hm = _____ dam = _____ m
   f. _____ m = 1 km   m. 1 km = _____ m = _____ dm = _____ cm = _____ mm
   g. _____ km = 1 m   n. 1 mm = _____ m = _____ km

5. Express each of the following in meters:
   a. 5 km 2 hm 9 dam 1 m = _____ m
   b. 9 m 4 dm 8 cm 7 mm = _____ m
   c. 9 m 9 cm 6 mm = _____ m
   d. 4 km 3 m 6 cm 2 mm = _____ m
   e. 7 km 9 hm 1 dam 8 m 8 dm 3 cm 4 mm = _____ m

6. Change each of the following to millimeters:
   67 m;   3 km;   29 cm;   8.9 m;   2.3 cm

7. Change each of the following to centimeters:
   456 mm;   87 m;   5.9 dm;   9 km;   2.6 m

8. Change each of the following to decimeters:
   223 m;   670 cm;   7.3 km;   17.9 m;   5,676 mm

9. Change each of the following to meters:
   a. 67,545 mm;   2.876 km;   285 cm;   54 dam;   334 mm;
   b. 7.8 hm;   32.6 dm;   2,435 cm;   .43 km;   45.2 dam

10. Change each of the following to dekameters:
    5.2 km;   59.6 m;   7,113 cm;   90 m;   62 hm

116   CHAPTER 2

11. Change each of the following to hectometers:

    513 dam;   91.1 m;   6.8 km;   342.65 m;   50 km

12. Change each of the following to kilometers:

    4,089,000 mm;   228.6 dam;   76 hm;   4,539 m;   112,000 cm

13. Find the missing equivalent measurements:

	km	hm	dam	m	dm	cm	mm
a.	?	?	?	800	?	?	?
b.	4	?	?	?	?	?	?
c.	?	?	?	?	?	7,000	?
d.	?	?	?	?	?	?	9,500,000

14. Complete each of the following:

    a. 4 m 5 mm = _____ mm
    b. 2 m 45 cm = _____ cm
    c. 5 cm 2 mm = _____ mm
    d. 3 cm 7 mm = _____ mm
    e. 6 km 340 m = _____ m
    f. 5 m 3 cm = _____ cm

15. Complete each of the following:

    a. 4 m 8 cm 9 mm = _____ mm
    b. 4 km 6 m 3 cm = _____ cm
    c. 3 m 6 dm 5 cm = _____ cm
    d. 6 m 7 cm 4 mm = _____ mm

16. Complete each of the following:

    a. 9 cm 3 mm = _____ cm
    b. 4 m 59 mm = _____ m
    c. 7 km 306 m = _____ km
    d. 5 m 8 cm = _____ m
    e. 3 cm 8 mm = _____ cm
    f. 8 km 9 m = _____ mm

17. Complete each of the following:

    a. 3 km 29 m 36 cm = _____ km
    b. 5 m 8 cm 2 mm = _____ m
    c. 9 km 5 m 8 cm = _____ km
    d. 7 m 6 cm 3 mm = _____ m

18. Complete each of the following:

    a. 4 cm 8 mm = _____ m
    b. 86 m 9 cm = _____ km
    c. 6 km 3 m 2 cm = _____ m
    d. 3 m 6 cm 5 mm = _____ cm

19. Sound travels in water at a speed of 1,450 meters per second. How many kilometers does it travel in 20 seconds?

20. If the acceleration of gravity is 9.8 meters per second per second, what is it in terms of millimeters per second per second?

Units of Measure and Measurement

21. How much longer is a metal rod that measures 1.4 meters than one that measures 96 centimeters?

22. Which measurement is greater:

    a. 5.5 m or 550 mm?   **c.** 4,000 mm or 8 m?   **e.** 10 m or 3,000 cm?
    b. 600 mm or 5 m?   **d.** 7 km or 6,000 m?   **f.** 7.2 cm or 85 mm?

23. Which measurement is smaller:

    a. 69 mm or 5.5 cm?   **c.** 4 km or 50,000 mm?   **e.** 90 cm or 7 m?
    b. 8 m or 5 km?   **d.** 7.56 m or 756 mm?   **f.** 4.3 km or 664 m?

24. Arrange the following measurements in order of size (longest first):

    3,400 m;   864,000 mm;   817,000 cm;   4.6 km

25. Arrange the following measurements in order of size (shortest first):

    5.67 km;   567 m;   56.7 mm   5,670,000 cm

2–2 MEASURE OF MASS OR WEIGHT—METRIC

Technically, the kilogram is a unit used to measure the *mass* of an object. However, in everyday use the word "weight" almost always means "mass." Therefore, for our purposes we will continue to speak of the kilogram as a unit of *weight*.

Since the weight of one (1) gram is so small, the *kilogram* is generally considered as the practical basic unit of weight. Note also that the weight of 1,000 kilograms is equivalent to one (1) metric ton (t).

To change a given number of units of one metric denomination to units of another metric denomination, we follow the procedures explained on pages 110–112. For short methods of computation, see sections 1–21 and 1–23.

EXERCISES

1. What metric unit of weight does each of these symbols represent?

    a. dg   **b.** hg   **c.** kg   **d.** mg   **e.** g   **f.** dag   **g.** cg

2. Complete each of the following:

   a. _____ cg = 1 mg
   b. _____ dg = 1 g
   c. _____ hg = 1 kg
   d. _____ cg = 1 dg
   e. _____ dg = 1 cg
   f. _____ g = 1 dag
   g. _____ dag = 1 hg
   h. _____ mg = 1 cg
   i. _____ hg = 1 dag
   j. _____ g = 1 dg
   k. _____ kg = 1 hg
   l. _____ dag = 1 g

3. Complete each of the following:

   a. _____ cg = 1 kg
   b. _____ g = 1 kg
   c. _____ dg = 1 kg
   d. _____ g = 1 mg
   e. _____ dag = 1 kg
   f. _____ g = 1 cg
   g. _____ cg = 1 g
   h. _____ mg = 1 g
   i. 1 mg = _____ g = _____ kg
   j. _____ kg = 1 g
   k. _____ kg = 1 mg
   l. _____ kg = 1 dag
   m. _____ kg = 1 cg
   n. _____ kg = 1 dg
   o. _____ hg = 1 kg
   p. _____ mg = 1 kg
   q. _____ kg = 1 hg
   r. 1 kg = _____ g = _____ mg

4. Change each of the following to milligrams:

   6.21 kg;   3.4 cg;   67 g;   4 cg;   .486 g

5. Change each of the following to centigrams:

   79 kg;   3.32 dg;   7.4 g;   29 mg;   117 g

6. Change each of the following to decigrams:

   .67 g;   56,900 mg;   4.8 kg;   54 cg;   5.36 g

7. Change each of the following to grams:

   810 dg;   70 dag;   230 mg;   8 kg;   45.6 cg

8. Change each of the following to dekagrams:

   23.2 cg;   87 kg;   7,000 g;   2.28 hg;   7,600 mg

9. Change each of the following to hectograms:

   432 dag;   50 kg;   3,000 g;   635.8 g;   1.1 kg

10. Change each of the following to kilograms:

    593.2 dag;   1,800,000 mg;   4,597 g;   23 hg;   397,000 cg

11. Change each of the following to metric tons:

    130 kg;   2,456,000 g;   6,000 kg;   54,760 kg;   83,000 hg

Units of Measure and Measurement

12. Find the missing equivalent weights:

	kg	g	cg	mg
a.	?	3,000	?	?
b.	3.6	?	?	?
c.	?	?	4,300	?
d.	?	?	?	600,000
e.	?	542	?	?

13. Complete each of the following:
    a. 3 cg 2 mg = __32__ mg
    b. 7 kg 210 g = _____ g
    c. 6 g 53 cg = _____ cg
    d. 1 g 76 mg = _____ mg
    e. 4 kg 570 g = _____ g
    f. 9 cg 2 mg = _____ mg

14. Complete each of the following:
    a. 3 g 9 dg 1 cg = _____ cg
    b. 7 g 8 cg 4 mg = _____ mg
    c. 6 kg 3 g 5 cg = _____ cg
    d. 7 kg 26 g 8 mg = _____ mg

15. Complete each of the following:
    a. 3 cg 1 mg = _____ cg
    b. 8 g 6 cg = _____ g
    c. 4 g 7 mg = _____ g
    d. 36 kg 170 g = _____ kg
    e. 9 kg 2 g = _____ kg
    f. 27 g 62 mg = _____ g

16. Complete each of the following:
    a. 9 g 2 cg 3 mg = _____ g
    b. 7 kg 4 g 1 cg = _____ kg
    c. 2 kg 8 g 5 mg = _____ kg
    d. 2 g 9 cg 5 mg = _____ g

17. Complete each of the following:
    a. 8 g 2 cg 7 mg = _____ cg
    b. 3 kg 7 g 1 cg = _____ g
    c. 6 cg 1 mg = _____ g
    d. 430 g 29 cg = _____ kg

18. If a person takes four 2 mg tablets each day, how many grams of this medicine are taken in one year?

19. Which weight is heavier:
    a. 8,500 cg or 6 kg?
    b. 78 mg or 8.9 g?
    c. 300 g or 1.8 kg?
    d. 4 kg or 4,000 g?
    e. 34 mg or 5.4 cg?
    f. 64 g or 270 mg?

20. Which weight is lighter:
    a. 7,300 g or 6.1 kg?
    b. 18 mg or 1.8 cg?
    c. 2.28 kg or 3,120 g?
    d. 3.26 g or 4,500 mg?
    e. 7,600 g or 8 kg?
    f. 37 cg or 2.1 g?

21. Arrange in order of weight (lightest first):
    5,200 g;   4.5 kg;   37,000 cg;   690,000 mg

22. Arrange in order of weight (heaviest first):
    556 g;   6,432,000 mg;   234.9 cg;   4.23 kg

2–3 MEASURE OF CAPACITY—METRIC

The units that measure capacity also measure volume. See section 2–5 on page 125 where units of cubic measure are studied.

To change a given number of units of one metric denomination to units of another metric denomination, we follow the procedures explained on pages 110–112. For short methods of computation, see sections 1–21 and 1–23.

EXERCISES

1. What metric unit of capacity does each of these symbols represent?

   a. daL   b. L   c. cL   d. hL   e. kL   f. dL   g. mL

2. Complete each of the following:

   a. _____ L = 1 dL   e. _____ hL = 1 daL   i. _____ mL = 1 cL
   b. _____ dL = 1 cL   f. _____ daL = 1 L   j. _____ hL = 1 kL
   c. _____ cL = 1 mL   g. _____ dL = 1 L   k. _____ daL = 1 hL
   d. _____ kL = 1 hL   h. _____ cL = 1 dL   l. _____ L = 1 daL

3. Complete each of the following:

   a. _____ L = 1 hL   e. 1 hL = _____ L = _____ mL
   b. _____ L = 1 mL   f. _____ mL = 1 L
   c. _____ kL = 1 L   g. _____ L = 1 kL
   d. _____ cL = 1 L   h. 1 dL = _____ L = _____ kL

4. Change each of the following to milliliters:

   .342 L;   81.2 cL;   8 cL;   3.1 L;   76 dL

5. Change each of the following to centiliters:

   30 mL;   2.9 L;   4 dL;   75.6 mL;   23 L

6. Change each of the following to deciliters:

   4.37 L;   400 cL;   4,433 mL;   6 kL;   6 daL

7. Change each of the following to liters:

   89.03 kL;   56 cL;   22,600 mL;   2 hL;   97 dL

Units of Measure and Measurement

8. Change each of the following to dekaliters:

   12,448 mL;   65.6 L;   445 L;   5.9 hL;   227 dL

9. Change each of the following to hectoliters:

   27 daL;   4,670,000 mL;   34.21 kL;   7.5 kL;   345 L

10. Change each of the following to kiloliters:

    226 daL;   6,578 L;   119 hL;   73.2 hL;   32 L

11. Find the missing equivalent capacities:

	L	dL	cL	mL
a.	5	?	?	?
b.	?	869	?	?
c.	?	?	?	300
d.	4.2	?	?	?
e.	?	?	454.3	?

12. Complete each of the following:

    a. 5 cL 2 mL = _____ mL
    b. 8 L 3 cL = _____ cL
    c. 2 L 9 mL = _____ mL
    d. 7 kL 300 L = _____ L
    e. 2 L 21 mL = _____ mL
    f. 34 L 17 cL = _____ cL

13. Complete each of the following:

    a. 5 L 2 cL 9 mL = _____ mL
    b. 7 L 62 cL 4 mL = _____ mL

14. Complete each of the following:

    a. 8 cL 1 mL = _____ cL
    b. 3 L 5 cL = _____ L
    c. 9 kL 28 L = _____ kL
    d. 6 L 8 mL = _____ L
    e. 4 L 59 cL = _____ L
    f. 3 cL 7 mL = _____ cL

15. Complete each of the following:

    a. 2 L 8 cL 3 mL = _____ L
    b. 7 L 1 cL 2 mL = _____ L

16. Complete each of the following:

    a. 9 L 4 cL 2 mL = _____ cL
    b. 2 L 5 cL 3 mL = _____ cL

17. Tony mixed 25 cL of a weed-killer solution with 375 cL of water. How many liters of the mixture did he have?

18. Which capacity is smaller:

    a. 5.8 cL or 6 mL?
    b. 73.4 mL or .5 L?
    c. 4.5 L or 5,600 mL?
    d. 34 dL or 34 daL?
    e. 6.9 L or 2,000 cL?
    f. 15 cL or 42 mL?

19. Arrange the following capacities in order of size (largest capacity first):

   .648 L;   662 mL;   6.59 dL;   65.7 cL

20. Arrange the following capacities in order of size (smallest capacity first):

   16,846 mL;   44.8 dL;   23 L;   958 cL

2–4 MEASURE OF AREA—METRIC

We have found in linear measure that ten (10) of any metric unit is equivalent to one (1) of the next higher metric unit. However, in the square measure one hundred (100) of any metric unit is equivalent to one (1) of the next higher unit. As we see in the following table, in the metric measure of area the exponent ^{2} is used to represent the word "square."

100 square millimeters (mm^2) = 1 square centimeter (cm^2)

100 square centimeters (cm^2) = 1 square decimeter (dm^2)

100 square decimeters (dm^2) = 1 square meter (m^2)

100 square meters (m^2) = 1 square dekameter (dam^2)

100 square dekameters (dam^2) = 1 square hectometer (hm^2)

100 square hectometers (hm^2) = 1 square kilometer (km^2)

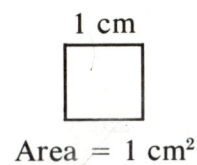

1 cm

Area = 1 cm^2

Special names are sometimes used. *Centare* may be used instead of square meter, *are* instead of square dekameter, and *hectare* instead of square hectometer. The hectare is used considerably. Thus,

   100 centares = 1 are (a)
   100 ares   = 1 hectare (ha)
   100 hectares = 1 square kilometer

Units of Measure and Measurement

To change from one metric unit of square measure to another metric unit of square measure, we follow the procedures on pages 110–112.

EXERCISES

1. What unit of metric square measure does each of these symbols represent?
   a. ha   **b.** mm²   **c.** dam²   **d.** dm²   **e.** m²   **f.** cm²

2. **a.** How many square meters are in 1 hectare?
   b. How many square centimeters are in 1 square meter?
   c. How many square meters are in 1 square kilometer?
   d. How many square millimeters are in 1 square meter?
   e. How many hectares are in 1 square kilometer?

3. Change each of the following to square millimeters:

   56.2 cm²;   .37 m²;   8 cm²;   3.74 m²;   2.2 dm²

4. Change each of the following to square centimeters:

   .78 m²;   6,700 mm²;   500 mm²;   98 dm²;   3.1 m²

5. Change each of the following to square decimeters:

   4.7 m²;   533.1 cm²;   487 cm²;   70 m²;   613,000 mm²

6. Change each of the following to square meters (or centares):

   537,000 mm²;   .516 km²;   34 km²;   27 dm²;   36,000 cm²

7. Change each of the following to square dekameters (or ares):

   2.28 m²;   6.3 hectares;   11,400 m²;   3,000 m²;   36 hectares

8. Change each of the following to hectares (or square hectometers):

   42 km²;   787.1 m²;   2.6 km²;   23,000 m²;   39.8 dam²

9. Change each of the following to square kilometers:

   4,320,000 m²;   638,259 m²;   411 hectares;   3,000 m²;   26.7 hectares

10. A plot of land measures 30 hectares. What part of a square kilometer is it? How many square meters does it measure?

2-5 MEASURE OF VOLUME—METRIC

The *volume,* also called capacity, is generally the number of units of cubic measure contained in a given space. However, the units of capacity (see section 2-3 on page 121) also are used to measure volume. Refer to the relationship given below between the cubic decimeter and the liter.

In linear measure ten (10) of any metric unit is equivalent to one (1) of the next higher metric unit. In the square measure one hundred (100) of any metric unit is equivalent to one (1) of the next higher unit. However, in the cubic measure one thousand (1,000) of any metric unit is equivalent to one (1) of the next higher unit. As shown below, in the metric measure of volume the exponent ^{3} is used to represent the word "cubic."

Volume = 1 cm^3

1,000 cubic millimeters (mm^3) = 1 cubic centimeter (cm^3)
1,000 cubic centimeters (cm^3) = 1 cubic decimeter (dm^3)
1,000 cubic decimeters (dm^3) = 1 cubic meter (m^3)

To change from one metric unit of cubic measure to another metric unit of cubic measure, we follow the procedures as outlined on pages 110-112.

Also note the following relationships:

The volume of one cubic decimeter has the same capacity as one (1) liter. Since 1 cubic decimeter is equivalent to 1,000 cubic centimeters (cm^3), and 1 liter is equivalent to 1,000 milliliters (mL), then the volume of 1 cubic centimeter (cm^3) has the same capacity as 1 milliliter (mL).

A gram is the weight of one (1) cubic centimeter (or 1 milliliter) of water at a temperature of 4 degrees Celsius, and a kilogram is the weight of 1,000 cubic centimeters (or 1 liter) of water at a temperature of 4 degrees Celsius.

MISCELLANEOUS EQUIVALENTS

1 liter = 1 cubic decimeter (dm^3) = 1,000 cubic centimeters (cm^3)
1 milliliter (mL) = 1 cubic centimeter (cm^3)
1 liter of water weighs 1 kilogram (kg)
1 milliliter (mL) or cubic centimeter (cm^3) of water weighs 1 gram (g)

EXERCISES

1. What unit of metric cubic measure does each of the following symbols represent?

    a. dam³   b. mm³   c. km³   d. cm³   e. m³   f. dm³   g. hm³

2. a. How many cubic centimeters are in 1 cubic hectometer?
    b. How many cubic millimeters are in 1 cubic centimeter?
    c. How many cubic centimeters are in 1 cubic meter?

3. Change each of the following to cubic millimeters:

    45.6 dm³;   63.9 m³;   7 cm³;   41 m³;   7.7 cm³

4. Change each of the following to cubic centimeters:

    7.9 m³;   540 mm³;   67 m³;   2.12 dm³;   6,557 mm³

5. Change each of the following to cubic decimeters:

    43,343 mm³;   39.7 cm³;   34 m³;   5,890 cm³;   91.4 m³

6. Change each of the following to cubic meters:

    47,800 cm³;   432 dm³;   100,000 cm³;   34,000 dm³;   4,670,000 mm³

7. a. 39 milliliters of liquid will occupy a space of how many cubic centimeters?
    b. A space of 227 cubic centimeters will hold how many milliliters of liquid?
    c. 7 liters of liquid will occupy a space of how many cubic centimeters?
    d. A space of 34,700 cubic centimeters will hold how many liters of liquid?
    e. A space of 67.8 cubic decimeters will hold how many liters of liquid?
    f. 5.6 liters of water weigh how many grams?
    g. .117 liter of water weighs how many grams?
    h. 62 centiliters of water weigh how many grams?
    i. 223 milliliters of water weigh how many grams?
    j. 9 liters of water weigh approximately how many kilograms?
    k. How many cubic decimeters are occupied by 43 kilograms of water?

l. How many liters of water are in a container if the water weighs 646 grams?

m. How many milliliters of water weigh 45 grams?

n. How many liters of water are in a container if the water weighs 11 kilograms?

o. Water weighing 2.6 kilograms fills a space of how many cubic centimeters?

p. Water weighing 37 grams occupies a space of how many cubic centimeters?

8. How much space is occupied by each of the following capacities?
   a. In cm^3: 61 cL;   5 dL;   75.5 mL;   9 mL;   .34 L
   b. In dm^3: 247 cL; 5L;   23 dL;   29.5 L; 6.61 L

9. Find the capacity that will fill each of the following volumes:
   a. In mL: 400 mm^3; 2 dm^3;   .07 m^3;   79 cm^3; 4.5 cm^3
   b. In cL: 2.7 dm^3; 2.22 cm^3; 4.54 m^3;   8 cm^3;   5,300 mm^3
   c. In L: 9.8 m^3;   27,000 cm^3; 128,000 mm^3; 4 dm^3; 32.1 dm^3

10. Find the weight of each of the following volumes or capacities of water:
    a. In kg: 3.65 L;   348 cm^3; 2,300 cm^3; 7 dm^3;   64 L
    b. In g: 7.6 mL;   4.98 cL;   5,800 mm^3; 71 cm^3; .121 L

11. Find the volume or capacity occupied by each of the following weights of water:
    a. In cm^3: 760 mg;   3.83 g;   3 kg;   70 g;   .6 kg
    b. In L: .008 kg;   3,430 mg;   87 kg;   4,700 g;   248 g
    c. In mL: 7,930 mg; 6,700 cg;   2 g;   833.1 g;   4.5 kg
    d. In cL: 29,400 cg; 377 g;   560 mg;   22 g;   9.3 kg

12. If an aquarium occupies a space of 7,200 cubic centimeters, how many liters of water will fill it? What is the weight of the water when the aquarium is full?

13. An oil tank has a volume of 616 cubic meters. How many liters of oil can it hold?

14. If it takes 180,000 liters of water to fill a swimming pool, how many kilograms does the water weigh when the pool is two-thirds full?

Units of Measure and Measurement

CUSTOMARY SYSTEM

Since the complete changeover to the metric system is a gradual one, we will find the customary units of measure still being used. Consequently the following sections deal with the customary units of measure for possible use in our everyday affairs.

We find that the conversion factors used in the customary system consist of many different numbers such as 12; 3; 36; 5,280; 1,760; 16; 2,000; etc. When we change from a small customary unit of measure to a larger unit, we sometimes get complicated answers. For example:

> 49 inches changed to yards is $1\frac{13}{36}$ yards.
>
> 39 ounces changed to pounds is $2\frac{7}{16}$ pounds.
>
> 8,000 feet changed to miles is $1\frac{17}{33}$ miles.
>
> 5,175 pounds changed to short tons is $2\frac{47}{80}$ tons.

However, we have seen that the conversion factors used in the metric system are all some power of ten. When we change from any unit of measure to another unit in the metric system, the computation is quick and easy because we work with the decimal system. For example:

> 46.2 millimeters = 4.62 centimeters
> 825 centimeters = 8.25 meters
> 9,573 milligrams = 9.573 grams
> 367 centiliters = 3.67 liters
> 9.45 meters = 9,450 millimeters
> 18.1 kilograms = 18,100 grams

In the following sections (2–6 to 2–11 inclusive), to change a given number of units of one customary denomination to units of another customary denomination, we follow the procedures that are explained on pages 110–112. See page 548 for tables of measure.

Fundamental Operations with Denominate Numbers

Any numbers which are expressed in terms of units of measure are called *denominate numbers*.

Addition

To add denominate numbers, we arrange like units in columns, then add each column. Where the sum of any column is greater than the number of units that make the next larger unit, it is simplified as illustrated in the model.

> Add:
>
> 4 ft. 10 in.
> 2 ft. 6 in.
> 6 ft. 16 in. = 7 ft. 4 in.
>
> Answer, 7 ft. 4 in.

Subtraction

To subtract denominate numbers, we arrange like units under each other, then subtract, starting from the right. When the number of units in the subtrahend is greater than the number of corresponding units in the minuend, we take one of the next larger units in the minuend and change it to an equivalent number of smaller units to permit subtraction as shown in the model.

> Subtract:
>
> 8 hr. 15 min. = 7 hr. 75 min.
> 2 hr. 50 min. = 2 hr. 50 min.
>   5 hr. 25 min.
>
> Answer, 5 hr. 25 min.

Multiplication

To multiply denominate numbers, we multiply each unit by the multiplier. Where the product of any column is greater than the number of units that make the next larger unit, it is simplified as shown in the model.

> Multiply:
>
> 3 yd. 8 in.
>   5
> 15 yd. 40 in. = 16 yd. 4 in.
>
> Answer, 16 yd. 4 in.

Division

To divide denominate numbers, we divide each unit by the divisor. If the unit is not exactly divisible, we change the remainder to the next smaller unit and combine with the given number of the smaller unit to form the next partial dividend.

> Divide:
>
>   4 lb. 11 oz.
> 4)18 lb. 12 oz.
>   16 lb.
>   2 lb. 12 oz. = 44 oz.
>   44 oz.
>
> Answer, 4 lb. 11 oz.

Units of Measure and Measurement

2-6 MEASURE OF LENGTH—CUSTOMARY

EXERCISES

1. Find the number of inches in:
   - **a.** 4 yd. 21 in.;   8 ft. 7 in.;   2 ft. 8 in.;   2 yd. 10 in.
   - **b.** $2\frac{7}{8}$ yd.;   7 yd.; $5\frac{5}{8}$ yd.;   36 yd.
   - **c.** $\frac{2}{3}$ ft.;   11 ft.;   3 ft.; $7\frac{1}{4}$ ft.

2. Find the number of feet in:
   - **a.** 76 in.;   288 in.;   1,044 in.;   936 in.
   - **b.** $\frac{1}{4}$ mi.;   23 mi.;   7 mi.; $5\frac{3}{8}$ mi.
   - **c.** 11 yd. 3 ft.;   3 mi. 760 ft.;   54 yd. 2 ft.;   4 rd. 5 ft.
   - **d.** 4 yd.; $7\frac{1}{3}$ yd.;   17 yd.; $\frac{1}{2}$ yd.
   - **e.** 5 rd.;   62 rd.; $\frac{1}{4}$ rd.; $7\frac{3}{4}$ rd.

3. Find the number of yards in:
   - **a.** 288 in.;   396 in.;   828 in.;   1872 in.
   - **b.** 12 ft.;   81 ft.;   57 ft.;   138 ft.
   - **c.** 3 mi.;   18 mi.; $\frac{1}{4}$ mi.; $4\frac{3}{8}$ mi.
   - **d.** 7 rd.; $\frac{1}{4}$ rd.; $2\frac{1}{2}$ rd.;   11 rd.
   - **e.** 2 mi. 440 yd.;   6 rd. 4 yd.;   12 rd. 1 yd.;   14 mi. 880 yd.

4. Find the number of rods in:
   - **a.** 44 yd.;   66 ft.; $82\frac{1}{2}$ ft.;   49.5 yd.
   - **b.** $5\frac{1}{4}$ mi.;   16 mi.;   7 mi.; $11\frac{1}{8}$ mi.

5. Find the number of miles in:
   - **a.** 15,840 ft.;   21,120 ft.;   68,640 ft.;   58,080 ft.
   - **b.** 12,320 yd.;   8,800 yd.;   15,840 yd.;   5,280 yd.

6. **a.** What part of a foot is:

   4 in.;   1 in.;   9 in.;   3 in.;   8 in.;   6 in.;   10 in.?

   b. What part of a yard is:

   9 in.;   12 in.;   29 in.;   30 in.;   1 ft.; $2\frac{1}{2}$ ft.;   2 ft.?

   c. What part of a mile is:

   440 yd.;   1,320 ft.;   110 yd.;   1,100 yd.;   3,300 ft.;   1,056 ft.?

7. Add and simplify:

1 ft. 7 in.	4 yd. 14 in.	2 mi. 800 yd.	5 yd. 2 ft. 11 in.
9 in.	3 yd. 31 in.	1 mi. 950 yd.	4 yd. 1 ft. 7 in.
		4 mi. 200 yd.	2 yd. 1 ft. 4 in.

8. Subtract:

6 yd. 1 ft. 7 in.	9 mi. 760 ft.	5 yd. 12 in.	6 ft.
3 yd. 2 ft. 9 in.	1,440 ft.	4 yd. 18 in.	4 ft. 8 in.

9. Multiply and simplify:

6 yd. 6 in.	3 mi. 352 yd.	4 ft. 2 in.	3 yd. 1 ft. 8 in.
6	7	5	4

10. Divide:

6) 31 yd. 6 in.   4) 24 mi. 352 yd.   3) 11 ft. 3 in.   2) 7 yd. 4 ft. 8 in.

2–7 MEASURE OF WEIGHT—CUSTOMARY

EXERCISES

1. Find the number of ounces in:
- **a.** 4 lb.,   15 lb., $2\frac{1}{4}$ lb., $5\frac{1}{2}$ lb.
- **b.** 6 lb. 5 oz.,   3 lb. 7 oz.,   4 lb. 1 oz.,   5 lb. 13 oz.

2. Find the number of pounds in:
- **a.** 8 s.t.,   30 s.t., $4\frac{1}{2}$ s.t., $11\frac{3}{5}$ s.t.
- **b.** 11 l.t.,   26 l.t., $1\frac{1}{2}$ l.t., $7\frac{3}{4}$ l.t.
- **c.** 2 s.t. 250 lb.,   4 l.t. 600 lb.,   11 s.t. 100 lb.,   7 l.t. 1,200 lb.
- **d.** 96 oz.,   336 oz.,   144 oz.,   544 oz.

3. Find the number of short tons in: 4,000 lb.,   30,000 lb.,   9,000 lb.

4. Find the number of long tons in: 6,720 lb.,   24,640 lb.,   15,680 lb.

5.
- **a.** What part of a pound is: 6 oz.,   11 oz.,   8 oz.,   4 oz.
- **b.** What part of a short ton is: 100 lb.,   250 lb.,   1,500 lb.,   1,000 lb.
- **c.** What part of a long ton is: 1,400 lb.,   560 lb.,   1,120 lb.,   840 lb.

Units of Measure and Measurement

6. Add and simplify:

3 s.t. 1,100 lb.	5 lb. 2 oz.	3 lb. 4 oz.	2 l.t. 900 lb.
1 s.t. 1,000 lb.	2 lb. 9 oz.	6 lb. 11 oz.	4 l.t. 900 lb.
		5 lb. 14 oz.	3 l.t. 650 lb.

7. Subtract:

| 11 lb. 3 oz. | 7 lb. 14 oz. | 12 lb. | 5 s.t. 100 lb. |
| 6 lb. 6 oz. | 5 lb. 11 oz. | 9 lb. 8 oz. | 2 s.t. 400 lb. |

8. Multiply and simplify:

| 6 lb. 4 oz. | 5 s.t. 250 lb. | 5 lb. 12 oz. | 1 lb. 7 oz. |
| 3 | 8 | 4 | 7 |

9. Divide:

4)16 lb. 12 oz.   6)24 s.t. 1,200 lb.   5)11 lb. 9 oz.   9)3 lb. 6 oz.

2–8 LIQUID MEASURE—CUSTOMARY

EXERCISES

1. Find the number of ounces in:

   a. 3 pt.,   23 pt., $2\frac{1}{4}$ pt., $4\frac{1}{2}$ pt.

   b. 5 qt.,   21 qt., $6\frac{3}{4}$ qt., $\frac{1}{2}$ qt.

   c. 3 gal.,   7 gal., $\frac{1}{2}$ gal., $2\frac{3}{4}$ gal.

   d. 2 gal. 7 oz.,   3 pt. 2 oz.,   4 qt. 11 oz.,   1 gal. 12 oz.

2. Find the number of pints in:

   a. 6 gal.,   11 gal., $2\frac{3}{4}$ gal., $\frac{1}{2}$ gal.

   b. 13 qt., $1\frac{1}{2}$ qt.,   23 qt., $7\frac{1}{4}$ qt.

   c. 328 oz.,   56 oz.,   184 oz.,   712 oz.

   d. 5 qt. 1 pt.,   9 qt. 1 pt.,   17 qt. 1 pt.   8 qt. 1 pt.

3. Find the number of quarts in:

   a. 6 gal.,   21 gal., $4\frac{3}{4}$ gal., $6\frac{1}{2}$ gal.

   b. 2 gal. 3 qt.,   11 gal. 3 qt.,   7 gal. 1 qt.,   5 gal. 2 qt.

   c. 16 pt.,   78 pt.,   7 pt.,   33 pt.

   d. 352 oz.,   96 oz.,   256 oz.,   192 oz.

4. Find the number of gallons in:

   a. 14 qt.,   68 qt.,   36 qt.,   15 qt.

   b. 112 pt.,   74 pt.,   23 pt.,   11 pt.

   c. 384 oz.,   224 oz.,   128 oz.,   544 oz.

5. a. What part of a pint is:

   7 oz.,   12 oz.,   13 oz.,   4 oz.,   11 oz.,   8 oz.?

   b. What part of a quart is:

   $\frac{1}{4}$ pt.,   4 oz.,   1 pt.,   20 oz., $1\frac{1}{2}$ pt.,   16 oz.?

   c. What part of a gallon is:

   5 pt., $3\frac{1}{2}$ qt.,   48 oz.,   2 qt., $1\frac{1}{2}$ pt.,   16 oz.?

6. Add and simplify:

     2 qt. 10 oz.   5 gal. 1 qt.   2 gal. 2 qt.   4 pt. 7 oz.
     1 qt. 9 oz.   4 gal. 1 qt.   3 gal. 1 qt.   2 pt. 11 oz.
                                                4 gal. 3 qt.   5 pt. 14 oz.

7. Subtract:

     6 gal. 2 qt.   4 qt. 3 oz.   3 pt. 8 oz.   11 gal.
     5 gal. 1 qt.   2 qt. 11 oz.   9 oz.   7 gal. 3 qt.

8. Multiply and simplify:

     2 pt. 5 oz.   2 gal. 3 qt.   4 qt. 7 oz.   5 gal. 1 qt.
            7   5   4   9

9. Divide:

   4)4 qt. 24 oz.   3)16 gal. 2 qt.   2)4 pt. 14 oz.   2)7 gal. 2 qt.

10. How many ounces of water must be added to:

    a. A 12-oz. can of frozen cranberry concentrate to make 3 pints of cranberry juice?

    b. A 20-oz. can of grape concentrate to make $2\frac{1}{2}$ quarts of grape juice?

    c. A 6-oz. can of grapefruit concentrate to make 2 pints of grapefruit juice?

    d. A 1-pt. can of orange concentrate to make $\frac{1}{2}$ gallon of orange juice?

Units of Measure and Measurement

2-9 DRY MEASURE—CUSTOMARY

EXERCISES

1. Find the number of pints in:

    8 qt.,   35 qt.,   7½ qt.,   5 qt. 1 pt.

2. Find the number of quarts in:

    a. 3 pk.,   21 pk.,   5¾ pk.,   11 pk. 3 qt.
    b. 36 pt.,   95 pt.,   3 pt.,   50 pt.

3. Find the number of pecks in:

    a. 5½ bu.,   14 bu.,   48 bu.,   8¼ bu.
    b. 19 bu. 1 pk.,   6 bu. 3 pk.,   12 qt.,   48 qt.

4. Find the number of bushels in: 10 pk.,   15 pk.,   56 pk.,   112 pk.

5. a. What part of a bushel is: 3 pk.,   1 pk.,   2 pk.,   8 qt.?
   b. What part of a peck is: 2 qt.,   7 qt.,   4 qt.,   3 qt.?

6. Add and simplify:

    2 bu. 2 pk.   1 pk. 6 qt.   3 pk. 2 qt.   2 bu. 1 pk.
    1 bu. 3 pk.   3 pk. 5 qt.   2 pk. 5 qt.   5 bu. 2 pk.
    4 bu. 3 pk.   4 pk. 6 qt.

7. Subtract:

    4 bu.   4 pk. 6 qt.   7 bu. 1 pk.   1 pk. 6 qt.
    3 bu. 1 pk.   1 pk. 3 qt.   4 bu. 3 pk.   7 qt.

8. Multiply and simplify:

    5 bu. 3 pk.   2 pk. 5 qt.   6 bu. 1 pk.   3 pk. 3 qt.
         5   2   6   4

9. Divide:

    3)12 pk. 3 qt.   3)14 bu. 1 pk.   2)4 pk. 2 qt.   2)7 bu. 2 pk.

2–10 MEASURE OF AREA—CUSTOMARY

EXERCISES

Change:
1. To square inches: 11 sq. ft., 3 sq. yd., 21 sq. ft., 7 sq. yd.
2. To square feet: 2 acres, 12 sq. yd., 35 sq. rd., 1,728 sq. in.
3. To square yards: 16 acres, 54 sq. ft., 2 sq. mi., 189 sq. ft., 15,552 sq. in.
4. To square rods: 242 sq. yd., 17 acres, 1,452 sq. yd., 75 acres.
5. To acres: 27 sq. mi., 29,040 sq. yd., 3,680 sq. rd., 8,960 sq. rd.
6. To square miles: 16,640 acres, 1,638,400 sq. rd., 11,520 acres, 5,760 acres.
7. What part of a square foot is: 18 sq. in.? 90 sq. in.? 80 sq. in.?
8. What part of a square yard is: 3 sq. ft.? 2 sq. ft.? 864 sq. in.?
9. What part of a square mile is: 160 acres? 38,720 sq. yd.? 34,848 sq. ft.?

2–11 MEASURE OF VOLUME—CUSTOMARY

EXERCISES

Change:
1. To cubic inches: 20 cu. ft., 5 cu. yd., 55 cu. ft.
2. To cubic inches: 47 cu. yd., $2\frac{1}{4}$ cu. ft., 32 cu. yd.
3. To cubic feet: 23 cu. yd., 10,368 cu. in., 214 cu. yd.
4. To cubic feet: 13,824 cu. in., 75 cu. yd., 55,296 cu. in.
5. To cubic yards: 297 cu. ft., 513,216 cu. in., 1,377 cu. ft.
6. To cubic yards: 326,592 cu. in., 1,512 cu. ft., 979,776 cu. in.
7. What part of a cubic foot is: 72 cu. in.? 432 cu. in.? 216 cu. in.?
8. What part of a cubic yard is: $\frac{3}{64}$ cu. ft.? 9 cu. ft.? 12 cu. ft.?
9. What part of a cubic foot is: 144 cu. in.? 192 cu. in.? 864 cu. in.?
10. What part of a cubic yard is: 15 cu. ft.? 21 cu. ft.? $4\frac{1}{2}$ cu. ft.?

Units of Measure and Measurement

2–12 VOLUME, CAPACITY, AND WEIGHT RELATIONSHIPS—CUSTOMARY

See page 549 for table of equivalents.

EXERCISES

1. Find the volume in cubic feet equal to a capacity of 195 gallons.
2. Find the capacity in gallons equal to a volume of 32 cubic feet.
3. Find the volume in cubic inches equal to a capacity of 34 gallons.
4. Find the capacity in gallons equal to a volume of 3,696 cu. in.
5. What volume in cubic feet do 2,250 lb. of fresh water occupy?
6. What is the weight in pounds of 75 cu. ft. of fresh water?
7. What volume in cubic feet do 704 lb. of sea water occupy?
8. What is the weight in pounds of 112 cu. ft. of sea water?
9. Find the volume in cubic feet equal to a capacity of 110 bushels.
10. Find the capacity in bushels equal to a volume of 180 cu. ft.
11. Find the equivalent:

    a. Volume in cubic feet:
    9,000 lb. of fresh water,   3,584 lb. of sea water

    b. Volume in cubic inches:
    3 gal.,   7 gal., $12\frac{1}{2}$ gal.

    c. Volume in cubic feet:
    116 bu.,   405 gal.,   8,850 gal.

    d. Weight in pounds:
    50 cu. ft. of sea water,   115 cu. ft. of fresh water

    e. Capacity in bushels:
    70 cu. ft.,   515 cu. ft.,   28 cu. ft.

    f. Capacity in gallons:
    5,313 cu. in.,   76 cu. ft.,   125 cu. ft.

2–13 MEASURE OF TIME

See pages 110–112 for directions to change units of time and page 129 for directions to compute with these units. The table of measure is found on page 549.

EXERCISES

1. Find the number of seconds in:
   a. 22 min.,   15 min., $2\frac{2}{3}$ min., $9\frac{3}{4}$ min.
   b. 2 hr.,   16 hr., $5\frac{1}{2}$ hr., $\frac{1}{3}$ hr.
   c. 1 hr. 45 min.,   2 min. 11 sec.,   23 min. 28 sec.,   2 hr. 11 min. 39 sec.

2. Find the number of minutes in:
   a. 2 hr.,   27 hr., $7\frac{1}{4}$ hr., $4\frac{1}{6}$ hr.
   b. 5 hr. 23 min.,   2 hr. 46 min.,   19 hr. 51 min.,   22 hr. 10 min.
   c. 660 sec.,   520 sec.,   2,460 sec.,   1,530 sec.

3. Find the number of hours in:
   a. 3 da.,   8 da., $6\frac{1}{8}$ da., $9\frac{1}{3}$ da.
   b. 11 da. 5 hr.,   4 da. 11 hr.,   1 da. 1 hr.,   16 da. 23 hr.
   c. 3,900 min.,   2,520 min.,   720 min.,   1,890 min.

4. Find the number of days in:
   a. 4 wk.,   27 wk.,   104 wk., $7\frac{2}{7}$ wk.
   b. 2 yr.,   10 yr.,[1] $3\frac{3}{4}$ yr., $6\frac{1}{4}$ yr.
   c. 5 wk. 2 da.,   8 wk. 6 da.,   5 yr. 23 da.,   1 yr. 333 da.
   d. 288 hr.,   168 hr.,   60 hr.,   104 hr.

5. Find the number of months in:
   a. 4 yr.,   9 yr., $\frac{1}{4}$ yr., $1\frac{1}{3}$ yr.
   b. 7 yr. 6 mo.,   3 yr. 10 mo.,   10 yr. 1 mo.,   1 yr. 11 mo.

---

[1] Where 4 or more years are involved, use 366 days for every fourth year.

Units of Measure and Measurement

6. Find the number of weeks in:

    21 yr., $6\frac{1}{2}$ yr.,   2 yr. 34 wk.,   7 yr. 49 wk.

7. Find the number of years in:

   a. 672 mo.,   60 mo.,   42 mo.,   144 mo.
   b. 572 wk.,   468 wk.,   130 wk.,   247 wk.
   c. 1,825 da.,   1,022 da.,   4,015 da.,   2,555 da.

8. a. What part of a minute is:

    25 sec.,   20 sec.,   30 sec.,   10 sec.?

   b. What part of an hour is:

    10 min.,   15 min.,   24 min.,   6 min.?

   c. What part of a day is:

    18 hr.,   22 hr.,   11 hr.,   20 hr.?

   d. What part of a week is:

    $1\frac{3}{4}$ da.,   2 da.,   6 da., $5\frac{1}{4}$ da.?

   e. What part of a year is:

    4 mo.,   13 wk.,   6 mo.,   219 da.?

9. Add and simplify:

   3 wk. 3 da.   27 hr. 19 min.   3 hr. 11 min. 41 sec.   11 yr. 6 mo.
   4 wk. 5 da.   32 hr. 45 min.   1 hr. 8 min. 23 sec.   5 yr. 7 mo.
                                          6 hr. 51 min. 9 sec.   1 yr. 2 mo.

10. Subtract:

   4 da. 21 hr.   45 min.   6 yr. 9 mo.   4 hr. 5 min. 9 sec.
   2 da. 8 hr.   23 min. 11 sec.   11 mo.   1 hr. 38 min. 39 sec.

11. Multiply and simplify:

   6 yr. 4 mo.   7 hr. 16 min.   6 da. 5 hr.   4 hr. 12 min. 15 sec.
         2   5   6   5

12. Divide:

   4)24 yr. 8 mo.   5)11 da. 11 hr.   6)21 hr. 36 min.   9)2 hr. 21 min. 18 sec.

2–14 MEASURE OF ANGLES AND ARCS

See pages 110–112 for directions to change units measuring angles and arcs and page 129 for directions to compute with these units. The table of measure is found on page 549.

EXERCISES

Change:

1. To minutes: 23°, 11°34′, 46°, 2,160″

2. To seconds: 31′, 2°, 13′17″, 5°46′52″

3. To degrees: 1,380′, 2,640′, 15,120″, 15,120′

4. Add and simplify:

   115°34′14″   56°45′   6°18′23″   112° 9′51″
    52° 8′33″   11°54′   14° 9′35″   4°53′ 6″
                                       29°57′22″   64°11′41″

5. Subtract:

   45°   134°54′   19°46′22″   90°
   25°15′   72°14′   4°51′ 9″   15°31′25″

6. Multiply and simplify:

   14°15′   62°55′   19°34′32″   3°19′44″
      6   2   4   12

7. Divide:

   3)36°24′   12)8°36′   7)51°34′   12)37°29′24″

2–15 RATES OF SPEED

We change the given units to the required units (in the metric model: kilometer to 1,000 meters and hour to 3,600 seconds; in the customary model: mile to 5,280 ft. and hour to 3,600 sec.), then we perform the necessary arithmetical operations. A *knot* is one (1) nautical mile per hour. A nautical mile is approximately equivalent to 6,080 ft. or 1.15 statute miles or 1.85 km. A kilometer is approximately equal to .54 nautical mile. See table of measure on page 549.

Note that in the metric system rates of speed are abbreviated in the following ways: meters per second (m/s); kilometers per hour (km/h); centimeters per second (cm/s); millimeters per minute (mm/min); and so on.

Units of Measure and Measurement

> | METRIC | CUSTOMARY |
> |---|---|
> | Change 90 km/h to m/s: | Change 45 m.p.h. to ft. per sec.: |
> | $90 \text{ km/h} = \dfrac{90 \text{ km}}{1 \text{ h}}$ | $45 \text{ m.p.h.} = \dfrac{45 \text{ mi.}}{1 \text{ hr.}}$ |
> | $\dfrac{90 \times 1{,}000}{3{,}600} = 25 \text{ m/s}$ | $\dfrac{45 \times 5{,}280}{3{,}600} = 66 \text{ ft. per sec.}$ |
> | *Answer,* 25 m/s | *Answer,* 66 ft. per sec. |

EXERCISES

Change, finding answers to the nearest hundredth wherever necessary:

1. To meters per second: 70 km/h;   1,240 mm/min;   100 cm/s
2. To meters per minute: 60 km/h;   115 mm/s;   120 cm/min
3. To centimeters per second: 440 mm/s;   24 m/min;   40 km/h
4. To kilometers per hour: 100,000 mm/s;   120 m/s;   3,300 cm/s
5. To centimeters per minute: 2 km/h;   5 m/h;   1,200 mm/s
6. To feet per second: 12 knots;   250 stat. m.p.h.;   100 stat. m.p.h.
7. To statute miles per hour: 300 ft. per sec.;   200 knots;   6,600 ft. per min.
8. To nautical miles per hour: 45 knots;   700 ft. per sec.;   95 stat. m.p.h.
9. To knots: 21.7 naut. m.p.h.;   112 stat. m.p.h.;   117 ft. per sec.
10. To feet per minute: 6,900 ft. per sec.;   225 stat. m.p.h.;   105 knots

2–16 TEMPERATURE

The three thermometer scales that are used to measure temperature are the *Celsius,* the *Fahrenheit,* and the *Kelvin.* The Celsius scale (formerly called centigrade) is based on 100 divisions, each called a *degree.* On the Celsius scale the freezing point of water is indicated as 0° and the boiling point is 100°. On the Fahrenheit scale the freezing point and the boiling point of water are indicated as 32° and 212° respectively for a 180° interval. It now appears that the Celsius scale will soon replace the Fahrenheit scale in the United States.

The Kelvin temperature scale, used in SI measurement, is related to the Celsius scale. One degree Celsius is exactly equal to one *kelvin* (the name used to mean degree Kelvin). The reading of a specific temperature on the Kelvin scale is approximately 273 kelvins more than its reading on the Celsius scale.

To change a Celsius temperature reading to a kelvin reading, we add 273° to the Celsius reading or use the formula:
$$K = C + 273$$
To change a kelvin temperature reading to a Celsius reading, we subtract 273° from the kelvin reading or use the formula:
$$C = K - 273.$$
To change a Celsius temperature reading to a Fahrenheit reading, we add 32° to nine-fifths of the Celsius temperature reading or use the formula:
$$F = \tfrac{9}{5}C + 32$$
To change a Fahrenheit temperature reading to a Celsius reading, we subtract 32° from the Fahrenheit temperature reading and take five-ninths of this answer or use the formula:
$$C = \tfrac{5}{9}(F - 32)$$

EXERCISES

1. Change each of the following Celsius temperature readings to a corresponding kelvin reading:

   a. 50°C   b. 212°C   c. 18°C   d. −23°C   e. 100°C

2. Change each of the following kelvin temperature readings to a corresponding Celsius reading:

   a. 115 K   b. 274 K   c. 800 K   d. 373 K   e. 411 K

3. Change each of the following Fahrenheit readings to a corresponding Celsius temperature reading:

   a. 27°F   b. 85°F   c. 173°F   d. 400°F   e. 223°F

4. Change each of the following Celsius temperature readings to a corresponding Fahrenheit temperature reading:

   a. 200°C   b. 45°C   c. 0°C   d. 2°C   e. 76°C

5. The average body temperature for humans is 98.6°F. What is the corresponding Celsius temperature?

2–17 TWENTY-FOUR HOUR CLOCK

To tell time on a 24-hour clock, a 4-digit numeral is used, the first two digits indicating the hour and the last two minutes. 1 P.M. is thought of as the 13th hour, 2 P.M. the 14th hour, etc. Thus 1200 is added to the given digits when expressing time from 1 P.M. to 12 midnight. The numeral 12 representing the hour is replaced by two zeros when expressing the time from midnight to 1 A.M.

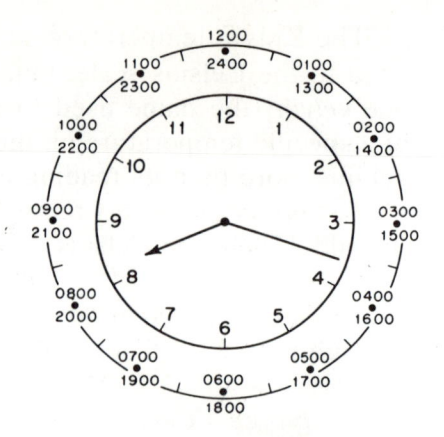

| 7:26 A.M. = 0726 | 2:51 P.M. = 1451 |
| 12:19 A.M. = 0019 | 12:03 P.M. = 1203 |

EXERCISES

1. Using the 24-hour system express each of the following:

   a. 6:21 A.M.
   b. 1:56 A.M.
   c. 4:00 A.M.
   d. 7:00 P.M.
   e. 3:30 P.M.
   f. 11:07 P.M.
   g. 9:13 A.M.
   h. 12:45 P.M.
   i. 1:04 A.M.
   j. 8:27 P.M.
   k. 12:05 A.M.
   l. 5:15 P.M.

2. Express each of the following in A.M. or P.M. time:

   a. 0230
   b. 1146
   c. 0300
   d. 1552
   e. 2330
   f. 1645
   g. 1829
   h. 1056
   i. 0001
   j. 0202
   k. 1927
   l. 1208

3. Find the difference in times between:
   a. 0141 and 0910   b. 0233 and 1509   c. 1040 and 2227

4. How long did it take an airplane to fly from Tampa to New York City if it left Tampa at 2219 and arrived in New York City at 0007 the next day?

5. A merchant ship left port at 0020 and arrived at its destination at 2000. A second ship traveled the same route, leaving port at 1045 and arriving at 0639 the next day. Which ship was slower, and how much slower was it?

2-18 TIME ZONES

There are four standard *time belts* in the United States, excluding Alaska and Hawaii: Eastern (EST), Central (CST), Mountain (MST), and Pacific (PST). Central time is one hour earlier than Eastern time, Mountain time is one hour earlier than Central time, and Pacific time is one hour earlier than Mountain time. The meridians at 75°, 90°, 105°, and 120° west longitude are used to determine the time in these zones.

Alaska has four time zones determined by the meridians at 120°, 135°, 150°, and 165° west longitude. Hawaii uses the time zone of the meridian at 150° west longitude. Parts of Canada east of Maine are in another time zone, the Atlantic Standard Time Zone, which is one hour later than Eastern Standard Time.

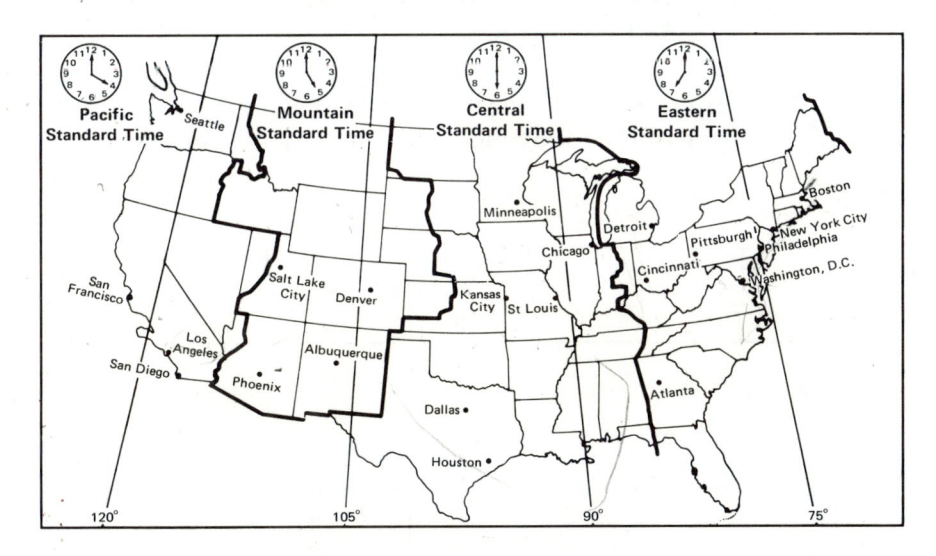

EXERCISES

1. In what time zone is your city located?
2. If it is 2 P.M. in the Pacific Time Zone, what time is it in:
    a. Eastern Zone?
    b. Mountain Zone?
    c. Central Zone?

Units of Measure and Measurement

3. If it is 6 A.M. in Phoenix, what time is it in:
   a. Boston?
   b. Minneapolis?
   c. Chicago?
   d. Denver?
   e. Detroit?
   f. Seattle?
   g. St. Louis?
   h. New York?
   i. Your city?

4. If a nationwide program is telecast at 9:30 P.M. from Los Angeles, what time is it seen in:
   a. Kansas City?
   b. Philadelphia?
   c. Albuquerque?
   d. San Diego?
   e. Houston?
   f. Your city?

5. How long did it take an airplane to fly from Washington to San Francisco if it left Washington at 9:45 A.M. (EST) and arrived in San Francisco at 11:20 A.M. (PST)?

6. If you place a telephone call at exactly noon in Dallas to Atlanta, what time is it in Atlanta?

MEASUREMENT

2–19 PRECISION AND ACCURACY IN MEASUREMENT

Measurement is never exact; it is approximate. The same length measured on each of the following scales shows a measurement of:

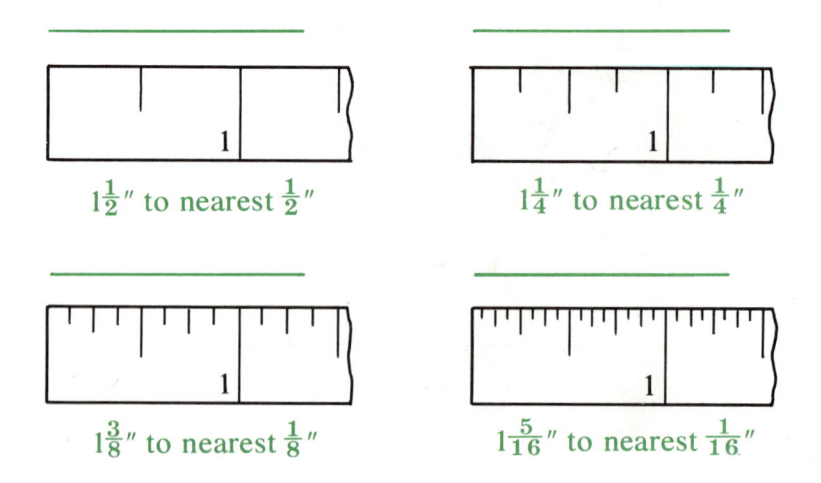

$1\frac{1}{2}''$ to nearest $\frac{1}{2}''$

$1\frac{1}{4}''$ to nearest $\frac{1}{4}''$

$1\frac{3}{8}''$ to nearest $\frac{1}{8}''$

$1\frac{5}{16}''$ to nearest $\frac{1}{16}''$

Observe that $1\frac{1}{2}''$ is precise to the nearest $\frac{1}{2}''$; $1\frac{1}{4}''$ is precise to the nearest $\frac{1}{4}''$; $1\frac{3}{8}''$ is precise to the nearest $\frac{1}{8}''$; $1\frac{5}{16}''$ is precise to the nearest $\frac{1}{16}''$. The smaller the unit, the more precise is the measurement. The measurement $1\frac{5}{16}''$ is the most precise of the measurements shown on the opposite page.

Precision is the closeness to the true measurement. It is determined by the smallest unit of measure used to make the measurement.

The precision of a measurement named by a numeral for a whole number ending in zeros is indicated by an underlined zero as shown below. (Also see section 2–20.)

49,00$\underline{0}$ kilometers is precise to the nearest kilometer.

49,0$\underline{0}$0 kilometers is precise to the nearest 10 kilometers.

49,$\underline{0}$00 kilometers is precise to the nearest 100 kilometers.

49,000 miles is precise to the nearest 1,000 miles.

49,000.0 miles is precise to the nearest tenth of a mile.

In measuring the length on the opposite page, $1\frac{5}{16}''$, the unit used was $\frac{1}{16}$ of an inch. The real length is between $1\frac{9}{32}''$ and $1\frac{11}{32}''$. The greatest possible error between the measurement $1\frac{5}{16}''$ and the true measurement can only be $\frac{1}{32}''$ since the *greatest possible error* is one-half of the unit used to measure.

The *relative error* is the ratio of the greatest possible error to the measurement. The *accuracy* of the measurement is determined by the relative error. The smaller the relative error, the more accurate is the measurement.

The measurement $1\frac{5}{16}''$ has a greatest possible error of $\frac{1}{32}''$.

The ratio of $\frac{1}{32}''$ to $1\frac{5}{16}''$ is $\frac{1}{42}$.

The relative error is $\frac{1}{42}$.

$\frac{1}{32}$ to $1\frac{5}{16}$

$\frac{1}{32} \div 1\frac{5}{16}$

$\frac{1}{32} \times \frac{16}{21} = \frac{1}{42}$

Units of Measure and Measurement

Sometimes to make quick comparisons the relative error is expressed as a percent. To do this we multiply the relative error by 100. (See section 1–29.)

The allowance for error in measurement is called *tolerance*. A required measurement of 5.9 inches with a tolerance of .25 inch allowed in both directions would be indicated as 5.9″ ± .25″ and would represent any measurement from 5.65 inches to 6.15 inches inclusive.

EXERCISES

1. To what unit of measure are these measurements precise?
   a. $3\frac{3}{8}$ in. d. $1\frac{3}{32}$ in. g. 1.17 min. j. .55 in. m. 4 hr. 12 min.
   b. $7\frac{1}{4}$ hr. e. $\frac{1}{6}$ yd. h. .022 mm k. 7.3 lb. n. 5 ft. 6 in.
   c. $11\frac{1}{2}$ pt. f. 2.4 mi. i. 4.009 g l. 2 km o. 2 kg 4 g

2. Find the unit of measure or precision of each of the following measurements:
   a. 16 mi. c. 23,0̲00 km e. 1,070,0̲00 mi. g. 6,783 cm
   b. 440 sec. d. 245 hr. f. 110̲,000 L h. 12,00̲0,000 kg

3. Which measurement in each of the following is more precise?
   a. $2\frac{1}{8}$ ft. or 2 ft. 5 in.? d. 1.119 m or 9.88 m?
   b. 5 lb. 3 oz. or $4\frac{1}{2}$ lb.? e. 1.00 g or 12.4 g?
   c. $10\frac{1}{4}$ in. or $6\frac{3}{8}$ in.? f. $3\frac{1}{8}$ mi. or $\frac{1}{4}$ mi.?

4. Find the greatest possible error in each of these measurements:
   a. $2\frac{1}{8}$ lb. f. $1\frac{1}{4}$ hr. k. 55 cm p. 14.7 g
   b. $5\frac{3}{4}$ lb. g. 440 yd. l. 9 mi. q. 1̲000 km
   c. $1\frac{3}{16}$ lb. h. 175 mi. m. 0.001 ft. r. 240,0̲00 ft.
   d. $\frac{5}{32}$ lb. i. $2\frac{23}{64}$ in. n. 17.3 km s. 1,01̲0,000 cm
   e. $9\frac{13}{16}$ in. j. 2.2 L o. 600 mi. t. 2 ft. 5 in.

5. Find the relative error in each of the measurements given in Exercise 4.

6. Find the relative error as a percent for each of the following measurements:

   a. $4\frac{3}{4}$ in.   b. 37 mi.   c. 1.2 kg   d. $1\frac{1}{3}$ min.   e. 70 cm

7. Which measurement in each of the following is more accurate?

   a. 27 m or 2.7 cm?   c. 1.001 g or 150 kg?   e. 27.1 in. or 2.25 in.?
   b. 250,000 ft. or $2\frac{1}{2}$ in.?   d. $5\frac{3}{8}$ in. or $\frac{3}{16}$ in.?   f. $3\frac{1}{4}$ lb. or $2\frac{1}{2}$ lb.?

8. Which is more precise: 1.6 m or 0.03 m? Which is more accurate?

9. Arrange the following measurements in order of precision, least precise first; and also in order of accuracy, least accurate first:

   a. 6 cm,   60 cm,   0.0060 cm,   0.060 cm,   6.0 cm,   0.06 cm
   b. 3.1 in.,   6.25 cm,   350 m,   6.114 ft.

10. What range of measurements does each of the following represent?

    a. $1\frac{1}{4}'' \pm \frac{1}{16}''$
    b. $2.3'' \pm .05''$
    c. 2.0 mm $\pm$ .01 mm
    d. 23.52 mm $\pm$ .005 mm
    e. $12.3'' \pm .09''$
    f. $1.345'' \pm .0002''$
    g. $5'' \pm .001''$
    h. $\frac{1}{2}'' \pm \frac{3}{64}''$

2–20 SIGNIFICANT DIGITS

Digits are *significant* in an approximate number when they indicate the precision which is determined by the value of the place of the last significant digit on the right.

> In 87.39, the significant digit 9 indicates precision to the nearest hundredth.
> In 31,0<u>0</u>0, the underlined 0 indicates precision to the nearest ten.

The digits described below in items (1) to (4) inclusive are significant, those described in items (5) and (6) are not significant except as noted in (5).

(1) All non-zero digits are significant.

> 912 has 3 significant digits.

Units of Measure and Measurement

(2) Zeros located between non-zero digits are significant.

> 5,006 has 4 significant digits.

(3) When a decimal or mixed decimal ends in zeros, these zeros are significant.

> 89.370 has 5 significant digits.

(4) When numbers are expressed in scientific notation, all the digits in the first factor are significant.

> In 2.61×10^4, digits 2, 6, and 1 are significant.

(5) When a number ends in zeros, these zeros are not significant unless they are specified as being significant or indicated as significant by a line drawn under them.

> 74,000 has 2 significant digits (7,4)
> 74,000 has 3 significant digits (7,4,0)
> 74,000 has 5 significant digits (7,4,0,0,0)
> 6,010 has 3 significant digits (6,0,1)

(6) In a decimal fraction (when the number is between 0 and 1) the zeros immediately following the decimal point are not significant.

> .00493 has 3 significant digits (4,9,3) but
> 4.0093 has 5 significant digits (4,0,0,9,3).
> .0040 has 2 significant digits (4,0).

EXERCISES

Determine the number of significant digits in each of the following examples:

1. 3	**15.** 11,099	**29.** 96.2
2. 24,000	**16.** 34,500	**30.** 11.05
3. 1.1109	**17.** .019	**31.** 100.0
4. 140	**18.** .60008	**32.** 76.00
5. .0101	**19.** 81.990	**33.** 1.20×10^8
6. 1,001	**20.** 107	**34.** 3,400,506
7. .90	**21.** 2.034	**35.** 135,000,000
8. 58	**22.** 23.04	**36.** 268,000,0̲00
9. 0.01	**23.** 203.4	**37.** 1.007×10^2
10. 5,070,000	**24.** 0.002034	**38.** 34.2066
11. 0.0010	**25.** 1,9̲00	**39.** .056700
12. 7,209,000	**26.** 5,600	**40.** 74.5̲0
13. 7.1×10^6	**27.** .000008	**41.** 8.79×10^7
14. 68,0̲00	**28.** 5.0×10^3	**42.** .060

2–21 APPROXIMATE NUMBERS

Numbers may be exact or approximate. Numbers used to count are exact; numbers used to measure are approximate. Estimated numbers are approximate numbers. An *approximate number* is a number that is almost equal to the true number.

Since measurement is approximate, we may be required to compute with approximate numbers arising through measurement. A measurement of 4.2 inches could mean any measurement from 4.15 inches to 4.25 inches. When written as 4.2″ ± .05″, it shows the measurement plus and minus the greatest possible error. The result of a computation with approximate numbers cannot be more accurate than the least accurate approximate number involved in the computation.

There are several methods of computing with approximate numbers. Observe the alternative methods on the next page.

Units of Measure and Measurement

Addition

To add approximate numbers, we add as we usually do, then round the sum using the unit of the least precise addend.

Subtraction

To subtract approximate numbers, we subtract as we usually do, then round the difference using the unit of the less precise of the given numbers.

Multiplication

To multiply approximate numbers, we multiply as we usually do, then round the product so that it contains the same number of significant digits as the factor having the smaller number of significant digits.

Division

To divide approximate numbers, we divide as we usually do, then round the quotient so that it contains the same number of significant digits as there are in either the dividend or divisor, whichever is less.

Add:

$$\begin{array}{r} 3.14 \\ 1.752 \\ +5.8 \\ \hline 10.692 \end{array} = 10.7$$

Subtract:

$$\begin{array}{r} 7.573 \\ -1.43 \\ \hline 6.143 \end{array} = 6.14$$

Multiply:

$$\begin{array}{r} 1.92 \\ 2.4 \\ \hline 768 \\ 3\ 84 \\ \hline 4.608 \end{array} = 4.6$$

Divide:

$$4.1_\wedge \overline{)9.7_\wedge 58} \quad \begin{array}{r} 2.38 = 2.4 \end{array}$$

$$\begin{array}{r} 8\ 2 \\ \hline 1\ 5\ 5 \\ 1\ 2\ 3 \\ \hline 3\ 28 \\ 3\ 28 \\ \hline \cdots \end{array}$$

Alternate Methods

To add or subtract, we first round the given approximate numbers to the least precise number, then perform the required operation.

To multiply when one of two factors contains more significant digits than the other, we round the factor which has more significant digits so that it contains only one more significant digit than the other. Then we multiply and round the product so that it contains the same number of significant digits as the factor having the smaller number of significant digits.

EXERCISES

Compute the following approximate numbers as indicated:

1. Add:
   a. 27 + 2.08 + .3
   b. 9.8 + 3.34
   c. .91 + 38 + 38.3
   d. 5.507 + 4.5 + 6.32
   e. .113 + .0084 + .07

2. Subtract:
   a. 23 − 12.8
   b. 1.06 − .009
   c. 3.3534 − 2.30007
   d. 34.1 − 19.56
   e. 45.628 − 9.52

3. Multiply:
   a. 11.76 × 4.3237
   b. 89 × 4.2
   c. .0075 × 4.33
   d. 3.25 × 1.679
   e. 2.4 × 3.5

4. Divide:
   a. .046)$\overline{87.6768}$
   b. 5)$\overline{45.955}$
   c. 2.5)$\overline{114.72}$
   d. 1.17)$\overline{246.3579}$
   e. 51)$\overline{9.87}$

ESTIMATING ANSWERS

For each of the following select your nearest estimate:

1. 23 × 61 is approximately: 1,000;   1,400;   14,000;   150
2. $2\frac{1}{8} \times 5\frac{1}{2}$ is approximately: 10;   20;   12;   8
3. 8% of 204 is approximately: 16,000; 1,500;   20;   16
4. 760 divided by 24 is approximately: 3,100;   30;   150;   50
5. .095 divided by .3 is approximately: 320;   29;   .003;   .32
6. 6,023 divided by 50 is approximately: 120;   12,000; 3,000;   30
7. 6 divided by $\frac{7}{8}$ is approximately: 5;   3;   7;   10
8. 33% of 630 is approximately: 20;   200;   210;   2,000
9. 199 times 503 is approximately: 1,000;   10,000; 100,000; 1,000,000
10. .07 times 97 is approximately: 68;   7;   8;   72

Units of Measure and Measurement

11. 30% of 810 is approximately: 2,500;   24,000;   430;   240
12. $\frac{2}{3}$ of 286 is approximately: 100;   150;   185;   210
13. 614 times 59 is approximately: 3,600;   36,000;   12,000;   1,200
14. 509 divided by .7 is approximately: 490;   250;   1,100;   750
15. 3% of $6.95 is approximately: $2; $.20; $.15; $1
16. 126,000 divided by 197 is approximately: 63,900;   640;   6.4;   6,400
17. $3\frac{1}{2}$ times $3\frac{15}{16}$ is approximately: 14;   11;   21;   18
18. 40% of 99,000 is approximately: 40,000;   4,000;   400;   25,000
19. $\frac{4}{5}$ of 494 is approximately: 620;   62;   4,000;   400
20. .39 times 811 is approximately: 320;   3,200;   290;   2,900
21. 119,123 divided by 396 is approximately: 40,000;   300;   30,000;   400
22. $7\frac{1}{4}$ times $8\frac{5}{6}$ is approximately: 56;   95;   65;   100
23. 8% of $29.95 is approximately: $3; $2.25; $9; $1.65
24. .8 divided by .019 is approximately: 40;   35;   400;   160
25. 118 divided by $1\frac{7}{8}$ is approximately: 100;   200;   6.5;   65
26. $5\frac{1}{8}$ times 16 is approximately: 3;   80;   100;   75
27. $37\frac{1}{2}$% of 9,023 is approximately: 38,000;   3,800;   33,000;   3,300
28. $\frac{15}{16}$ divided by 4 is approximately: $\frac{1}{4}$; $\frac{1}{8}$; $\frac{3}{4}$; $\frac{1}{2}$
29. 131 times 68 is approximately: 1,000;   5,000;   10,000;   15,000
30. 130,119 divided by 213 is approximately: 680;   68;   600;   6,000
31. 13 divided by .06 is approximately: 780;   78;   22;   220
32. 5.8 times 11 is approximately: 54;   35;   65;   66
33. $26\frac{1}{2}$ divided by $2\frac{7}{8}$ is approximately: 7;   8;   9;   11
34. $33\frac{1}{3}$% of 17,448 is approximately: 581;   5,800;   5,000;   50,000
35. $9\frac{1}{3}$ times $6\frac{1}{6}$ is approximately: 49;   53;   70;   59

CHAPTER REVIEW

1. Change:
   a. 52.7 m to mm
   b. 208 mm to cm
   c. 36 km to m
   d. 1,455 cm to m
   e. 4,600 m to km

   (2–1)

2. Change:
   a. 11.6 kg to g
   b. 82 mg to cg
   c. 48 g to mg
   d. 579 g to kg
   e. 9.3 cg to mg

   (2–2)

3. Complete:
   a. 15 L = _____ cL
   b. 6.4 cL = _____ mL
   c. 967 mL = _____ L
   d. 8.1 L = _____ dL
   e. .58 L = _____ mL

   (2–3)

4. Complete:
   a. 9.8 cm² = _____ mm²
   b. 5,300 m² = _____ hectares
   c. 672 cm² = _____ m²
   d. 23.95 km² = _____ hectares
   e. 350 hectares = _____ km²

   (2–4)

5. a. Change:
   (1) 41.56 cm³ to mm³
   (2) 60.89 cm³ to dm³
   (3) 75.2 m³ to dm³
   (4) 9.3 m³ to cm³

   (2–5)

   b. 50.6 liters of water occupies a space of _____ cm³ and weighs _____ kg. (2–5)

   c. How many liters of water will fill a tank if it occupies a space of 20,000 dm³? (2–5)

   d. 704 milliliters of water weigh how many grams? (2–5)

6. How many inches are in $4\frac{3}{4}$ yards? (2–6)

7. What part of a pound is 10 ounces? (2–7)

Units of Measure and Measurement

8. How many liquid ounces are in $1\frac{1}{2}$ quarts? (2–8)
9. Change 19 pecks to bushels. (2–9)
10. Change 306 sq. ft. to sq. yd. (2–10)
11. Change $28\frac{1}{3}$ cu. yd. to cu. ft. (2–11)
12. What part of the day is 16 hours? (2–13)
13. Multiply 32°42′ by 15. Simplify your answer. (2–14)
14. Change 110 m/s to km/h (2–15)
15. The room thermostat is set at 65°F. What would it be on the Celsius scale? What is it in kelvins? (2–16)
16. Express as A.M. or P.M.:   a. 0400   b. 2145 (2–17)
17. If it is 9 A.M. in Houston, what time is it in Savannah? In San Diego? (2–18)
18. a. Find the greatest possible error and the relative error in each of the following measurements: (2–19)

    (1) 12 g   (2) 3.4 mm   (3) 4,500 km   (4) $2\frac{5}{6}$ hr.   (5) $5\frac{1}{4}$ in.

    b. Which is more precise, 71,000 m or .005 mm? Which is more accurate? (2–19)

    c. Arrange the following measurements in order of precision, most precise first; and also in order of accuracy, most accurate first: (2–19)

    (1) .3 cm   (2) 3 cm   (3) .03 cm   (4) 30 cm   (5) .003 cm

19. How many significant digits are in each of the following? (2–20)

    a. 4.0027   c. 6,008,000   e. 2.93×10^9
    b. .00427   d. 5,981   f. 20.035

20. Compute the approximate numbers as indicated: (2–21)

    a. 25 + .59 + 1.4   c. 6.2×5.8
    b. 48.3 − .067   d. $6\overline{)2.045}$

ACHIEVEMENT TEST

1. Add: (1–2)
   92,776
   9,468
   43,584
   6,865
   25,379

2. Subtract: (1–3)
   843,179
   232,089

3. Multiply: (1–4)
   807
   689

4. Divide: (1–5)
   184)91,080

5. Add: (1–12)
   $\frac{1}{2} + \frac{5}{6}$

6. Subtract: (1–13)
   $11 - 1\frac{2}{3}$

7. Multiply: (1–14)
   $12 \times 2\frac{3}{8}$

8. Divide: (1–15)
   $14 \div \frac{2}{5}$

9. Add: (1–18)
   .6 + .006 + .06

10. Subtract: (1–19)
    7.5 − .69

11. Multiply: (1–20)
    .014 × .07

12. Divide: (1–22)
    1.5 ÷ .03

13. Find 9% of $560.85 to the nearest cent. (1–30, 1–33, 1–34)
14. What percent of 105 is 42? (1–31, 1–33, 1–34)
15. 27 is 15% of what number? (1–32, 1–33, 1–34)
16. Find the square root of 556,516. (1–36)
17. Write all the even prime numbers. (1–37, 1–39)
18. a. What is the greatest common factor of 24, 60, and 96? (1–38)
    b. What is the least common multiple of 12, 18, and 30? (1–40)
19. Which distance is shorter: 6,750 m or 7.1 km? (2–1)
20. Which weight is heavier: .084 g or 56 mg? (2–2)
21. How many cL are in 62.3 liters? (2–3)

Units of Measure and Measurement

22. How many m² are in 12 hectares? (2–4)

23. a. Change 376 dm³ to m³. (2–5)

    b. 4,050 liters of water weighs _____ kg and occupies a space of _____ dm³. (2–5)

24. a. An altitude of 26,400 ft. is how many miles high? (2–6)

    b. How many sq. yd. are contained in a floor space of 225 sq. ft.? (2–10)

    c. The capacity of a certain truck is 5 cu. yd. What is this in cubic feet? (2–11)

25. Divide 16°6′ by 12. (2–14)

26. Change 90 km/h to m/s. (2–15)

27. a. What part of a year is 4 months? (2–13)

    b. Express as A.M. or P.M.: (1) 1320 (2) 0900 (2–17)

    c. If it is 3 P.M. in Buffalo, what time is it in Washington, D.C.? In New Orleans? In Phoenix? In Kansas City? (2–18)

28. Which is warmer, a temperature of 20°C or 60°F? (2–16)

29. Which measurement is more precise, 85 mm or 8.5 mm? Which is more accurate? (2–19)

30. How many significant digits are in each of the following? (2–20)

    a. 8,400   **b.** 5,003   **c.** .0010   **d.** 7.2×10^8 **e.** .006

3

CHAPTER 3
Positive and Negative Numbers

INVENTORY TEST

The numeral at the end of each problem indicates the section where explanatory material may be found.

1. Which of the following numerals name positive numbers? Which name negative numbers? (3–1)

   $^-3.25 \quad ^+78 \quad ^-\frac{5}{6} \quad 0 \quad ^+.562 \quad ^-200 \quad ^-3\frac{7}{10}$

2. Which of the following numerals name integers? Which name rational numbers? Which name irrational numbers? Which name real numbers? (3–1 to 3–2)

   $^-39 \quad ^+\frac{7}{12} \quad ^-\sqrt{85} \quad ^+5\frac{2}{3} \quad ^-.61 \quad ^+\frac{40}{8}$

3. Find the value of: **a.** $|^-12|$ **b.** $|^+9|$ (3–3)
4. Draw the graph of $^-3, ^-1, 0, ^+2, ^+4$ on a number line. (3–4)
5. Which of the following sentences are true? (3–5)

   a. $^-6 > ^-2$ **c.** $^-4 \not> ^+3$ **e.** $^+2 < ^+5$
   b. $0 < ^-5$ **d.** $^+1 \not< ^-1$ **f.** $^-7 \not< 0$

6. If a tail wind of 25 km/h is represented by $^+25$ km/h, how can a head wind of 30 km/h be represented? (3–6)
7. Use the number line as a scale to draw a vector that illustrates each of the following movements. Write the numeral that is represented by each of these vectors. (3–7)

   a. From $^+4$ to $^-3$ **b.** From $^-2$ to $^+2$

8. **a.** What is the opposite of $^-11$? (3–8)
   b. What is the additive inverse of $^+7$?
   c. What is the opposite of the opposite of $^-\frac{3}{4}$?
   d. Write symbolically: The opposite of negative ten is positive ten.

9. Add on the number line, using vectors: (3–9)
   a. $^+4 + {}^-9$
   b. $^-2 + {}^-5$
   c. $^-4 + {}^+6$
   d. $^+7 + {}^+1$

10. Add as indicated: (3–10)
    a. $^-15 + {}^+21$
    b. $^-32 + {}^+17$
    c. $^-19 + {}^-14$
    d. $^+26 + {}^-50$

11. Add as indicated: (3–11)
    a. $^+2.6 + {}^+.35$
    b. $\frac{^-5}{6} + {}^+3\frac{2}{3}$
    c. $^+.839 + {}^-.84$
    d. $^-4\frac{3}{5} + {}^-3\frac{1}{2}$

12. Subtract on the number line, using vectors: (3–12)
    a. $^-8 - {}^+6$
    b. $^-11 - {}^-5$
    c. $^+1 - {}^-6$
    d. $^+2 - {}^+10$

13. Subtract as indicated: (3–13)
    a. $^-50 - {}^+37$
    b. $^-25 - {}^-19$
    c. $12 - 41$
    d. $0 - {}^-7$

14. Subtract as indicated: (3–14)
    a. $\frac{^-2}{3} - \frac{^-5}{12}$
    b. $^-.008 - {}^+3.5$
    c. $^+2\frac{1}{6} - {}^-3\frac{1}{2}$
    d. $6.1 - 9$

15. Multiply as indicated: (3–15)
    a. $^-12 \times {}^-6$
    b. $^+23 \times {}^-34$
    c. $(4)(^-7)(^-3)$
    d. $(^-1)(5)(^-3)(^-2)$

16. Multiply as indicated: (3–16)
    a. $\frac{^+5}{8} \times \frac{^-3}{10}$
    b. $^-4\frac{1}{2} \times {}^-1\frac{5}{6}$
    c. $^-.49 \times {}^+2.71$
    d. $^-.58 \times {}^-.005$

17. Divide as indicated: (3–17)
    a. $^+20 \div {}^-4$
    b. $^-18 \div {}^-1$
    c. $^-54 \div 6$
    d. $^-9 \div {}^-9$

18. Divide as indicated: (3–18)
    a. $\frac{^-3}{4} \div \frac{^+9}{10}$
    b. $^-2\frac{13}{16} \div {}^-2\frac{1}{4}$
    c. $^+4.576 \div {}^-.04$
    d. $^-.5 \div {}^-.005$

Positive and Negative Numbers

19. Perform the indicated operation: (3–19)

    a. $(-5) + (-4)$ **c.** $(-1)(-3)$
    b. $(-7) - (-7)$ **d.** $(-63) \div (+9)$

20. a. Does the commutative property hold for the: (3–10, 3–11, 3–13 to 3–18)

    (1) subtraction of integers?
    (2) addition of rational numbers?
    (3) multiplication of integers?
    (4) division of rational numbers?
    (5) multiplication of rational numbers?
    (6) division of integers?
    (7) subtraction of rational numbers?
    (8) addition of integers?

  b. What is the multiplicative inverse of: (3–16)

    (1) $^-2$?   (2) $^+11$?   (3) $^-\frac{4}{5}$?   (4) $^+\frac{7}{3}$?   (5) $^-2\frac{1}{4}$?

EXTENSION OF THE NUMBER SYSTEM

When we add any two whole numbers, we find that the sum is always a whole number. Thus, the whole numbers are said to be *closed* under the operation of addition.

However, when we subtract one whole number from another whole number, the difference is not always a whole number.

For example:

In each of the following subtractions the difference is a whole number.

$$\begin{array}{ccccc} 5 & 5 & 5 & 5 & 5 \\ -1 & -2 & -3 & -4 & -5 \\ \hline 4 & 3 & 2 & 1 & 0 \end{array}$$

But in each of the following subtractions the difference cannot be a whole number.

$$\begin{array}{ccccc} 5 & 5 & 5 & 5 & 5 \\ -6 & -7 & -8 & -9 & -10 \\ \hline ? & ? & ? & ? & ? \end{array}$$

Thus we see that the whole numbers are not closed under the operation of subtraction. In order to be able to subtract in problems like those above, let us now enlarge our number system.

3–1 INTEGERS

The numerals naming whole numbers may be arranged in a definite order on a number line so that they correspond one-to-one with points on the number line to the right of the point labeled 0.

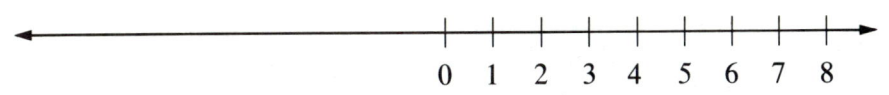

Now let us extend the number line to the left of the point marked 0. Using the interval between 0 and 1 as the unit of measure, we locate equally spaced points to the left of 0. The first new point is labeled ⁻1, read "negative one"; the second point ⁻2, read "negative two"; the third point ⁻3 etc.

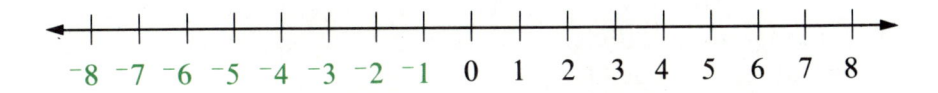

Positive and Negative Numbers 161

The numbers corresponding to the points to the left of 0 are *negative numbers* and their numerals are identified by the raised dash (⁻) symbols that precede the digits. The numbers corresponding to the points to the right of 0 are *positive numbers* and their numerals may either contain no sign or for emphasis are identified by the raised (⁺) symbols that precede the digits, in which case the number line would look like this:

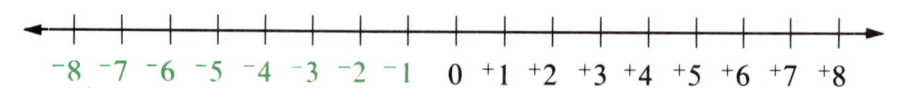

```
◄──┼──┼──┼──┼──┼──┼──┼──┼──┼──┼──┼──┼──┼──┼──┼──┼──►
  ⁻8 ⁻7 ⁻6 ⁻5 ⁻4 ⁻3 ⁻2 ⁻1  0 ⁺1 ⁺2 ⁺3 ⁺4 ⁺5 ⁺6 ⁺7 ⁺8
```

Positive and negative numbers are sometimes called *signed numbers* or *directed numbers*. The numeral ⁻8 names the number *negative eight*. The numeral ⁺7 or simply 7 names the number *positive seven*. 0 is neither positive nor negative. All numbers greater than 0 are positive numbers.

Examination of the number line below will show that the points that correspond to ⁺4 and ⁻4 fall on opposite sides of the point marked 0 but are the same distance from 0.

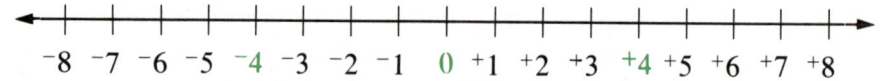

```
◄──┼──┼──┼──┼──┼──┼──┼──┼──┼──┼──┼──┼──┼──┼──┼──┼──►
  ⁻8 ⁻7 ⁻6 ⁻5 ⁻4 ⁻3 ⁻2 ⁻1  0 ⁺1 ⁺2 ⁺3 ⁺4 ⁺5 ⁺6 ⁺7 ⁺8
```

A pair of numbers, one positive and the other negative, such as ⁺4 and ⁻4, which have the same absolute value (see section 3–3) are called *opposites*. ⁻4 is the opposite of ⁺4 and ⁺4 is the opposite of ⁻4. Each is the opposite of the other. The opposite of zero is zero.

The group of numbers which consists of all the whole numbers and their opposites is called the *integers*. Accordingly, both . . . , ⁻2, ⁻1, 0, 1, 2, . . . and . . . , ⁻2, ⁻1, 0, ⁺1, ⁺2, . . . describe the *integers*. Observe that the three dots are used at both ends since the integers continue in both directions without ending. Thus, 1, 2, 3, . . . describes the *positive integers* (or the natural numbers); 0, 1, 2, . . . describes the *non-negative integers* (or the whole numbers); and . . . , ⁻3, ⁻2, ⁻1, 0 describes the *non-positive integers*.

The *even integers* consist of the integers that are divisible by two (2). The even integers are described in the following way: . . . , ⁻4, ⁻2, 0, 2, 4, ... The *odd integers* consist of all the integers that are not divisible by two (2). The *odd integers* are described by . . . , ⁻3, ⁻1, 1, 3, ...

EXERCISES

1. Read or write in words, each of the following:

   a. ⁻8   c. ⁻6   e. ⁻23   g. ⁺62
   b. ⁺10   d. ⁺9   f. ⁺49   h. ⁻115

2. Write symbolically:

   a. Positive six
   b. Negative four
   c. Positive two
   d. Negative one
   e. Positive twenty
   f. Negative fifteen
   g. Negative seventeen
   h. Positive sixty-one

3. What number is the opposite of each of the following?

   a. ⁺9   c. 0   e. ⁻16   g. ⁻82
   b. ⁻3   d. ⁺10   f. ⁺35   h. ⁻101

4. a. Does ⁻5 name a whole number? An integer?
   b. Does 14 name a whole number? An integer?

3–2 RATIONAL NUMBERS; REAL NUMBERS

Numerals naming the fractional numbers of arithmetic may be arranged in a definite order on the number line to the right of the point labeled 0. There is a point on the number line corresponding to each fractional number.

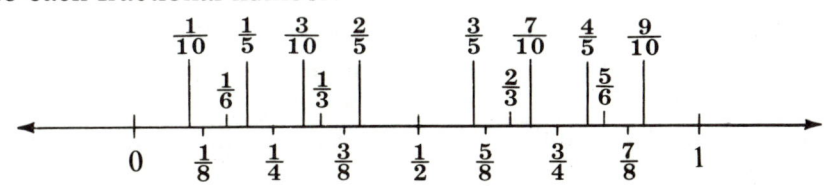

Examination of the number line below will show that there are fractional subdivisions to the left of zero matching those to the right of zero with each point of division corresponding to a fractional number. Since there are an infinite number of positive and negative fractional numbers, only a few are indicated. However, there is a point on the number line corresponding to each positive and negative fractional number.

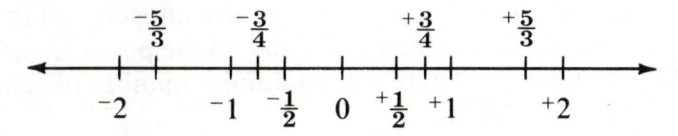

Positive and Negative Numbers

163

> The numeral $^+\frac{3}{4}$ or $^+\left(\frac{3}{4}\right)$ or just $\frac{3}{4}$
> names the number *positive three-fourths*.
>
> The numeral $^-\frac{1}{2}$ or $^-\left(\frac{1}{2}\right)$
> names the number *negative one-half*.

Observe in numerals like $^-\frac{1}{2}$ or $^+\frac{3}{4}$ where parentheses are not used the $^-$ or $^+$ symbol belongs to the entire fraction although it may appear that the symbol belongs to the numerator alone.

Each positive fractional number has an opposite negative fractional number and each negative fractional number has an opposite positive fractional number.

> $^-\frac{4}{5}$ is the opposite of $^+\frac{4}{5}$.
>
> $^+1\frac{5}{8}$ is the opposite of $^-1\frac{5}{8}$.

All the fractional numbers and their opposites are called the *rational numbers*. This includes the integers since each integer may be named in fraction form. For example $^-5 = \frac{^-5}{1}$, $^+3 = \frac{^+3}{1}$, and $0 = \frac{0}{1} = \frac{0}{2} = \frac{0}{3}$, etc. A rational number may be described as a number named by a numeral that expresses a quotient of two integers with division by zero excluded.

An *irrational number* is a number that is both a non-terminating and non-repeating decimal. An irrational number cannot be named by a quotient of two integers with division by zero excluded. All the positive and negative irrational numbers are called the *irrational numbers*.

All the rational numbers and all the irrational numbers are called the *real numbers*. They include all the integers, all the positive and negative fractional numbers, and all the positive and negative irrational numbers. There are an infinite number of real numbers in our number system.

EXERCISES

1. Read, or write in words, each of the following:
   a. $^-\frac{3}{5}$ c. $^-3\frac{1}{2}$ e. $^-6.2$ g. $^-\frac{19}{5}$
   b. $^+\frac{7}{6}$ d. $^+.75$ f. $^+\frac{1}{15}$ h. $^-.003$

2. Write symbolically:
   a. Negative three-tenths
   b. Positive one-fourth
   c. Negative six-hundredths
   d. Positive one and five-tenths
   e. Negative five and sixty-one hundredths
   f. Positive one hundred twenty-two thousandths

3. What number is the opposite of each of the following?
   a. $^+\frac{11}{12}$ c. $^+.01$ e. $^-4\frac{2}{5}$ g. $^+6\frac{3}{8}$
   b. $^-\frac{5}{3}$ d. $^-2.17$ f. $^+.004$ h. $^-51.6$

4. a. Is $^-\frac{7}{10}$ a whole number? An integer? A rational number? A real number?
   b. Is $^+5\frac{1}{6}$ a whole number? An integer? A rational number? A real number?
   c. Is $^+.35$ a whole number? An integer? A rational number? A real number?
   d. Is $^-6.7$ a whole number? An integer? A rational number? A real number?
   e. Is $^-3$ a whole number? An integer? A rational number? A real number?
   f. Is $^-\sqrt{21}$ a whole number? An integer? A rational number? A real number?

5. Which of the following numerals name integers? Which name rational numbers? Which name irrational numbers? Which name real numbers?
   a. $^-12$; $^+.05$; $^-\frac{5}{8}$; $^-\sqrt{31}$; $\frac{8}{2}$; $^-6\frac{2}{3}$
   b. $^+\frac{7}{12}$; $^-100$; $^-1\frac{3}{4}$; $^-3.02$; $^+\sqrt{19}$; $^-\frac{24}{8}$
   c. $^-.115$; $^-\frac{1}{16}$; $^+\sqrt{36}$; $^-\frac{49}{7}$; $^-62$; $^+7\frac{2}{3}$
   d. $^+19$; $^-\sqrt{5}$; $^-\frac{13}{15}$; $^-.59$; $^-2\frac{3}{8}$; $^+\frac{22}{7}$

Positive and Negative Numbers

3-3 ABSOLUTE VALUE

The *absolute value* of any number is the value of the corresponding arithmetic number which has no sign. The absolute value of $^-3$ is 3 and the absolute value of $^+3$ is 3. The absolute value of 0 is 0.

A pair of vertical bars | | is the symbol used to designate absolute value.

$|^-5|$ is read "the absolute value of negative 5."

EXERCISES

1. Read, or write in words, each of the following:
   a. $|15|$
   b. $|^-3|$
   c. $|^+22|$
   d. $|^-\frac{7}{16}|$
   e. $|^-.06|$
   f. $|^+3\frac{2}{6}|$

2. Write symbolically, using the absolute value symbol:
   a. The absolute value of negative nine
   b. The absolute value of positive four-fifths
   c. The absolute value of negative seven and four-ninths
   d. The absolute value of positive five-thousandths

3. Find the value of each of the following:
   a. $|^-8|$
   b. $|^+4|$
   c. $|^-.7|$
   d. $|^+5|$
   e. $|0|$
   f. $|^-\frac{3}{8}|$
   g. $|^-1\frac{2}{3}|$
   h. $|^+19|$
   i. $|^-27|$
   j. $|^-.41|$
   k. $|^+\frac{16}{5}|$
   l. $|^-200|$

In each of the following first find the absolute value of each number as required, then apply the necessary operation to obtain the answer:

4. a. $|^-5| + |^-6|$
   b. $|^+\frac{1}{3}| + |^-\frac{1}{3}|$
   c. $|^-1\frac{1}{2}| + |^+2\frac{3}{4}|$
   d. $|^-.62| + |^-1.4|$

166 CHAPTER 3

5. **a.** $|^-16| - |^+5|$
   b. $|^{+}\tfrac{5}{6}| - |^{-}\tfrac{2}{3}|$
   c. $|^-4\tfrac{1}{4}| - |^-3\tfrac{1}{8}|$
   d. $|^+1.7| - |^-.4|$

6. **a.** $|^+9| \times |^-7|$
   b. $|^{+}\tfrac{2}{3}| \times |^{-}\tfrac{7}{10}|$
   c. $|^+2\tfrac{3}{7}| \times |^-10|$
   d. $|^-3.14| \times |^+.02|$

7. **a.** $|^-51| \div |^+17|$
   b. $|^{+}\tfrac{3}{4}| \div |^{-}\tfrac{2}{3}|$
   c. $|^-1\tfrac{3}{4}| \div |^-5\tfrac{1}{4}|$
   d. $|^+6.5| \div |^-.01|$

3–4 THE REAL NUMBER LINE; GRAPH OF A GROUP OF NUMBERS

The *real number line* is the complete collection of points which corresponds to all the real numbers. The real number line is endless in both directions and only a part of it is shown at any one time. There are an infinite number of points on the real number line. However, there is one and only one point that corresponds to each real number and one and only one real number that corresponds to each point on the real number line.

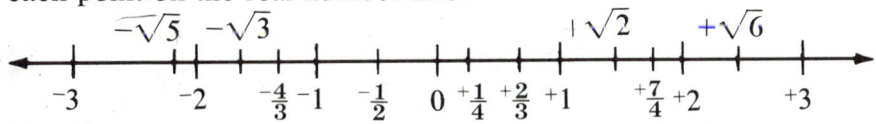

Usually the real number line is labeled only with the numerals naming the integers.

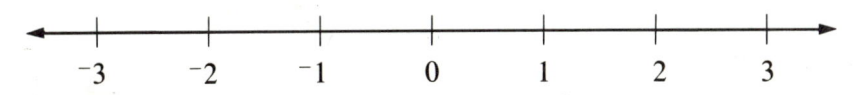

Each point on the number line is called the *graph* of the number to which it corresponds and each number is called the *coordinate* of the related point on the line. Capital letters are generally used to identify particular points.

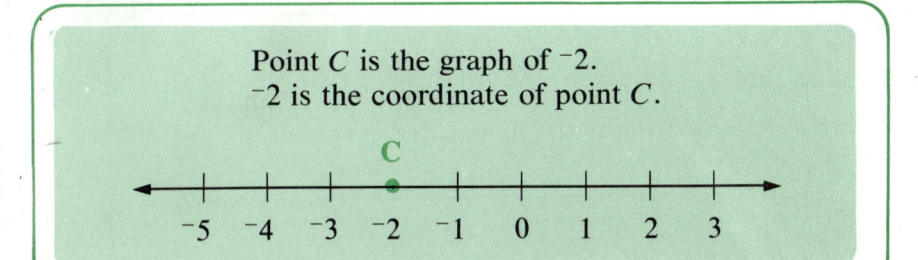

> Point C is the graph of $^-2$.
> $^-2$ is the coordinate of point C.

Positive and Negative Numbers

Thus, the *graph of a number* is a point on the number line whose coordinate is the number.

The *graph of a group of numbers* is made up of the points on the number line whose coordinates are the numbers.

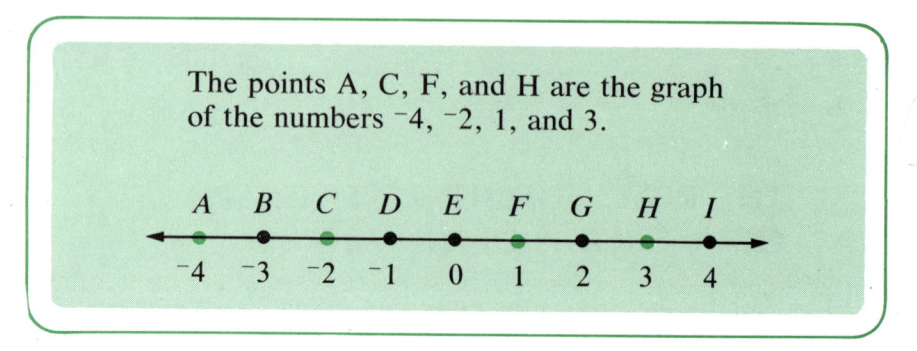

> The points A, C, F, and H are the graph of the numbers ⁻4, ⁻2, 1, and 3.
>
> ```
>  A    B    C    D    E    F    G    H    I
> ←●────●────●────●────●────●────●────●────●→
> ⁻4   ⁻3   ⁻2   ⁻1    0    1    2    3    4
> ```

To draw the graph of a group of numbers, we first draw an appropriate number line and locate the point or points whose coordinate or coordinates are listed in the given numbers. We use heavy solid or colored dots to indicate these points.

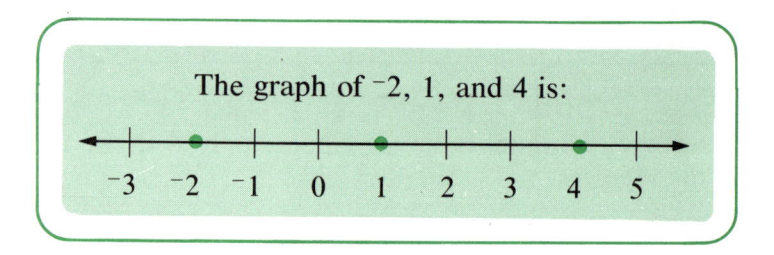

> The graph of ⁻2, 1, and 4 is:
>
> ```
> ←──┼────●────┼────┼────●────┼────┼────●────┼──→
>   ⁻3   ⁻2   ⁻1    0    1    2    3    4    5
> ```

Sometimes a restricted number line is required. To construct this kind of number line, we draw a straight line and, using a convenient unit of measure, we locate the required points of division. These points of division are then properly labeled.

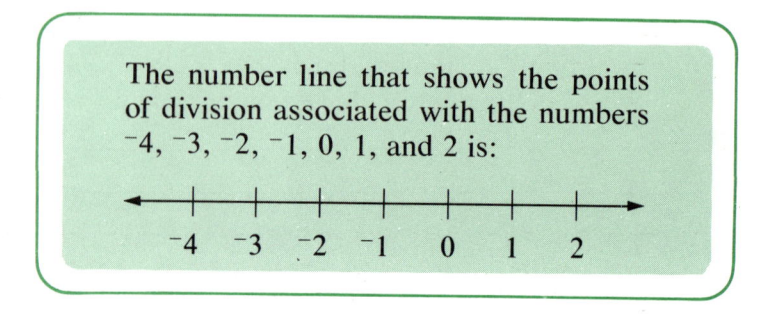

> The number line that shows the points of division associated with the numbers ⁻4, ⁻3, ⁻2, ⁻1, 0, 1, and 2 is:
>
> ```
> ←──┼────┼────┼────┼────┼────┼────┼──→
>   ⁻4   ⁻3   ⁻2   ⁻1    0    1    2
> ```

EXERCISES

1.
```
    L M N P Q R S T U V A B C D E F G H I J K
<---•-•-•-•-•-•-•-•-•-•-•-•-•-•-•-•-•-•-•-•-•--->
   -10 -9 -8 -7 -6 -5 -4 -3 -2 -1 0 +1 +2 +3 +4 +5 +6 +7 +8 +9 +10
```

   a. What number corresponds to point *I*? *M*? *F*? *U*? *Q*? *G*? *R*? *D*? *B*?
   b. What letter labels the point corresponding to each of the following numbers?

   +4   -4   +2   -1   -7   0   +9   -3   -8   +7

2. Draw the graph of each of the following numbers on a number line:

   a. -4
   b. -6 and +2
   c. -7, -3, 0, and +1
   d. -2, -1, 2, and 3
   e. -4, -3, 0, -1, 4, and 6
   f. -3, +5, and -2
   g. +2, -4, +3, -7, +8, -3, and -2
   h. -8, 0, 2, -5, +3, 7, -3, and 6
   i. -5, -4, -3, . . . , 2
   j. -9, -7, -5, . . . , 3

3. Draw the graph of each of the following on a number line:

   a. all integers less than +3 and greater than -3
   b. all integers greater than -2 and less than 0
   c. all integers less than +1 and greater than -1

4. Write the coordinates of which each of the following is the graph:

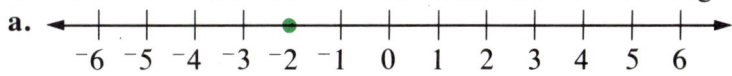

   a.
   b.
   c.
   d.
   e.
   f.
   g.

Positive and Negative Numbers   169

3-5 COMPARING INTEGERS

The number line may be drawn horizontally or vertically.

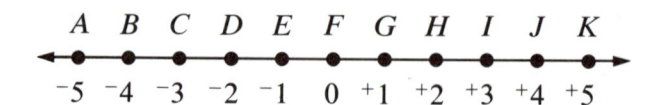

```
    A   B   C   D   E   F   G   H   I   J   K
←———•———•———•———•———•———•———•———•———•———•———•———→
   ⁻5  ⁻4  ⁻3  ⁻2  ⁻1   0  ⁺1  ⁺2  ⁺3  ⁺4  ⁺5
```

L ● +5
M ● +4
N ● +3
P ● +2
Q ● +1
R ● 0
S ● ⁻1
T ● ⁻2
U ● ⁻3
V ● ⁻4
W ● ⁻5

On the horizontal number line, of two numbers the number corresponding to the point on the line *farther to the right* is the greater number.

On the vertical number line, the numbers *above* zero are positive and those *below* zero are negative. Any number corresponding to a point is greater than any number corresponding to a point located below it.

EXERCISES

1. On the horizontal number line, which point corresponds to the greater number:
   a. Point K or point C?
   b. Point B or point G?
   c. Point F or point D?
   d. Point I or point A?

2. On the vertical number line, which point corresponds to the greater number:
   a. Point M or point S?
   b. Point V or point Q?
   c. Point S or point T?
   d. Point W or point R?

3. On the horizontal number line is the point corresponding to ⁻1 to the left or to the right of the point corresponding to ⁻4? Which number is greater, ⁻1 or ⁻4?

4. On the vertical number line is the point corresponding to ⁻5 above or below the point corresponding to ⁻2? Which number is greater, ⁻2 or ⁻5?

5. Which of the following sentences are true?
   a. ⁺5 > ⁺7
   b. ⁻4 > ⁻6
   c. ⁺3 < 0
   d. ⁺2 < ⁻5
   e. 0 > ⁻3
   f. ⁻3 > ⁻3
   g. ⁻2 < ⁺2
   h. ⁻3 ≮ ⁻5
   i. ⁻4 < ⁻4
   j. ⁺6 < ⁻8
   k. ⁻3 ≯ ⁻1
   l. ⁻2 ≮ 0
   m. ⁺5 > ⁻70
   n. ⁻10 ≮ ⁻5
   o. ⁺25 ≯ ⁻40
   p. 0 < ⁻1

6. Rewrite each of the following and insert between the two numerals the symbol (=, <, or >) which will make the sentence true:

   a. ⁻5 ? ⁻2   d. ⁻7 ? ⁻7   g. ⁻3 ? ⁻4
   b. ⁺8 ? ⁺6   e. ⁺15 ? ⁻21   h. 0 ? ⁻7
   c. ⁻7 ? ⁺4   f. ⁻8 ? 0   i. ⁻15 ? ⁺1

7. Name the following numbers in order of size (smallest first):

   a. ⁻3, ⁺7, ⁻11, ⁺15, ⁻8, ⁺4, 0, ⁺11, ⁻2, ⁻1
   b. ⁺6, ⁻5, ⁻7, 0, ⁺4, ⁻11, ⁺1, ⁻3, ⁻6, ⁺2

8. Name the following numbers in order of size (greatest first):

   a. ⁻1, ⁺8, ⁻9, ⁺3, 0, ⁻5, ⁺5, ⁻10, ⁺15, ⁻6
   b. ⁺5, ⁻6, ⁺1, 0, ⁺11, ⁻4, ⁺2, ⁺7, ⁻1, ⁻12

9. Which is greater:

   a. The absolute value of ⁻5 or the absolute value of ⁺2?
   b. The absolute value of ⁺9 or the absolute value of ⁻9?

10. Which has the greater opposite number?

    a. ⁺8 or ⁺4   b. ⁻3 or ⁻5   c. ⁻5 or ⁺7   d. ⁺2 or ⁻4

3–6 OPPOSITE MEANINGS

Positive and negative numbers are used in science, statistics, weather reports, stock reports, sports, and many other fields to express opposite meanings or directions. For example:

If ⁺3% indicates an *increase* of 3% in the cost of living, ⁻4% would indicate a 4% *decrease* in the cost of living.

EXERCISES

1. If ⁺21° means 21 degrees above zero, what does ⁻15° mean?
2. If ⁻47 kilometers indicates 47 kilometers south, what does ⁺16 kilometers indicate?
3. If $35 earned is represented by ⁺$35, how can $28 spent be represented?

Positive and Negative Numbers

4. If the deficiency of 29.3 mm of precipitation is indicated by ⁻29.3 mm, how can the excess of 18.4 mm of precipitation be indicated?

5. If ⁻$500 means the loss of $500 in a transaction, what does ⁺$425 mean?

6. If ⁺1$\frac{1}{4}$ points represents a 1$\frac{1}{4}$ point gain in a stock, what does ⁻2 points represent?

7. If ⁻12 amperes represents a discharge of 12 amperes of electricity, how can a charge of 6 amperes of electricity be represented?

8. If ⁺55° indicated 55° east longitude, how can 42° west longitude be indicated?

9. If ⁻11 weeks represents 11 weeks ago, what does ⁺8 weeks represent?

10. If ⁺1,000 km indicates 1,000 kilometers above sea level, how can 325 kilometers below sea level be indicated?

3–7 OPPOSITE DIRECTIONS; VECTORS

Positive and negative numbers, used as directed numbers, indicate movements in opposite directions. A movement to the right of a particular point is generally considered as moving in a positive direction from the point and a movement to the left of a point as moving in a negative direction from the point.

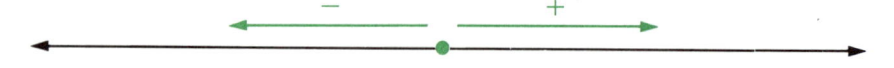

The sign in the numeral naming a directed number indicates the direction, and the absolute value of the directed number represents the magnitude (distance in units) of the movement.

> ⁺5 means moving 5 units to the right.
>
> ⁻2 means moving 2 units to the left.

An arrow that represents a directed line segment is called a *vector*. It is used to picture the size and the direction of the movement that is indicated by a signed number. The absolute value of the signed number indicates the length of the arrow and the sign indicates the direction.

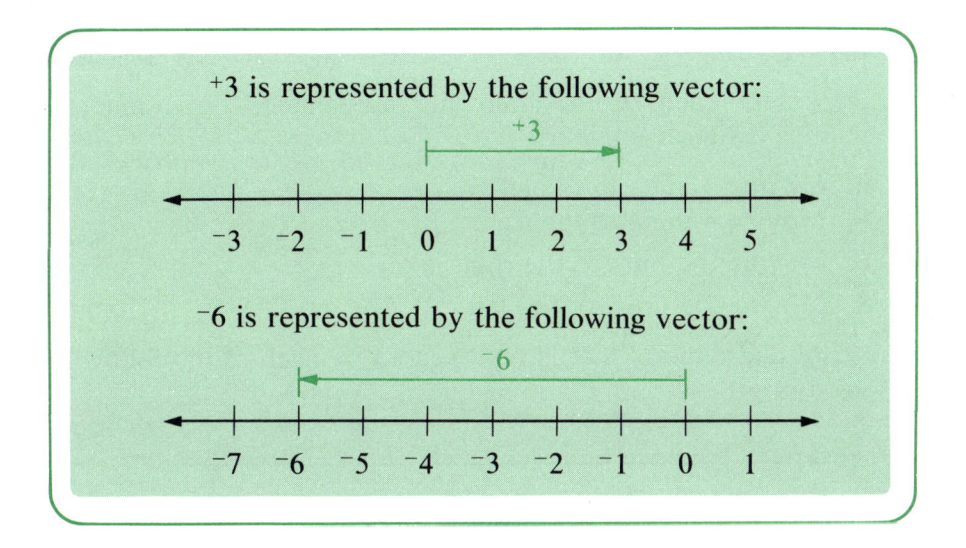

⁺3 is represented by the following vector:

```
          +3
    |—————————▶|
◀—+———+———+———+———+———+———+———+———+—▶
 ⁻3  ⁻2  ⁻1   0   1   2   3   4   5
```

⁻6 is represented by the following vector:

```
              ⁻6
    |◀—————————————————|
◀—+———+———+———+———+———+———+———+———+—▶
 ⁻7  ⁻6  ⁻5  ⁻4  ⁻3  ⁻2  ⁻1   0   1
```

To determine the numeral that describes the movement from one point to another point on the number line, we draw the vector between the points with the arrowhead pointing in the direction of the movement. We use the number of units of length in the vector as the absolute value of the numeral. We select the sign according to the direction of the vector. If it points to the right, the sign is positive (+); if it points to the left, the sign is negative (−).

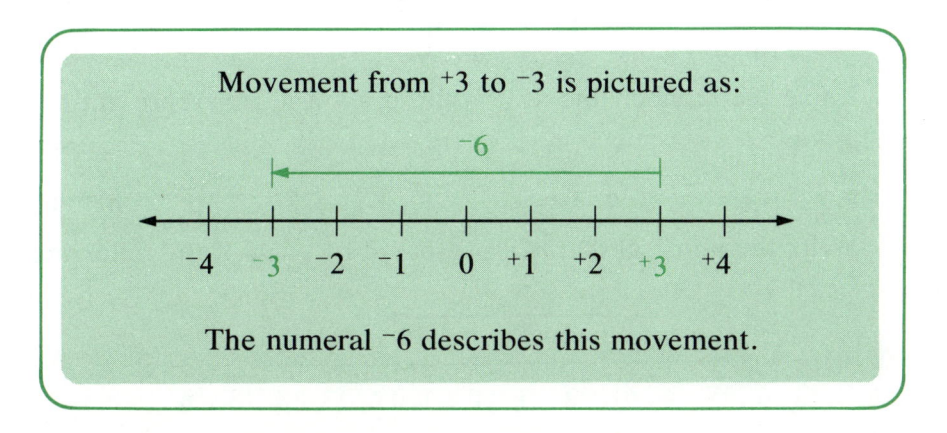

Movement from ⁺3 to ⁻3 is pictured as:

```
              ⁻6
    |◀————————————————|
◀—+———+———+———+———+———+———+———+———+—▶
 ⁻4  ⁻3  ⁻2  ⁻1   0  ⁺1  ⁺2  ⁺3  ⁺4
```

The numeral ⁻6 describes this movement.

Positive and Negative Numbers

EXERCISES

1. A movement of how many units of distance and in what direction is represented by each of the following numbers?

   a. ⁻8
   b. ⁺12
   c. ⁻3
   d. ⁺4
   e. 15
   f. ⁻3$\frac{1}{2}$
   g. ⁻2.25
   h. ⁺$\frac{3}{8}$

2. Represent each of the following movements by a numeral naming a signed number:

   a. Moving 3 units to the left
   b. Moving 4 units to the right
   c. Moving 9$\frac{1}{2}$ units to the right
   d. Moving 7.5 units to the left

3. Write the numeral that is represented by each of the following vectors:

   a. 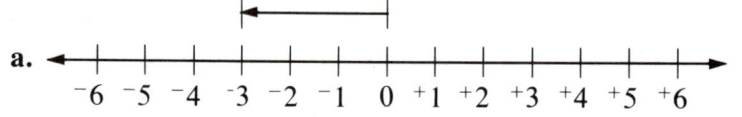
   b. 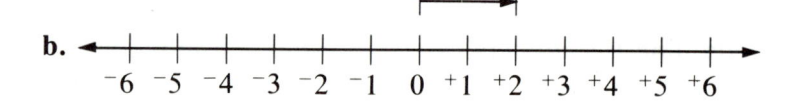
   c. 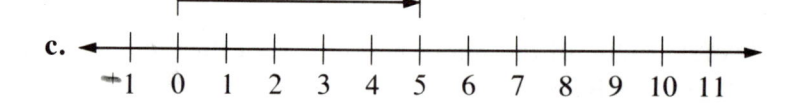
   d.

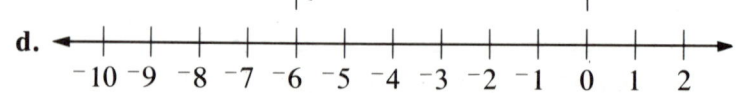

4. Using the number line as a scale, draw a vector representing:

   a. ⁻2
   b. ⁺7
   c. ⁻8
   d. 6
   e. ⁺2$\frac{1}{3}$
   f. ⁻3.5
   g. ⁻11
   h. ⁺4$\frac{3}{4}$

5. Write the numeral that is represented by each of the following vectors:

   a.

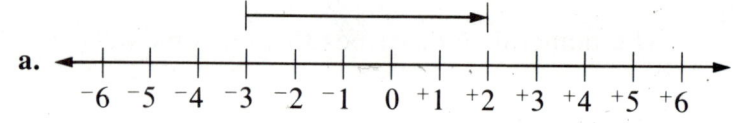

b.

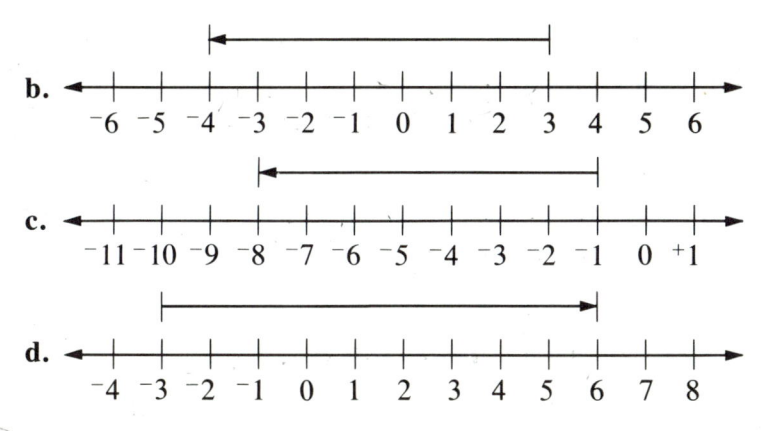

c.

d.

6. Use the number line as a scale to draw a vector that illustrates each of the following movements. Write the numeral that is represented by each of these vectors.

a. From 0 to ⁺6
b. From 0 to ⁻8
c. From ⁻6 to 0
d. From ⁺3 to 0
e. From ⁺4 to ⁺8
f. From ⁺10 to ⁺1
g. From ⁻1 to ⁻6
h. From ⁻4 to ⁻2
i. From ⁺2 to ⁻2
j. From ⁻7 to ⁺10
k. From ⁺11 to ⁻10
l. From ⁺3 to ⁺13
m. From ⁻3 to ⁻6
n. From ⁻9 to ⁻5
o. From ⁺1 to ⁻4
p. From ⁻11 to ⁺11
q. From ⁺8 to ⁻1
r. From ⁺9 to ⁺1

3–8 ADDITIVE INVERSE OF A NUMBER; MORE ABOUT OPPOSITES

Sometimes the *opposite of a number* is called the *additive inverse of the number*.

When the sum of two numbers is zero, each addend is said to be the *additive inverse* of the other addend. We shall see in sections 3–9 and 3–10 that the sum of a number and its opposite is zero.

⁻6 is the opposite of or the additive inverse of ⁺6 (or 6).
⁺4 (or 4) is the opposite of or the additive inverse of ⁻4.

Positive and Negative Numbers

A centered dash symbol is used in algebra to indicate the opposite of a number.

> The expression $-(7)$ and sometimes -7 are read as:
>   "The opposite of seven."
>
> The sentence $-(^-7) = {}^+7$ is read as:
>   "The opposite of negative seven is positive seven."
>
> The sentence $-7 = {}^-7$ is read as:
>   "The opposite of seven is negative seven."

EXERCISES

1. What number is the additive inverse of each of the following numbers;
   a. $^-2$ b. $^+48$ c. 0 d. $^-\frac{5}{24}$ e. 2.08 f. $^+\frac{3}{4}$ g. $^-7\frac{3}{5}$ h. $^-\frac{7}{3}$

2. What number is the opposite of each of the following numbers?
   a. $^+10$ b. $^-6$ c. $^-\frac{4}{5}$ d. $^-.26$ e. $^+\frac{14}{9}$ f. $^-3\frac{3}{4}$ g. 0 h. $^+15$

3. Read, or write in words, each of the following:
   a. $^-8$
   b. $-(^+6)$
   c. $-\left(^-\frac{3}{4}\right)$
   d. $-(^-3)$
   e. $-(^-2) = {}^+2$
   f. $-(^+10) = {}^-10$
   g. $-15 = {}^-15$
   h. $-(^-7.4) = 7.4$

4. Write symbolically, using the "opposite symbol":
   a. The opposite of thirty-two
   b. The opposite of negative eight
   c. The opposite of positive two
   d. The opposite of positive thirteen
   e. The opposite of negative ten
   f. The opposite of eleven is negative eleven
   g. The opposite of negative three is positive three
   h. The opposite of negative seventy-one is positive seventy-one
   i. The opposite of positive twenty is negative twenty
   j. The opposite of the opposite of sixty-three is sixty-three

5. a. Find the opposite of the opposite of: $^+2$; $^-11$; $^+\frac{7}{12}$; $^-.8$; $^-1\frac{2}{3}$
   b. What is always true about the opposite of the opposite of a number?

COMPUTATION WITH POSITIVE AND NEGATIVE NUMBERS

3–9 ADDITION ON THE NUMBER LINE

If we move 6 units to the right from the point marked 0, we reach the point whose coordinate is $^+6$. If we then move 2 units to the right of this point, we reach the point whose coordinate is $^+8$. The sum of these two movements is the same as a single movement of 8 units to the right.

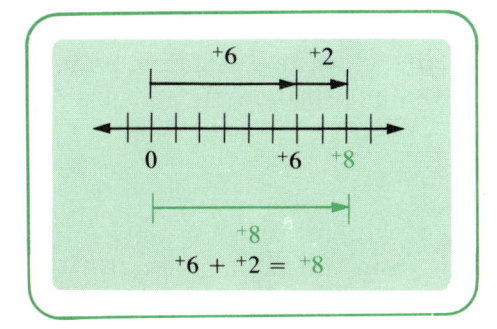

Observe in the illustration that a vector is drawn to represent the sum and each addend. To avoid confusion, we draw the vectors representing the addends above the number line and the vector representing the sum below the number line. The vector representing the first addend begins at the point 0 and the second vector representing the second addend is drawn from the point reached by the first vector. The resultant vector that represents the sum is drawn from the point 0 to the point reached by the vector representing the second addend.

If we move 6 units to the left from the point marked 0, we reach the point whose coordinate is $^-6$. If we then move 2 units to the left of this point, we reach the point whose coordinate is $^-8$. The sum of these two movements is the same as a single movement of 8 units to the left. Observe how vectors are used to illustrate this.

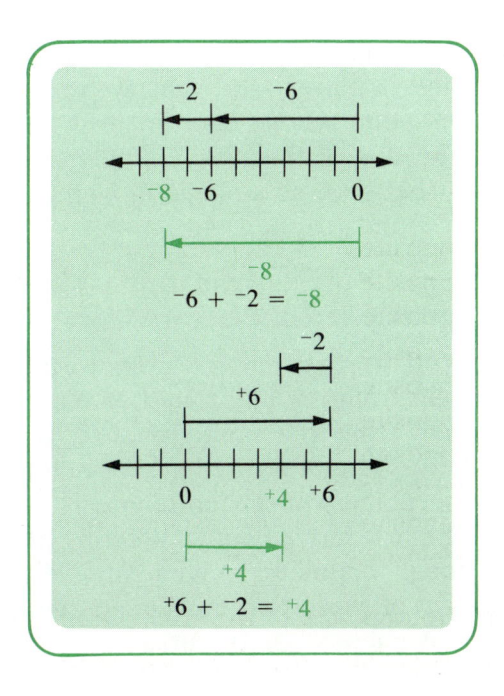

If we first move 6 units to the right from the point marked 0 and then from the point reached we move 2 units to the left, we reach the point whose coordinate is $^+4$. The sum of these two movements is the same as a single movement of 4 units to the right.

Positive and Negative Numbers

If we first move 6 units to the right from the point marked 0 and then from the point reached we move 8 units to the left, we reach the point whose coordinate is ⁻2. The sum of these two movements is the same as a single movement of 2 units to the left.

If we first move 6 units to the left from the point marked 0 and then from the point reached we move 2 units to the right, we reach the point whose coordinate is ⁻4. The sum of these two movements is the same as a single movement of 4 units to the left.

If we first move 6 units to the left from the point marked 0 and then from the point reached we move 8 units to the right, we reach the point whose coordinate is ⁺2. The sum of these two movements is the same as a single movement of 2 units to the right.

If we first move 4 units to the right from the point marked 0 and then from the point reached we move 4 units to the left, we reach the point whose coordinate is 0. The first movement is offset by an equal opposite movement. The sum of these two movements is a movement of zero units.

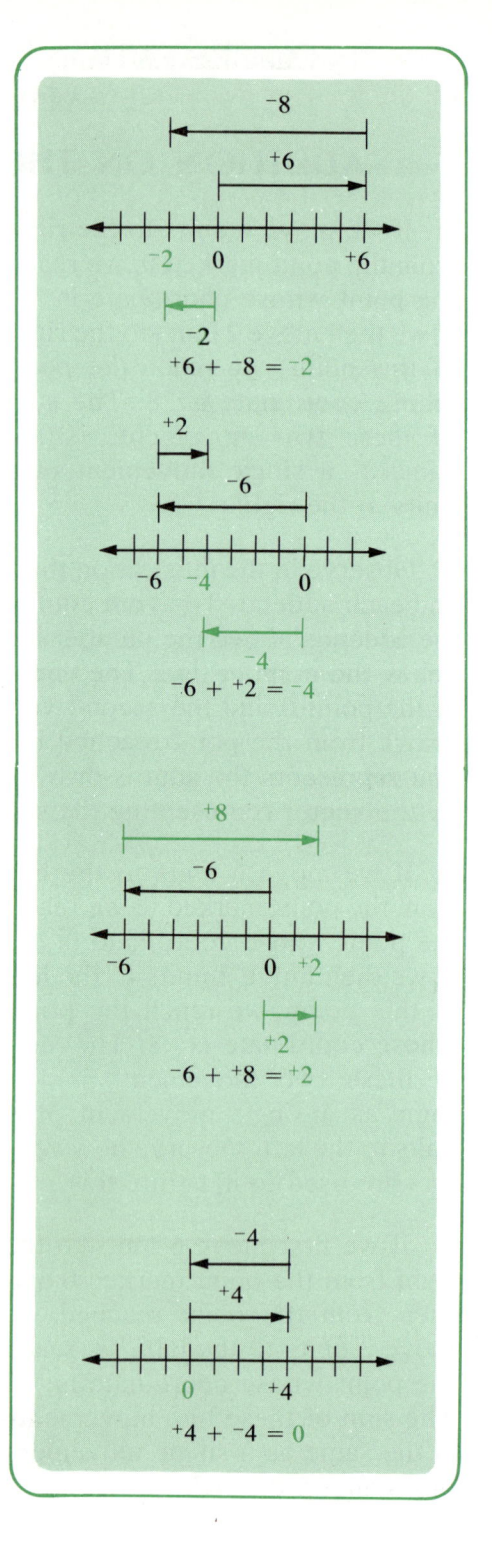

We may add rational numbers on the number line. For example:

If we first move $2\frac{1}{2}$ units to the right from the point marked 0 and then from the point reached we move $5\frac{1}{2}$ units to the left, we reach a point whose coordinate is $^-3$. The sum of these two movements is the same as a single movement of 3 units to the left.

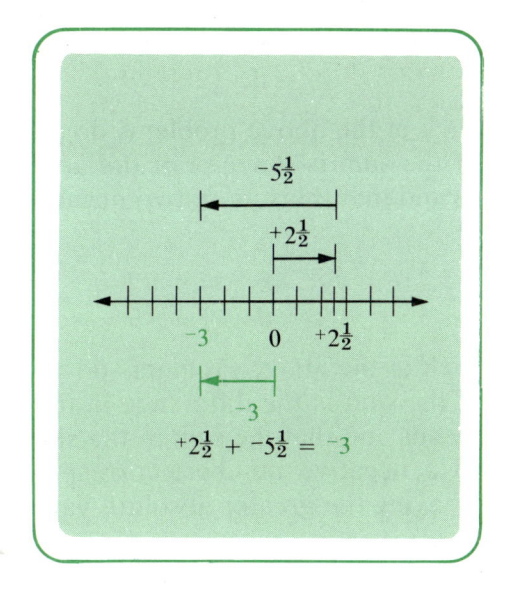

$$^+2\tfrac{1}{2} + {}^-5\tfrac{1}{2} = {}^-3$$

Observe that when vectors are used, the sum of two positive numbers or the sum of two negative numbers is represented by a vector whose length is equal to the sum of the lengths of the vectors representing the two addends. This resultant vector has a positive direction when the two addends are both positive numbers and a negative direction when the two addends are both negative numbers.

However, the sum of a positive number and a negative number or the sum of a negative number and a positive number is represented by a vector whose length is equal to the difference in the lengths of the vectors representing the two addends and its direction is determined by the direction of the longer of the two vectors representing the two addends.

EXERCISES

In each of the following add as indicated, using the number line to find the sum:

1. a. $^+1 + {}^+4$ c. $^+3 + {}^+3$ e. $^+8 + {}^+2$ g. $^+2 + {}^+5$
   b. $^+5 + {}^+2$ d. $^+6 + {}^+2$ f. $^+4 + {}^+4$ h. $^+9 + {}^+4$

In each of the above problems do you see that the absolute value of the sum is the sum of the absolute values of the two addends and that the sum of two positive integers is a positive integer?

Positive and Negative Numbers

2. a. $^-4 + {}^-2$ c. $^-1 + {}^-8$ e. $^-6 + {}^-4$ g. $^-2 + {}^-3$
   b. $^-3 + {}^-3$ d. $^-2 + {}^-9$ f. $^-4 + {}^-1$ h. $^-6 + {}^-6$

In each of the above problems do you see that the absolute value of the sum is the sum of the absolute values of the two addends and that the sum of two negative integers is a negative integer?

3. a. $^+5 + {}^-2$ c. $^+4 + {}^-7$ e. $^+6 + {}^-2$ g. $^+10 + {}^-6$
   b. $^+7 + {}^-3$ d. $^+1 + {}^-9$ f. $^+1 + {}^-6$ h. $^+2 + {}^-11$

In each of the above problems do you see that the absolute value of the sum is the difference in the absolute values of the two addends and that the sign in the sum, when we add a positive and a negative number, corresponds to the sign of the addend having the greater absolute value?

4. a. $^-8 + {}^+2$ c. $^-1 + {}^+4$ e. $^-9 + {}^+12$ g. $^-10 + {}^+8$
   b. $^-4 + {}^+8$ d. $^-5 + {}^+8$ f. $^-3 + {}^+1$ h. $^-7 + {}^+12$

In each of the above problems do you see that the absolute value of the sum is the difference in the absolute values of the two addends and that the sign in the sum, when we add a negative number and a positive number, corresponds to the sign of the addend having the greater absolute value?

5. a. $^+4 + {}^-4$ c. $^+10 + {}^-10$ e. $^-3 + {}^+3$ g. $^+8 + {}^-8$
   b. $^+5 + {}^-5$ d. $^+2 + {}^-2$ f. $^+7 + {}^-7$ h. $^+9 + {}^-9$

In each of the above problems do you see that the sum of an integer and its opposite is zero?

6. a. $0 + {}^+9$ c. $0 + {}^-4$ e. $^-10 + 0$ g. $0 + {}^-7$
   b. $^+3 + 0$ d. $^+2 + 0$ f. $^+8 + 0$ h. $0 + {}^-4$

In each of the above problems do you see that the sum of any non-zero integer and zero is the non-zero integer?

7. a. $^+2\frac{1}{2} + {}^+\frac{1}{2}$ c. $^+4\frac{1}{2} + {}^-5$ e. $^-6\frac{2}{5} + 0$ g. $^-2\frac{1}{4} + {}^{-\frac{5}{8}}$
   b. $^-3\frac{1}{7} + {}^-3\frac{1}{7}$ d. $^-4\frac{2}{3} + {}^-5\frac{1}{3}$ f. $^-2\frac{1}{2} + {}^-2\frac{1}{2}$ h. $^-4 + {}^+6\frac{1}{3}$

In the above problems do you see that the same procedures are used to add rational numbers that are used to add integers?

8. a. $^+2 + {}^+3$ e. $^+7 + {}^+10$ i. $^-12 + {}^+9$ m. $^+2 + {}^-2$
   b. $^+8 + {}^-3$ f. $0 + {}^-8$ j. $^+1 + {}^+8$ n. $^-5 + {}^-5$
   c. $^-3 + {}^+3$ g. $^-6 + {}^-4$ k. $^+1\frac{1}{6} + {}^{-\frac{2}{3}}$ o. $^-2 + {}^+10$
   d. $^-1 + {}^-5$ h. $^-4 + {}^+2$ l. $^-2\frac{1}{4} + {}^-3\frac{1}{2}$ p. $^+6 + {}^-13$

3-10 ADDITION OF INTEGERS; PROPERTIES

Our observations in adding on the number line lead us to conclude that:

To add two positive integers, we find the sum of their absolute values and prefix the numeral with a positive sign. Observe that addition may be indicated horizontally or vertically.

In $^+6 + {}^+2 = {}^+8$, the absolute value of the sum (8) is the sum of the absolute values of the addends (6 + 2) and the sign is the common sign, $^+$ in this case.

To add two negative integers, we find the sum of their absolute values and prefix the numeral with a negative sign.

In $^-6 + {}^-2 = {}^-8$, the absolute value of the sum (8) is the sum of the absolute values of the addends (6 + 2) and the sign is the common sign, $^-$ in this case.

To add a positive integer and a negative integer or *a negative integer and a positive integer,* we subtract the smaller absolute value from the greater absolute value of the two addends and prefix the difference with the sign of the number having the greater absolute value.

In $^+6 + {}^-2 = {}^+4$ the absolute value of the sum (4) is equal to the difference in the absolute values of the addends (6 − 2) and the sign of the sum ($^+$) is the same as the sign of the addend with the greater absolute value (the sign of $^+6$).

Horizontally:

$^+6 + {}^+2 = {}^+(6 + 2) = {}^+8$

Answer, $^+8$

Vertically:

$^+6$
$^+2$
―――
$^+8$

Horizontally:

$^-6 + {}^-2 = {}^-(6 + 2) = {}^-8$

Answer, $^-8$

Vertically:

$^-6$
$^-2$
―――
$^-8$

Horizontally:

$^+6 + {}^-2 = {}^+(6 − 2) = {}^+4$

Answer, $^+4$

Vertically:

$^+6$
$^-2$
―――
$^+4$

Positive and Negative Numbers

In $^+6 + {}^-8 = {}^-2$, the absolute value of the sum (2) is equal to the difference in the absolute values of the addends (8 − 6) and the sign of the sum (⁻) is the same as the sign of the addend with the greater absolute value (the sign of $^-8$).

In $^-6 + {}^+2 = {}^-4$, the absolute value of the sum (4) is equal to the difference in the absolute values of the addends (6 − 2) and the sign of the sum (⁻) is the same as the sign of the addend with the greater absolute value (the sign of $^-6$).

In $^-6 + {}^+8 = {}^+2$, the absolute value of the sum (2) is equal to the difference in the absolute values of the addends (8 − 6) and the sign of the sum (⁺) is the same as the sign of the addend with the greater absolute value (the sign of $^+8$).

The sum of any integer and its opposite is zero.

The sum of any non-zero integer and zero is the non-zero integer.

The numeral naming a positive integer does not require the positive (⁺) sign.

To add three or more integers, we either add successively as shown in the model or we add first the positive numbers, then the negative numbers, and finally the two answers.

Horizontally:
$^+6 + {}^-8 = {}^-(8 − 6) = {}^-2$

Answer, $^-2$

Vertically:
$^+6$
$^-8$
―――
$^-2$

Horizontally:
$^-6 + {}^+2 = {}^-(6 − 2) = {}^-4$

Answer, $^-4$

Vertically:
$^-6$
$^+2$
―――
$^-4$

Horizontally:
$^-6 + {}^+8 = {}^+(8 − 6) = {}^+2$

Answer, $^+2$

Vertically:
$^-6$
$^+8$
―――
$^+2$

$^+4 + {}^-4 = 0$ *Answer, 0*
$^-7 + {}^+7 = 0$ *Answer, 0*

$^-9 + 0 = {}^-9$ *Answer, $^-9$*
$0 + {}^+2 = {}^+2$ *Answer, $^+2$*

$3 + {}^-11 = {}^-8$ *Answer, $^-8$*
$^-1 + 7 = 6$ *Answer, 6*

$^+1 + {}^-4 + {}^+2$
$= {}^-4 + {}^+3$
$= {}^-1$ *Answer, $^-1$*

EXERCISES

Add as indicated:

1. a. $^+4 + {^+6}$ b. $^+34 + {^+6}$ c. $^+17 + {^+41}$ d. $^+117 + {^+88}$
2. a. $^-5 + {^-10}$ b. $^-19 + {^-49}$ c. $^-36 + {^-55}$ d. $^-53 + {^-41}$
3. a. $^+10 + {^-3}$ b. $^+32 + {^-19}$ c. $^+22 + {^-17}$ d. $^+51 + {^-42}$
4. a. $^+2 + {^-8}$ b. $^+5 + {^-15}$ c. $^+56 + {^-72}$ d. $^-69 + {^+84}$
5. a. $^-3 + {^+1}$ b. $^-32 + {^+15}$ c. $^-41 + {^+25}$ d. $^-103 + {^+95}$
6. a. $^-1 + {^+7}$ b. $^-19 + {^+50}$ c. $^-19 + {^+64}$ d. $^-77 + {^+110}$
7. a. $0 + {^+5}$ b. $0 + {^-14}$ c. $0 + {^+31}$ d. $^-38 + 0$
8. a. $^+13 + {^-13}$ b. $^-36 + {^+36}$ c. $^+82 + {^-82}$ d. $^-44 + {^+44}$
9. a. $6 + {^-12}$ b. $^-41 + 17$ c. $27 + {^-35}$ d. $^-42 + 90$

10. a. $^+2 + {^+1} + {^+5}$ d. $^+21 + {^-17} + {^-4}$ g. $^-2 + {^+3} + {^-6} + {^+3}$
    b. $^-7 + {^+8} + {^+1}$ e. $^+18 + {^+9} + {^-22}$ h. $^+10 + {^+15} + {^-2} + {^-6}$
    c. $^-12 + {^-5} + {^-27}$ f. $^-20 + {^-45} + {^+54}$ i. $^+16 + {^-22} + {^-34} + {^+48}$

11. a. $^-15 + {^-6}$ e. $^-54 + {^+63}$ i. $^-28 + {^+75} + {^-46}$
    b. $^-9 + {^+14}$ f. $^+23 + {^-23}$ j. $^+17 + {^-6} + {^-35}$
    c. $^+20 + {^-10}$ g. $^-89 + {^-79}$ k. $^+73 + {^-29} + {^-57} + {^+7}$
    d. $^+54 + {^+65}$ h. $^-22 + 0$ l. $^-44 + {^+18} + {^-26} + {^+6}$

12. a. $(^+6 + {^+4}) + {^-2}$ c. $(^-12 + {^+15}) + {^-7}$ e. $^-18 + (^+39 + {^-16})$
    b. $^+7 + (^-8 + {^-6})$ d. $^-11 + (^-6 + {^-24})$ f. $(^+24 + {^-30}) + {^+24}$

13. Add:

a.
$^+4$	$^-6$	$^-18$	$^+43$	$^-36$	$^+32$	$^-19$	$^+12$
$^+9$	$^-16$	$^+9$	$^-29$	$^+36$	0	$^-19$	$^-51$

b.
$^+17$	$^-62$	$^-29$	$^+85$	$^+34$	0	$^-29$	$^-18$
$^-51$	$^+35$	$^-29$	$^+117$	$^-5$	$^-65$	$^+56$	$^-1$

c.
$^-45$	$^-27$	$^+53$	$^-66$	$^+38$	$^+71$	0	$^-202$
$^-29$	$^+0$	$^-61$	$^+52$	$^-38$	$^-19$	$^-76$	$^+83$

d.
			$^+8$	$^-60$	$^-7$	$^+104$	$^-34$
$^+2$	$^-13$	$^-28$	$^-6$	$^-8$	$^+32$	$^-78$	$^-43$
$^-8$	$^+7$	$^+52$	$^+2$	$^+15$	$^-56$	$^-32$	$^-57$
$^+4$	$^-3$	$^-18$	$^-1$	$^+27$	$^+31$	$^+29$	$^-85$

Positive and Negative Numbers

14. Find the missing numbers:

   a. $^+4 + {}^-8 = ?$
   b. $^-12 + {}^+9 = \square$
   c. $^+11 + {}^-21 = n$
   d. $^-6 + {}^-15 = \square$
   e. $^-16 + {}^+16 = n$
   f. $^+33 + {}^+44 = ?$
   g. $^-16 + {}^+9 = n$
   h. $^-28 + {}^-46 = n$
   i. $^+35 + {}^-17 = n$

15. Find the missing numbers:

   a. $? + {}^+7 = {}^+9$
   b. $\square + {}^-3 = {}^-10$
   c. $n + {}^-3 = {}^-3$
   d. $\square + {}^-18 = {}^-4$
   e. $? + {}^+22 = {}^+5$
   f. $n + 31 = {}^+7$
   g. $n + {}^+4 = {}^-17$
   h. $n + {}^-42 = 0$
   i. $n + {}^-57 = {}^-20$

16. Find the missing numbers:

   a. $^+6 + ? = {}^+13$
   b. $^-3 + \square = {}^-8$
   c. $^+7 + n = {}^-2$
   d. $^-12 + \square = {}^-5$
   e. $^+26 + n = 0$
   f. $^-16 + ? = {}^-4$
   g. $^+26 + n = {}^+17$
   h. $^-46 + n = {}^-46$
   i. $^-40 + n = {}^+13$

Properties

In the following exercises we determine the properties that hold for the integers under the operation of addition.

1. **Commutative Property**

   a. What do we mean when we say that the commutative property holds for the addition of a certain group of numbers?

   b. In each of the following find the two sums by adding on the number line using vectors. Compare the two sums by comparing their vectors.

   (1) $^+1 + {}^+4; {}^+4 + {}^+1.$ Does $^+1 + {}^+4 = {}^+4 + {}^+1?$
   (2) $^-5 + {}^+2; {}^+2 + {}^-5.$ Does $^-5 + {}^+2 = {}^+2 + {}^-5?$

   c. In each of the following add as indicated, then compare the sums:

   (1) $^+6 + {}^+5; {}^+5 + {}^+6.$ Does $^+6 + {}^+5 = {}^+5 + {}^+6?$
   (2) $^-4 + {}^-2; {}^-2 + {}^-4.$ Does $^-4 + {}^-2 = {}^-2 + {}^-4?$
   (3) $^+8 + {}^-3; {}^-3 + {}^+8.$ Does $^+8 + {}^-3 = {}^-3 + {}^+8?$
   (4) $^-6 + {}^+4; {}^+4 + {}^-6.$ Does $^-6 + {}^+4 = {}^+4 + {}^-6?$

   d. Can you find two integers for which the commutative property of addition does not hold?

   e. Can we say that the commutative property holds for the addition of integers?

2. Associative Property
   a. What do we mean when we say that the associative property holds for the addition of a certain group of numbers?
   b. In each of the following find the two sums by adding on the number line using vectors. Compare the two sums by comparing their vectors.
      (1) $(^-5 + {}^+6) + {}^+2$; $^-5 + ({}^+6 + {}^+2)$.
         Does $(^-5 + {}^+6) + {}^+2 = {}^-5 + ({}^+6 + {}^+2)$?
      (2) $({}^+7 + {}^-3) + {}^-4$; ${}^+7 + ({}^-3 + {}^-4)$.
         Does $({}^+7 + {}^-3) + {}^-4 = {}^+7 + ({}^-3 + {}^-4)$?
   c. In each of the following add as indicated, then compare the sums:
      (1) $({}^-4 + {}^-7) + {}^+12$; ${}^-4 + ({}^-7 + {}^+12)$.
         Does $({}^-4 + {}^-7) + {}^+12 = {}^-4 + ({}^-7 + {}^+12)$?
      (2) $({}^+3 + {}^-12) + {}^-2$; ${}^+3 + ({}^-12 + {}^-2)$.
         Does $({}^+3 + {}^-12) + {}^-2 = {}^+3 + ({}^-12 + {}^-2)$?
   d. Can you find three integers for which the associative property of addition does not hold?
   e. Can we say that the associative property holds for the addition of integers?

3. Closure
   a. What do we mean when we say that a certain group of numbers is closed under the operation of addition?
   b. Is the sum of each of the following an integer?

      (1) ${}^+5 + {}^-8$ (4) ${}^-6 + {}^+2$ (7) ${}^+7 + 0$
      (2) ${}^-4 + {}^-4$ (5) $0 + {}^-12$ (8) ${}^-14 + {}^+8$
      (3) ${}^+6 + {}^+10$ (6) ${}^-10 + {}^+10$ (9) ${}^+30 + {}^-30$

   c. Can you find two integers for which the sum is not an integer?
   d. Is the sum of two integers always an integer?
   e. Can we say that the integers are closed under the operation of addition?

4. Additive Identity
   a. What do we mean when we say that a certain group of numbers has an additive identity?

Positive and Negative Numbers

 b. What is the sum of each of the following?

  (1) ⁻6 + 0   (3) ⁺7 + 0   (5) 0 + ⁺14   (7) ⁺10 + 0
  (2) 0 + ⁺4   (4) 0 + ⁻2   (6) ⁻12 + 0   (8) 0 + ⁻3

 c. In each case in problem **4b**, is the sum of the integer and zero the integer itself?
 d. Can you find any integer which when added to zero does not have the given integer as the sum?
 e. Does the set of integers have an additive identity? If so, what is it?

5. Additive Inverse

 a. What do we mean by an additive inverse of an integer?
 b. What is the sum of each of the following?

  (1) ⁻3 + ⁺3   (4) ⁻4 + ⁺4   (7) ⁺6 + ⁻6
  (2) ⁺6 + ⁻6   (5) 0 + 0   (8) ⁻105 + ⁺105
  (3) ⁺12 + ⁻12   (6) ⁺25 + ⁻25   (9) ⁻53 + ⁺53

 c. Are we adding in each case in problem **5b** an integer and its additive inverse? What is the sum in each case?
 d. Can you find any integer which when added to its additive inverse does not have zero as the sum?
 e. Does each integer have an additive inverse?

3–11 ADDITION OF RATIONAL NUMBERS; PROPERTIES

 Examination of addition on the number line reveals that, to add positive and negative fractional numbers, we follow the same procedures as we do when we add integers. (See sections 3–9 and 3–10.) Since in the addition of rational numbers we sometimes add the absolute values and at other times subtract the absolute values, we should review at this time both the addition and the subtraction of the fractional numbers of arithmetic. (See sections 1–12 and 1–13.) Although the model problems below only illustrate common fractions, rational numbers may also be named by decimal fraction numerals. Thus sections 1–18 and 1–19 should also be reviewed at this time.

 To add two positive rational numbers, we find the sum of their absolute values and prefix the numeral with a positive sign.

 To add two negative rational numbers, we find the sum of their absolute values and prefix the numeral with a negative sign.

Add: $\frac{-1}{8} + \frac{-5}{8} = ?$

$$\frac{-1}{8} + \frac{-5}{8} = -\left(\left|\frac{-1}{8}\right| + \left|\frac{-5}{8}\right|\right)$$

$$= -\left(\frac{1}{8} + \frac{5}{8}\right) = -\left(\frac{1+5}{8}\right) = \frac{-6}{8} = \frac{-3}{4}$$

or simply,
Vertically:
$$\frac{-1}{8}$$
$$\frac{-5}{8}$$
$$\frac{-6}{8} = \frac{-3}{4}$$

Horizontally:
$$\frac{-1}{8} + \frac{-5}{8} = \frac{-1 + -5}{8} = \frac{-6}{8} = \frac{-3}{4}$$

Answer, $\frac{-3}{4}$

To add a positive rational number and a negative rational number or *a negative rational number and a positive rational number, subtract the smaller absolute value from the greater absolute value of the two addends and prefix the difference with the sign of the number having the greater absolute value.*

Add: $\frac{+2}{3} + \frac{-1}{2} = ?$

$$\frac{+2}{3} + \frac{-1}{2} = +\left(\left|\frac{+2}{3}\right| - \left|\frac{-1}{2}\right|\right) = +\left(\frac{2}{3} - \frac{1}{2}\right)$$

$$= +\left(\frac{2}{3} \times \frac{2}{2} - \frac{1}{2} \times \frac{3}{3}\right) = +\left(\frac{4}{6} - \frac{3}{6}\right) = \frac{+1}{6}$$

or simply,
Vertically:
$$\frac{+2}{3} = \frac{+2}{3} \times \frac{2}{2} = \frac{+4}{6}$$
$$\frac{-1}{2} = \frac{-1}{2} \times \frac{3}{3} = \frac{-3}{6}$$
$$\frac{+1}{6}$$

Horizontally:
$$\frac{+2}{3} + \frac{-1}{2} = \frac{+2 \times 2}{3 \times 2} + \frac{-1 \times 3}{2 \times 3}$$
$$= \frac{+4}{6} + \frac{-3}{6} = \frac{+4 + -3}{6} = \frac{+1}{6}$$

Answer, $\frac{+1}{6}$

Positive and Negative Numbers

Observe in the model problem below that sometimes it is necessary to interchange the absolute values to make this subtraction possible.

Add: $\dfrac{^+1}{4} + \dfrac{^-5}{8} = ?$

$\dfrac{^+1}{4} + \dfrac{^-5}{8} = ^-\left(\left|\dfrac{^-5}{8}\right| - \left|\dfrac{^+1}{4}\right|\right)^* = ^-\left(\dfrac{5}{8} - \dfrac{1}{4}\right)$

$= ^-\left(\dfrac{5}{8} - \dfrac{1}{4} \times \dfrac{2}{2}\right) = ^-\left(\dfrac{5}{8} - \dfrac{2}{8}\right) = \dfrac{^-3}{8}$

*(Since the smaller absolute value happens to be in the first addend in this problem, we interchange absolute values in order to subtract.)

or simply,

Vertically:

$\dfrac{^+1}{4} = \dfrac{^+2}{8}$

$\dfrac{^-5}{8} = \dfrac{^-5}{8}$

$\overline{} \quad \dfrac{^-3}{8}$

Horizontally:

$\dfrac{^+1}{4} + \dfrac{^-5}{8} = \dfrac{^+2}{8} + \dfrac{^-5}{8} = \dfrac{^+2 + ^-5}{8} = \dfrac{^-3}{8}$

Answer, $\dfrac{^-3}{8}$

The sum of any rational number and its opposite is zero.

The sum of any non-zero rational number and zero is the non-zero rational number.

The numeral naming a positive rational number does not require the positive (⁺) sign.

To add three or more rational numbers, we either add successively or first add the positive numbers, then the negative numbers, and finally the two answers.

$^+3\tfrac{1}{2} + ^-4\tfrac{1}{4} + ^+1\tfrac{5}{6} \quad = ^+3\tfrac{6}{12} + ^-4\tfrac{3}{12} + ^+1\tfrac{10}{12}$

$\phantom{^+3\tfrac{1}{2} + ^-4\tfrac{1}{4} + ^+1\tfrac{5}{6} \quad} = ^+3\tfrac{6}{12} + ^+1\tfrac{10}{12} + ^-4\tfrac{3}{12}$

$\phantom{^+3\tfrac{1}{2} + ^-4\tfrac{1}{4} + ^+1\tfrac{5}{6} \quad} = ^+4\tfrac{16}{12} + ^-4\tfrac{3}{12}$

Answer, $^+1\tfrac{1}{12} \quad = ^+\tfrac{13}{12} = ^+1\tfrac{1}{12}$

EXERCISES

Add as indicated:

1. a. $^+\frac{1}{10} + ^+\frac{3}{10}$ b. $^+\frac{1}{4} + ^+\frac{7}{12}$ c. $^+2\frac{1}{4} + ^+2\frac{5}{6}$ d. $^+4\frac{9}{16} + ^+7\frac{3}{10}$
2. a. $^+.6 + ^+.7$ b. $^+4.6 + ^+1.5$ c. $^+.018 + ^+.04$ d. $^+30.41 + ^+3.568$
3. a. $^-\frac{3}{5} + ^-\frac{1}{5}$ b. $^-\frac{2}{3} + ^-\frac{1}{2}$ c. $^-3\frac{7}{8} + ^-1\frac{5}{8}$ d. $^-8\frac{5}{6} + ^-6\frac{7}{16}$
4. a. $^-.74 + ^-.6$ b. $^-.06 + ^+.16$ c. $^-16 + ^-1.8$ d. $^-6.92 + ^-74.9$
5. a. $^+\frac{5}{6} + ^-\frac{1}{3}$ b. $^+3\frac{1}{2} + ^-2\frac{1}{4}$ c. $^+4\frac{5}{8} + ^-\frac{9}{16}$ d. $^+5\frac{1}{2} + ^-7\frac{5}{12}$
6. a. $^+1.5 + ^-2.6$ b. $^+.8 + ^-1.2$ c. $^+.439 + ^-.19$ d. $^+18.7 + ^-1.87$
7. a. $^-\frac{3}{25} + ^-\frac{9}{100}$ b. $^-7\frac{1}{10} + ^+1\frac{2}{5}$ c. $^-5\frac{1}{3} + ^+7\frac{2}{9}$ d. $^-2\frac{3}{10} + ^+2\frac{5}{16}$
8. a. $^-7.4 + ^+4.7$ b. $^-.81 + ^+.8$ c. $^-.006 + ^+.0031$ d. $^-2.26 + ^+22.3$
9. a. $^+2\frac{1}{3} + ^-2\frac{1}{3}$ b. $^-\frac{9}{16} + ^+\frac{9}{16}$ c. $^+5\frac{1}{4} + ^-5\frac{1}{4}$ d. $^-3\frac{11}{12} + ^+3\frac{11}{12}$
10. a. $^-4.1 + ^+4.1$ b. $^+.6 + ^-.6$ c. $^-.019 + ^+.019$ d. $^+7.153 + ^-7.153$
11. a. $^-6\frac{7}{8} + 0$ b. $0 + ^+7\frac{1}{4}$ c. $^-3.7 + 0$ d. $0 + ^+.044$
12. a. $^-8\frac{2}{3} + 5\frac{1}{4}$ b. $8\frac{1}{3} + ^-2\frac{5}{6}$ c. $^-.52 + 1.3$ d. $4.72 + ^-2.142$
13. a. $^+\frac{5}{6} + ^-\frac{1}{3} + ^-\frac{7}{9}$ c. $^-3\frac{1}{6} + ^-2\frac{7}{8} + ^-1\frac{3}{4}$ e. $^-4.19 + ^-.408 + ^+62.3$
    b. $^+1\frac{3}{5} + ^-7\frac{1}{2} + ^+4\frac{7}{10}$ d. $^-2\frac{1}{4} + ^+1\frac{2}{3} + ^-\frac{5}{8}$ f. $^+.206 + ^-.7242 + ^+.54$
14. a. $(^-\frac{5}{6} + ^-\frac{2}{3}) + ^-\frac{11}{12}$ d. $^-4\frac{7}{12} + (^-6\frac{1}{4} + ^+2\frac{3}{16})$ g. $(^-.6 + ^+.72) + ^-.1$
    b. $^+4\frac{1}{3} + (^+3\frac{1}{2} + ^+7\frac{3}{4})$ e. $^+3 + (^-1\frac{5}{6} + ^-\frac{4}{5})$ h. $^-3.6 + (^-.85 + ^-2.3)$
    c. $^+\frac{9}{16} + (^-1\frac{5}{8} + ^+2\frac{1}{4})$ f. $(^-10\frac{7}{8} + ^+5\frac{1}{2}) + ^-4\frac{1}{6}$ i. $^+7.16 + (^-81.3 + ^+.685)$

15. Add:

$^-\frac{13}{15}$	$^+\frac{3}{4}$	$^-\frac{3}{9}$	$^+2\frac{1}{2}$	$^+5\frac{7}{12}$	$^-4\frac{9}{10}$	$^-3\frac{7}{16}$
$^-\frac{7}{15}$	$^-\frac{7}{8}$	$^+\frac{5}{18}$	$^-\frac{3}{8}$	$^+4\frac{1}{3}$	$^+6\frac{5}{6}$	$^-8\frac{5}{12}$

16. Add:

$^+\frac{3}{8}$	$^-\frac{9}{16}$	$^+\frac{3}{4}$	$^-7\frac{5}{6}$	$^+5\frac{1}{2}$	$^+2\frac{1}{2}$	$^-2\frac{13}{16}$
$^-\frac{1}{2}$	$^-\frac{11}{12}$	$^-\frac{5}{16}$	$^+9\frac{3}{8}$	$^-3\frac{2}{3}$	$^-6\frac{7}{8}$	$^-5\frac{5}{7}$
$^+\frac{2}{3}$	$^-\frac{1}{4}$	$^-\frac{7}{8}$	$^-3\frac{7}{12}$	$^-1\frac{3}{5}$	$^+4\frac{3}{4}$	$^+2\frac{11}{12}$

17. Add:

			$^-21.4$	$^-6.39$	$^-.031$	$^-7.4$
$^-7.8$	$^+.115$	$^-6.88$	$^+17.5$	$^+4.01$	$^-.046$	$^+6.13$
$^+3.9$	$^-.217$	$^-2.01$	$^-11.3$	$^+3.95$	$^+.012$	$^-1.819$

Positive and Negative Numbers

18. Find the missing numbers:

    a. $^-\!\tfrac{5}{6} + {}^-\!\tfrac{1}{6} = ?$
    b. $^-\!\tfrac{2}{3} + {}^+\!\tfrac{1}{4} = n$
    c. $^-3\tfrac{1}{3} + {}^-2\tfrac{5}{12} = \square$
    d. $^+3\tfrac{3}{4} + {}^+2\tfrac{1}{8} = n$
    e. $^-1\tfrac{11}{12} + {}^+2\tfrac{9}{16} = n$
    f. $^+9\tfrac{2}{5} + {}^-7\tfrac{3}{10} = n$

19. Find the missing numbers:

    a. $^-.8 + {}^-.4 = \square$
    b. $^+.16 + {}^-3.4 = n$
    c. $^-4.3 + {}^+3 = ?$
    d. $^+5.2 + {}^-5.2 = n$
    e. $^+6.1 + {}^-.61 = n$
    f. $^-.716 + {}^+.82 = n$

20. Find the missing numbers:

    a. $? + {}^+\!\tfrac{1}{4} = {}^+2\tfrac{3}{8}$
    b. $n + {}^-\!\tfrac{3}{5} = {}^+\!\tfrac{7}{10}$
    c. $\square + {}^-3\tfrac{1}{3} = {}^-7\tfrac{5}{6}$
    d. $n + {}^-7\tfrac{5}{7} = 0$
    e. $n + {}^+2\tfrac{2}{3} = {}^+\!\tfrac{7}{12}$
    f. $n + {}^-7\tfrac{3}{4} = {}^-6\tfrac{2}{3}$

21. Find the missing numbers:

    a. $? + {}^+.5 = {}^+1.2$
    b. $n + {}^-3.4 = {}^-6.1$
    c. $n + {}^+8.7 = {}^+2.74$
    d. $\square + {}^-11 = {}^-18.6$
    e. $n + {}^-.374 = {}^-.374$
    f. $n + {}^+6.14 = {}^-3.894$

22. Find the missing numbers:

    a. $^-\!\tfrac{1}{3} + \square = {}^-\!\tfrac{1}{6}$
    b. $^+2\tfrac{1}{4} + n = {}^-1\tfrac{1}{2}$
    c. $^+3\tfrac{1}{3} + ? = {}^+6\tfrac{5}{6}$
    d. $^-3\tfrac{9}{13} + ? = {}^+3\tfrac{9}{13}$
    e. $^-2\tfrac{1}{5} + n = {}^-8\tfrac{3}{10}$
    f. $^-9\tfrac{1}{5} + n = {}^+4$

23. Find the missing numbers:

    a. $^+.9 + ? = {}^+2.5$
    b. $^-1.3 + n = {}^-.2$
    c. $^-4.1 + \square = 0$
    d. $^+.6 + n = {}^+.18$
    e. $^-3.9 + n = {}^+4.1$
    f. $^+.788 + n = {}^-.0002$

Properties

In the following exercises we determine the properties that hold for the rational numbers under the operation of addition.

1. Commutative Property

    a. What do we mean when we say that the commutative property holds for the addition of a certain group of rational numbers?

    b. In each of the following find the two sums by adding on the number line, using vectors. Compare the two sums by comparing their vectors.

    (1) $^+\!\tfrac{1}{4} + {}^-\!\tfrac{1}{2}; \ {}^-\!\tfrac{1}{2} + {}^+\!\tfrac{1}{4}.$

      Does $^+\!\tfrac{1}{4} + {}^-\!\tfrac{1}{2} = {}^-\!\tfrac{1}{2} + {}^+\!\tfrac{1}{4}?$

    (2) $^-3\tfrac{1}{4} + {}^-1\tfrac{1}{2}; \ {}^-1\tfrac{1}{2} + {}^-3\tfrac{1}{4}.$

      Does $^-3\tfrac{1}{4} + {}^-1\tfrac{1}{2} = {}^-1\tfrac{1}{2} + {}^-3\tfrac{1}{4}?$

c. In each of the following add as indicated, then compare the sums:
   (1) $^-\frac{1}{6} + {}^+\frac{2}{3}$; $^+\frac{2}{3} + {}^-\frac{1}{6}$.
   Does $^-\frac{1}{6} + {}^+\frac{2}{3} = {}^+\frac{2}{3} + {}^-\frac{1}{6}$?
   (2) $^+3\frac{3}{8} + {}^+1\frac{5}{8}$; $^+1\frac{5}{8} + {}^+3\frac{3}{8}$.
   Does $^+3\frac{3}{8} + {}^+1\frac{5}{8} = {}^+1\frac{5}{8} + {}^+3\frac{3}{8}$?
   (3) $^+3\frac{1}{3} + {}^-4\frac{11}{12}$; $^-4\frac{11}{12} + {}^+3\frac{1}{3}$.
   Does $^+3\frac{1}{3} + {}^-4\frac{11}{12} = {}^-4\frac{11}{12} + {}^+3\frac{1}{3}$?
   (4) $^-2.17 + {}^-.436$; $^-.436 + {}^-2.17$.
   Does $^-2.17 + {}^-.436 = {}^-.436 + {}^-2.17$?

d. Can you find two rational numbers for which the commutative property of addition does not hold?

e. Can we say that the commutative property holds for the addition of rational numbers?

2. Associative Property

   a. What do we mean when we say that the associative property holds for the addition of a certain group of numbers?

   b. In each of the following find the two sums by adding on the number line, using vectors. Compare the two sums by comparing their vectors.
   (1) $(^-\frac{1}{2} + {}^-\frac{1}{2}) + {}^-\frac{1}{4}$; $^-\frac{1}{2} + (^-\frac{1}{2} + {}^-\frac{1}{4})$.
   Does $(^-\frac{1}{2} + {}^-\frac{1}{2}) + {}^-\frac{1}{4} = {}^-\frac{1}{2} + (^-\frac{1}{2} + {}^-\frac{1}{4})$?
   (2) $(^+2\frac{2}{5} + {}^-3\frac{1}{5}) + {}^+\frac{2}{5}$; $^+2\frac{2}{5} + (^-3\frac{1}{5} + {}^+\frac{2}{5})$.
   Does $(^+2\frac{2}{5} + {}^-3\frac{1}{5}) + {}^+\frac{2}{5} = {}^+2\frac{2}{5} + (^-3\frac{1}{5} + {}^+\frac{2}{5})$?

   c. In each of the following add as indicated, then compare the sums:
   (1) $(^-3\frac{1}{2} + {}^+2\frac{3}{8}) + {}^-4\frac{1}{4}$; $^-3\frac{1}{2} + (^+2\frac{3}{8} + {}^-4\frac{1}{4})$.
   Does $(^-3\frac{1}{2} + {}^+2\frac{3}{8}) + {}^-4\frac{1}{4} = {}^-3\frac{1}{2} + (^+2\frac{3}{8} + {}^-4\frac{1}{4})$?
   (2) $(^-6.4 + {}^+3.31) + {}^+.92$; $^-6.4 + ({}^+3.31 + {}^+.92)$.
   Does $(^-6.4 + {}^+3.31) + {}^+.92 = {}^-6.4 + ({}^+3.31 + {}^+.92)$?

   d. Can you find three rational numbers for which the associative property of addition does not hold?

   e. Can we say that the associative property holds for the addition of rational numbers?

3. **Closure**
   a. What do we mean when we say that a certain group of numbers is closed under the operation of addition?
   b. Is the sum of each of the following a rational number?

   (1) $^+\frac{7}{8} + ^+\frac{3}{8}$ (4) $^-4\frac{1}{5} + 0$ (7) $^-.381 + ^-4.92$

   (2) $^-\frac{5}{6} + ^+\frac{2}{3}$ (5) $^-3\frac{2}{5} + ^+2\frac{10}{13}$ (8) $^+.06 + ^-.006$

   (3) $^-\frac{1}{6} + ^-\frac{6}{15}$ (6) $^+3\frac{1}{3} + ^-5\frac{1}{4}$ (9) $^-7.05 + ^+3.4$

   c. Can you find two rational numbers for which the sum is not a rational number?
   d. Is the sum of two rational numbers always a rational number?
   e. Can we say that the rational numbers are closed under the operation of addition?

4. **Additive Identity**
   a. What do we mean when we say that a certain group of numbers has an additive identity?
   b. What is the sum of each of the following?

   (1) $^+\frac{5}{6} + 0$ (3) $0 + ^+4.2$ (5) $^-\frac{12}{5} + 0$ (7) $0 + ^-.7$

   (2) $0 + ^-2\frac{1}{2}$ (4) $^-.12 + 0$ (6) $0 + ^+5.15$ (8) $^-8\frac{11}{14} + 0$

   c. In each case in problem **4b,** is the sum of the rational number and zero the rational number itself?
   d. Can you find any rational number which when added to zero does not have the given rational number as the sum?
   e. Do the rational numbers have an additive identity? If so, what is it?

5. **Additive Inverse**
   a. What do we mean by an additive inverse of a rational number?
   b. What is the sum of each of the following?

   (1) $^-\frac{9}{10} + ^+\frac{9}{10}$ (4) $^-5.44 + ^+5.44$ (7) $^+11\frac{2}{3} + ^-11\frac{2}{3}$

   (2) $^+3\frac{1}{3} + ^-3\frac{1}{3}$ (5) $^-\frac{11}{6} + ^+\frac{11}{6}$ (8) $^-7.5 + ^+7.5$

   (3) $^+.8 + ^-.8$ (6) $^+.003 + ^-.003$ (9) $^-3.14 + ^+3.14$

   c. Are we adding in each case in problem **5b** a rational number and its additive inverse? What is the sum in each case?
   d. Can you find a rational number which when added to its additive inverse does not have zero as the sum?
   e. Does each rational number have an additive inverse?

3-12 SUBTRACTION ON THE NUMBER LINE

Since subtraction is the inverse operation of addition, to subtract we find the addend which when added to the given addend (subtrahend) will equal the given sum (minuend).

$$^+6 - {}^-2 = ?$$
is thought of as
$$^-2 + ? = {}^+6$$

Using this idea of inverse operation, to subtract on the number line we locate the points reached by the movement represented by the given sum and by the given addend, moving in both cases from the point marked 0.

We draw (1) a vector below the number line to represent the movement indicated by the given sum and (2) a vector above the number line to represent the movement indicated by the given addend.

To find the missing addend, we draw (3) a vector *from* the point reached by the vector representing the given addend *to* the point reached by the vector representing the sum. The length of this vector is the absolute value and the direction of the vector indicates the sign of the missing addend.

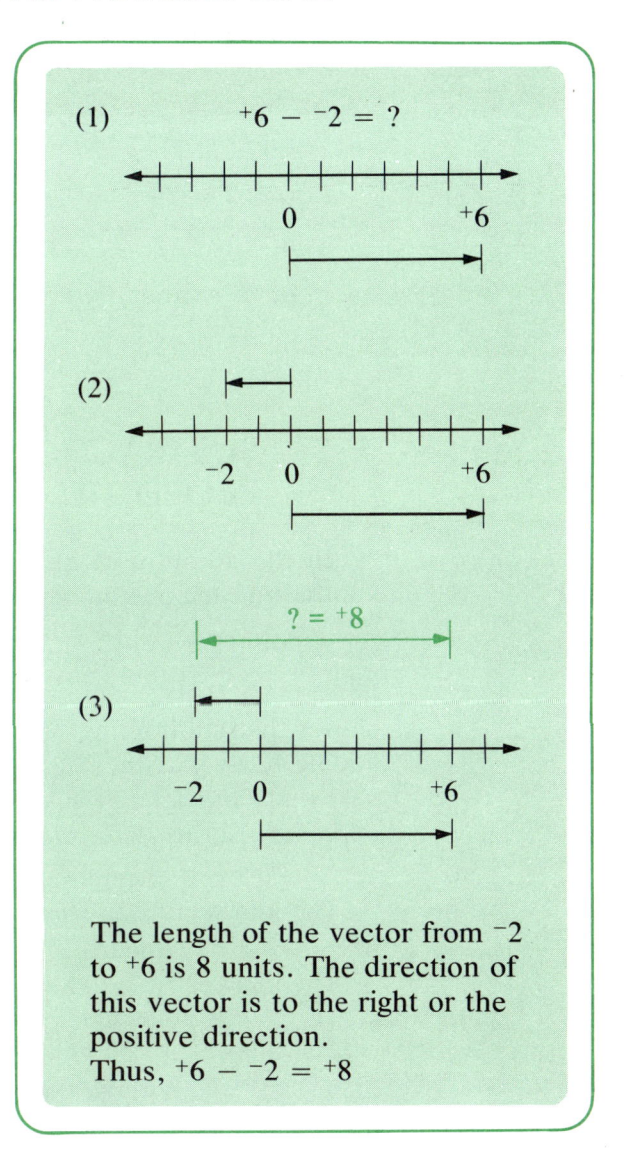

(1) $^+6 - {}^-2 = ?$

(2)

(3)

The length of the vector from $^-2$ to $^+6$ is 8 units. The direction of this vector is to the right or the positive direction.
Thus, $^+6 - {}^-2 = {}^+8$

EXERCISES

In each of the following subtract as indicated on the number line, using vectors to find the difference:

1. a. $^+5 - {}^+2$ **b.** $^+7 - {}^+3$ **c.** $^+8 - {}^+1$ **d.** $^+12 - {}^+4$

Positive and Negative Numbers 193

2. a. $^+2 - {}^+6$ b. $^+1 - {}^+7$ c. $^+4 - {}^+9$ d. $^+3 - {}^+5$
3. a. $^-7 - {}^-2$ b. $^-9 - {}^-3$ c. $^-8 - {}^-4$ d. $^-6 - {}^-2$
4. a. $^-5 - {}^-9$ b. $^-2 - {}^-8$ c. $^-1 - {}^-10$ d. $^-3 - {}^-9$
5. a. $^+9 - {}^-2$ b. $^+4 - {}^-7$ c. $^+6 - {}^-4$ d. $^+8 - {}^-9$
6. a. $^-8 - {}^+1$ b. $^-2 - {}^+4$ c. $^-5 - {}^+5$ d. $^-7 - {}^+10$
7. a. $^-7 - {}^-7$ b. $^+2 - {}^+2$ c. $^-9 - {}^-9$ d. $^+5 - {}^+5$
8. a. $0 - {}^-5$ b. $0 - {}^+6$ c. $^-1 - 0$ d. $^+10 - 0$
9. a. $^-1\frac{1}{2} - {}^+2\frac{1}{2}$ b. $^-7 - {}^-8\frac{1}{4}$ c. $^+6\frac{1}{4} - {}^-5\frac{3}{4}$ d. $^-3\frac{1}{3} - {}^-3\frac{1}{3}$

3–13 SUBTRACTION OF INTEGERS; PROPERTIES

In the addition of integers we learned that $^+6 + {}^+2 = {}^+8$. In subtracting integers on the number line we found that $^+6 - {}^-2 = {}^+8$.

A comparison of $^+6 - {}^-2 = {}^+8$
and $^+6 + {}^+2 = {}^+8$

shows that subtracting $^-2$ from $^+6$ gives the same answer ($^+8$) as adding $^+2$ to $^+6$. Since $^+2$ is the opposite of $^-2$, we may say in this case that *subtracting a number* gives the same answer as *adding its opposite*. Let us determine if this is true in other cases.

a. $^+3 - {}^+5 = ?$ means "What number added to $^+5$ equals $^+3$?" The answer is $^-2$. Thus, $^+3 - {}^+5 = {}^-2$.
b. $^-6 - {}^+2 = ?$ means "What number added to $^+2$ equals $^-6$?" The answer is $^-8$. Thus, $^-6 - {}^+2 = {}^-8$.
c. $^-2 - {}^-7 = ?$ means "What number added to $^-7$ equals $^-2$?" The answer is $^+5$. Thus, $^-2 - {}^-7 = {}^+5$.
d. $0 - {}^-8 = ?$ means "What number added to $^-8$ equals 0?" The answer is $^+8$. Thus, $0 - {}^-8 = {}^+8$.

a. $^+3 - {}^+5 = {}^-2$ c. $^-2 - {}^-7 = {}^+5$
   $^+3 + {}^-5 = {}^-2$ $^-2 + {}^+7 = {}^+5$

b. $^-6 - {}^+2 = {}^-8$ d. $0 - {}^-8 = {}^+8$
   $^-6 + {}^-2 = {}^-8$ $0 + {}^+8 = {}^+8$

We conclude that both subtracting a number and adding the opposite of the number which is to be subtracted give the same answer.

Thus, *to subtract any integer from another integer,* we add to the minuend the opposite (or additive inverse) of the integer which is to be subtracted. The problem first could be rewritten, arranged horizontally or vertically. For example:

Subtract: $^-7 - {}^+2 = ?$

Horizontally:

$^-7 - {}^+2 = {}^-7 + {}^-2 = {}^-9$

Answer, $^-9$

Vertically:

Subtract: Add:
$^-7$ $^-7$
$\underline{{}^+2}$ → $\underline{{}^-2}$
             $^-9$

Subtract: $^-1 - {}^-5 = ?$

Horizontally:

$^-1 - {}^-5 = {}^-1 + {}^+5 = {}^+4$

Answer, $^+4$

Vertically:

Subtract: Add:
$^-1$ $^-1$
$\underline{{}^-5}$ → $\underline{{}^+5}$
             $^+4$

Subtract: $0 - {}^+3 = ?$

Horizontally:

$0 - {}^+3 = 0 + {}^-3 = {}^-3$

Answer, $^-3$

Vertically:

Subtract: Add:
0 0
$\underline{{}^+3}$ → $\underline{{}^-3}$
             $^-3$

The numeral naming a positive number does not require the positive ($^+$) sign.

$$7 - 12 = 7 + {}^-12 = {}^-5$$

Answer, $^-5$

Positive and Negative Numbers

EXERCISES

Subtract as indicated:

1. a. $^+5 - {^+3}$ b. $^+10 - {^+4}$ c. $^+31 - {^+18}$ d. $^+39 - {^+19}$
2. a. $^+2 - {^+7}$ b. $^+8 - {^+31}$ c. $^+14 - {^+56}$ d. $^+35 - {^+70}$
3. a. $^-12 - {^-10}$ b. $2 - 28$ c. $19 - 54$ d. $67 - 100$
4. a. $^-3 - {^-7}$ b. $^-61 - {^-28}$ c. $0 - {^-61}$ d. $^-76 - {^-103}$
5. a. $0 - {^-6}$ b. $^+92 - {^+92}$ c. $^-56 - 0$ d. $^-29 - {^-29}$
6. a. $^+14 - {^-6}$ b. $^-12 - {^-40}$ c. $^-94 - {^+38}$ d. $^-110 - {^-42}$
7. a. $6 - 11$ b. $0 - {^+24}$ c. $^-27 - {^-65}$ d. $0 - {^+108}$
8. a. $^-24 - {^+18}$ b. $^+27 - {^-39}$ c. $^+60 - {^+38}$ d. $^+51 - {^-51}$
9. a. $^+9 - 0$ b. $^-17 - 0$ c. $^-15 - {^-15}$ d. $^+31 - 0$
10. a. $^-12 - {^-12}$ b. $^-48 - {^+65}$ c. $^-73 - {^-36}$ d. $^-169 - {^+250}$
11. a. $(^-8 - {^+5}) - {^+9}$ c. $^-23 - ({^+10} - {^+15})$ e. $({^+11} - {^-6}) - {^+14}$
    b. $({^+9} - {^+12}) - {^-6}$ d. $^+17 - ({^-4} - {^-20})$ f. $^-2 - ({^-43} - {^+43})$

12. Subtract in each of the following:

a. $\begin{array}{r} ^-26 \\ +{^+19} \\ \hline \end{array}$ $\begin{array}{r} 0 \\ +{^+10} \\ \hline \end{array}$ $\begin{array}{r} ^+17 \\ +{^+28} \\ \hline \end{array}$ $\begin{array}{r} ^+65 \\ -{^-35} \\ \hline \end{array}$ $\begin{array}{r} ^+18 \\ +{^+40} \\ \hline \end{array}$ $\begin{array}{r} ^-32 \\ +{^+32} \\ \hline \end{array}$ $\begin{array}{r} ^-87 \\ +{^+87} \\ \hline \end{array}$ $\begin{array}{r} 93 \\ 200 \\ \hline \end{array}$

b. $\begin{array}{r} 8 \\ 43 \\ \hline \end{array}$ $\begin{array}{r} ^-97 \\ +{^+39} \\ \hline \end{array}$ $\begin{array}{r} ^+7 \\ -{^-38} \\ \hline \end{array}$ $\begin{array}{r} 29 \\ 52 \\ \hline \end{array}$ $\begin{array}{r} ^-59 \\ -{^-59} \\ \hline \end{array}$ $\begin{array}{r} ^-85 \\ +{^+114} \\ \hline \end{array}$ $\begin{array}{r} ^-42 \\ -{^-81} \\ \hline \end{array}$ $\begin{array}{r} 0 \\ -34 \\ \hline \end{array}$

c. $\begin{array}{r} ^-40 \\ -{^-15} \\ \hline \end{array}$ $\begin{array}{r} ^+62 \\ +{^+75} \\ \hline \end{array}$ $\begin{array}{r} ^+9 \\ -{^-56} \\ \hline \end{array}$ $\begin{array}{r} ^-99 \\ 0 \\ \hline \end{array}$ $\begin{array}{r} ^-45 \\ -{^-78} \\ \hline \end{array}$ $\begin{array}{r} ^-91 \\ -{^-39} \\ \hline \end{array}$ $\begin{array}{r} ^-18 \\ -{^-72} \\ \hline \end{array}$ $\begin{array}{r} ^-54 \\ +{^+6} \\ \hline \end{array}$

13. Find the missing numbers:
    a. $^+5 - {^+10} = ?$
    b. $^-18 - {^-11} = \square$
    c. $^-32 - {^+43} = n$
    d. $0 - {^-61} = \square$
    e. $^+16 - {^-16} = ?$
    f. $^-40 - {^-40} = n$
    g. $^-23 - {^-54} = n$
    h. $^+100 - {^-75} = n$
    i. $^+35 - {^+71} = n$

14. Find the missing numbers:
    a. $? - {^+4} = {^+8}$
    b. $\square - {^-12} = {^-18}$
    c. $n - {^-1} = {^+7}$
    d. $\square - {^+5} = {^-12}$
    e. $? - {^-14} = 0$
    f. $n - 8 = {^-4}$
    g. $n - {^-14} = 10$
    h. $^-47 - n = {^-52}$
    i. $n - {^+21} = {^-19}$

15. Find the missing numbers:
    a. $^-5 - \square = {^-13}$
    b. $^+12 - ? = {^-3}$
    c. $^-5 - n = {^+11}$
    d. $^-23 - ? = 0$
    e. $0 - n = {^-17}$
    f. $^-25 - \square = {^+15}$
    g. $^+200 - n = {^+245}$
    h. $^-51 - n = {^-52}$
    i. $^+61 - n = {^-34}$

Properties

In the following exercises we determine the properties that hold for the integers under the operation of subtraction.

1. **Commutative Property**
   a. If the commutative property holds for the subtraction of integers, what would have to be true?
   b. In each of the following subtract as indicated, then compare answers:
      (1) $^+8 - ^+6$; $^+6 - ^+8$. Does $^+8 - ^+6 = ^+6 - ^+8$?
      (2) $^-18 - ^-23$; $^-23 - ^-18$. Does $^-18 - ^-23 = ^-23 - ^-18$?
      (3) $^+5 - ^-7$; $^-7 - ^+5$. Does $^+5 - ^-7 = ^-7 - ^+5$?
      (4) $^-6 - ^+2$; $^+2 - ^-6$. Does $^-6 - ^+2 = ^+2 - ^-6$?
   c. Does the commutative property hold for the subtraction of integers?

2. **Associative Property**
   a. If the associative property holds for the subtraction of integers, what would have to be true?
   b. In each of the following subtract as indicated, then compare answers:
      (1) $(^-12 - ^+9) - ^-7$; $^-12 - (^+9 - ^-7)$.
         Does $(^-12 - ^+9) - ^-7 = ^-12 - (^+9 - ^-7)$?
      (2) $(^-4 - ^-8) - ^+3$; $^-4 - (^-8 - ^+3)$.
         Does $(^-4 - ^-8) - ^+3 = ^-4 - (^-8 - ^+3)$?
   c. Does the associative property hold for the subtraction of integers?

3. **Closure**
   a. What do we mean when we say that a certain group of numbers is closed under the operation of subtraction?
   b. When you subtract in each of the following, is the difference in each case an integer?
      (1) $^-9 - ^-2$ (4) $^+10 - ^-10$ (7) $3 - 8$
      (2) $^+6 - ^+11$ (5) $0 - ^+6$ (8) $^+4 - ^-12$
      (3) $^-2 - ^+5$ (6) $^-3 - 0$ (9) $^-5 - ^-5$
   c. Can you find two integers for which the difference is not an integer?
   d. Is the difference of two integers always an integer?
   e. Can we say that integers are closed under the operation of subtraction?

Positive and Negative Numbers

3-14 SUBTRACTION OF RATIONAL NUMBERS; PROPERTIES

Examination of subtraction on the number line in section 3-12 and the development in section 3-13, that subtracting a number and adding the opposite of the number which is to be subtracted both give the same answer, lead us to conclude that:

To subtract any rational number from another rational number, we add to the minuend the opposite (or additive inverse) of the rational number which is to be subtracted. The problem first could be rewritten, arranged horizontally or vertically. For example:

Subtract: $\frac{^+3}{7} - \frac{^+4}{7} = ?$

Horizontally:

$$\frac{^+3}{7} - \frac{^+4}{7} = \frac{^+3}{7} + \frac{^-4}{7} = \frac{^+3 + ^-4}{7} = \frac{^-1}{7}$$

Answer, $\frac{^-1}{7}$

Vertically:

Subtract: Add:
$\frac{^+3}{7}$ $\frac{^+3}{7}$
$\frac{^+4}{7}$ → $\frac{^-4}{7}$
                  $\frac{^-1}{7}$

Subtract: $^-2\frac{1}{3} - ^+1\frac{5}{9} = ?$

Horizontally:

$^-2\frac{1}{3} - ^+1\frac{5}{9} = ^-2\frac{1}{3} + ^-1\frac{5}{9}$
$= ^-2\frac{3}{9} + ^-1\frac{5}{9}$
$= ^-3\frac{8}{9}$

Answer, $^-3\frac{8}{9}$

Vertically:

Subtract: Add:
$^-2\frac{1}{3}$ $^-2\frac{1}{3} = ^-2\frac{3}{9}$
$^+1\frac{5}{9}$ → $^-1\frac{5}{9} = ^-1\frac{5}{9}$
                     $^-3\frac{8}{9}$

Subtract: $0 - ^-7.6 = ?$

Horizontally:

$0 - ^-7.6 = 0 + ^+7.6 = ^+7.6$

Answer, $^+7.6$

Vertically:

Subtract: Add:
0   0
$^-7.6$ → $^+7.6$
            $^+7.6$

EXERCISES

Subtract as indicated:

1. a. $^{+}\frac{3}{4} - {}^{+}\frac{1}{4}$ b. $^{+}\frac{2}{3} - {}^{+}\frac{7}{9}$ c. $^{+}3\frac{5}{6} - {}^{+}1\frac{3}{5}$ d. $^{+}7\frac{1}{4} - {}^{+}4\frac{2}{3}$

2. a. $^{+}\frac{1}{8} - {}^{+}\frac{7}{8}$ b. $^{+}\frac{7}{12} - {}^{+}\frac{5}{4}$ c. $^{+}2\frac{1}{3} - {}^{+}4\frac{3}{5}$ d. $^{+}\frac{13}{2} - {}^{+}1\frac{1}{3}$

3. a. $^{+}.61 - {}^{+}.57$ b. $^{+}.7 - {}^{+}.07$ c. $^{+}1.5 - {}^{+}.17$ d. $^{+}.857 - {}^{+}.85$

4. a. $^{+}.71 - {}^{+}1.3$ b. $^{+}.53 - {}^{+}.6$ c. $^{+}.37 - {}^{+}3.7$ d. $^{+}.1576 - {}^{+}.157$

5. a. $^{-}\frac{7}{8} - {}^{-}\frac{1}{8}$ b. $^{-}3\frac{1}{3} - {}^{-}2\frac{1}{4}$ c. $^{-}\frac{3}{5} - {}^{-}\frac{2}{5}$ d. $^{-}6\frac{1}{2} - {}^{-}4\frac{11}{12}$

6. a. $^{-}\frac{1}{12} - {}^{-}\frac{5}{12}$ b. $^{-}\frac{2}{3} - {}^{-}\frac{4}{5}$ c. $^{-}2\frac{1}{6} - {}^{-}3\frac{3}{8}$ d. $^{-}5\frac{2}{3} - {}^{-}6\frac{11}{16}$

7. a. $^{-}9.5 - {}^{-}2.3$ b. $^{-}8 - {}^{-}.72$ c. $^{-}.108 - {}^{-}.0092$ d. $^{-}35.6 - {}^{-}.907$

8. a. $^{-}.06 - {}^{-}.3$ b. $^{-}3.7 - {}^{-}10.5$ c. $^{-}7.87 - {}^{-}37.2$ d. $^{-}.0003 - {}^{-}.2$

9. a. $^{+}\frac{12}{15} - {}^{-}\frac{2}{15}$ b. $^{+}\frac{7}{10} - {}^{-}\frac{1}{8}$ c. $^{+}4\frac{1}{2} - {}^{-}2\frac{2}{3}$ d. $^{+}2\frac{11}{12} - {}^{-}5\frac{3}{4}$

10. a. $^{+}.24 - {}^{-}.2$ b. $^{+}7.8 - {}^{-}.104$ c. $^{+}5 - {}^{-}.05$ d. $^{+}.316 - {}^{-}.14$

11. a. $^{-}\frac{3}{5} - {}^{+}\frac{1}{2}$ b. $^{-}\frac{7}{18} - {}^{+}\frac{5}{9}$ c. $^{-}7\frac{1}{8} - {}^{+}6\frac{3}{4}$ d. $^{-}12\frac{1}{2} - {}^{+}8\frac{9}{10}$

12. a. $^{-}.68 - {}^{+}.135$ b. $^{-}8.573 - {}^{+}1.158$ c. $^{-}8.36 - {}^{+}.8$ d. $^{-}3.8 - {}^{+}6$

13. a. $^{-}\frac{7}{8} - 0$ b. $^{+}3\frac{1}{4} - 0$ c. $^{+}3.9 - 0$ d. $^{-}.028 - 0$

14. a. $0 - {}^{-}\frac{2}{3}$ b. $0 - {}^{+}7\frac{13}{15}$ c. $^{-}.7 - {}^{-}.7$ d. $0 - {}^{-}4.15$

15. a. $^{-}\frac{11}{12} - {}^{-}\frac{11}{12}$ b. $^{+}6\frac{2}{3} - {}^{+}6\frac{2}{3}$ c. $0 - {}^{+}.624$ d. $^{+}6.713 - {}^{+}6.173$

16. a. $\frac{5}{9} - \frac{8}{9}$ b. $3\frac{3}{16} - 8$ c. $.67 - 1.5$ d. $.081 - .14$

17. a. $\left({}^{-}\frac{1}{3} - {}^{-}\frac{1}{4}\right) - {}^{-}\frac{1}{8}$ c. $\left({}^{-}1\frac{2}{3} - {}^{-}2\frac{1}{5}\right) - {}^{+}3\frac{3}{4}$ e. $({}^{+}.08 - {}^{-}.5) - {}^{-}.304$

    b. $^{+}5\frac{1}{2} - \left({}^{-}2\frac{5}{7} - {}^{-}3\frac{11}{14}\right)$ d. $^{-}4.6 - ({}^{+}9.3 - {}^{+}6.3)$ f. $^{+}.521 - ({}^{+}.81 - {}^{+}.1)$

18. Subtract in each of the following:

a.	$^{+}\frac{3}{5}$	$^{-}\frac{1}{3}$	$^{+}\frac{5}{8}$	$^{-}3\frac{3}{5}$	$^{-}6$	$^{+}5\frac{3}{5}$	$3\frac{3}{16}$
	$^{+}\frac{1}{5}$	$^{-}\frac{5}{6}$	$^{-}\frac{2}{5}$	$^{+}\frac{13}{15}$	$^{-}2\frac{9}{10}$	$^{-}2\frac{1}{4}$	$7\frac{2}{3}$
b.	$^{-}6\frac{1}{2}$	$3\frac{7}{8}$	$^{+}5\frac{1}{3}$	$6\frac{1}{3}$	$^{-}10\frac{7}{32}$	$^{+}6\frac{7}{12}$	$^{-}1\frac{9}{16}$
	$^{+}4\frac{5}{6}$	$7\frac{11}{12}$	$^{-}3\frac{2}{5}$	$9\frac{2}{3}$	$^{-}2\frac{3}{4}$	$^{-}5\frac{13}{16}$	$^{-}2\frac{5}{8}$
c.	$^{-}.7$	$^{+}1.5$	$.024$	$^{-}6.384$	$^{-}11.38$	3.98	$^{-}.78$
	$^{-}.8$	$^{-}.6$	$.03$	$^{+}1.42$	$^{-}.9$	6	$^{-}.9924$

Positive and Negative Numbers

19. Find the missing numbers:

   a. $^+\frac{3}{7} - {}^+\frac{1}{7} = ?$
   b. $^-\frac{11}{12} - {}^-\frac{37}{60} = n$
   c. $^+2\frac{1}{2} - {}^+\frac{3}{8} = \square$
   d. $^+3\frac{7}{8} - {}^-1 = n$
   e. $^-7\frac{1}{3} - {}^-8\frac{5}{12} = n$
   f. $3\frac{5}{8} - 8\frac{3}{16} = n$

20. Find the missing numbers:

   a. $^+.5 - {}^-.1 = \square$
   b. $^-7.8 - {}^-3.8 = n$
   c. $\square - {}^-3\frac{1}{4} = {}^-5\frac{1}{2}$
   d. $n - 6\frac{7}{12} = 4\frac{5}{6}$
   e. $^+6.07 - {}^-9.18 = n$
   f. $^-.7 - {}^-3.48 = n$

21. Find the missing numbers:

   a. $? - {}^-\frac{7}{16} = {}^+\frac{15}{16}$
   b. $n - {}^+\frac{5}{8} = {}^-\frac{1}{4}$
   c. $n - {}^-2\frac{1}{4} = {}^-6\frac{1}{2}$
   d. $? - {}^-7\frac{5}{16} = 2\frac{7}{8}$
   e. $n - 9\frac{7}{8} = 11$
   f. $n - 1\frac{2}{3} = \frac{11}{15}$

22. Find the missing numbers:

   a. $? - {}^-\frac{1}{12} = {}^+\frac{7}{12}$
   b. $^-\frac{7}{10} - n = {}^+\frac{3}{5}$
   c. $^-7\frac{1}{2} - ? = {}^-3\frac{3}{8}$
   d. $12 - n = {}^-\frac{5}{6}$
   e. $^-4\frac{11}{16} - n = {}^-3\frac{2}{3}$
   f. $^-8\frac{5}{12} - n = 6\frac{3}{4}$

23. Find the missing numbers:

   a. $^-.5 + \square = {}^+.8$
   b. $n - {}^+2.6 = {}^-1.8$
   c. $n - {}^-.68 = {}^-1.4$
   d. $12 - n = {}^-\frac{5}{6}$
   e. $n - {}^-.125 = {}^-.001$
   f. $n - 20.8 = 5.43$

Properties

In the following exercises we determine the properties that hold for the rational numbers under the operation of subtraction.

1. Commutative Property

   a. If the commutative property holds for the subtraction of rational numbers, what would have to be true?
   b. In each of the following subtract as indicated, then compare answers:

   (1) $^-\frac{3}{4} - {}^+\frac{1}{2}$; $^+\frac{1}{2} - {}^-\frac{3}{4}$. Does $^-\frac{3}{4} - {}^+\frac{1}{2} = {}^+\frac{1}{2} - {}^-\frac{3}{4}$?

   (2) $^-2\frac{5}{6} - {}^-3\frac{1}{3}$; $^-3\frac{1}{3} - {}^-2\frac{5}{6}$. Does $^-2\frac{5}{6} - {}^-3\frac{1}{3} = {}^-3\frac{1}{3} - {}^-2\frac{5}{6}$?

   (3) $^+5 - {}^+\frac{11}{15}$; $^+\frac{11}{15} - {}^+5$. Does $^+5 - {}^+\frac{11}{15} = {}^+\frac{11}{15} - {}^+5$?

   (4) $^+7.6 - {}^-8.4$; $^-8.4 - {}^+7.6$. Does $^+7.6 - {}^-8.4 = {}^-8.4 - {}^+7.6$?

   c. Does the commutative property hold for the subtraction of rational numbers?

2. **Associative Property**
   a. If the associative property holds for the subtraction of rational numbers, what would have to be true?
   b. In each of the following subtract as indicated, then compare answers:
      (1) $(^{+}\!\frac{5}{6} - ^{-}\!\frac{7}{12}) - ^{-}\!\frac{1}{3}$; $^{+}\!\frac{5}{6} - (^{-}\!\frac{7}{12} - ^{-}\!\frac{1}{3})$.

      Does $(^{+}\!\frac{5}{6} - ^{-}\!\frac{7}{12}) - ^{-}\!\frac{1}{3} = ^{+}\!\frac{5}{6} - (^{-}\!\frac{7}{12} - ^{-}\!\frac{1}{3})$?

      (2) $(^{-}2\frac{1}{3} - ^{-}1\frac{1}{9}) - ^{+}\!\frac{5}{6}$; $^{-}2\frac{1}{3} - (^{-}1\frac{1}{9} - ^{+}\!\frac{5}{6})$.

      Does $(^{-}2\frac{1}{3} - ^{-}1\frac{1}{9}) - ^{+}\!\frac{5}{6} = ^{-}2\frac{1}{3} - (^{-}1\frac{1}{9} - ^{+}\!\frac{5}{6})$?
   c. Does the associative property hold for the subtraction of rational numbers?

3. **Closure**
   a. What do we mean when we say that a certain group of numbers is closed under the operation of subtraction?
   b. When you subtract in each of the following, is the difference in each case a rational number?

      (1) $^{+}\!\frac{3}{5} - ^{-}\!\frac{7}{15}$ (4) $^{-}4\frac{6}{7} - 0$ (7) $^{+}2\frac{2}{3} - ^{+}1\frac{1}{5}$

      (2) $^{-}\!\frac{3}{10} - ^{+}2\frac{1}{2}$ (5) $^{-}.04 - ^{-}.002$ (8) $0 - ^{-}3\frac{3}{4}$

      (3) $^{-}2\frac{1}{3} - ^{-}4\frac{1}{2}$ (6) $^{+}7\frac{5}{12} - ^{-}7\frac{5}{12}$ (9) $^{-}4.5 - ^{+}.14$
   c. Can you find two rational numbers for which the difference is not a rational number?
   d. Can we say that the rational numbers are closed under the operation of subtraction?

3–15 MULTIPLICATION OF INTEGERS; PROPERTIES

To develop the procedure of determining the product of two positive integers (like $^{+}3 \times ^{+}4$), or of two negative integers (like $^{-}3 \times ^{-}4$), or of a positive integer and a negative integer (like $^{+}3 \times ^{-}4$), or of a negative integer and a positive integer (like $^{-}3 \times ^{+}4$), let us analyze the statements shown on the following page:

Positive and Negative Numbers

Representing a bank deposit of $6 by $^+6$, a withdrawal of $6 by $^-6$, 5 weeks from now by $^+5$, and 5 weeks ago by $^-5$, let us solve the following using signed numbers.

a. If we deposit in our bank account $6 each week for 5 weeks, how will our bank account 5 weeks from now compare with our present bank account?

There will be $30 more in our bank account 5 weeks from now.

$$^+5 \times {^+6} = {^+30}$$

b. If we deposited in our bank account $6 each week for the past 5 weeks, how did our bank account 5 weeks ago compare with our present bank account?

There was $30 less in the account 5 weeks ago.

$$^-5 \times {^+6} = {^-30}$$

c. If we withdraw from our bank account $6 each week for the next 5 weeks, how will our bank account 5 weeks from now compare with our present bank account?

There will be $30 less in the account 5 weeks from now.

$$^+5 \times {^-6} = {^-30}$$

d. If we withdrew from our bank account $6 each week for the past 5 weeks, how did our bank account 5 weeks ago compare with our present bank account?

There was $30 more in the account 5 weeks ago.

$$^-5 \times {^-6} = {^+30}$$

From the above it appears that the product of the two positive integers is a positive integer, the product of two negative integers is a positive integer, and the product of a positive integer and a negative integer or a negative integer and a positive integer is a negative integer. Let us check this further.

Multiplication may be thought of as repeated addition. Four times three, expressed as 4×3, is the sum of four 3's: $(3 + 3 + 3 + 3)$ or 12. The expression four times three may be written as 4×3, $4 \cdot 3$, $(4) \times (3)$, $(4)3$, or $(4)(3)$. Observe that the $\times$ symbol may be replaced by a raised dot or the factors may be written next to each other within parentheses without any multiplication symbol.

Since a numeral without a sign represents a positive number, 4 × 3 may be expressed as ⁺4 × ⁺3 or (⁺4)(⁺3).
Thus, 4 × 3 or ⁺4 × ⁺3 = ⁺3 + ⁺3 + ⁺3 + ⁺3 = ⁺12.
That is: ⁺4 × ⁺3 = ⁺12.
This is illustrated on the number line as follows:

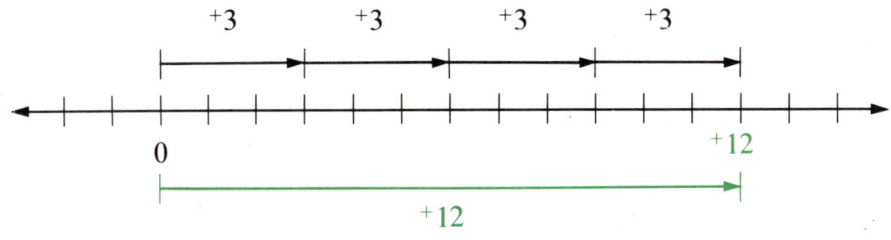

Thus, the product of two positive integers is a positive integer whose absolute value is the product of the absolute values of the two given factors.

The product of zero and any integer is zero.

⁺2 × 0 = 0   ⁻5 × 0 = 0
0 × ⁻1 = 0   0 × ⁺6 = 0

The indicated product ⁺4 × ⁻3 means four ⁻3's, or the sum of ⁻3 + ⁻3 + ⁻3 + ⁻3, which is ⁻12.
That is: ⁺4 × ⁻3 = ⁻12
This is illustrated on the number line as follows:

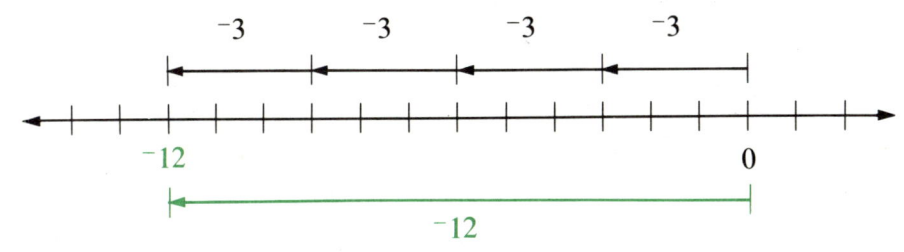

Let us examine the following sequence:

Observe that each time the second factor decreases by 1, the product decreases by 5.
Thus, the product ⁺5 × ⁻1 should be 5 less than 0, the product of ⁺5 × 0. This is ⁻5.
The product of ⁺5 × ⁻2 should be 5 less than ⁻5, the product of ⁺5 × ⁻1. This is ⁻10.

⁺5 × ⁺3 = ⁺15
⁺5 × ⁺2 = ⁺10
⁺5 × ⁺1 = ⁺5
⁺5 × 0 = 0
⁺5 × ⁻1 = ⁻5
⁺5 × ⁻2 = ⁻10
⁺5 × ⁻3 = ⁻15

Positive and Negative Numbers

The product of $^+5 \times ^-3$ should be 5 less than $^-10$, the product of $^+5 \times ^-2$. This is $^-15$.

Thus, $^+5 \times ^-3 = ^-15$.

This also illustrates that $^+5 \times ^-3 = ^-15$.

Thus, the product of a positive integer and a negative integer is a negative integer whose absolute value is the product of the absolute values of the two given factors.

Let us examine the following sequence:

Observe that each time the first factor decreases by 1, the product decreases by 2.

Thus, the product of $^-1 \times ^+2$ should be 2 less than 0, the product of $0 \times ^+2$. This is $^-2$.

The product of $^-2 \times ^+2$ should be 2 less than $^-2$, the product of $^-1 \times ^+2$. This is $^-4$.

The product of $^-3 \times ^+2$ should be 2 less than $^-4$, the product of $^-2 \times ^+2$. This is $^-6$.

$^+4 \times ^+2 = ^+8$
$^+3 \times ^+2 = ^+6$
$^+2 \times ^+2 = ^+4$
$^+1 \times ^+2 = ^+2$
$0 \times ^+2 = 0$
$^-1 \times ^+2 = ^-2$
$^-2 \times ^+2 = ^-4$
$^-3 \times ^+2 = ^-6$
$^-4 \times ^+2 = ^-8$

The product of $^-4 \times ^+2$ should be 2 less than $^-6$, the product of $^-3 \times ^+2$. This is $^-8$.

Thus, $^-4 \times ^+2 = ^-8$.

This illustrates that $^-4 \times ^+2 = ^-8$.

Thus, the product of a negative integer and a positive integer is a negative integer whose absolute value is the product of the absolute values of the two given factors.

Let us examine the following sequence:

Observe that each time the second factor decreases by 1, the product increases by 3.

Thus, the product of $^-3 \times ^-1$ should be 3 more than 0, the product of $^-3 \times 0$. This is $^+3$.

The product of $^-3 \times ^-2$ should be 3 more than $^+3$, the product of $^-3 \times ^-1$. This is $^+6$.

The product $^-3 \times ^-3$ should be 3 more than $^+6$, the product of $^-3 \times ^-2$. This is $^+9$.

$^-3 \times ^+3 = ^-9$
$^-3 \times ^+2 = ^-6$
$^-3 \times ^+1 = ^-3$
$^-3 \times 0 = 0$
$^-3 \times ^-1 = ^+3$
$^-3 \times ^-2 = ^+6$
$^-3 \times ^-3 = ^+9$

Thus, $^-3 \times ^-3 = ^+9$.

This illustrates that $^-3 \times ^-3 = ^+9$.

Thus, the product of two negative integers is a positive integer whose absolute value is the product of the absolute values of the two given factors.

Summarizing:

To multiply two positive integers or two negative integers, we find the product of their absolute values and prefix the numeral for this product with a positive sign.

Multiply: $^+6 \times {^+5} = ?$ Multiply: $^-3 \times {^-8} = ?$

$^+6 \times {^+5} = {^+(6 \times 5)} = {^+30}$ $^-3 \times {^-8} = {^+(3 \times 8)} = {^+24}$

or $^+6$
$\ \underline{^+5}$
$\ ^+30$

or $^-3$
$\ \underline{^-8}$
$\ ^+24$

Answer, $^+30$ *Answer*, $^+24$

To multiply a positive integer by a negative integer or a negative integer by a positive integer, we find the product of their absolute values and prefix the numeral for this product with a negative sign.

Multiply: $^+7 \times {^-4} = ?$ Multiply: $^-6 \times {^+8} = ?$

$^+7 \times {^-4} = {^-(7 \times 4)} = {^-28}$ $^-6 \times {^+8} = {^-(6 \times 8)} = {^-48}$

or $^+7$
$\ \underline{^-4}$
$\ ^-28$

or $^-6$
$\ \underline{^+8}$
$\ ^-48$

Answer, $^-28$ *Answer*, $^-48$

$^-7 \times 9 = {^-63}$ *Answer*, $^-63$.

The numeral naming a positive integer does not require the positive ($^+$) sign.

$0 \times {^-7} = 0$ *Answer*, 0
$^+4 \times 0 = 0$ *Answer*, 0

The product of zero and any integer is zero.

Positive and Negative Numbers

To multiply three or more factors, we multiply successively. Observe that an odd number of negative factors produces a negative number as the product, while an even number of negative factors produces a positive number as the product, provided zero is not a factor. In each case the absolute value of the product is equal to the product of the absolute values of the given factors.

> Multiply:
>
> $(^-3)(^+1)(^-5)(^-2) = {}^-30$
>
> *Answer,* $^-30$
>
> $(^-4)(^-2)(^-1)(^-3) = {}^+24$
>
> *Answer,* $^+24$

EXERCISES

Multiply as indicated:

1. a. $^+7 \times {}^+5$ b. $^+7 \times {}^+9$ c. $^+5 \times {}^+15$ d. $^+36 \times {}^+56$
2. a. $^-6 \times {}^-4$ b. $^-4 \times {}^-14$ c. $^-20 \times {}^-16$ d. $^-1 \times {}^-115$
3. a. $^+6 \times {}^-7$ b. $^+5 \times {}^-16$ c. $^+21 \times {}^-100$ d. $^+56 \times {}^-50$
4. a. $^-5 \times {}^+6$ b. $^-6 \times {}^+13$ c. $^-11 \times {}^+28$ d. $^-61 \times {}^+27$
5. a. $^-4 \times 4$ b. $^-10 \times 5$ c. $6 \times {}^-49$ d. $32 \times {}^-16$
6. a. $0 \times {}^-17$ b. $^+26 \times 0$ c. $^-18 \times 0$ d. $0 \times {}^+29$
7. a. $(^-2)(^-1)$ f. $(^-5)(^-4)$ k. $(^-12)(^-12)$ p. $^-5(^-7)$
   b. $(^-7)(^+3)$ g. $(0)(^-2)$ l. $(^-35)(30)$ q. $^-8(12)$
   c. $(^+13)(^-2)$ h. $(^-12)(5)$ m. $(105)(^-61)$ r. $6(^-7)$
   d. $(^-5)(^-10)$ i. $(^-4)(^-13)$ n. $(^-49)(^-100)$ s. $14(^-4)$
   e. $(^+6)(^+6)$ j. $(7)(^-8)$ o. $(^-275)(^-189)$ t. $^-14(^-10)$
8. Multiply in each of the following:

   a. $\begin{array}{r}^-5\\\underline{^-2}\end{array}$ $\begin{array}{r}^+7\\\underline{^-4}\end{array}$ $\begin{array}{r}^-8\\\underline{^+8}\end{array}$ $\begin{array}{r}^+5\\\underline{^+6}\end{array}$ $\begin{array}{r}^-12\\\underline{^-4}\end{array}$ $\begin{array}{r}^+12\\\underline{^-7}\end{array}$ $\begin{array}{r}^-1\\\underline{^-5}\end{array}$ $\begin{array}{r}^-15\\\underline{^+3}\end{array}$

   b. $\begin{array}{r}^+6\\\underline{^-11}\end{array}$ $\begin{array}{r}^+18\\\underline{^-8}\end{array}$ $\begin{array}{r}0\\\underline{^-21}\end{array}$ $\begin{array}{r}^-35\\\underline{^+1}\end{array}$ $\begin{array}{r}16\\\underline{^-8}\end{array}$ $\begin{array}{r}^-59\\\underline{24}\end{array}$ $\begin{array}{r}^-31\\\underline{^-46}\end{array}$ $\begin{array}{r}^-200\\\underline{^-200}\end{array}$

9. a. $(^-3 \times ^-7) \times {}^+3$ c. $^+5 \times (^-3 \times {}^+7)$ e. $(^-3 \times ^-3) \times ^-3$

    b. $(^+7 \times ^-4) \times ^-1$ d. $^-4 \times (^-8 \times ^-2)$ f. $^+6 \times (^-4 \times ^-3)$

10. a. $^+8 \times ^-6 \times {}^+4$ c. $^-8 \times ^-8 \times ^-8$ e. $^-2 \times {}^+5 \times ^-1 \times ^-3$

    b. $^+6 \times ^-4 \times ^-5$ d. $^-12 \times {}^+6 \times ^-1$ f. $^-3 \times ^-3 \times ^-3 \times ^-3$

11. a. $(^-3)(2)(^-1)$ c. $(2)(5)(^-4)(3)$ e. $(5)(^-1)(2)(^-3)(4)$

    b. $(^-2)(^-2)(^-2)$ d. $(4)(^-6)(3)(^-2)$ f. $(^-4)(^-3)(^-1)(^-3)(^-2)$

12. Find the missing numbers:

    a. $^-8 \times ^-7 = ?$ d. $7 \times ^-10 = \square$ g. $10 \times ^-2 = n$

    b. $6 \times ^-4 = \square$ e. $^-14 \times ^-8 = ?$ h. $^-38 \times 6 = n$

    c. $^-7 \times 5 = n$ f. $^-13 \times 7 = n$ i. $^-4 \times 32 = n$

13. Find the missing numbers:

    a. $4 \times ? = ^-12$ d. $^-2 \times \square = 12$ g. $^-25 \times n = ^-100$

    b. $6 \times n = ^-30$ e. $8 \times ? = ^-64$ h. $6 \times n = 36$

    c. $^-14 \times \square = ^-42$ f. $^-7 \times n = 0$ i. $^-3 \times n = 36$

14. Find the missing numbers:

    a. $? \times 5 = ^-20$ d. $\square \times ^-9 = 72$ g. $n \times ^-8 = 56$

    b. $\square \times ^-2 = 20$ e. $n \times ^-9 = ^-45$ h. $n \times 3 = ^-66$

    c. $n \times ^-7 = ^-42$ f. $? \times 12 = 36$ i. $n \times ^-25 = ^-100$

Properties

In the following exercises we determine the properties that hold for the integers under the operation of multiplication.

1. Commutative Property

    a. What do we mean when we say that the commutative property holds for the multiplication of a certain group of numbers?

    b. In each of the following multiply as indicated, then compare the products:

        (1) $^+8 \times {}^+4$; $^+4 \times {}^+8$. Does $^+8 \times {}^+4 = {}^+4 \times {}^+8$?
        (2) $^-5 \times ^-6$; $^-6 \times ^-5$. Does $^-5 \times ^-6 = ^-6 \times ^-5$?
        (3) $^+13 \times ^-2$; $^-2 \times {}^+13$. Does $^+13 \times ^-2 = ^-2 \times {}^+13$?
        (4) $^-7 \times {}^+7$; $^+7 \times ^-7$. Does $^-7 \times {}^+7 = {}^+7 \times ^-7$?

Positive and Negative Numbers

c. Can you find two integers for which the commutative property of multiplication does not hold?
d. Can we say that the commutative property holds for the multiplication of integers?

2. Associative Property

   a. What do we mean when we say that the associative property holds for the multiplication of a certain set of numbers?
   b. In each of the following multiply as indicated, then compare the products:

      (1) $(^+6 \times {}^-4) \times {}^-2$; $^+6 \times ({}^-4 \times {}^-2)$.
          Does $(^+6 \times {}^-4) \times {}^-2 = {}^+6 \times ({}^-4 \times {}^-2)$?
      (2) $(^-3 \times {}^-5) \times {}^+9$; $^-3 \times ({}^-5 \times {}^+9)$.
          Does $(^-3 \times {}^-5) \times {}^+9 = {}^-3 \times ({}^-5 \times {}^+9)$?

   c. Can you find three integers for which the associative property of multiplication does not hold?
   d. Can we say that the associative property holds for the multiplication of integers?

3. Distributive Property

   a. What do we mean when we say that the distributive property of multiplication over addition holds for the integers?
   b. In each of the following perform the indicated operations, then compare answers:

      (1) $^+5 \times ({}^-8 + {}^+4)$; $(^+5 \times {}^-8) + (^+5 \times {}^+4)$.
          Does $^+5 \times ({}^-8 + {}^+4) = (^+5 \times {}^-8) + (^+5 \times {}^+4)$?
      (2) $^-2 \times ({}^+7 + {}^-8)$; $(^-2 \times {}^+7) + (^-2 \times {}^-8)$.
          Does $^-2 \times ({}^+7 + {}^-8) = (^-2 \times {}^+7) + (^-2 \times {}^-8)$?

   c. Can you find any integers for which multiplication is not distributed over addition?
   d. Can we say that the distributive property of multiplication over addition holds for the set of integers?

4. Closure

   a. What do we mean when we say that a certain group of numbers is closed under the operation of multiplication?
   b. Is the product of each of the following an integer?

      (1) $^-5 \times {}^+3$ (4) $^+4 \times {}^+10$ (7) $0 \times {}^+4$
      (2) $^+12 \times {}^-6$ (5) $^-9 \times 0$ (8) $^-13 \times {}^-13$
      (3) $^-5 \times {}^-5$ (6) $^-1 \times {}^-16$ (9) $^-32 \times {}^+1$

c. Can you find two integers for which the product is not an integer?
d. Is the product of two integers always an integer?
e. Can we say that the integers are closed under the operation of multiplication?

5. **Multiplicative Identity**

    a. What do we mean when we say that a certain group of numbers has a multiplicative identity?
    b. What is the product of each of the following?

    (1) $^-8 \times 1$ (3) $^+12 \times 1$ (5) 0×1 (7) $1 \times\ ^+25$
    (2) $1 \times\ ^+6$ (4) $1 \times\ ^-27$ (6) $1 \times\ ^-1$ (8) $^-200 \times 1$

    c. In each case in problem **5b** is the product of the integer and one (1) the integer itself?
    d. Can you find any integer which when multiplied by one (1) does not have the given integer as the product?
    e. Do the integers have a multiplicative identity? If so, what is it?

3–16 MULTIPLICATION OF RATIONAL NUMBERS; PROPERTIES

We follow the same procedures to multiply rational numbers as we do to multiply integers. See section 3–15. It may be necessary to review at this time the multiplication of the rational numbers of arithmetic (common fractions, mixed numbers, and decimal fractions). See sections 1–14 and 1–20.

To multiply two positive rational numbers or *two negative rational numbers,* we find the product of their absolute values and prefix the numeral for this product with a positive sign.

Multiply: $^+\frac{1}{4} \times\ ^+\frac{3}{7} = ?$
$^+\frac{1}{4} \times\ ^+\frac{3}{7} =\ ^+(\frac{1}{4} \times \frac{3}{7}) =\ ^+\frac{3}{28}$ *Answer,* $^+\frac{3}{28}$

Multiply: $^-\frac{2}{3} \times\ ^-2\frac{1}{4} = ?$
$^-\frac{2}{3} \times\ ^-2\frac{1}{4} =\ ^+(\frac{2}{3} \times 2\frac{1}{4}) =\ ^+(\frac{2}{3} \times \frac{9}{4}) =\ ^+\frac{18}{12} =\ ^+1\frac{1}{2}$
                                               Answer, $^+1\frac{1}{2}$

To multiply a positive rational number by a negative rational number or *a negative rational number by a positive rational number*, we find the product of their absolute values and prefix the numeral for this product with a negative sign. The numeral naming a positive rational number does not require the positive (+) sign.

> Multiply: $^+1\frac{1}{2} \times {}^-2\frac{1}{3} = ?$
>
> $^+1\frac{1}{2} \times {}^-2\frac{1}{3} = {}^-\left(1\frac{1}{2} \times 2\frac{1}{3}\right) = {}^-\left(\frac{3}{2} \times \frac{7}{3}\right) = {}^-\frac{7}{2} = {}^-3\frac{1}{2}$
>
> Answer, $^-3\frac{1}{2}$
>
> Multiply: $^-1.5 \times {}^+.3 = ?$
>
> $^-1.5 \times {}^+.3 = {}^-(1.5 \times .3) = {}^-.45$
>
> Answer, $^-.45$

The product of zero and any rational number is zero.

> $0 \times {}^-3\frac{3}{5} = 0$ Answer, 0
>
> $^+4.013 \times 0 = 0$ Answer, 0

To multiply three or more factors, we multiply successively. The absolute value of the product is equal to the product of the absolute values of the given factors. An odd number of negative factors produces a negative number as the product, while an even number of negative factors produces a positive number as the product, provided zero is not a factor.

> $(^-.4)(^-7)(^-.06) = {}^-.168$ Answer, $^-.168$
>
> $\left(\frac{-3}{5}\right)\left(\frac{+1}{4}\right)\left(\frac{-5}{6}\right) = {}^+\frac{1}{8}$ Answer, $^+\frac{1}{8}$

EXERCISES

Multiply as indicated:

1. a. $^{+}\frac{1}{3} \times ^{+}\frac{4}{5}$ b. $^{+}\frac{5}{6} \times ^{+}24$ c. $^{+}\frac{3}{5} \times ^{+}4\frac{1}{10}$ d. $^{+}3\frac{1}{2} \times ^{+}1\frac{7}{12}$
2. a. $^{+}.7 \times ^{+}.5$ b. $^{+}51 \times ^{+}.03$ c. $^{+}1.2 \times ^{+}2.5$ d. $^{+}.275 \times ^{+}.36$
3. a. $^{-}\frac{4}{5} \times ^{-}\frac{15}{16}$ b. $^{-}12 \times ^{-}\frac{11}{16}$ c. $^{-}\frac{2}{3} \times ^{-}1\frac{1}{8}$ d. $^{-}2\frac{2}{3} \times ^{-}3\frac{3}{4}$
4. a. $^{-}.03 \times ^{-}.2$ b. $^{-}.4 \times ^{-}50$ c. $^{-}.61 \times ^{-}.88$ d. $^{-}.002 \times ^{-}.002$
5. a. $^{+}\frac{3}{4} \times ^{-}\frac{1}{3}$ b. $^{+}32 \times ^{-}2\frac{3}{8}$ c. $^{+}5\frac{1}{4} \times ^{-}\frac{13}{16}$ d. $^{+}3\frac{3}{5} \times ^{-}6\frac{2}{3}$
6. a. $^{+}.115 \times ^{-}.9$ b. $^{+}100 \times ^{-}.003$ c. $^{+}.01 \times ^{-}.004$ d. $^{+}5.1 \times ^{-}83.6$
7. a. $^{-}\frac{7}{8} \times ^{+}\frac{8}{7}$ b. $^{-}5\frac{1}{3} \times ^{+}2$ c. $^{-}4\frac{2}{5} \times ^{+}\frac{3}{8}$ d. $^{-}4\frac{1}{4} \times ^{+}3\frac{1}{3}$
8. a. $^{-}.006 \times ^{+}.004$ b. $^{-}24 \times ^{+}7.5$ c. $^{-}4.13 \times ^{+}.42$ d. $^{-}1.04 \times ^{+}.004$
9. a. $9 \times ^{-}\frac{2}{3}$ b. $1\frac{2}{5} \times 20$ c. $^{-}.6 \times 5.4$ d. $.03 \times ^{-}.829$
10. a. $0 \times ^{-}\frac{5}{6}$ b. $^{+}7\frac{2}{3} \times 0$ c. $^{-}.59 \times 0$ d. $0 \times ^{+}7.25$
11. a. $(^{-}\frac{1}{3})(^{-}\frac{1}{3})$ f. $(^{-}1\frac{1}{3})(^{-}3\frac{1}{2})$ k. $(^{-}.2)(^{-}.2)$ p. $\frac{5}{12}(^{-}2\frac{1}{4})$
    b. $(^{-}\frac{3}{5})(\frac{5}{6})$ g. $(24)(^{-}2\frac{1}{8})$ l. $(.7)(^{-}.16)$ q. $\frac{^{-}9}{16}(80)$
    c. $(\frac{1}{3})(^{-}\frac{7}{9})$ h. $(\frac{9}{10})(^{-}5\frac{1}{4})$ m. $(^{-}5.3)(.46)$ r. $\frac{^{-}2}{5}(^{-}6\frac{2}{3})$
    d. $(36)(^{-}2\frac{5}{6})$ i. $(^{-}1\frac{7}{14})(^{-}3\frac{1}{3})$ n. $(^{-}.006)(^{-}.09)$ s. $^{-}.8(8.25)$
    e. $(^{-}\frac{7}{10})(^{-}60)$ j. $(^{-}1\frac{5}{7})(^{-}2\frac{5}{8})$ o. $(3.5)(^{-}.208)$ t. $3.2(^{-}.001)$

12. Multiply in each of the following:

    $\begin{array}{c}^{-}.03\\\underline{^{-}.3}\end{array}$ $\begin{array}{c}^{-}51\\\underline{.08}\end{array}$ $\begin{array}{c}.3\\\underline{^{-}.41}\end{array}$ $\begin{array}{c}^{-}.005\\\underline{^{-}.001}\end{array}$ $\begin{array}{c}^{-}300\\\underline{^{-}.004}\end{array}$ $\begin{array}{c}^{-}1.1\\\underline{^{-}.11}\end{array}$ $\begin{array}{c}^{-}.3187\\\underline{1.5}\end{array}$

13. a. $(^{+}\frac{2}{5} \times ^{-}\frac{1}{2}) \times ^{+}\frac{5}{6}$ c. $(^{-}1\frac{1}{2} \times 3\frac{1}{8}) \times ^{-}1\frac{1}{2}$ e. $(.08 \times ^{-}4.3) \times ^{-}.039$
    b. $^{+}\frac{3}{8} \times (^{-}\frac{5}{6} \times ^{+}\frac{2}{5})$ d. $^{-}4\frac{1}{6} \times (^{+}3\frac{1}{5} \times ^{-}1\frac{3}{10})$ f. $^{-}2.6 \times (^{-}.15 \times ^{-}8.2)$

14. a. $^{-}\frac{1}{4} \times \frac{1}{2} \times ^{-}\frac{1}{3}$ c. $^{-}2\frac{1}{2} \times ^{-}\frac{5}{8} \times ^{-}4\frac{1}{5}$ e. $.06 \times ^{-}3.4 \times .5$
    b. $^{-}\frac{4}{5} \times ^{-}\frac{5}{6} \times ^{-}\frac{3}{8}$ d. $^{-}1\frac{4}{5} \times 3\frac{1}{7} \times ^{-}6\frac{2}{3}$ f. $^{-}1.15 \times ^{-}.003 \times ^{-}.014$

15. a. $(\frac{3}{8})(^{-}\frac{1}{2})(\frac{4}{9})$ c. $(^{-}1\frac{2}{5})(10)(^{-}\frac{12}{15})$ e. $(^{-}.4)(^{-}.2)(^{-}.1)$
    b. $(^{-}\frac{5}{6})(\frac{3}{10})(^{-}\frac{2}{3})$ d. $(^{-}1\frac{1}{4})(^{-}\frac{5}{8})(2\frac{1}{3})(^{-}24)$ f. $(^{-}1.3)(^{-}.32)(^{-}.25)(^{-}.006)$

Positive and Negative Numbers

16. Find the missing numbers:

a. $\frac{-2}{3} \times \frac{-2}{3} = ?$
b. $\frac{3}{4} \times \frac{-1}{3} = n$
c. $\frac{-7}{15} \times {}^-80 = \square$
d. ${}^-1\frac{3}{5} \times 3\frac{1}{8} = n$
e. ${}^-2\frac{1}{2} \times {}^-4\frac{7}{8} = n$
f. $8\frac{1}{6} \times {}^-10 = n$

17. Find the missing numbers:

a. $.2 \times {}^-.01 = \square$
b. ${}^-.004 \times {}^-.004 = n$
c. ${}^-.14 \times 50 = ?$
d. $2.55 \times {}^-.122 = n$
e. ${}^-.0051 \times {}^-.06 = n$
f. ${}^-.0016 \times 1.4 = n$

Properties

In the following exercises we determine the properties that hold for the rational number under the operation of multiplication.

1. **Commutative Property**

   a. What do we mean when we say that the commutative property holds for the multiplication of a certain group of numbers?

   b. In each of the following multiply as indicated, then compare the products:

   (1) $\frac{-5}{6} \times \frac{-1}{4}; \frac{-1}{4} \times \frac{-5}{6}$.
   Does $\frac{-5}{6} \times \frac{-1}{4} = \frac{-1}{4} \times \frac{-5}{6}$?

   (2) ${}^-1\frac{1}{3} \times {}^+15; {}^+15 \times {}^-1\frac{1}{3}$.
   Does ${}^-1\frac{1}{3} \times {}^+15 = {}^+15 \times {}^-1\frac{1}{3}$?

   (3) ${}^+2\frac{3}{4} \times {}^+1\frac{1}{5}; {}^+1\frac{1}{5} \times {}^+2\frac{3}{4}$.
   Does ${}^+2\frac{3}{4} \times {}^+1\frac{1}{5} = {}^+1\frac{1}{5} \times {}^+2\frac{3}{4}$?

   (4) ${}^+7.6 \times {}^-.05; {}^-.05 \times {}^+7.6$.
   Does ${}^+7.6 \times {}^-.05 = {}^-.05 \times {}^+7.6$?

   c. Can you find two rational numbers for which the commutative property of multiplication does not hold?

   d. Can we say that the commutative property holds for the multiplication of rational numbers?

2. **Associative Property**

   a. What do we mean when we say that the associative property holds for the multiplication of a certain group of numbers?

b. In each of the following multiply as indicated, then compare the products:

(1) $(\frac{-3}{4} \times \frac{-1}{3}) \times \frac{+5}{8}$; $\frac{-3}{4} \times (\frac{-1}{3} \times \frac{+5}{8})$.

Does $(\frac{-3}{4} \times \frac{-1}{3}) \times \frac{+5}{8} = \frac{-3}{4} \times (\frac{-1}{3} \times \frac{+5}{8})$?

(2) $(^{+}3\frac{1}{2} \times ^{+}1\frac{2}{5}) \times ^{-}2\frac{1}{4}$; $^{+}3\frac{1}{2} \times (^{+}1\frac{2}{5} \times ^{-}2\frac{1}{4})$.

Does $(^{+}3\frac{1}{2} \times ^{+}1\frac{2}{5}) \times ^{-}2\frac{1}{4} = ^{+}3\frac{1}{2} \times (^{+}1\frac{2}{5} \times ^{-}2\frac{1}{4})$?

c. Can you find three rational numbers for which the associative property of multiplication does not hold?

d. Can we say that the associative property holds for the multiplication of rational numbers?

3. Distributive Property

a. What do we mean when we say that the distributive property of multiplication over addition holds for the rational numbers?

b. In each of the following perform the indicated operations, then compare answers:

(1) $\frac{-1}{3} \times (\frac{+5}{9} + \frac{-1}{6})$; $(\frac{-1}{3} \times \frac{+5}{9}) + (\frac{-1}{3} \times \frac{-1}{6})$.

Does $\frac{-1}{3} \times (\frac{+5}{9} + \frac{-1}{6}) = (\frac{-1}{3} \times \frac{+5}{9}) + (\frac{-1}{3} \times \frac{-1}{6})$?

(2) $^{+}2\frac{1}{2} \times (^{-}1\frac{1}{4} + ^{+}1\frac{1}{3})$; $(^{+}2\frac{1}{2} \times ^{-}1\frac{1}{4}) + (^{+}2\frac{1}{2} \times ^{+}1\frac{1}{3})$.

Does $^{+}2\frac{1}{2} \times (^{-}1\frac{1}{4} + ^{+}1\frac{1}{3}) = (^{+}2\frac{1}{2} \times ^{-}1\frac{1}{4}) + (^{-}2\frac{1}{2} \times ^{+}1\frac{1}{3})$?

c. Can you find any rational numbers for which multiplication is not distributed over addition?

d. Can we say that the distributive property of multiplication over addition holds for the rational numbers?

4. Closure

a. What do we mean when we say that a certain group of numbers is closed under the operation of multiplication?

b. Is the product of each of the following a rational number?

(1) $\frac{+5}{6} \times \frac{+2}{3}$ (4) $\frac{-5}{14} \times ^{-}28$ (7) $^{+}0.7 \times ^{-}6.9$

(2) $^{+}1\frac{1}{3} \times \frac{-7}{9}$ (5) $^{-}1 \times ^{+}8\frac{1}{5}$ (8) $^{-}2.81 \times ^{-}4.02$

(3) $^{-}3\frac{3}{4} \times ^{+}2\frac{1}{9}$ (6) $0 \times ^{-}6\frac{13}{14}$ (9) $^{-}8.2 \times ^{+}.003$

c. Can you find two rational numbers for which the product is not a rational number?

Positive and Negative Numbers

d. Is the product of two rational numbers always a rational number?
e. Can we say that the rational numbers are closed under the operation of multiplication?

5. Multiplicative Identity

   a. What do we mean when we say that a certain group of numbers has a multiplicative identity?
   b. What is the product of each of the following?
      (1) $1 \times {}^+\frac{3}{4}$ (3) $1 \times {}^-3\frac{1}{6}$ (5) ${}^-9\frac{1}{5} \times 1$ (7) $1 \times {}^-5.41$
      (2) ${}^-\frac{10}{11} \times 1$ (4) ${}^+5\frac{3}{4} \times 1$ (6) $1 \times {}^+7\frac{1}{4}$ (8) ${}^+30.6 \times 1$
   c. In each case in problem **5b**, is the product of the rational number and one (1) the rational number itself?
   d. Can you find any rational number which when multiplied by one (1) does not have the given rational number as the product?
   e. Do the rational numbers have a multiplicative identity? If so, what is it?

6. Multiplicative Inverse

   a. What do we mean by a multiplicative inverse of a rational number?
   b. What is the multiplicative inverse of:
         8? ${}^+\frac{1}{8}$? ${}^-\frac{3}{10}$? ${}^+\frac{12}{7}$? ${}^-2\frac{1}{2}$?
   c. Is each of the following given indicated products a product of a rational number and its multiplicative inverse? Multiply each of the following:
      (1) ${}^+\frac{2}{5} \times {}^+\frac{5}{2}$ (4) ${}^+\frac{3}{2} \times {}^+\frac{2}{3}$ (7) ${}^+10 \times {}^+\frac{1}{10}$
      (2) ${}^-\frac{1}{4} \times {}^-4$ (5) ${}^+2\frac{1}{6} \times {}^+\frac{6}{13}$ (8) ${}^-\frac{5}{8} \times {}^-1\frac{3}{5}$
      (3) ${}^-\frac{7}{9} \times {}^-\frac{9}{7}$ (6) ${}^-3\frac{1}{3} \times {}^-\frac{3}{10}$ (9) ${}^+\frac{7}{22} \times {}^+3\frac{1}{7}$
   d. In each case is the product of the two non-zero rational numbers one (1)?
   e. Can you find any non-zero rational number that does not have a multiplicative inverse?
   f. Does each non-zero rational number have a multiplicative inverse?

3–17 DIVISION OF INTEGERS; PROPERTIES

Since division is the inverse operation of multiplication, to divide we find the factor (quotient) which multiplied by the given factor (divisor) will equal the given product (dividend).

> To divide $^+15$ by $^+3$ means to find the number which multiplied by $^+3$ will equal $^+15$.
> That is: $^+15 \div {}^+3 = ?$ becomes $^+3 \times ? = {}^+15$.
> Since $^+3 \times {}^+5 = {}^+15$.
> then $^+15 \div {}^+3 = {}^+5$.
> $^+15 \div {}^+3 = {}^+5$ is also written as $\dfrac{^+15}{^+3} = {}^+5$

Thus, the quotient of a positive integer divided by a positive integer is a positive number whose absolute value is the quotient of the absolute values of the two given integers.

> To divide $^-15$ by $^-3$ means to find the number which multiplied by $^-3$ will equal $^-15$.
> That is: $^-15 \div {}^-3 = ?$ becomes $^-3 \times ? = {}^-15$.
> Since $^-3 \times {}^+5 = {}^-15$,
> then $^-15 \div {}^-3 = {}^+5$.
> $^-15 \div {}^-3 = {}^+5$ is also written as $\dfrac{^-15}{^-3} = {}^+5$

Thus, the quotient of a negative integer divided by a negative integer is a positive number whose absolute value is the quotient of the absolute values of the two given integers.

> To divide $^-15$ by $^+3$ means to find the number which multiplied by $^+3$ will equal $^-15$.
> That is: $^-15 \div {}^+3 = ?$ becomes $^+3 \times ? = {}^-15$.
> Since $^+3 \times {}^-5 = {}^-15$,
> then $^-15 \div {}^+3 = {}^-5$.
> $^-15 \div {}^+3 = {}^-5$ is also written as $\dfrac{^-15}{^+3} = {}^-5$

Positive and Negative Numbers

Thus, the quotient of a negative integer divided by a positive integer is a negative number whose absolute value is the quotient of the absolute values of the two given integers.

> To divide $^+15$ by $^-3$ means to find the number which multiplied by $^-3$ will equal $^+15$.
> That is: $^+15 \div {}^-3 = ?$ becomes $^-3 \times ? = {}^+15$.
> Since $^-3 \times {}^-5 = {}^+15$,
> then $^+15 \div {}^-3 = {}^-5$.
> $^+15 \div {}^-3 = {}^-5$ is also written as $\frac{^+15}{^-3} = {}^-5$

Thus, the quotient of a positive integer divided by a negative integer is a negative number whose absolute value is the quotient of the absolute values of the two given integers.

Summarizing:

To divide a positive integer by a positive integer or *a negative integer by a negative integer,* we divide their absolute values and prefix the numeral for this quotient with a positive sign.

> Divide: $^+26 \div {}^+2 = ?$
> $^+26 \div {}^+2 = {}^+(26 \div 2) = {}^+13$
> or $\frac{^+26}{^+2} = {}^+13$
>
> Answer, $^+13$
>
> Divide: $^-35 \div {}^-7 = ?$
> $^-35 \div {}^-7 = {}^+(35 \div 7) = {}^+5$
> or $\frac{^-35}{^-7} = {}^+5$
>
> Answer, $^+5$

To divide a positive integer by a negative integer or *a negative integer by a positive integer,* we divide their absolute values and prefix the numeral for this quotient with a negative sign.

> Divide: $^-40 \div {}^+8 = ?$
> $^-40 \div {}^+8 = {}^-(40 \div 8) = {}^-5$
> or $\frac{^-40}{^+8} = {}^-5$
>
> Answer, $^-5$
>
> Divide: $^+16 \div {}^-4 = ?$
> $^+16 \div {}^-4 = {}^-(16 \div 4) = {}^-4$
> or $\frac{^+16}{^-4} = {}^-4$
>
> Answer, $^-4$

Any non-zero integer divided by itself is one.

> Divide: $^-5 \div {^-5} = ?$
>
> $^-5 \div {^-5} = {^+}(5 \div 5) = {^+}1$ or $\frac{^-5}{^-5} = {^+}1$
>
> Answer, $^+1$

Zero divided either by a positive integer or by a negative integer is zero. We cannot divide a positive integer or a negative integer by zero.

> $0 \div {^+7} = 0$ $\frac{0}{^-7} = 0$
>
> Answer, 0   Answer, 0

The numeral naming a positive number does not require the positive ($^+$) sign.

> $^-18 \div 9 = {^-2}$ or $\frac{^-18}{9} = {^-2}$
>
> Answer, $^-2$

Sometimes when an integer is divided by a non-zero integer, the quotient is not an integer but a rational number.

> Divide: $7 \div {^-3} = ?$
>
> $7 \div {^-3} = {^-}(7 \div 3) = \frac{^-7}{3} = {^-2}\frac{1}{3}$
>
> Answer, $^-2\frac{1}{3}$
>
> Divide: $^-6 \div {^-15} = ?$
>
> $^-6 \div {^-15} = {^+}(6 \div 15) = \frac{^+6}{15} = \frac{^+2}{5}$
>
> Answer, $\frac{^+2}{5}$

Positive and Negative Numbers

EXERCISES

Divide as indicated:

1. a. $^+15 \div {^+3}$ b. $^+66 \div {^+6}$ c. $^+54 \div {^+6}$ d. $^+80 \div {^+20}$
2. a. $^-36 \div {^-9}$ b. $^-80 \div {^-10}$ c. $^-32 \div {^-8}$ d. $^-98 \div {^-7}$
3. a. $^-42 \div {^+6}$ b. $^-84 \div {^+6}$ c. $^-121 \div {^+11}$ d. $^-68 \div {^+17}$
4. a. $^+45 \div {^-9}$ b. $^+56 \div {^-7}$ c. $^+17 \div {^-1}$ d. $^+125 \div {^-25}$
5. a. $0 \div {^-4}$ b. $0 \div {^+13}$ c. $0 \div {^-18}$ d. $0 \div {^+40}$
6. a. $^-60 \div 10$ b. $42 \div {^-6}$ c. $^-52 \div 13$ d. $121 \div {^-11}$
7. a. $^-8 \div {^-8}$ b. $^-14 \div 14$ c. $18 \div {^-18}$ d. $^-90 \div {^-90}$
8. a. $8 \div {^-3}$ b. $^-16 \div {^-7}$ c. $^-28 \div 12$ d. $^-50 \div {^-8}$
9. a. $^-2 \div {^-7}$ b. $^-3 \div {^+18}$ c. $8 \div {^-20}$ d. $^-48 \div {^-56}$

10. Divide in each of the following:

    a. $\dfrac{^-16}{^-4}$ $\dfrac{^+24}{^-6}$ $\dfrac{^-48}{^+6}$ $\dfrac{^-75}{^-15}$ $\dfrac{^+60}{^-10}$ $\dfrac{^+99}{^+11}$ $\dfrac{^-92}{^+23}$ $\dfrac{^-250}{^-25}$

    b. $\dfrac{^-18}{^+3}$ $\dfrac{^+72}{^-8}$ $\dfrac{^-37}{^+37}$ $\dfrac{^-84}{^+7}$ $\dfrac{^+96}{^+6}$ $\dfrac{^-75}{^-15}$ $\dfrac{^+150}{^-15}$ $\dfrac{^-72}{^+72}$

    c. $\dfrac{30}{^-6}$ $\dfrac{^-117}{9}$ $\dfrac{^-45}{^-5}$ $\dfrac{^-330}{15}$ $\dfrac{128}{^-16}$ $\dfrac{19}{^-19}$ $\dfrac{^-88}{^-8}$ $\dfrac{675}{^-75}$

    d. $\dfrac{^-60}{5}$ $\dfrac{^-22}{2}$ $\dfrac{84}{^-6}$ $\dfrac{^-160}{8}$ $\dfrac{48}{^-3}$ $\dfrac{^-78}{^-6}$ $\dfrac{195}{^-15}$ $\dfrac{^-540}{^-60}$

11. a. $(^-16 \div {^+4}) \div {^-2}$ c. $(^-32 \div {^-8}) \div {^-4}$ e. $^-80 \div (^-12 \div {^-6})$
    b. $^+50 \div (^+6 \div {^-3})$ d. $^-60 \div (^-24 \div {^+6})$ f. $(^-63 \div {^+7}) \div {^+3}$

12. Find the missing numbers:

    a. $^-21 \div 7 = ?$ d. $^-48 \div {^-6} = \square$ g. $108 \div {^-9} = n$
    b. $32 \div {^-4} = \square$ e. $^-84 \div 9 = ?$ h. $^-95 \div {^-17} = n$
    c. $\dfrac{6}{^-1} = n$ f. $\dfrac{^-42}{^-6} = n$ i. $\dfrac{^-165}{11} = n$

13. Find the missing numbers:

    a. $? \div 3 = {^-6}$ d. $\square \div {^-7} = {^-11}$ g. $n \div {^-17} = 0$
    b. $\square \div {^-1} = 17$ e. $? \div 5 = {^-1}$ h. $n \div {^-8} = {^-96}$
    c. $\dfrac{n}{^-6} = 4$ f. $\dfrac{n}{7} = {^-8}$ i. $\dfrac{n}{^-3} = 21$

Properties

In the following exercises we determine the properties that hold for the integers under the operation of division.

1. **Commutative Property**
   a. If the commutative property holds for the division of integers, what would have to be true?
   b. In each of the following divide as indicated, then compare quotients:

   (1) $^+16 \div {}^+4;\ {}^+4 \div {}^+16.$ Does $^+16 \div {}^+4 = {}^+4 \div {}^+16?$
   (2) $^-20 \div {}^-5;\ {}^-5 \div {}^-20.$ Does $^-20 \div {}^-5 = {}^-5 \div {}^-20?$
   (3) $^+49 \div {}^-7;\ {}^-7 \div {}^+49.$ Does $^+49 \div {}^-7 = {}^-7 \div {}^+49?$
   (4) $^-72 \div {}^+8;\ {}^+8 \div {}^-72.$ Does $^-72 \div {}^+8 = {}^+8 \div {}^-72?$

   c. Does the commutative property hold for the division of integers?

2. **Associative Property**
   a. If the associative property holds for the division of integers, what would have to be true?
   b. In each of the following divide as indicated, then compare quotients:

   (1) $(^+27 \div {}^-9) \div {}^-3;\ {}^+27 \div ({}^-9 \div {}^-3).$
   Does $(^+27 \div {}^-9) \div {}^-3 = {}^+27 \div ({}^-9 \div {}^-3)?$
   (2) $(^-60 \div {}^-6) \div {}^+2;\ {}^-60 \div ({}^-6 \div {}^+2).$
   Does $(^-60 \div {}^-6) \div {}^+2 = {}^-60 \div ({}^-6 \div {}^+2)?$

   c. Does the associative property hold for the division of integers?

3. **Distributive Property**
   a. If the distributive property of division over addition holds for the integers, what would have to be true?
   b. In each of the following perform the indicated operations, then compare answers:

   (1) $^+36 \div ({}^+13 + {}^-4);\ (^+36 \div {}^+13) + ({}^+36 \div {}^-4).$
   Does $^+36 \div ({}^+13 + {}^-4) = ({}^+36 \div {}^+13) + ({}^+36 \div {}^-4)?$
   (2) $({}^-20 + {}^-12) \div {}^-4;\ ({}^-20 \div {}^-4) + ({}^-12 \div {}^-4).$
   Does $({}^-20 + {}^-12) \div {}^-4 = ({}^-20 \div {}^-4) + ({}^-12 \div {}^-4)?$

   c. Is division distributed over addition in problem **3b** (1)? In problem **3b** (2)? When is division distributed over addition in the integers?

4. Closure

   a. What do we mean when we say that a certain group of numbers is closed under the operation of division?

   b. When you divide in each of the following, is the quotient in each case an integer?

   (1) $^-15 \div {}^-5$ (4) $^-1 \div {}^+4$ (7) $^+7 \div {}^-12$
   (2) $^+6 \div {}^+8$ (5) $^+15 \div {}^-24$ (8) $^-27 \div {}^-1$
   (3) $0 \div {}^-4$ (6) $^-64 \div {}^+16$ (9) $^-24 \div {}^+16$

   c. Can you find two integers for which the quotient is not an integer provided that division by zero is excluded?

   d. Can we say that the integers are closed under the operation of division provided that division by zero is excluded?

3–18 DIVISION OF RATIONAL NUMBERS; PROPERTIES

We follow the same procedures to divide rational numbers as we do when we divide integers. See section 3–17. However, it may be necessary to review at this time the division of the rational numbers of arithmetic (common fractions, mixed numbers, and decimal fractions). See sections 1–15 and 1–22.

To divide a positive rational number by a positive rational number or *a negative rational number by a negative rational number,* we divide their absolute values and prefix the numeral for this quotient with a positive sign.

Divide: $\frac{-3}{4} \div \frac{-6}{7} = ?$

$\frac{-3}{4} \div \frac{-6}{7} = {}^+\left(\frac{3}{4} \div \frac{6}{7}\right) = {}^+\left(\frac{3}{4} \times \frac{7}{6}\right) = {}^+\frac{7}{8}$

Answer, $^+\frac{7}{8}$

To divide a positive rational number by a negative rational number or *a negative rational number by a positive rational number,* we divide their absolute values and prefix the numeral for this quotient with a negative sign.

> Divide $^-2\frac{1}{4} \div {}^+1\frac{1}{2} = ?$
>
> $^-2\frac{1}{4} \div {}^+1\frac{1}{2} = {}^-(2\frac{1}{4} \div 1\frac{1}{2})$
> $= {}^-(\frac{9}{4} \div \frac{3}{2}) = {}^-(\frac{9}{4} \times \frac{2}{3}) = {}^-\frac{3}{2} = {}^-1\frac{1}{2}$
>
> Answer, $^-1\frac{1}{2}$

Any non-zero rational number divided by itself is one (1).

> Divide: $^-.5 \div {}^-.5 = ?$
>
> $^-.5 \div {}^-.5 = {}^+(.5 \div .5) = {}^+1$ or $\frac{^-.5}{^-.5} = {}^+1$
>
> Answer, $^+1$

Zero divided either by a positive rational number or by a negative rational number is zero. We cannot divide a positive rational number or a negative rational number by zero.

> $0 \div {}^-3\frac{1}{3} = 0$ $0 \div {}^-7.6 = 0$
> Answer, 0   Answer, 0

The numeral naming a positive rational number does not require the positive (+) sign.

EXERCISES

Divide as indicated:

1. a. $^+\frac{5}{16} \div {}^+\frac{3}{8}$ b. $^+16 \div {}^+\frac{2}{5}$ c. $^+3\frac{1}{2} \div {}^+21$ d. $^+8\frac{1}{3} \div {}^+\frac{5}{6}$
2. a. $^+0.6 \div {}^+0.3$ b. $^+18 \div {}^+0.6$ c. $^+6.4 \div {}^+0.16$ d. $^+0.002 \div {}^+0.0005$
3. a. $^-\frac{5}{6} \div {}^-\frac{5}{12}$ b. $^-\frac{4}{5} \div {}^-2$ c. $^-1\frac{7}{8} \div {}^-\frac{5}{32}$ d. $^-3\frac{1}{3} \div {}^-\frac{5}{6}$
4. a. $^-5.4 \div {}^-0.3$ b. $^-3.9 \div {}^-3$ c. $^-0.006 \div {}^-0.03$ d. $^-28 \div {}^-0.07$

Positive and Negative Numbers

5. a. $\frac{-4}{5} \div \frac{+1}{4}$ b. $^-6 \div \frac{+1}{5}$ c. $\frac{-7}{12} \div \frac{+1}{15}$ d. $^-2\frac{3}{4} \div {^+2\frac{1}{16}}$

6. a. $^-0.54 \div {^+0.06}$ b. $^-0.0032 \div {^+0.8}$ c. $^-13 \div {^+5.2}$ d. $^-0.168 \div {^+28}$

7. a. $\frac{+3}{8} \div \frac{-3}{4}$ b. $\frac{+8}{9} \div {^-3}$ c. $^+8\frac{3}{4} \div {^-5}$ d. $^+2\frac{1}{4} \div {^+1\frac{11}{16}}$

8. a. $^+7.2 \div {^-1.2}$ b. $^+3 \div {^-1.2}$ c. $^+6.405 \div {^-0.05}$ d. $^+0.0018 \div {^-0.6}$

9. a. $0 \div \frac{-13}{16}$ b. $0 \div {^-2\frac{3}{4}}$ c. $0 \div {^-0.9}$ d. $0 \div {^-3.7}$

10. a. $\frac{-3}{5} \div \frac{3}{4}$ b. $3\frac{3}{4} \div {^-4\frac{3}{5}}$ c. $^-3.6 \div 0.04$ d. $96 \div {^-1.6}$

11. a. $\frac{-4}{5} \div \frac{-4}{5}$ b. $^-6\frac{7}{12} \div 6\frac{7}{12}$ c. $0.58 \div {^-0.58}$ d. $^-3.415 \div {^-3.415}$

12. Divide in each of the following:

   a. $\dfrac{\frac{-2}{3}}{\frac{5}{6}}$ $\dfrac{^-6}{\frac{-9}{10}}$ $\dfrac{\frac{3}{4}}{^-15}$ $\dfrac{\frac{-11}{12}}{^-1\frac{1}{6}}$ $\dfrac{3\frac{1}{8}}{^-3\frac{3}{4}}$ $\dfrac{\frac{-5}{12}}{1\frac{2}{3}}$ $\dfrac{^-6\frac{1}{4}}{^-3\frac{1}{3}}$

   b. $\dfrac{^-42.5}{0.5}$ $\dfrac{1.21}{^-11}$ $\dfrac{^-8}{^-1.6}$ $\dfrac{^-0.1}{0.5}$ $\dfrac{^-0.03}{^-0.006}$ $\dfrac{8.03}{^-8.03}$ $\dfrac{^-0.005}{^-0.15}$

13. a. $\left(\frac{-5}{8} \div \frac{-9}{16}\right) \div \frac{+5}{6}$ c. $\left(^-3\frac{1}{3} \div {^+2\frac{1}{2}}\right) \div {^+2\frac{1}{4}}$ e. $(1.8 \div {^-0.3}) \div {^-0.2}$

    b. $\frac{-4}{5} \div \left(\frac{-3}{4} \div \frac{-7}{8}\right)$ d. $^-3\frac{1}{8} \div \left(^-15 \div {^-6\frac{2}{3}}\right)$ f. $^-12.6 \div (^-5.4 \div {^-0.03})$

14. Find the missing numbers:

    a. $\frac{-9}{16} \div \frac{-7}{8} = ?$ d. $\frac{-5}{6} \div \frac{-5}{6} = \square$ g. $3\frac{7}{16} \div {^-5} = n$

    b. $\frac{9}{10} \div \frac{-7}{12} = \square$ e. $^-2\frac{1}{10} \div \frac{3}{4} = ?$ h. $^-3\frac{1}{9} \div {^-2\frac{1}{3}} = n$

    c. $\dfrac{^-3}{\frac{1}{2}} = n$ f. $\dfrac{4}{^-3\frac{1}{5}} = n$ i. $\dfrac{^-1\frac{1}{8}}{^-6\frac{3}{4}} = n$

15. Find the missing numbers:

    a. $^-0.9 \div {^-0.3} = \square$ d. $0.4 \div {^-0.16} = ?$ g. $55.8 \div {^-0.23} = n$

    b. $^-0.04 \div 7 = ?$ e. $^-0.406 \div {^-5.8} = \square$ h. $^-0.003 \div 0.003 = n$

    c. $\dfrac{8.4}{^-0.6} = n$ f. $\dfrac{^-8}{0.2} = n$ i. $\dfrac{^-0.4}{^-0.01} = n$

Properties

In the following exercises we determine the properties that hold for the rational numbers under the operation of division.

1. Commutative Property

    a. If the commutative property holds for the division of rational numbers, what would have to be true?

b. In each of the following divide as indicated, then compare quotients:
(1) $^+\frac{2}{3} \div {}^-\frac{1}{3}$; $^-\frac{1}{3} \div {}^+\frac{2}{3}$. Does $^+\frac{2}{3} \div {}^-\frac{1}{3} = {}^-\frac{1}{3} \div {}^+\frac{2}{3}$?
(2) $^-2\frac{1}{4} \div {}^-\frac{1}{2}$; $^-\frac{1}{2} \div {}^-2\frac{1}{4}$. Does $^-2\frac{1}{4} \div {}^-\frac{1}{2} = {}^-\frac{1}{2} \div {}^-2\frac{1}{4}$?
(3) $^-4\frac{1}{2} \div {}^+2\frac{1}{4}$; $^+2\frac{1}{4} \div {}^-4\frac{1}{2}$. Does $^-4\frac{1}{2} \div {}^+2\frac{1}{4} = {}^+2\frac{1}{4} \div {}^-4\frac{1}{2}$?
(4) $^+.6 \div {}^+.03$; $^+.03 \div {}^+.6$. Does $^+.6 \div {}^+.03 = {}^+.03 \div {}^+.6$?

c. Does the commutative property hold for the division of rational numbers?

2. Associative Property
    a. If the associative property holds for the division of rational numbers, what would have to be true?
    b. In each of the following divide as indicated, then compare quotients:
    (1) $(^-\frac{7}{8} \div {}^+\frac{1}{4}) \div {}^-\frac{1}{2}$; $^-\frac{7}{8} \div ({}^+\frac{1}{4} \div {}^-\frac{1}{2})$.
        Does $(^-\frac{7}{8} \div {}^+\frac{1}{4}) \div {}^-\frac{1}{2} = {}^-\frac{7}{8} \div ({}^+\frac{1}{4} \div {}^-\frac{1}{2})$?
    (2) $(^+2\frac{1}{9} \div {}^-1\frac{1}{6}) \div {}^-3\frac{1}{3}$; $^+2\frac{1}{9} \div ({}^-1\frac{1}{6} \div {}^-3\frac{1}{3})$.
        Does $(^+2\frac{1}{9} \div {}^-1\frac{1}{6}) \div {}^-3\frac{1}{3} = {}^+2\frac{1}{9} : ({}^-1\frac{1}{6} : {}^-3\frac{1}{3})$?

    c. Does the associative property hold for the division of rational numbers?

3. Distributive Property
    a. If the distributive property of division over addition holds for the rational numbers, what would have to be true?
    b. In each of the following perform the indicated operations, then compare answers:
    (1) $^-\frac{3}{5} \div ({}^+\frac{6}{15} + {}^-\frac{1}{3})$; $(^-\frac{3}{5} \div {}^+\frac{6}{15}) + ({}^-\frac{3}{5} \div {}^-\frac{1}{3})$.
        Does $^-\frac{3}{5} \div ({}^+\frac{6}{15} + {}^-\frac{1}{3}) = ({}^-\frac{3}{5} \div {}^+\frac{6}{15}) + ({}^-\frac{3}{5} \div {}^-\frac{1}{3})$?
    (2) $(^+2\frac{2}{3} + {}^-4\frac{1}{4}) \div {}^-1\frac{1}{2}$; $(^+2\frac{2}{3} \div {}^-4\frac{1}{4}) + ({}^-2\frac{2}{3} \div {}^-1\frac{1}{2})$.
        Does $(^+2\frac{2}{3} + {}^-4\frac{1}{4}) \div {}^-1\frac{1}{2} = ({}^+2\frac{2}{3} \div {}^-4\frac{1}{4}) + ({}^-2\frac{2}{3} \div {}^-1\frac{1}{2})$?

    c. Is division distributed over addition in problem **3b (1)**? In problem **3b (2)**? When is division distributed over addition in the rational numbers?

4. Closure
    a. What do we mean when we say that a certain group of numbers is closed under the operation of division?

Positive and Negative Numbers

b. When you divide in each of the following, is the quotient in each case a rational number?

(1) $\frac{^-5}{7} \div \frac{^-2}{7}$ (4) $-3\frac{1}{3} \div +2\frac{1}{2}$ (7) $+2.16 \div ^-.06$

(2) $\frac{^+1}{4} \div \frac{^-3}{8}$ (5) $\frac{^+11}{14} \div -2\frac{3}{7}$ (8) $^+.08 \div ^+.14$

(3) $0 \div \frac{^-7}{15}$ (6) $-2\frac{7}{10} \div -2$ (9) $^-.12 \div ^-.1$

c. Can you find two rational numbers for which the quotient is not a rational number provided that division by zero is excluded?

d. Can we say that the rational numbers are closed under the operation of division provided that division by zero is excluded?

3–19 COMPUTATION—USING NUMERALS WITH CENTERED SIGNS

The number that is the *opposite of six,* named by the numeral -6, and the number *negative six,* named by the numeral $^-6$ are one and the same number. Since there is no need to have two symbols that differ so slightly to name the same number, to simplify matters we shall discard the use of the raised sign and henceforth use only numerals with centered signs to name numbers. The numeral -6 represents both the *opposite of six* and the number *negative six.* The numeral $+6$ or 6 names the number *positive six.*

While both sign locations are correct in naming signed numbers, the numerals having the signs centered are more generally used.

To compute with signed numbers named by numerals with centered signs, we follow the same procedures as with numbers named by numerals having raised signs.

Also, to simplify an algebraic expression such as $7 - 11 + 6 - 3$, we may take it to mean $(+7) + (-11) + (+6) + (-3)$ where the given signs tell you which numbers are positive and which are negative but the operation is considered to be addition.

$$+7 - 11 + 6 - 3 = (+7) + (-11) + (+6) + (-3) = -1$$

EXERCISES

1. Add:

   a. $+7 \atop +5$ $\quad -6 \atop -7$ $\quad +7 \atop -4$ $\quad +2 \atop -7$ $\quad -9 \atop +3$ $\quad -11 \atop +14$ $\quad -3 \atop -8$ $\quad +3 \atop -13$ $\quad -17 \atop +16$ $\quad -8 \atop +25$

   b. $+6 \atop -6$ $\quad -8 \atop +8$ $\quad -4 \atop +4$ $\quad 0 \atop +13$ $\quad -4 \atop 0$ $\quad +\frac{3}{5} \atop +\frac{1}{5}$ $\quad +\frac{5}{8} \atop +\frac{7}{8}$ $\quad -\frac{2}{3} \atop -\frac{3}{4}$ $\quad -0.4 \atop +1.7$ $\quad -3.86 \atop -1.14$

2. Add as indicated:

   a. $(+4) + (-8)$
   b. $(-5) + (+7) + (-3)$
   c. $[(-8) + (-12)] + (-2)$
   d. $(+2) + (+6) + (-8)$
   e. $(-13) + (+6) + (+8) + (-2)$
   f. $[(+16) + (-9)] + [(+3) + (-8)]$

3. Simplify:

   a. $9 + 3$
   b. $7 - 8$
   c. $-3 + 4$
   d. $-6 - 9$
   e. $8 - 14$
   f. $-3 - 6$
   g. $8 - 3 - 4$
   h. $11 + 9 - 8$
   i. $15 - 8 + 3$
   j. $6 - 9 + 2$
   k. $5 - 7 - 4$
   l. $-3 + 6 - 2$
   m. $12 - 3 + 5 - 11$
   n. $-4 + 6 + 3 - 6$
   o. $2 - 9 + 1 - 3 + 4$
   p. $4 + 3 - 8 + 7 - 6$

4. Subtract:

   a. $+32 \atop +27$ $\quad +7 \atop +8$ $\quad 12 \atop 4$ $\quad 3 \atop 11$ $\quad -6 \atop -3$ $\quad -3 \atop -13$ $\quad +8 \atop -5$ $\quad -2 \atop +14$ $\quad -9 \atop +2$ $\quad +11 \atop -11$

   b. $+6 \atop +6$ $\quad -3 \atop -3$ $\quad 0 \atop +5$ $\quad 8 \atop 0$ $\quad 0 \atop -2$ $\quad +\frac{7}{8} \atop -\frac{1}{8}$ $\quad -.15 \atop .28$ $\quad -2\frac{1}{2} \atop +\frac{2}{3}$ $\quad +2.08 \atop +1.62$ $\quad -3\frac{1}{2} \atop +2\frac{1}{2}$

5. Subtract as indicated:

   a. $(-7) - (+9)$
   b. $(-6) - (-2)$
   c. $(0) - (9)$
   d. $(+8) - (-4)$
   e. $(8 - 6) - (3 - 9)$
   f. $(7 - 12) - (5 - 8)$
   g. $(-8) + (-4) - (-14)$
   h. $(+5) - (-7) - (+5)$
   i. $[(-12) - (+7)] - (-3)$

6. a. From -9 subtract 5.
   b. Subtract 8 from 0.
   c. Take 13 from 8.
   d. From -2 take -2.
   e. Subtract 13 from 6.
   f. From 7 take -5.

7. Multiply:

   a. $7 \atop 5$ $\quad +8 \atop +7$ $\quad -6 \atop -6$ $\quad -7 \atop -9$ $\quad -6 \atop 7$ $\quad -5 \atop +5$ $\quad +8 \atop -6$ $\quad +14 \atop -3$ $\quad -1 \atop +1$ $\quad -11 \atop -11$

Positive and Negative Numbers

b. | 0 | −9 | 0 | +6 | 0 | +0.2 | +0.05 | $-2\frac{1}{2}$ | $-4\frac{3}{4}$ | −2.5 |
    | 8 | 0 | −3 | 0 | +4 | 0.3 | +0.64 | −4 | $+3\frac{1}{3}$ | −0.03 |

8. Multiply as indicated:
   a. (−8) × (−7)
   b. (+7) · (−6)
   c. (−3)(+3)
   d. −6(8 − 12)
   e. (+4)(−0.9)
   f. $-\frac{3}{4}(-8)$
   g. (−9)(0)(−3)
   h. [(+15)(−2)] × (−3)
   i. (−8) × [(−3)(−4)]
   j. −7(3 − 8) − (4 − 5)
   k. (−4)(−3)(+2)(−3)
   l. (−4)(−3)(−1)(−5)

9. Find the value of each of the following:
   a. (−3)²
   b. (+6)²
   c. (−5)³
   d. (+5)⁴
   e. (−1)⁵
   f. (−3)⁴
   g. (−2)⁶
   h. (+4)³

10. Divide as indicated:
    a. $\frac{+16}{+4}$ $\frac{-20}{-4}$ $\frac{+27}{-9}$ $\frac{-56}{+7}$ $\frac{-51}{-17}$ $\frac{-63}{-9}$ $\frac{-90}{+30}$ $\frac{+42}{+6}$ $\frac{-60}{+12}$ $\frac{-75}{-5}$
    b. $\frac{-7}{-7}$ $\frac{+16}{-16}$ $\frac{-31}{+31}$ $\frac{-7}{-1}$ $\frac{-15}{+1}$ $\frac{+8}{-1}$ $\frac{0}{+4}$ $\frac{0}{-8}$ $\frac{0}{+6}$ $\frac{0}{-5}$

CHAPTER REVIEW

1. Which of the following numerals name positive numbers? Which name negative numbers? (3–1)

   ⁻81   ⁺3.9   0 $\frac{-4}{5}$ $\frac{+48}{12}$ ⁻.642 $+7\frac{3}{8}$

2. Which of the following numerals name integers? Which name rational numbers? Which name irrational numbers? Which name real numbers? (3–1) (3–2)

   $\frac{+13}{16}$ ⁻.7   ⁺15 $-\sqrt{22}$ $+3\frac{5}{6}$ $\frac{-18}{3}$

3. a. Find the value of: (1) |⁺38| (2) $\left|\frac{-11}{12}\right|$ (3–3)

   b. Which is greater: The absolute value of ⁻6 or the absolute value of ⁺4? (3–3)

4. Draw the graph of ⁻4, ⁻1, 3, and 4 on a number line. (3–4)

5. Write the coordinates of which the following is the graph: (3–4)

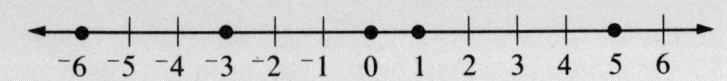

   $\leftarrow\bullet\,+\,+\,\bullet\,+\,+\,\bullet\,\bullet\,+\,+\,+\,\bullet\,+\,\rightarrow$
   $\quad\,^-6\,\,^-5\,\,^-4\,\,^-3\,\,^-2\,\,^-1\,\,0\,\,1\,\,2\,\,3\,\,4\,\,5\,\,6$

6. Which of the following sentences are true? (3–5)
   a. $^-10 < ^-3$ b. $^-9 > 0$ c. $^+4 \not< ^-8$ d. $^-2 \not> ^+1$ e. $^+3 > ^+5$

7. If 25 meters to the left is indicated by $^-25$ meters, how can 63 meters to the right be indicated? (3–6)

8. Using the number line as the scale, draw the vector and write the numeral represented by this vector that illustrates a movement from: (3–7)
   a. 0 to $^-5$ b. $^-4$ to $^+2$ c. $^-3$ to $^-7$ d. $^+6$ to $^-1$

9. a. What is the opposite of $^-.28$? (3–8)
   b. What is the additive inverse of $^-1\frac{7}{8}$?
   c. What is the opposite of the opposite of $^+43$?
   d. Write symbolically: The opposite of positive nineteen is negative nineteen.
   e. What is the value of: (1.) $-(^-1.7) = ?$ (2.) $-(^+51) = ?$
   f. Which has the greater opposite number: $^-3$ or $^-7$?

10. Add on the number line, using vectors: (3–9)
    a. $^+3 + ^+2$ b. $^-4 + ^+1$ c. $^+6 + ^-8$ d. $^-2 + ^-7$

11. Add as indicated: (3–10, 3–19)
    a. $^-6 + ^+5$
    b. $^-7 + ^-9$
    c. $(-8) + (+11)$
    d. $(+10) + (-10)$
    e. $(-1) + (-4) + (+6) + (-3)$
    f. Simplify: $6 - 3 + 5 - 9 + 1 - 7 - 2 + 10 - 4$

12. Add as indicated: (3–11, 3–19)
    a. $^+\frac{3}{4} + ^-\frac{3}{8}$
    b. $^-.36 + ^+.4$
    c. $(^+2\frac{1}{2}) + (^-5\frac{2}{3})$
    d. $(^-5.4) + (^-.91)$

13. Subtract on the number line, using vectors: (3–12)
    a. $^-4 - ^+2$
    b. $^-3 - ^-6$
    c. $^+2 - ^-5$
    d. $^+1 - ^+7$

Positive and Negative Numbers

14. Subtract as indicated: (3–13, 3–19)

    a. $^-26 - {^+17}$
    b. $^-7 - {^-8}$
    c. $10 - 15$
    d. $(+9) - (-4)$

15. Subtract as indicated: (3–14, 3–19)

    a. $^+\frac{3}{5} - {^-\frac{2}{3}}$
    b. $^-.04 - {^+.135}$
    c. $\left(-4\frac{3}{4}\right) - \left(-6\frac{1}{2}\right)$
    d. $8.2 - 11$

16. Multiply as indicated: (3–15, 3–19)

    a. $^-3 \times {^-9}$
    b. $^+8 \times {^-6}$
    c. $(+5)(-1)$
    d. $(-7)(-2)$
    e. $(-1)(-3)(+2)(-5)(-4)$

17. Multiply as indicated: (3–16, 3–19)

    a. $^-1\frac{2}{3} \times {^-1\frac{1}{2}}$
    b. $^+.56 \times {^-.09}$
    c. $\left(-\frac{9}{20}\right)\left(+\frac{5}{6}\right)$
    d. $(-1.38)(-2.5)$

18. Divide as indicated: (3–17, 3–19)

    a. $^+60 \div {^-12}$
    b. $^-42 \div {^-7}$
    c. $(18) \div (-2)$
    d. $(-15) \div (-15)$

19. Divide as indicated: (3–18, 3–19)

    a. $^-\frac{3}{8} \div {^-\frac{5}{6}}$
    b. $^+6.56 \div {^-.08}$
    c. $\left(-2\frac{1}{4}\right) \div \left(+3\frac{1}{2}\right)$
    d. $(-10.5) \div (-.1)$

20. a. Does the associative property hold for the: (3–10, 3–11, 3–13 to 3–18)

    (1) multiplication of integers?
    (2) subtraction of rational numbers?
    (3) division of integers?
    (4) addition of rational numbers?
    (5) division of rational numbers?
    (6) addition of integers?
    (7) multiplication of rational numbers?
    (8) subtraction of integers?

  b. What is the multiplicative inverse of: (3–16)

    (1) $^-100$?   (2) $^+\frac{1}{12}$?   (3) $^-\frac{2}{3}$?   (4) $^+\frac{11}{4}$?   (5) $^-3\frac{1}{2}$?

ACHIEVEMENT TEST

The numeral at the end of each problem indicates the section where explanatory material may be found.

1. Add:

    39,969
    45,784
    92,638
    25,709
    63,175   (1–2)

2. Subtract:

    526,003
    405,369   (1–3)

3. Multiply:

    3,859
      487   (1–4)

4. Divide:

    793)672,464   (1–5)

5. Add:

    $1\frac{3}{4} + \frac{2}{5} + 3\frac{9}{10}$ (1–12)

6. Subtract:

    $16\frac{5}{8} - 9$ (1–13)

7. Multiply:

    $3\frac{1}{7} \times 4\frac{3}{8}$ (1–14)

8. Divide:

    $10\frac{4}{5} \div 2\frac{7}{10}$ (1–15)

9. Add:

    .983 + 62.7 + 4.51   (1–18)

10. Subtract:

    $85 − $3.79   (1–19)

11. Multiply:

    .003 × 500   (1–20)

12. Divide:

    12 ÷ .8   (1–22)

13. Find 11.9% of $6,457.   (1–30, 1–33, 1–34)
14. What percent of 85 is 51?   (1–31, 1–33, 1–34)
15. 6% of what number is $1,500?   (1–32, 1–33, 1–34)
16. Find the square root of 88,209.   (1–36)
17. What is the greatest common factor of 48, 80, and 144?   (1–38)
18. Find the least common multiple of 18, 27, and 63.   (1–40)

Positive and Negative Numbers

19. Which of the following statements are true and what property is illustrated by each of the true statements? (1–44)

    a. 20 × 5 = 5 × 20
    b. 20 − 5 = 5 − 20
    c. 20 ÷ 5 = 5 ÷ 20
    d. 20 + 5 = 5 + 20
    e. (60 − 10) − 2 = 60 − (10 − 2)
    f. (60 × 10) × 2 = 60 × (10 × 2)
    g. (60 + 10) + 2 = 60 + (10 + 2)
    h. (60 ÷ 10) ÷ 2 = 60 ÷ (10 ÷ 2)
    i. 8 × (2 + 4) = (8 × 2) + (8 × 4)
    j. 8 ÷ (2 + 4) = (8 ÷ 2) + (8 ÷ 4)

20. Change:

    a. 10,000 m to km   (2–1)   c. 128.5 cm³ to dm³   (2–5)
    b. 43 m² to cm²   (2–4)

21. Change:

    a. $4\frac{5}{8}$ yd. to in.   (2–6)   c. 810 cu. ft. to cu. yd. (2–11)
    b. 37 sq. ft. to sq. in.   (2–10)

22. Change:

    a. 58.4 kg to g   (2–2)   c. 68.5 mg to g   (2–2)
    b. .726 L to cL   (2–3)

23. Which is colder, a temperature of 25°C or 80°F?   (2–16)

24. Which is faster, a speed of 20 m/s or 60 km/h?   (2–15)

25. a. What part of a minute is 48 seconds?   (2–13)

    b. Write in A.M. or P.M. time: (1) 1750   (2) 0003   (2–17)

    c. If it is 10 P.M. in Salt Lake City, what time is it in Pittsburgh? Minneapolis? Portland, Oregon?   (2–18)

26. Which is more precise, 19 cm or .19 cm? Which is more accurate?   (2–19)

27. How many significant digits are in each of the following?   (2–20)

    a. 3,060   b. 91,000   c. .008   d. 7.4 × 10⁵   e. 12.090

28. Which of the following numerals name positive numbers? Which name negative numbers? (3–1)

   $^{-}.375$ $^{+}54$ $^{-}\frac{2}{3}$ $^{+}.181$ $^{-}2\frac{3}{4}$ $^{-}200$

29. Which of the following numerals name integers? Which name rational numbers? Which name irrational numbers? Which name real numbers? (3–1, 3–2)

   $^{-}.6$ $^{+}3\frac{1}{2}$ $^{-}17$ $^{+}\frac{32}{4}$ $^{-}\sqrt{41}$ $^{-}\frac{9}{10}$

30. a. Are all the integers closed under the operation of addition? Subtraction? Multiplication? Division? (3–10 to 3–18)

   b. What is the additive identity for the integers? For the rational numbers?

   c. What is the multiplicative identity for the integers? For the rational numbers?

31. On a number line, draw the graph of: $-5, -2, 0, 1$, and 4. (3–4)

32. Write the coordinates of which the following is the graph: (3–4)

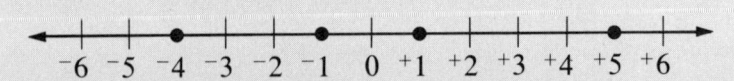

33. Which of the following sentences are true? (3–5)

   a. $^{-}10 < ^{-}4$ b. $^{-}7 > 0$ c. $^{-}12 \not< ^{-}12$ d. $^{-}3 \not> ^{+}2$

34. a. What is the opposite of $^{-}25$? (3–8)

   b. What is the additive inverse of $^{+}.3$?

   c. What is the value of: (1) $-(^{+}48) = ?$ (2) $-\left(^{-}4\frac{1}{2}\right) = ?$

35. a. Find the value of: (3–3)

   (1) $|^{+}62| = ?$

   (2) $|^{-}.33| = ?$

   b. Find the multiplicative inverse of: (3–16)

   (1) $^{+}\frac{1}{10}$

   (2) $^{-}8$

   (3) $^{-}\frac{5}{6}$

Positive and Negative Numbers   231

36. Using the number line as a scale, draw the vector and write the numeral represented by this vector that illustrates a movement: (3–7)

    a. From ⁻5 to ⁺4

    b. From ⁺3 to ⁻3

37. Add as indicated: (3–11, 3–19)

    a. ⁺7 + ⁻9

    b. ⁻3 + ⁻1

    c. $(-14) + (+14)$

    d. $(+8) + (-5)$

    e. $(-2) + (-5) + (+8) + (+6) + (-7)$

    f. Simplify: $10 - 8 + 1 - 5 - 6 + 11 - 3 + 1 - 4$

38. Subtract as indicated: (3–14, 3–19)

    a. ⁺3 − ⁻4   **c.** $8 - 10$

    b. 0 − ⁺6   **d.** $(+7) - (-3)$

39. Multiply as indicated: (3–15, 3–19)

    a. ⁻4 × ⁻5   **c.** $(+3)(-3)$

    b. ⁻1 × ⁺6   **d.** $(-3)(-1)(+5)(-2)(-3)$

40. Divide as indicated: (3–16, 3–19)

    a. ⁻48 ÷ ⁻8   **c.** $(+27) \div (-3)$

    b. ⁻30 ÷ ⁺5   **d.** $(-18) \div (-18)$

4

CHAPTER 4
Algebra

LANGUAGE OF ALGEBRA

4–1 SYMBOLS—REVIEW

SYMBOLS OF OPERATION

In algebra we continue to use the operational symbols of arithmetic.

The signs of operation may assume any one of several meanings, as follows:

The *addition* symbol or plus symbol $+$ means sum, add, more than, increased by, exceeded by.

The *subtraction* symbol or minus symbol $-$ means difference, subtract, take away, less, less than, decreased by, diminished by.

The *multiplication* symbol $\times$ or the raised dot $(\cdot)$ means product, multiply, times.

The *division* symbol $\div$ or $\overline{)}$ means quotient, divide. The fraction bar, as in $\frac{a}{b}$, is generally used in algebra to indicate division. The word "over" is sometimes used to express the relationship of the numerator to the denominator.

Other symbols used in both algebra and arithmetic include exponents, the square root symbol, parentheses, and the verbs of mathematical sentences such as $=$, $\neq$, $>$, $\not>$, $<$, $\not<$, $\geq$, $\not\geq$, $\leq$, and $\not\leq$.

EXPONENT

An *exponent* is the small numeral written to the upper right (superscript) of the base. When it names a natural number, it tells how many times the factor is being used in multiplication. 8^2 represents 8×8 and is read "the square of eight" and 8^3 represents $8 \times 8 \times 8$ and is read "the cube of eight." The factor that is being repeated is called the *base*. The number 8^4 uses 8 as the base and 4 as the exponent. Numbers such as 8^2, 8^3, 8^4, 8^5, etc. are called *powers* of 8.

SQUARE ROOT

The *square root* symbol $\sqrt{}$ written over the numeral indicates the square root of the corresponding number. The *square root* of a number is that number which when multiplied by itself produces the given number.

PARENTHESES

Parentheses ( ) are generally used to group together two or more numerals so that they are treated as a single quantity. Sometimes parentheses are used to set off a numeral from another so that the meaning will not be misunderstood. (9 + 5) may be read as "the quantity nine plus five."

CONSTANT

A numeral is sometimes called a *constant* because it names a definite number.

VARIABLE

A *variable* is a letter (small letter, capital letter, letter with a subscript such as b_1, read "b sub one," or letter with a prime mark such as S', read "S prime") or a frame (such as □, ▭, ○, and △) or a blank which holds a place open for a number. Sometimes the first letter of a key word is used as the variable. A variable may represent any number, but under certain conditions it represents a specific number or numbers.

$-x$ means the opposite of the variable x and not negative x or minus x. $-x$ is not necessarily a negative number. When x is a positive number, $-x$ is a negative number; but when x is a negative number, $-x$ is a positive number.

EXERCISES

1. Read, or write in words, each of the following:

   a. $19 - 8$
   b. 10×36
   c. $\frac{21}{7}$
   d. $51 + 47$
   e. 15^2
   f. $\sqrt{62}$
   g. $(11) \times (72)$
   h. $(32 - 29)$
   i. 56^3
   j. $6 \cdot 45$
   k. 44^2
   l. $\sqrt{102}$
   m. $(72 + 54)$
   n. $(9 \times 8) \times 12$
   o. $4 + (3 + 11)$

Algebra

2. What symbol represents the variable in each of the following?

   a. $\square - 9 = 31$ e. $T' - 32 = 18$

   b. $28 + ? = 56$ f. $\frac{n}{3} = 8$

   c. $x + 7 = 31$ g. $15 + 29 = \triangle$

   d. $9y = 54$ h. $48 = b_2 - 11$

3. In each of the following, what is the base? What is the exponent? How many times is the base being used as a factor?

   a. 6^{11} e. 21^{16}
   b. 10^7 f. 1^{100}
   c. 3^9 g. 2^8
   d. 15^{10} h. 25^2

4–2 MATHEMATICAL PHRASES OR MATHEMATICAL EXPRESSIONS

A *numerical expression* or *numerical phrase* consists of a single numeral with or without operational symbols, like 15 or 2^5, or two or more numerals with operational symbols, like $6 + 5$; $63 - 27$; 9×23; $54 \div 9$; $8 \times (2 + 9)$; etc.

An *algebraic expression* or *algebraic phrase* may be a numerical expression as described above or an expression containing one or more variables joined by operational symbols like d; $4a - b$; $7x^2 + 5x - 2$; etc.

Both numerical expressions and algebraic expressions are *mathematical expressions*.

In an algebraic expression no multiplication symbol is necessary when the factors are two letters (variables) or a numeral and a letter. In the latter case the numeral always precedes the letter or the variable. This numeral is usually called the *numerical coefficient* of the variable. a times b may be expressed as $a \times b$, $a \cdot b$, ab (preferred), $(a)(b)$, $a(b)$, or $(a)b$. In $6y^4$, the 6 is the numerical coefficient of y^4.

To write algebraic expressions, we write numerals, variables, and operational symbols as required in the proper order.

The sum of x and six. *Answer*, $x + 6$

The product of b and five. *Answer*, $5b$

The square of the distance (d). *Answer*, d^2

The difference between the base (b) and the altitude (a). *Answer*, $b - a$

The circumference (c) divided by pi (π). *Answer*, $\dfrac{c}{\pi}$

The square root of the area (A). *Answer*, $\sqrt{A}$

Eight times the sum of sixteen and y. *Answer*, $8(16 + y)$

EXERCISES

1. Write each of the following as a numerical expression:
   a. Eighteen added to ten.
   b. From fifteen subtract nine.
   c. Twenty times seventeen.
   d. Sixty divided by twelve.
   e. The sum of seven and five.
   f. The difference between sixty and twenty-one.
   g. The product of six and eight.
   h. The quotient of fourteen divided by seven.
   i. The square of ten.
   j. The cube of twenty-one.
   k. The square root of six.
   l. Nine times the sum of five and two.

2. Write each of the following as an algebraic expression:
   a. c added to b.
   b. From m subtract r.
   c. t times x.
   d. a divided by d.
   e. The sum of p and i.
   f. The difference between l and m.
   g. The product of b and h.
   h. The quotient of d divided by r.

Algebra

i. The square of V.
j. The cube of s.
k. The square root of A.
l. Two times the sum of l and w.
m. The product of s and six.
n. Twenty less than x.

3. Write each of the following as an algebraic expression:
   a. The sum of angles E, F, G, and H.
   b. 180° decreased by angle B.
   c. The product of the length (l), width (w), and height (h).
   d. The square root of the difference between the squares of the hypotenuse (h) and the base (b).
   e. Six times the square of the side (s).
   f. 273° more than the Celsius temperature reading (C).
   g. Twice the product of pi and the radius (r) times the sum of the radius (r) and the height (h).
   h. The product of force (F) and distance (d).
   i. The sum of a, b, and c divided by 3.
   j. The number of teeth (t) of the driving gear multiplied by the number of revolutions it makes per minute (R).
   k. The gross weight (W) divided by the wing area (A).
   l. The number of articles (n) multiplied by the price of one article (p).
   m. The true course ($T.C.$) increased by west variation ($W.V.$).
   n. The electromotive force (E) divided by the resistance (R).
   o. The product of the weight of the body (W) and the square of the velocity (v) divided by the product of the acceleration of gravity (g) and the radius of the circle (r).
   p. The profit (p) divided by the selling price (s).
   q. The sum of the wing drag (D_w) and the parasite drag (D_p).
   r. The quotient of the atomic weight ($A.W.$) divided by the equivalent weight ($E.W.$).
   s. The span of the wing (s) multiplied by the chord (c).
   t. Pi times radius (r) times sum of radius and slant height (l).

4. Read, or write in words, each of the following:
   a. $x + 10$
   b. $8n$
   c. $\dfrac{a}{15}$
   d. b^2
   e. $z - 5$
   f. $\sqrt{t}$
   g. $90 - A$
   h. ab
   i. $R_1 + R_2$
   j. xyz
   k. $2\pi r^2$
   l. $\dfrac{d}{t}$
   m. $\tfrac{1}{6}\pi d^3$
   n. $h(b + b')$
   o. $16(y - 1)$

4–3 MATHEMATICAL SENTENCES—OPEN SENTENCES; EQUATIONS; INEQUALITIES

Sentences such as:

"Some number a increased by five is equal to eighteen"
expressed in symbols as $a + 5 = 18$

or "Seven times each number n is less than thirty"
expressed in symbols as $7n < 30$

or "Each number x decreased by nine is greater than twelve"
expressed in symbols as $x - 9 > 12$

are *mathematical sentences*. A mathematical sentence that contains a variable is called an *open sentence*. $a + 5 = 18$, $7n < 30$, and $x - 9 > 12$ are open sentences. A verbal (word) sentence may be expressed as an equivalent algebraic sentence and vice-versa.

An open sentence that has the equality sign = as its verb is called an *equation*. The sentence $a + 5 = 18$ is an equation.

If the sentence uses $\neq$, $>$, $<$, $\not>$, $\not<$, $\geq$, $\leq$, $\not\geq$, or $\not\leq$ as its verb, it is called an *inequality*. The sentences $7n < 30$ and $x - 9 > 12$ are inequalities.

Symbolic verbs used in mathematical sentences include:

$=$ meaning "is equal to."
$\neq$ meaning "is not equal to."
$<$ meaning "is less than."
$\not<$ meaning "is not less than."
$>$ meaning "is greater than."
$\not>$ meaning "is not greater than."
$\leq$ or $\leqq$ meaning "is less than or equal to."
$\not\leq$ or $\not\leqq$ meaning "is not less than or not equal to."
$\geq$ or $\geqq$ meaning "is greater than or equal to."
$\not\geq$ or $\not\geqq$ meaning "is not greater than or not equal to."

Observe that the symbol $\not<$ is equivalent to symbol $\geq$; symbol $\not>$ is equivalent to symbol $\leq$; symbol $\not\leq$ is equivalent to symbol $>$; and symbol $\not\geq$ is equivalent to symbol $<$.

An open sentence is not a statement. We cannot tell whether it is true or false. It is only after we substitute a number for the variable that the open sentence becomes a statement. Then we can determine whether the sentence is true or false.

Algebra

The sentence "$5y \geq 20$" is the shortened form of "$5y > 20$ or $5y = 20$" and is read "Five times each number y is greater than or equal to twenty."

The sentence "$9 > b > 4$" is the shortened form of "$9 > b$ and $b > 4$" and is read "Nine is greater than each number b which is greater than four" or "Each number b is less than nine and greater than four."

The sentence "$2 \leq x \leq 6$" is read "Two is less than or equal to each number x which is less than or equal to six" or "Each number x is greater than or equal to two and less than or equal to six."

EXERCISES

1. Read, or write in words, each of the following:

   a. $n < 15$
   b. $5c > 30$
   c. $b - 8 > 11$
   d. $x + 9 < 20$
   e. $4t \not< 36$
   f. $3x + 5 \neq x - 10$
   g. $8n \not> 21 - 6n$
   h. $\dfrac{d}{9} = 14$
   i. $y - 7 < 4y + 6$
   j. $11b - b \neq 25 - 2b$
   k. $16a + 5 \not> 8a - 3$
   l. $m \geq 39$
   m. $12x \leq 72$
   n. $b + 1 \geq 6$
   o. $3n \not\leq 54$
   p. $7c - 1 \not\geq 3c + 4$
   q. $3 < x < 8$
   r. $7 > n > 1$
   s. $5 < a < 32$
   t. $17 > 4c > 0$
   u. $2 \leq y \leq 56$
   v. $14 \geq d \geq 10$
   w. $3 < r \leq 8$
   x. $1 \leq b < 15$

2. Write each of the following as an open sentence symbolically:

   a. Some number r decreased by five is equal to twenty-one.
   b. Each number n increased by ten is greater than sixteen.
   c. Four times each number t plus one is not equal to ten.
   d. Four times each number x is less than forty-eight.
   e. Nine times each number b is not greater than eighty-one.
   f. Each number y increased by one is not less than fourteen.

g. Each number *a* divided by six is greater than or equal to twenty.
h. Eight times each number *d* decreased by seven is less than or equal to nine.
i. Ten times each number *s* is greater than thirty and less than seventy.
j. Each number *c* is less than eleven and greater than six.

4–4 FORMULAS

A special kind of equation, called a *formula,* is a mathematical rule expressing the relationship of two or more quantities by means of numerals, variables, and operating symbols. The formula contains algebraic expressions.

To express mathematical and scientific principles as formulas, we write numerals, operating symbols, and letters (variables) representing the given quantities in the required order to show the relationship between quantities. A quantity may be represented by the first letter of a key word.

Write as a formula:

The net price (n) is equal to the list price (l) decreased by the discount (d).

Answer, $n = l - d$

To translate a formula to a word statement, we write a word rule stating the relationship expressed by the formula.

Translate:

$i = prt$ where i = interest, p = principal, r = annual rate of interest, and t = time in years.

Answer, The interest equals the principal times the annual rate of interest times the time in years.

Algebra

EXERCISES

1. Express each of the following as a formula:

   a. The area of a rectangle (A) is equal to the product of the length (l) and width (w).

   b. The sum of angles A, B, and C of triangle ABC is 180°.

   c. The circumference of a circle (c) is equal to twice the product of pi and the radius (r).

   d. The central angle of a regular polygon (a) equals 360° divided by the number of sides (n).

   e. The perimeter of a rectangle (p) is twice the sum of the length (l) and width (w).

   f. The area of a circle (A) is equal to one-fourth the product of pi and the square of the diameter (d).

   g. The hypotenuse (h) of a right triangle is equal to the square root of the sum of the squares of the altitude (a) and the base (b).

   h. The volume of a sphere (V) is four-thirds the product of pi and the cube of the radius (r).

   i. The distance (d) a freely falling body drops is one-half the product of the acceleration due to gravity (g) and the square of the time of falling (t).

   j. The temperature reading on the Fahrenheit scale (F) is equal to nine-fifths of the reading on the Celsius scale (C) increased by 32°.

   k. The displacement of the piston (D) equals the area of the piston (A) times the stroke (s).

   l. The pitch of a roof (p) is equal to the rise (r) divided by the span (s).

   m. The cutting speed (s) of a bandsaw in feet per minute is equal to pi times the diameter (d) in feet times the number of revolutions per minute (R).

   n. The capital (C) of a business is the difference between the assets (A) and the liabilities (L).

o. The selling price (*s*) is equal to the sum of the cost (*c*) and the profit (*p*).

p. The rate of discount (*r*) is equal to the discount (*d*) divided by the list price (*l*).

q. The tip speed of a propeller (*T.S.*) is equal to pi times the diameter (*d*) times the number of revolutions per second (*N*).

r. The horsepower (*H.P.*) required for the wing of an airplane equals the product of the drag of the wing (*D*) and the velocity (*V*) divided by 550.

s. The lift (*L*) of an airplane wing is the product of the lift coefficient (*C*), half of the air density (*d*), the wing area (*A*), and the square of the air speed or velocity (*V*).

t. The aspect ratio (*r*) of a wing of an airplane is equal to the span of the wing (*s*) divided by the chord (*c*).

2. Express each of the following formulas as a word statement:

   a. $d = rt$ where d = distance traveled, r = average rate of speed, and t = time of travel.

   b. $A = p + i$ where A = amount, p = principal, and i = interest.

   c. $C = K - 273°$ where C = Celsius temperature reading and K = Kelvin temperature reading.

   d. $d = \dfrac{m}{v}$ where d = density, m = mass, and v = volume.

   e. $p = 2l + 2w$ where p = perimeter of rectangle, l = length, and w = width.

   f. $A = \pi r^2$ where A = area of circle, π = pi or 3.14, and r = radius.

   g. $I = \dfrac{E}{R}$ where I = current in amperes, E = electromotive force in volts, and R = resistance in ohms.

   h. $A = \dfrac{h}{2}(b_1 + b_2)$ where A = area of trapezoid, h = height, b_1 = lower base, and b_2 = upper base.

   i. $b = \sqrt{h^2 - a^2}$ where b = base of right triangle, h = hypotenuse, and a = altitude.

   j. $a = \dfrac{360°}{n}$ where a = central angle of a regular polygon and n = the number of sides.

Algebra

EVALUATION

4–5 ALGEBRAIC EXPRESSIONS

To evaluate an algebraic expression means to find the value of the algebraic expression which depends upon the values of the variables. If these values change, the value of the expression usually changes.

(1) *To evaluate an algebraic expression*, we copy the expression and substitute the given value for each variable. We then perform the necessary operations as indicated in the expression.
Observe that the value of each term (part of expression connected to other parts by either a plus or minus sign) is found and these values are combined.

(2) If the expression is a fraction, we simplify both the numerator and denominator separately, and then express the fraction in simplest terms.

(3) If there are parentheses in the expression, we find the value of the expressions in each set of parentheses, and then perform the necessary operations to get the answer.

(4) If there are exponents in the expression, we raise each quantity to the proper power usually before performing the other operations.

---

(1) Find the value of $2x + 3y$ when $x = 8$ and $y = 5$.

$$2x + 3y$$
$$= 2 \cdot 8 + 3 \cdot 5$$
$$= 16 + 15$$
$$= 31$$

Answer, 31

(2) Find the value of $\dfrac{a+b}{a-b}$ when $a = 6$ and $b = 3$.

$$\dfrac{a+b}{a-b} = \dfrac{6+3}{6-3}$$
$$= \dfrac{9}{3} = 3$$

Answer, 3

(3) Find the value of $2(l + w)$ when $l = 10$ and $w = 7$.

$$2(l + w)$$
$$= 2(10 + 7)$$
$$= 2(17)$$
$$= 34$$

Answer, 34

(4) Find the value of $a^2 + b^2$ when $a = 5$ and $b = 4$.

$$a^2 + b^2$$
$$= 5^2 + 4^2$$
$$= 25 + 16$$
$$= 41$$

Answer, 41

EXERCISES

Find the value of each of the following algebraic expressions:

When $a = 6$ and $c = 3$:

1. $a + c =$ ~~9~~
   6 3 8
2. $a - c =$ 3
   6 3
3. ac 9
4. $\dfrac{a}{c}$ $\dfrac{6}{3}$
5. $a + 9c$
   $6 + 12$

When $x = 12$, $y = 4$, and $z = 2$:

6. $3x + 5y$
   $15 + 7 = 22$
7. $4xyz - 3xy$
   $22 - 19 = 3$
8. $4xz - 8z - 11$
9. $\dfrac{4x - 5y}{3y + x}$

When $c = 10$, $n = -8$, and $y = -4$:

10. $4n + 5y$
11. $2c - 6n$
12. $\dfrac{2n}{3y}$
13. $cn + 2ny - 4y$

When $b = 8$, $c = 5$, and $x = 2$:

14. $b(c + x)$
15. $4b + (c - x)$
16. $5b + c(b + x)$
17. $(b + c)(b - x)$

When $m = 8$, $n = 4$, and $y = 1$:

18. $m(n - y)$
19. $3mn(m + y)$
20. $my + m(3m - 5n)$
21. $\dfrac{6(mn - 2y)}{11 + (m - ny)}$

When $a = -6$, $x = 4$, and $y = -4$:

22. $a - (x - y)$
23. $(a - x)(a + y)$
24. $\dfrac{3(y - 2a)}{x - 2(y + 2)}$
25. $a + y(x - a)$

When $a = 5$, $b = 4$, and $c = 2$:

26. $3a^2$
27. $b^3 - c^2$
28. $4a^2 + 3b^2$
29. $a^2 - 3a + 7$
30. $\dfrac{(a + b)^2}{a^2 + b^2}$

When $d = 2$, $n = 3$, and $x = 6$:

31. $n^2 - d^2$
32. $5x^2 - 2x + 9$
33. $8d^2 - 5dn$
34. $\dfrac{2n^2 + 5n + 6}{3n^2 - 7n + 2}$

When $b = -10$, $c = 5$, and $d = -2$:

35. $6d^2$
36. $c^2 - 2d^2$
37. $9b^3 - 4b^2 - 3b + 8$
38. $2b^2 - 3bd + 5d^2$

When $a = 10$, $b = 4$, $c = 3$, $d = 2\tfrac{1}{2}$, and $n = 1.5$:

39. $5a + 3c + 4d$
40. $a + b(b - n)$
41. $a^2 - 3ab + 2b^2$
42. $n^2 - \dfrac{n}{a}$

Algebra 245

4–6 FORMULAS

To find the required value, we copy the given formula, substitute the given values for the letters, and perform the necessary operations.

> Find the value of F when $C = 20$, using the formula $F = 1.8\,C + 32$:
> $F = 1.8\,C + 32$
> $F = 1.8 \times 20 + 32$
> $F = 36 + 32$
> $F = 68$ *Answer,* $F = 68$

EXERCISES

Find the value of:
1. p when $s = 8$, using the formula $p = 3\,s$.
2. A when $l = 12$ and $w = 7$, using the formula $A = lw$.
3. E when $I = 9$ and $R = 15$, using the formula $E = IR$.
4. W when $F = 90$ and $d = 8$, using the formula $W = Fd$.
5. A when $p = 125$ and $i = 19$, using the formula $A = p + i$.
6. K when $C = 38$, using the formula $K = C + 273$.
7. F when $C = 40$, using the formula $F = 1.8\,C + 32$.
8. v when $V = 15$, $g = 32$, and $t = 3$, using the formula $v = V + gt$.
9. B when $A = 53$, using the formula $B = 90 - A$.
10. C when $A = 9{,}800$ and $L = 1{,}250$, using the formula $C = A - L$.
11. b when $p = 23$ and $e = 8$, using the formula $b = p - 2\,e$.
12. C when $A = 42$ and $B = 76$, using the formula $C = 180 - A - B$.
13. a when $n = 8$, using the formula $a = \dfrac{360}{n}$.
14. r when $d = 171$ and $t = 9$, using the formula $r = \dfrac{d}{t}$.
15. I when $E = 110$ and $R = 22$, using the formula $I = \dfrac{E}{R}$.

16. W when $w = 75$, $l = 12$, and $L = 6$, using the formula $W = \frac{wl}{L}$.

17. A when $d = 50$, using the formula $A = .7854 \, d^2$.
18. s when $t = 7$, using the formula $s = 16 \, t^2$.
19. W when $I = 6$ and $R = 15$, using the formula $W = I^2R$.

20. V when $\pi = \frac{22}{7}$, $r = 14$, and $h = 35$, using the formula $V = \pi r^2 h$.
21. A when $p = 140$, $r = .06$, and $t = 5$, using the formula $A = p(1 + rt)$.
22. B when $A = 83$ and $C = 28$, using the formula $B = 180 - (A + C)$.
23. l when $a = 6$, $n = 9$, and $d = 3$, using the formula $l = a + (n - 1)d$.
24. C when $F = 50$, using the formula $C = \frac{5}{9}(F - 32)$.

25. E when $I = 7$, $r = 16$, $R = 23$, using the formula $E = Ir + IR$.

26. k when $m = 14$ and $v = 5$, using the formula $k = \frac{1}{2} mv^2$.

27. A when $\pi = \frac{22}{7}$, $r = 7$, and $h = 18$, using the formula $A = 2\pi r(r + h)$.
28. K when $C = -15$, using the formula $K = C + 273$.

29. C when $F = -4$, using the formula $C = \frac{5}{9}(F - 32)$.

30. I when $n = 12$, $E = 110$, $R = 24$, and $r = 3$, using the formula $I = \frac{nE}{R + nr}$.

OPERATIONS

A *monomial* is a number or a variable or a product of variables or a product of a number and a variable or variables. Examples of monomials are 6, b, x^2y^3z, $9c$, and $-5\,a^4x$.

A *polynomial* is a monomial or a sum of monomials. The following expressions are examples of polynomials: $7 + n$, $4x - 5$, $2b^2 - 5ab + 8b^2$, and $4a^3 - 2a^2 + a - 9$.

Like terms are terms which have a common factor of identical variables raised to identical powers. For example, $9\,b^2d^3$ and $-7\,b^2d^3$ are like terms; but x and x^2 are not like terms.

Observe that an unexpressed numerical coefficient of 1 is understood in both x meaning $1\,x$ or $+1\,x$ and $-x$ meaning $-1\,x$. Of course $0\,x = 0$.

Algebra

4–7 ADDITION

Monomials

To add like terms, we use the distributive property of multiplication over addition, or briefly:

We find the algebraic sum of the numerical coefficients and prefix it to their common factor of variables.

To add unlike terms like $5b$ to $3a$, we indicate their sum as $3a + 5b$ since it cannot be expressed as a single term.

The sum of $2x$ and $5x$:

$$2x + 5x = (2+5)x = 7x$$

or
$$\begin{array}{r} 2x \\ 5x \\ \hline 7x \end{array}$$
Answer, $7x$

$$\begin{array}{r} -3a^3 \\ -3a^3 \\ \hline -6a^3 \end{array}$$
Answer, $-6a^3$

$$\begin{array}{r} +7xy \\ -8xy \\ \hline -xy \end{array}$$
Answer, $-xy$

EXERCISES

1. Add:

a.
$+3x$	$-7a$	$+6b$	$-4cd$	$+8n$	$-x$	$+6c$
$+2x$	$-5a$	$-8b$	$-cd$	$-13n$	$-x$	$-6c$

b.
$-1.4d$	$-4\frac{2}{3}xy$	$-6y^2$	$-8c^3d$	$7b^4$	$2a^3bc^2$	$-2x^2$
$+2.3d$	$+2\frac{1}{3}xy$	$+5y^2$	$-3c^3d$	$-b^4$	$4a^3bc^2$	$-2x^2$

c.
$-5x$	$-2bc$	$-8mn$	$2x^2y$	$-a$	$+6c$	$+7a^2x^3$
$+6x$	$-3bc$	$-3mn$	$-4x^2y$	$+2a$	$-7c$	$-4a^2x^3$
$-8x$	$+5bc$	$-mn$	$6x^2y$	$-a$	$-9c$	$+3a^2x^3$
				$+3a$	$+10c$	$-8a^2x^3$

2. Add as indicated:

a. $(6c) + (-9c)$
b. $(-2d) + (-d)$
c. $(-x) + (5x)$
d. $(-8y^3) + (-4y^3)$
e. $(-a) + (+a) + (-a)$
f. $(-4b) + (-b) + (5b)$
g. $(2xy) + (-5xy) + (-8xy)$
h. $(-7d^4) + (4d^4) + (-8d^4) + (-d^4)$
i. $(-m^2x) + (m^2x) + (3m^2x) + (-9m^2x)$
j. $(3a^2b^2) + (-5a^2b^2) + (-6a^2b^2) + (4a^2b^2)$

3. Simplify:

   a. $3n + 7n + 4n$
   b. $2a - 8a + a - 3a$
   c. $-d - 2d - 3d - d$
   d. $5b + 6b - 8b - 9b + b$
   e. $4x - x - 2x + 3x - 5x$
   f. $6ax - 3ax + ax - 2ax$
   g. $-9bc - 3bc + 2bc + 4bc$
   h. $-3c^2 + 4c^2 - c^2 - 10c^2$
   i. $8a^2b^2 - a^2b^2 - 5a^2b^2 - 4a^2b^2 + 2a^2b^2$
   j. $10y^3 - 3y^3 - 8y^3 + y^3 - 8y^3$

4. Add:

   a. $4a$ and $5x$ b. $7c$ and $-5d$ c. $9x^2$ and $2x$ d. $-4bx$ and $-2cx$

Polynomials

To add two or more polynomials, we write the polynomials so that like terms are under each other and we add each column.

Polynomials should be expressed in descending or ascending order of the powers of the variable.

$x^2 + 5 - 3x$ is written as $x^2 - 3x + 5$
  or $5 - 3x + x^2$

$$\begin{array}{r} 4a^2 - ab + 3b^2 \\ 3a^2 - 7ab - 4b^2 \\ \hline 7a^2 - 8ab - b^2 \end{array}$$

Answer, $7a^2 - 8ab - b^2$

EXERCISES

1. Add:

   a. $\begin{array}{r} 3x + 2 \\ 7x + 6 \end{array}$ $\begin{array}{r} 5a + 8b \\ 9a + b \end{array}$ $\begin{array}{r} 7c - d \\ c - 4d \end{array}$

   b. $\begin{array}{r} 4x^2 - 3y^2 \\ -2x^2 + 3y^2 \end{array}$ $\begin{array}{r} -7ax - 2y \\ -4ax + y \end{array}$ $\begin{array}{r} -3ac^3 + 9ac^2 \\ +3ac^3 - 9ac^2 \end{array}$

   c. $\begin{array}{r} 9a - 4b + 5c \\ 3a + 4b - 5c \end{array}$ $\begin{array}{r} 7a^2 - 3ab - 4b^2 \\ 2a^2 - 5ab + b^2 \end{array}$ $\begin{array}{r} 6m^2 - mn \\ 3mn + 2n^2 \end{array}$

   d. $\begin{array}{r} 2a - 3b \\ -4a + 7b \\ 2a - 5b \end{array}$ $\begin{array}{r} 3b^2 - 5b - 8 \\ 3b^2 - 7b - 9 \\ -5b^2 + 2b - 4 \end{array}$ $\begin{array}{r} x^2 - 4xy - y^2 \\ -5x^2 - 9xy + 2y^2 \\ -4x^2 - xy - y^2 \end{array}$

2. Add as indicated:

   a. $(x + 5) + (x + 2)$
   b. $(c - 3d) + (4c - 5d)$

Algebra   249

  c. $(x^2 - 2y^2) + (y^2 - x^2)$
  d. $(m + 3n) + (3m - n) + (2m - 5n)$
  e. $(b^2 - 4bc) + (2bc - c^2) + (c^2 - b^2)$
  f. $(a^2 - 5a + 9) + (2a^2 - a - 7) + (4 - 2a - 3a^2)$

3. Simplify:

  a. $6x + 3y + 2x + 7y$
  b. $3a - 1 + 4 - 5a$
  c. $2c^2 - 3c + c^2$
  d. $7x^2 - 4x + 8 + 9x - 4 - 2x^2$
  e. $4c^2 - 5cd - d^2 + 8c^2 - 4d^2 - 9cd - c^2 + d^2$
  f. $8b^3 - 2b^2 + b - 3 - 2b^2 - 10b^3 - b + 7 - b^2 - 1$

4–8 SUBTRACTION

Monomials

To subtract like terms, we use the distributive property of multiplication with respect to subtraction or, briefly:

We add the additive inverse (or opposite) of the subtrahend to the minuend.

To subtract unlike terms, like subtracting $3x$ from $-5b$, we indicate the difference as $-5b - 3x$ since it cannot be expressed as a single term.

From $4ab$ take $7ab$:

$(4ab) - (7ab) = (4 - 7)ab = -3ab$

or

 Subtract: Add:
  $4ab$ $+4ab$
  $7ab$ → $-7ab$
        $\overline{-3ab}$

Answer, $-3ab$

EXERCISES

1. Subtract:

 a. $\begin{array}{r}7x\\6x\end{array}$ $\begin{array}{r}2a\\8a\end{array}$ $\begin{array}{r}+5x^2y\\+7x^2y\end{array}$ $\begin{array}{r}-6y\\-2y\end{array}$ $\begin{array}{r}-2b^3\\-11b^3\end{array}$ $\begin{array}{r}+9b\\-3b\end{array}$ $\begin{array}{r}+\ ax\\-4ax\end{array}$

 b. $\begin{array}{r}-5abc\\+3abc\end{array}$ $\begin{array}{r}-2r^2s\\+10r^2s\end{array}$ $\begin{array}{r}-8c\\-7c\end{array}$ $\begin{array}{r}0\\+3cd\end{array}$ $\begin{array}{r}9b^2x\\0\end{array}$ $\begin{array}{r}+6mn\\+6mn\end{array}$ $\begin{array}{r}-4abc\\-4abc\end{array}$

2. Subtract as indicated:
   a. $(6m) - (-5m)$
   b. $(-3c^2x) - (-c^2x)$
   c. $(-8dy) - (-8dy)$
   d. $(-2cd^2) - (+3cd^2)$
   e. $(10abx) - (11abx)$
   f. $(a^2y^4) - (-a^2y^4)$

3. Do as indicated:
   a. From $2n$ subtract $7n$.
   b. Take $5y$ from $-9y$.
   c. From $8x$ take $3y$.
   d. Subtract $-2b$ from a.
   e. Subtract $-3x^2y^5$ from $-x^2y^5$.
   f. From 0 take $-4cx$.
   g. From $-m^2$ subtract $-mn$.
   h. Take $4x^2$ from 25.

Polynomials

To subtract one polynomial from another, we write the polynomials so that like terms are under each other and we subtract each column.

$$7x^2 - 4x + 6$$
$$2x^2 - 5x - 4$$
$$\overline{5x^2 + x + 10}$$

Answer, $5x^2 + x + 10$

EXERCISES

1. Subtract:

   a. $3c + 7$
      $\underline{c + 6}$

      $2x^2 - 3y^2$
      $\underline{8x^2 + 7y^2}$

      $7a^2 - 9a - 5$
      $\underline{2a^2 + 9a - 5}$

   b. $3cd + 6$
      $\underline{5c^2 - 5cd}$

      $a^2 - 7ab - 2b^2$
      $\underline{-6a^2 - 8ab + b^2}$

2. Subtract as indicated:
   a. $(6a - y) - (8a - 6y)$
   b. $(x - y) - (-a - x)$
   c. $(x^2 - 4x - 8) - (3x^2 - 5x + 7)$
   d. $(4m^2 - 5mn - 2n^2) - (4n^2 - 8mn - m^2)$

3. a. From $4a - 7b$ take $2a + 9b$.
   b. Subtract $9x^2 - 5x - 6$ from $3x^2 - 6x + 11$.
   c. Take $6c^2$ from $2c^2 - 5$.
   d. From $3b^3 - 4b^2 + 6$ subtract $9b - 8$.
   e. What must be added to $12y^2 - 8y + 1$ to get $8y^2 - 6y - 2$?
   f. From zero subtract $2x^2 - 3y^2$.
   g. From the sum of $8m^2 - 7m + 2$ and $3m^2 - 9m - 7$ subtract $12 - 11m^2$.
   h. Subtract the sum of $9a^2 + 5ab - 10b^2$ and $6b^2 - a^2$ from $4a^2 - 2ab$.
   i. From the sum of $4n^2 - 8n - 5$ and $8n + 5$ subtract the sum of $3n - 4$ and $9 - 6n - n^2$.

Algebra

4–9 MULTIPLICATION

Law of Exponents for Multiplication

Since $a^5 = a \cdot a \cdot a \cdot a \cdot a$ and $a^3 = a \cdot a \cdot a$
then $a^5 \cdot a^3 = \underline{a \cdot a \cdot a \cdot a \cdot a} \cdot \underline{a \cdot a \cdot a} = a^8$
Thus $a^5 \cdot a^3 = a^{5+3} = a^8$

In general, $a^m \cdot a^n = a^{m+n}$ where a represents any number and m and n are positive integers. Observe that the exponent of the variable in the product is the sum of the exponents of that variable in the given factors. This is not true for the product of two different variables. For example, the product of b^5 and c^3 is b^5c^3. Here the variables are written alphabetically.

Monomials

To multiply monomials, we use the commutative and associative properties and the law of exponents for multiplication, or, briefly:

We multiply the numerical coefficients as we do signed numbers and find the product of the variables using the law of exponents.

We prefix the numerical product to the product of the variables arranged in alphabetical order.

Note: $x = 1\,x^1$

Multiply $8\,c^3d^2$ by $3\,c^4d^6$:

$(8\,c^3d^2)(3\,c^4d^6) = (8 \cdot 3)(c^3 \cdot c^4)(d^2 \cdot d^6)$
$ = 24\,c^7\,d^8$

Answer, $24\,c^7d^8$

Multiply $-7\,m^4x^2$ by $+8\,m^2y^3$:

$-7\,m^4x^2$
$\underline{+8\,m^2y^3}$
$-56\,m^6x^2y^3$

Answer, $-56\,m^6x^2y^3$

EXERCISES

1. Write in shortened form using exponents:
   a. $c \cdot c \cdot c \cdot x \cdot x$
   b. $a \cdot a \cdot a \cdot y \cdot y \cdot y \cdot y$
   c. $b \cdot b \cdot b \cdot b \cdot b \cdot d \cdot d \cdot d \cdot d$
   d. $x \cdot x \cdot x \cdot x \cdot x \cdot x \cdot x \cdot x \cdot y$

2. Multiply:
   a. $n^5 \cdot n^7$
   b. $x^8 \cdot x^{11}$
   c. $t^5 \cdot t$
   d. $10^5 \times 10^8$
   e. $10^{12} \times 10^9$
   f. $6^2 \times 6^7$
   g. $b^9 \cdot a^{10}$
   h. $c^2 \cdot y$
   i. $n^4 \cdot d^5$
   j. $a^5x^2 \cdot ax^8$
   k. $c^7d^2 \cdot c^2d^5$
   l. $x^9y \cdot y^7$

3. Multiply:
   a. $\underline{\begin{array}{c} a^4 \\ a^7 \end{array}}$ $\underline{\begin{array}{c} m^{15} \\ m^{10} \end{array}}$ $\underline{\begin{array}{c} -c \\ -c \end{array}}$ $\underline{\begin{array}{c} 6\,a^6 \\ 4\,a^4 \end{array}}$ $\underline{\begin{array}{c} -5\,x \\ -6\,x \end{array}}$ $\underline{\begin{array}{c} +8\,m^3 \\ -8\,m^3 \end{array}}$ $\underline{\begin{array}{c} -2\,b^4 \\ +5\,b^9 \end{array}}$

   b. $\underline{\begin{array}{c} 6\,m^2n \\ 3\,mn^5 \end{array}}$ $\underline{\begin{array}{c} -4\,b^7c \\ 9\,b^4c^5 \end{array}}$ $\underline{\begin{array}{c} 8\,a^3b^2 \\ 6\,a^2b^8 \end{array}}$ $\underline{\begin{array}{c} -7\,x \\ -xy \end{array}}$ $\underline{\begin{array}{c} -y \\ -8 \end{array}}$ $\underline{\begin{array}{c} -1 \\ 4\,d^3 \end{array}}$ $\underline{\begin{array}{c} -cy \\ -5\,bx^9 \end{array}}$

   c. $\underline{\begin{array}{c} -s \\ 5\,t \end{array}}$ $\underline{\begin{array}{c} 6\,c \\ -\,cx \end{array}}$ $\underline{\begin{array}{c} -7\,x^3y^5 \\ -4\,x^2y^9 \end{array}}$ $\underline{\begin{array}{c} -3\,bcd \\ 8 \end{array}}$ $\underline{\begin{array}{c} -9\,ax^3 \\ -2\,a^4y^2 \end{array}}$ $\underline{\begin{array}{c} -10\,dt^5 \\ -10\,dt^5 \end{array}}$ $\underline{\begin{array}{c} 8\,m^7n^6r^8 \\ -m^3n^4r^8 \end{array}}$

4. Multiply as indicated:
   a. $7\,y \cdot y$
   b. $6\,ax^2 \cdot 2\,a^5x^3$
   c. $(-\dot{x})(-4\,x^6)$
   d. $(7\,c^4d^5)(-8\,c^3d^7)$
   e. $(-9\,b^2x)(-6\,x^5y^6)$
   f. $6(8\,a)$
   g. $-5\,x^3(x^4)$
   h. $-c(-d)$
   i. $-8\,a^4x^3(4\,a^5x^9)$
   j. $10\,b^7y^5(-3\,b^6y^8)$
   k. $4\,y^3 \cdot 3\,y^2 \cdot 2\,y^7$
   l. $5\,a \cdot 2\,b \cdot 9\,c$
   m. $(-b)(-b)(-b)$
   n. $(-3\,c^2)(-3\,c^2)(-3\,c^2)(-3\,c^2)$
   o. $(2\,x^2y)(-5\,x^3)(-3\,xy^2)(-y^7)$

5. Multiply as indicated:
   a. $(c^3y)^2$
   b. $(-5\,b^4x^3)^2$
   c. $(-3\,x)^3$
   d. $(-4\,m^3x^5)^4$
   e. $(7\,a^5c^6d)^4$
   f. $(-2\,b^2x^4y^8)^5$

Polynomials

To multiply a polynomial by a monomial, we use the distributive property by multiplying each term of the polynomial by the monomial.

$2\,a^3x(3\,a^2 + 5\,ax + 7\,x^2)$
$= (2\,a^3x \cdot 3\,a^2) + (2\,a^3x \cdot 5\,ax) + (2\,a^3x \cdot 7\,x^2)$
$= 6\,a^5x + 10\,a^4x^2 + 14\,a^3x^3$

Answer, $6\,a^5x + 10\,a^4x^2 + 14\,a^3x^3$

Algebra

EXERCISES

1. Multiply:

   a. $-5(x + 7)$

   b. $6(3a - 9b)$

   c. $y(y + 1)$

   d. $-a^3(a^2b + b^5)$

   e. $-6x^5y^2(-4x^2 - 3y^4)$

   f. $4c^2(6c^2 - 4c + 3)$

   g. $-9b^3d^2(2b^2 - 5bd - 6d^2)$

   h. $-3b^4(7b^2 - 6b + 9)$

   i. $-10m^7n^3(8m^4 + 3m^2n^2 - 5n^4)$

   j. $7ab^3x^5(4a^5b^2 - 6a^3x^6 + bx^7)$

2. Multiply:

$$\begin{array}{r} 3x - 7 \\ 4 \\ \hline \end{array} \qquad \begin{array}{r} 12b^2 - 5bc \\ -b \\ \hline \end{array} \qquad \begin{array}{r} 6a^2 + ax - 4x^2 \\ -2ax \\ \hline \end{array} \qquad \begin{array}{r} x^3 - 5xy + 6xy^2 - 8y^3 \\ 9x^2y^3 \\ \hline \end{array}$$

4–10 DIVISION

Law of Exponents for Division

Since $a^8 = a \cdot a \cdot a \cdot a \cdot a \cdot a \cdot a \cdot a$ and $a^3 = a \cdot a \cdot a$

then $\dfrac{a^8}{a^3} = \dfrac{\overset{1}{\cancel{a}} \cdot \overset{1}{\cancel{a}} \cdot \overset{1}{\cancel{a}} \cdot a \cdot a \cdot a \cdot a \cdot a}{\underset{1}{\cancel{a}} \cdot \underset{1}{\cancel{a}} \cdot \underset{1}{\cancel{a}}} = a^5$

Thus $\dfrac{a^8}{a^3} = a^{8-3} = a^5$

In general $a^m \div a^n = \dfrac{a^m}{a^n} = a^{m-n}$ where a represents any nonzero number and m and n are positive integers such that $m > n$. Observe that for any given variable its exponent in the quotient is equal to its exponent in the dividend (numerator) minus its exponent in the divisor (denominator).

Monomials

To divide monomials, we divide the numerical coefficients as we do signed numbers and find the quotient of the variables using the law of exponents for division. We prefix the numerical quotient to the product of the quotients of the variables arranged in alphabetical order.

Divide $-48\,b^9x^8$ by $6\,b^3x^4$:

$$\frac{-48\,b^9x^8}{6\,b^3x^4} = \frac{-48}{6} \cdot \frac{b^9}{b^3} \cdot \frac{x^8}{x^4}$$
$$= -8 \quad b^6 \quad x^4$$

Answer, $-8\,b^6x^4$

EXERCISES

1. Divide:

   a. x^7 by x^4
   b. d^9 by d^3
   c. y^{12} by y^2
   d. 10^8 by 10^2
   e. 10^{11} by 10^5
   f. 8^6 by 8^3
   g. m^9 by m^9
   h. b^6c^3 by c^2
   i. x^7y^{15} by xy^8
   j. a^5b^9 by a^4b^6
   k. $m^{12}x^8$ by m^5
   l. $r^{10}s^8t^4$ by rs^7t^2

2. Divide:

   a. $\dfrac{6\,a^{12}}{3}$ $\dfrac{-35\,x^8}{-7}$ $\dfrac{9\,b^5}{-1}$ $\dfrac{-8\,y^6}{-8}$ $\dfrac{54\,c^3d^9}{9\,c^2d^3}$ $\dfrac{-72\,m^6x^8}{-8\,m^2x^4}$ $\dfrac{+18\,b^3c}{-6\,b^2}$

   b. $\dfrac{+32\,gt^2}{-16\,g}$ $\dfrac{-36\,a^7x^8y}{4\,a^5x^2y}$ $\dfrac{-25\,b^6c^5}{-25\,b^2c}$ $\dfrac{80\,m^7x^3}{-5\,m^7x^3}$ $\dfrac{-x}{-1}$ $\dfrac{24\,c^6x^5y^3}{4\,c^4y^3}$ $\dfrac{-2\,m^2x^4}{2\,m^2x^4}$

3. Divide as indicated:

   a. $x^6 \overline{)-x^{12}}$
   b. $-3 \overline{)-12\,a^4b^5}$
   c. $-5\,c^2d \overline{)20\,c^7d^8}$
   d. $-6\,a^6 \overline{)-30\,a^{18}y^4}$
   e. $-2\,x^3y^4 \overline{)-8\,x^{10}y^6}$
   f. $(a^6b^9) \div (a^3b)$
   g. $(32\,b^5x^7) \div (-4\,x^2)$
   h. $(-72\,s^{10}) \div (-9\,s^2)$
   i. $(30\,b^8t^7) \div (-5\,b^5t^7)$

Polynomials

To divide a polynomial by a monomial, we use the distributive property by dividing each term of the polynomial by the monomial.

Divide $4\,c^9x^3 - 10\,c^6x^5 + 12\,c^4x^7$ by $-2\,c^4x^2$:

$$\frac{4\,c^9x^3 - 10\,c^6x^5 + 12\,c^4x^7}{-2\,c^4x^2} = \frac{4\,c^9x^3}{-2\,c^4x^2} + \frac{-10\,c^6x^5}{-2\,c^4x^2} + \frac{12\,c^4x^7}{-2\,c^4x^2}$$
$$= -2\,c^5x + 5\,c^2x^3 - 6\,x^5$$

Answer, $-2\,c^5x + 5\,c^2x^3 - 6\,x^5$

Algebra

EXERCISES

1. Divide:

   a. $\dfrac{24a + 16}{8}$ $\dfrac{6a - 12x}{-3}$ $\dfrac{14c - 7}{-7}$ $\dfrac{30b - 5}{5}$ $\dfrac{2m^2 + 7mn - n^2}{-1}$

   b. $\dfrac{x^4 - x^3}{-x}$ $\dfrac{8d^2 - d}{-d}$ $\dfrac{a^4b^3 - a^3b^4 + a^2b^5}{a^2b^2}$ $\dfrac{16x^5y^3 - 20x^4y^4 + 36x^3y^5}{-4x^3y^2}$

2. Divide as indicated:

   a. $-9\overline{)27x^4 - 18}$
   b. $5b^2\overline{)20b^4 - 15b^3 + 35b^2}$
   c. $-2x^2y^4\overline{)8x^{10}y^6 - 4x^6y^8 - 10x^4y^{12}}$
   d. $(8a - 12b) \div (-4)$
   e. $(12x^7 - 9x^5 + 6x^3) \div (-3x^2)$
   f. $(36a^9b^7 + 30a^8b^6 - 42a^5b^4) \div (6a^5b^3)$

4–11 ZERO EXPONENTS; NEGATIVE EXPONENTS

The law of exponents for division (see section 4–10) indicates that $a^m \div a^n = \dfrac{a^m}{a^n} = a^{m-n}$ where a represents any non-zero number and m and n are positive integers such that $m > n$. Here we use exponents which name natural numbers indicating how many times the factor is being used in multiplication as a repeated factor. We found that for any given variable its exponent in the quotient is equal to its exponent in the dividend (numerator) minus its exponent in the divisor (denominator).

For example: $\dfrac{7^9}{7^5} = 7^{9-5} = 7^4$ Answer, 7^4

and when $b \neq 0$, $\dfrac{b^{12}}{b^4} = b^{12-4} = b^8$ Answer, b^8

In order to give meaning to zero exponents and negative exponents, let us extend the law of exponents for division to include $\dfrac{a^m}{a^n}$ both for $m = n$ and $m < n$ when $a \neq 0$.

For example: $\dfrac{a^6}{a^6} = \dfrac{\overset{1}{\cancel{a}} \cdot \overset{1}{\cancel{a}} \cdot \overset{1}{\cancel{a}} \cdot \overset{1}{\cancel{a}} \cdot \overset{1}{\cancel{a}} \cdot \overset{1}{\cancel{a}}}{\underset{1}{\cancel{a}} \cdot \underset{1}{\cancel{a}} \cdot \underset{1}{\cancel{a}} \cdot \underset{1}{\cancel{a}} \cdot \underset{1}{\cancel{a}} \cdot \underset{1}{\cancel{a}}} = \dfrac{1}{1} = 1$

By the law of exponents, $\dfrac{a^6}{a^6} = a^{6-6} = a^0$

Since in general $\dfrac{a^m}{a^m} = a^{m-m} = a^0$ and $\dfrac{a^m}{a^m} = 1$, then we may define a^0 to equal 1 when a represents any non-zero number. That is: $a^0 = 1$.

The expression 0^0 is meaningless.

We have found that $\dfrac{a^8}{a^3} = \dfrac{\overset{1}{\cancel{a}} \cdot \overset{1}{\cancel{a}} \cdot \overset{1}{\cancel{a}} \cdot a \cdot a \cdot a \cdot a \cdot a}{\underset{1}{\cancel{a}} \cdot \underset{1}{\cancel{a}} \cdot \underset{1}{\cancel{a}}} = a^5$

and by the law of exponents that $\dfrac{a^8}{a^3} = a^{8-3} = a^5$ when $a \neq 0$.

Here we see that the exponent five (5) in the quotient a^5 indicates that the numerator has five (5) more of the repeated factor than the denominator.

In a similar manner, we find that $\dfrac{a^3}{a^8} = \dfrac{\overset{1}{\cancel{a}} \cdot \overset{1}{\cancel{a}} \cdot \overset{1}{\cancel{a}}}{\underset{1}{\cancel{a}} \cdot \underset{1}{\cancel{a}} \cdot \underset{1}{\cancel{a}} \cdot a \cdot a \cdot a \cdot a \cdot a} = \dfrac{1}{a^5}$

and by the law of exponents that $\dfrac{a^3}{a^8} = a^{3-8} = a^{-5}$.

That is: $a^{-5} = \dfrac{1}{a^5}$ when $a \neq 0$.

Here we see that the exponent negative five (-5) in the quotient a^{-5} indicates that the denominator has five (5) more of the repeated factor than the numerator.

Thus in general,

$a^{-1} = \dfrac{1}{a}$; $a^{-2} = \dfrac{1}{a^2}$; $a^{-3} = \dfrac{1}{a^3}$; $a^{-4} = \dfrac{1}{a^4}$; etc. when $a \neq 0$.

EXERCISES

1. Find the numerical value of each of the following:
    a. 10^0
    b. 3^0
    c. 100^0
    d. 19^0
    e. 35^0
    f. 147^0
    g. 500^0
    h. 1^0

2. Find the numerical value of each of the following:
    a. 3×10^0
    b. 8×10^0
    c. 5×10^0
    d. 4×6^0
    e. $5^0 \times 7^0$
    f. $12^0 \times 3$
    g. $9^0 \times 4^0$
    h. $16^0 \times 7$

3. Find the numerical value of each of the following when the variables are not zero:

   a. m^0
   b. d^0
   c. r^0
   d. $x^0 y^0$
   e. $(4c)^0$
   f. $10 n^0$
   g. $7 z^0$
   h. $(7z)^0$

4. Divide, then determine the numerical value when the variables are not zero:

   a. $\dfrac{b^4}{b^4}$
   b. $\dfrac{m^8}{m^8}$
   c. $\dfrac{t^9}{t^9}$
   d. $\dfrac{r^3 s^2}{r^3 s^2}$
   e. $\dfrac{ab^3 c^5}{ab^3 c^5}$
   f. $\dfrac{x^2 y^4 z^6}{x^2 y^4 z^6}$
   g. $\dfrac{b^9 c^6 d^7}{b^9 c^6 d^7}$
   h. $\dfrac{r^3 s t^7}{r^3 s t^7}$

5. Name each of the following, using negative exponents:

   a. $\dfrac{1}{10^3}$
   b. $\dfrac{1}{10^2}$
   c. $\dfrac{1}{10^1}$
   d. $\dfrac{1}{10^5}$
   e. $\dfrac{1}{10^8}$
   f. $\dfrac{1}{10^{10}}$
   g. $\dfrac{1}{3^4}$
   h. $\dfrac{1}{5^1}$
   i. $\dfrac{1}{8^6}$
   j. $\dfrac{1}{12^9}$
   k. $\dfrac{1}{7^5}$
   l. $\dfrac{1}{6^{14}}$

6. Name each of the following as a fraction, using a positive exponent:

   a. 10^{-1}
   b. 10^{-6}
   c. 10^{-4}
   d. 10^{-9}
   e. 10^{-12}
   f. 10^{-7}
   g. 8^{-5}
   h. 6^{-3}
   i. 12^{-2}
   j. 15^{-7}
   k. 100^{-15}
   l. 70^{-10}

7. Find the numerical value of each of the following:

   a. 10^{-2}
   b. 10^{-4}
   c. 10^{-3}
   d. 10^{-7}
   e. 10^{-10}
   f. 10^{-9}
   g. 2^{-2}
   h. 1^{-7}
   i. 3^{-4}
   j. 4^{-3}
   k. 5^{-6}
   l. 7^{-2}

8. Multiply, expressing each product as a power of the given base:

   a. $x^4 \cdot x^6$
   b. $x^{-4} \cdot x^{-6}$
   c. $b^3 \cdot b^{-3}$
   d. $c^{-8} \cdot c^5$
   e. $n^{-9} \cdot n^{-1}$
   f. $d^0 \cdot d^{-6}$
   g. $y^{-2} \cdot y^2$
   h. $a^{-1} \cdot a^{-1}$
   i. $w^{-3} \cdot w$
   j. $x^{-5} \cdot x^6$

9. Multiply, expressing first each product as a power of the given base, then find its numerical value:

   a. $10^3 \times 10^6$
   b. $10^{-1} \times 10^{-4}$
   c. $10^{-5} \times 10^3$
   d. $10^6 \times 10^{-2}$
   e. $10^{-11} \times 10^8$
   f. $10^{-5} \times 10^7$
   g. $10^0 \times 10$
   h. $10^{-2} \times 10^{-4}$
   i. $10^{-6} \times 10^6$
   j. $3^2 \times 3^{-3}$
   k. $5^{-7} \times 5^4$
   l. $6^{-2} \times 6^{-2}$
   m. $1^{12} \times 1^{-8}$
   n. $3^{-5} \times 3^9$
   o. $2^{-8} \times 2^{-1}$

10. Divide, expressing each quotient as a power of the given base:

   a. $\dfrac{n^2}{n^5}$
   b. $\dfrac{b^9}{b^{14}}$
   c. $\dfrac{r}{r^6}$
   d. $m^9 \div m^{17}$
   e. $d^5 \div d^5$
   f. $t^{12} \div t^{11}$
   g. $\dfrac{s^2}{s^3}$
   h. $\dfrac{a^8}{a^{15}}$
   i. $\dfrac{c}{c^{10}}$
   j. $b^4 \div b^{11}$
   k. $z \div z^{10}$
   l. $w^4 \div w^0$
   m. $\dfrac{x^5}{x^{15}}$
   n. $\dfrac{v^3}{v^{11}}$
   o. $\dfrac{n^0}{n^2}$

11. Divide, expressing first each quotient as a power of the given base, then find its numerical value:

   a. $\dfrac{10^2}{10^6}$
   b. $\dfrac{10^9}{10^{16}}$
   c. $\dfrac{10^{-2}}{10^2}$
   d. $\dfrac{10^{-8}}{10^{12}}$
   e. $\dfrac{10^7}{10^{-5}}$
   f. $\dfrac{10^6}{10^{-9}}$
   g. $\dfrac{10^{-5}}{10^{-6}}$
   h. $\dfrac{10^{-3}}{10^{-1}}$
   i. $\dfrac{10^0}{10^{-7}}$
   j. $\dfrac{10^{-16}}{10^{-7}}$
   k. $\dfrac{10^{11}}{10^0}$
   l. $\dfrac{10^{-6}}{10^{-6}}$
   m. $\dfrac{3^{-7}}{3^{-5}}$
   n. $\dfrac{2^8}{2^{-1}}$
   o. $\dfrac{5^0}{5^{-3}}$

12. Compute, expressing each answer as a decimal numeral:

   a. 3×10^0
   b. 8×10^{-2}
   c. 9×10^{-6}
   d. 5×10^{-5}
   e. 7×10^{-3}
   f. 6×10^{-4}
   g. 4×10^{-12}
   h. 9×10^{-15}
   i. 2×10^{-20}
   j. 6×5^0
   k. 7×2^{-2}
   l. 8×3^{-3}
   m. 5×9^{-1}
   n. 4×6^{-2}
   o. 2×7^{-4}

13. Show that:

   a. $\dfrac{1}{2^{-3}} = 2^3$
   b. $\dfrac{1}{10^{-7}} = 10^7$
   c. $\dfrac{a^{-3}}{b^{-4}} = \dfrac{b^4}{a^3}$
   d. $\dfrac{c^2 x^{-5}}{d^3 y^{-2}} = \dfrac{c^2 y^2}{d^3 x^5}$

In the following, simplify each answer by scientific notation:

14. Multiply:

   a. 2×10^5 by 3×10^8
   b. 2×10^{11} by 2×10^{15}
   c. 3×10^{12} by 5×10^{16}
   d. 9×10^{21} by 4×10^{14}
   e. 8×10^{19} by 6×10^{13}
   f. 7×10^{15} by 9×10^9

15. Divide:

   a. $\dfrac{6 \times 10^8}{3 \times 10^4}$
   b. $\dfrac{8 \times 10^{15}}{2 \times 10^7}$
   c. $\dfrac{3.4 \times 10^9}{1.7 \times 10^2}$
   d. $\dfrac{9.6 \times 10^{11}}{1.2 \times 10^7}$
   e. $\dfrac{5.4 \times 10^{25}}{1.8 \times 10^{14}}$
   f. $\dfrac{4.8 \times 10^{18}}{2.4 \times 10^6}$

Algebra

OPEN SENTENCES IN ONE VARIABLE

4–12 INTRODUCTION TO THE SOLUTION OF EQUATIONS IN ONE VARIABLE

To solve an equation in one variable means to find the number represented by the variable which, when substituted for the variable, will make the sentence true.

Any number that makes the sentence true is said to *satisfy* the equation and to be the *root* or the *solution* of the open sentence.

> The equation $n + 3 = 12$ indicates that some number n plus three equals twelve. When 9 is substituted for the variable n, the resulting sentence $9 + 3 = 12$ is a true sentence. Thus 9 is said to be the root of the equation $n + 3 = 12$.

All the numbers which satisfy the sentence are called the *solutions* of the sentence. Since there is only one number, 9, which satisfies the equation $n + 3 = 12$, then we say that the solution of the equation $n + 3 = 12$ is 9.

Checking an equation is testing whether some number is a solution of this equation by substituting the number for the variable in the equation. If the resulting sentence is true, the number is a solution; if not, the number is not a solution.

> 9 is a solution to $n + 3 = 12$.
> 5 is not a solution because $5 + 3 \neq 12$.

The expressions at the left and at the right of the equality sign in an equation are called *members* or *sides* of the equation.

> In the equation $n + 3 = 12$, the expression $n + 3$ is the left member and the expression 12 is the right member.

Equations which have exactly all the same solutions are called *equivalent equations*. The equations $n + 3 = 12$ and $n = 9$ are equivalent equations because they both have the same solution, 9. An equation is considered to be in its *simplest form* when one side contains only the variable and the other side is a constant term. The equation $n = 9$ is an equation in the simplest form.

EXERCISES

1. Which of the following numbers will make the sentence $x + 3 = 9$ true?
   a. $x = 7$
   b. $x = 3$
   c. $x = 5$
   d. $x = 6$
   e. $x = 12$
   f. $x = 9$

2. Which of the following numbers will make the sentence $n - 4 = 12$ true?
   a. $n = 8$
   b. $n = 3$
   c. $n = 4$
   d. $n = 18$
   e. $n = 16$
   f. $n = 12$

3. Which of the following numbers will make the sentence $6b = 30$ true?
   a. $b = 24$
   b. $b = 6$
   c. $b = \frac{1}{5}$
   d. $b = 5$
   e. $b = 36$
   f. $b = 30$

4. Which of the following numbers will make the sentence $\frac{c}{8} = 4$ true?
   a. $c = 2$
   b. $c = 4$
   c. $c = 32$
   d. $c = \frac{1}{2}$
   e. $c = 1$
   f. $c = 8$

5. Which of the following numbers is the root of the equation $r + 5 = 5$?
   a. $r = 10$
   b. $r = 1$
   c. $r = 0$
   d. $r = -1$
   e. $r = -10$
   f. $r = 5$

6. Which of the following numbers is the root of the equation $8n = 48$?
   a. $n = 40$
   b. $n = 56$
   c. $n = 6$
   d. $n = 14$
   e. $n = 26$
   f. $n = 48$

7. Which of the following numbers is the root of the equation $y - 5 = 1$?
   a. $y = 5$
   b. $y = -1$
   c. $y = 4$
   d. $y = 6$
   e. $y = 3$
   f. $y = 1$

Algebra

8. Which of the following numbers is the root of the equation $\frac{d}{2} = 6$?

   a. $d = 3$ c. $d = 8$ e. $d = \frac{1}{3}$
   b. $d = 4$ d. $d = 12$ f. $d = 6$

9. Which of the following numbers is a solution to the equation $s + 11 = 33$?

   a. 44   c. 4   e. 11
   b. 3   d. 22   f. 33

10. Which of the following numbers is a solution to the equation $\frac{m}{15} = 12$?

    a. 160   c. 180   e. 200
    b. 3   d. 27   f. 12

11. Which of the following numbers is a solution to the equation $9v = 54$?

    a. 45   c. 7   e. -9
    b. 63   d. 6   f. 54

12. Which of the following numbers is a solution to the equation $w - 8 = 8$?

    a. 0   c. -16 e. 16
    b. 88   d. 1   f. 8

13. a. Write the right member of the equation $12x - 9 = 31$.
    b. Write the left member of the equation $4y + 17 = 5$.

14. Which of the following equations have 7 as the solution? Which are equivalent equations?

    a. $n + 13 = 6$ c. $12n = 84$ e. $\frac{n}{7} = 1$
    b. $n - 3 = 4$ d. $n = 7$ f. $0 + n = 7$

15. Which of the following equations have 8 as the solution? Which are equivalent equations?
    Which of the equivalent equations is in the simplest form?

    a. $x + 7 = 15$ c. $x - 3 = 5$ e. $10x = 80$
    b. $\frac{x}{4} = 2$ d. $x = 8$ f. $x + 0 = 8$

4–13 INVERSE OPERATIONS; PROPERTIES OF EQUALITY

We have learned that *inverse operations undo each other.* That is: subtraction undoes addition, addition undoes subtraction, division undoes multiplication, and multiplication undoes division.

We found that:

(1) When we first add a number to the given number n and then subtract this number from the sum, we return to the given number n.

That is: $(n + 1) - 1 = n$; $(n + 2) - 2 = n$; $(n + 3) - 3 = n$; etc.

(2) When we first subtract a number from the given number n and then add this number to the answer, we return to the given number n.

That is: $(n - 1) + 1 = n$; $(n - 2) + 2 = n$; $(n - 3) + 3 = n$; etc.

(3) When we first multiply the given number n by a number and then we divide the product by this number, we return to the given number n.

That is: $(n \times 2) \div 2 = n$ or $\frac{2n}{2} = n$;

$(n \times 3) \div 3 = n$ or $\frac{3n}{3} = n$;

$(n \times 4) \div 4 = n$ or $\frac{4n}{4} = n$; etc.

(4) When we first divide the given number n by a number and then we multiply the quotient by this number, we return to the given number n.

That is: $(n \div 2) \times 2 = n$ or $\frac{n}{2} \times 2 = n$;

$(n \div 3) \times 3 = n$ or $\frac{n}{3} \times 3 = n$;

$(n \div 4) \times 4 = n$ or $\frac{n}{4} \times 4 = n$; etc.

The *properties of equality* which state that the results are equals when equals are increased or decreased or multiplied or divided by equals, with division by zero excluded are called *axioms*.

Algebra

These axioms allow us in each of the following cases to obtain an equation in the simplest form.

    a. When we subtract the same number from both sides of a given equation.

        For example: Given $n + 7 = 28$

        Subtracting 7 from each side: $(n + 7) - 7 = 28 - 7$

        we get: $n = 21$

    b. When we add the same number to both sides of a given equation.

        For example: Given: $n - 7 = 28$

        Adding 7 to each side: $(n - 7) + 7 = 28 + 7$

        we get: $n = 35$

    c. When we divide both sides of a given equation by the same non-zero number.

        For example: Given: $7n = 28$

        Dividing each side by 7: $\frac{7n}{7} = \frac{28}{7}$

        we get: $n = 4$

    d. When we multiply both sides of a given equation by the same non-zero number.

        For example: Given: $\frac{n}{7} = 28$

        Multiplying each side by 7: $7 \cdot \frac{n}{7} = 7 \cdot 28$

        we get: $n = 196$

EXERCISES

In each of the following, find the missing number, and where required, the missing operation:

1. a. $(15 + 6) - \Box = 15$ **b.** $(18 + 43) \; ? \; \Box = 18$ **c.** $(n + 5) \; ? \; \Box = n$

2. a. $(24 - 2) + ? = 24$ **b.** $(52 - 17) \; ? \; \Box = 52$ **c.** $(x - 9) \; ? \; \Box = x$

3. a. $(7 \times 3) \div \Box = 7$ **b.** $(26 \times 18) \; ? \; \Box = 26$ **c.** $(a \times 8) \; ? \; \Box = a$

4. a. $(11 \times 8) \div \Box = 8$ **b.** $(41 \times 72) \; ? \; \Box = 72$ **c.** $(10y) \; ? \; \Box = y$

5. a. $(12 \div 2) \times \square = 12$
   b. $(15 \div 4) ? \square = 15$
   c. $(d \div 5) ? \square = d$

6. a. $\frac{5}{6} \times ? = 5$
   b. $\frac{11}{4} \times ? = 11$
   c. $\frac{n}{8} \times ? = n$

7. a. $(r - 10) ? \square = r$
   c. $(x + 45) ? \square = x$
   e. $(17 m) ? \square = m$
   b. $(16 h) ? \square = h$
   d. $(y - 32) ? \square = y$
   f. $\frac{a}{50} ? \square = a$

8. In each of the following, what number do you subtract from both sides of the given equation to get an equivalent equation in simplest form? Also write this equivalent equation.

   a. $n + 9 = 19$
   b. $c + 3 = 11$
   c. $r + 36 = 53$
   d. $t + 90 = 117$

9. In each of the following, what number do you add to both sides of the given equation to get an equivalent equation in simplest form? Also write this equivalent equation.

   a. $w - 4 = 9$
   b. $x - 13 = 7$
   c. $m - 6 = 0$
   d. $p - 28 = 28$

10. In each of the following, by what number do you divide both sides of the given equation to get an equivalent equation in simplest form? Also write this equivalent equation.

    a. $11 c = 55$
    b. $42 a = 21$
    c. $7 c = 3$
    d. $16 y = 38$

11. In each of the following, by what number do you multiply both sides of the given equation to get an equivalent equation in simplest form? Also write this equivalent equation.

    a. $\frac{x}{6} = 12$
    b. $\frac{n}{8} = 2$
    c. $\frac{b}{16} = 13$
    d. $\frac{w}{9} = 9$

12. What operation with what number do you use on both sides of each of the following equations to get an equivalent equation in simplest form? Also write this equivalent equation.

    a. $8 c = 24$
    b. $n + 7 = 12$
    c. $y - 4 = 4$
    d. $9 d = 45$
    e. $\frac{w}{7} = 6$
    f. $13 r = 60$
    g. $m - 21 = 8$
    h. $z + 37 = 37$
    i. $\frac{n}{5} = 1$
    j. $18 t = 12$
    k. $h - 110 = 39$
    l. $3 y = 2$
    m. $r + 87 = 125$
    n. $50 = c - 19$
    o. $24 = \frac{b}{8}$
    p. $r - 83 = 0$
    q. $20 t = 75$
    r. $\frac{v}{12} = 9$
    s. $91 = p + 62$
    t. $6 = 18 t$

Algebra

4–14 SOLVING EQUATIONS IN ONE VARIABLE

To solve an equation in one variable, we transform the given equation to the simplest equivalent equation which has only the variable itself as one member and a constant naming a number as the other member. To make this transformation we use the properties of equality (axioms) with inverse operations or with the additive inverse or multiplicative inverse as required.

An equation may be compared to balanced scales. To keep the equation in balance, any change on one side of the equality sign must be balanced by an equal change on the other side of the equality sign.

Observe in the models of this page and on the following pages, that there are four basic types of equations:

Type I $n + 3 = 12$ **Type III** $3n = 12$

Type II $n - 3 = 12$ **Type IV** $\frac{n}{3} = 12$

They are solved both by inverse operations and by the additive inverse or the multiplicative inverse. When the inverse operation method is used, the operation indicated in the given equation by the variable and its connected constant is undone by performing the inverse operation on both sides of the equation.

To check whether the number found is a solution, we substitute this number for the variable in the given equation. If the resulting sentence is true, then the number is a solution. A check mark is sometimes used to indicate that the resulting sentence is true. In this section there will only be one solution to each equation.

Basic Type I

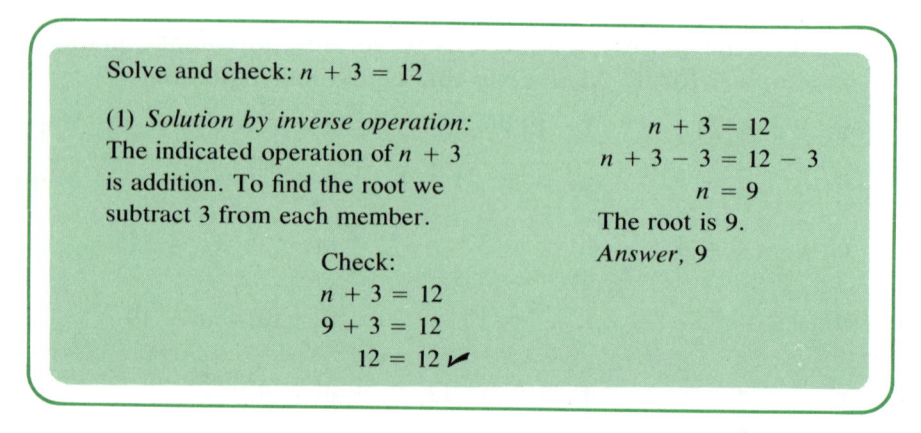

Solve and check: $n + 3 = 12$

(1) *Solution by inverse operation:*
The indicated operation of $n + 3$ is addition. To find the root we subtract 3 from each member.

$n + 3 = 12$
$n + 3 - 3 = 12 - 3$
$n = 9$
The root is 9.
Answer, 9

Check:
$n + 3 = 12$
$9 + 3 = 12$
$12 = 12$ ✓

Solve and check: $n + 3 = 12$

(2) *Solution by additive inverse:*
The additive inverse of $+3$ is -3.
We add the additive inverse to both members, using the addition axiom.

$n + 3 = 12$
$n + 3 + (-3) = 12 + (-3)$
$n = 9$
Answer, 9

Check:
$n + 3 = 12$
$9 + 3 = 12$
$12 = 12$ ✓

Observe that the equation $n + 3 = 12$ may also appear as $3 + n = 12$ or $12 = n + 3$ or $12 = 3 + n$. These are forms of the same equation. In each case, the constant 3 must be eliminated in order to get an equivalent equation in the simplest form. This is $n = 9$ when solving $n + 3 = 12$ or $3 + n = 12$; and $9 = n$ when solving $12 = n + 3$ or $12 = 3 + n$. The equation $9 = n$ may be rewritten as $n = 9$.

EXERCISES

Solve and check:

1. $x + 3 = 13$
2. $n + 29 = 54$
3. $b + 36 = 95$
4. $a + 75 = 120$
5. $11 + y = 34$
6. $8 + c = 27$
7. $29 + x = 46$
8. $51 + s = 69$
9. $21 = a + 7$
10. $65 = y + 19$
11. $38 = x + 38$
12. $80 = n + 42$
13. $8 = 5 + x$
14. $32 = 17 + c$
15. $55 = 26 + n$
16. $94 = 56 + T$
17. $m + 45 = 45$
18. $39 + d = 59$
19. $93 = y + 37$
20. $68 = 9 + b$
21. $b + \frac{3}{4} = 9$
22. $2\frac{7}{8} + x = 4\frac{1}{4}$
23. $y + .9 = 6$
24. $c + \$.05 = \$.83$
25. $a + 6 = 4$
26. $r + 5 = -2$
27. $7 + s = 5$
28. $-8 = x + 3$
29. $\$15 = a + \$2.75$
30. $y + 14 = 43$
31. $36 + x = 144$
32. $78 = 78 + r$
33. $12 = n + 5\frac{3}{8}$
34. $c + 5 = 2.3$
35. $6 = s + 3.4$
36. $m + 12 = 0$

Basic Type II

> Solve and check: $n - 3 = 12$
>
> (1) *Solution by addition axiom:* $\qquad\qquad n - 3 = 12$
> The indicated operation of $n - 3$ $\qquad n - 3 + 3 = 12 + 3$
> is subtraction. To find the root we $\qquad\qquad\qquad n = 15$
> add 3 to both members. $\qquad\qquad$ The root is 15.
> $\qquad\qquad\qquad\qquad\qquad\qquad\qquad\qquad$ Answer, 15
>
> (2) *Solution by additive inverse:* $\qquad\qquad n - 3 = 12$
> The additive inverse of -3 is $+3$. $\qquad n + (-3) = 12$
> We add the additive inverse to both $\qquad n + (-3) + (+3)$
> members using the addition axiom. $\qquad\qquad\qquad = 12 + (+3)$
> $\qquad\qquad\qquad\qquad\qquad\qquad\qquad\qquad n = 15$
>
> Check: $\qquad\qquad\qquad\qquad$ Answer, 15
> $n - 3 = 12$
> $15 - 3 = 12$
> $12 = 12$ ✓

Observe that the equation $n - 3 = 12$ may also appear as $12 = n - 3$ but not as $3 - n = 12$ or $12 = 3 - n$. When solving $n - 3 = 12$ or $12 = n - 3$, the constant 3 must be eliminated in each case in order to get an equivalent equation in the simplest form. This is $n = 15$ when solving $n - 3 = 12$, or $15 = n$ when solving $12 = n - 3$. The equation $15 = n$ may be rewritten as $n = 15$.

EXERCISES

Solve and check:

1. $a - 3 = 8$
2. $x - 12 = 9$
3. $b - 25 = 52$
4. $n - 60 = 19$

5. $15 = c - 4$
6. $6 = s - 26$
7. $64 = x - 33$
8. $72 = y - 85$

9. $y - 7 = 7$
10. $39 = t - 39$
11. $n - 24 = 0$
12. $0 = m - 93$

13. $x - \frac{1}{2} = \frac{3}{4}$
14. $a - 7\frac{1}{3} = 6$
15. $8 = y - 2\frac{3}{8}$
16. $5\frac{1}{4} = n - 5\frac{1}{4}$

17. $n - .9 = 2.6$
18. $54 = b - 1.5$
19. $8.3 = x - 6$
20. $x - 6 = -8$

21. $x - 21 = -35$
22. $y - 12 = -9$
23. $a - 4 = -4$
24. $a - 49 = 85$

25. $b - .3 = 4.7$
26. $31 = c - 20$
27. $x - 12 = 0$
28. $6\frac{1}{2} = y - \frac{7}{8}$

29. $n - 1\frac{1}{2} = -3\frac{1}{2}$
30. $a - \$.08 = \$.64$
31. $s - \$1.45 = \$.75$
32. $a - \$.27 = \$.15$

Basic Type III

Solve and check: $3n = 12$

(1) *Solution by division axiom:*
The indicated operation of $3n$ is multiplication. To find the root we divide both members by 3.

$3n = 12$
$\frac{3n}{3} = \frac{12}{3}$
$n = 4$
The root is 4.
Answer, 4

(2) *Solution by multiplicative inverse:*
The multiplicative inverse of 3 is $\frac{1}{3}$. We multiply both members by $\frac{1}{3}$, using the multiplication axiom.

$3n = 12$
$\frac{1}{3} \cdot 3n = \frac{1}{3} \cdot 12$
$n = 4$
Answer, 4

Check:
$3n = 12$
$3 \cdot 4 = 12$
$12 = 12$ ✓

Observe that the equation $3n = 12$ may also appear as $12 = 3n$. When solving $3n = 12$ or $12 = 3n$, the numerical coefficient 3 must be eliminated in each case in order to get an equivalent equation in simplest form. This is $n = 4$ when solving $3n = 12$, or $4 = n$ when solving $12 = 3n$. The equation $4 = n$ may be rewritten as $n = 4$.

EXERCISES

Solve and check:

1. $8n = 56$
2. $10x = 70$
3. $9y = 27$
4. $12a = 84$
5. $12 = 3a$
6. $54 = 6d$
7. $63 = 7n$
8. $60 = 15y$

9. $5x = 5$
10. $14 = 14b$
11. $6y = 0$
12. $0 = 9m$
13. $4n = 9$
14. $12x = 20$
15. $17 = 5c$
16. $26 = 8N$
17. $6y = 1$
18. $4 = 7T$

19. $8b = 6$
20. $10 = 18y$
21. $8x = .24$
22. $.2a = 16$
23. $9c = \$.45$
24. $\$.06n = \$1.32$
25. $\frac{1}{3}a = 27$
26. $\frac{5}{8}x = 35$
27. $40 = \frac{4}{5}b$
28. $2\frac{1}{2}c = 15$

29. $9x = -27$
30. $-6b = 54$
31. $-42 = -7y$
32. $8d = -48$
33. $-x = 8$
34. $\frac{2}{3}n = -18$
35. $14b = -4$
36. $-24y = -12$
37. $16T = 400$
38. $180 = 20x$

39. $19r = 19$
40. $16b = 30$
41. $25x = 15$
42. $1.05c = 420$
43. $1\frac{1}{4}n = 8\frac{3}{4}$
44. $15T = 8$
45. $9b = -54$
46. $-12x = 80$
47. $21 = 24c$
48. $.04p = 96$

Basic Type IV

Solve and check: $\frac{n}{3} = 12$

(1) *Solution by inverse operation:*

The indicated operation of $\frac{n}{3}$ is division. To find the root we multiply both members by 3.

$\frac{n}{3} = 12$
$3 \cdot \frac{n}{3} = 3 \cdot 12$
$n = 36$
The root is 36.
Answer, 36.

(2) *Solution by multiplicative inverse:*

Since $\frac{n}{3} = \frac{1}{3}n$, we use the multiplicative inverse of $\frac{1}{3}$, which is 3. We multiply both members by 3, using the multiplicative axiom.

$\frac{n}{3} = 12$
$3 \cdot \frac{n}{3} = 3 \cdot 12$
$n = 36$
Answer, 36

Check:

$\frac{n}{3} = 12$
$\frac{36}{3} = 12$
$12 = 12$ ✓

Observe that the equation $\frac{n}{3} = 12$ may also appear as $12 = \frac{n}{3}$. When solving $\frac{n}{3} = 12$ or $12 = \frac{n}{3}$, the denominator 3 must be eliminated in order to get an equivalent equation in simplest form. This is $n = 36$ when solving $\frac{n}{3} = 12$, or $36 = n$ when solving $12 = \frac{n}{3}$. The equation $36 = n$ may be rewritten as $n = 36$.

EXERCISES

Solve and check:

1. $\frac{x}{2} = 7$
2. $\frac{c}{5} = 3$
3. $\frac{y}{6} = 0$
4. $9 = \frac{a}{4}$
5. $\frac{x}{3} = 15$
6. $\frac{b}{12} = 4$
7. $16 = \frac{d}{20}$
8. $\frac{s}{10} = 10$
9. $\frac{c}{9} = 1.8$
10. $\frac{n}{1.04} = 60$
11. $\frac{1}{2} n = 27$
12. $\frac{1}{3} a = \$.54$
13. $\frac{t}{5} = -9$
14. $\frac{b}{-4} = 8$
15. $\frac{a}{18} = -6$
16. $\frac{d}{-3} = -15$
17. $\frac{n}{7} = 1$
18. $\frac{r}{2} = 24$
19. $\$1.25 = \frac{1}{8} m$
20. $\frac{1}{4} r = 9$
21. $\frac{b}{9} = -3$
22. $\frac{c}{-5} = 12$
23. $30 = \frac{n}{6}$
24. $\frac{y}{5} = 5$

The solution of more difficult equations may involve the use of more than one axiom or the combining of terms as follows:

Solve and check: $5n + 4 = 49$

$5n + 4 = 49$
$5n + 4 - 4 = 49 - 4$
$5n = 45$
$\frac{5n}{5} = \frac{45}{5}$
$n = 9$

Check:
$5n + 4 = 49$
$(5 \cdot 9) + 4 = 49$
$45 + 4 = 49$
$49 = 49$ ✓

Answer, 9

Algebra

Solve and check: $4x + 5x = 36$

$4x + 5x = 36$ Check:
$9x = 36$ $4x + 5x = 36$
$\dfrac{9x}{9} = \dfrac{36}{9}$ $(4 \cdot 4) + (5 \cdot 4) = 36$
$x = 4$ $16 + 20 = 36$
                        $36 = 36$ ✔

Answer, 4

Solve and check: $6y + 5 = 29 - 2y$

$6y + 5 = 29 - 2y$
$6y + (2y) + 5 = 29 + (-2y) + (2y)$
$8y + 5 = 29$
$8y + 5 + (-5) = 29 + (-5)$
$8y = 24$
$y = 3$

Check:
$6y + 5 = 29 - 2y$
$(6 \cdot 3) + 5 = 29 - (2 \cdot 3)$
$18 + 5 = 29 - 6$
$23 = 23$ ✔

Answer, 3

EXERCISES

Solve and check:

1. $3n + 7 = 31$
2. $9 + 8x = 57$
3. $71 = 5d + 6$
4. $80 = 8 + 9c$
5. $6y + 13 = 35$
6. $59 = 12m + 35$
7. $26 + 11b = 70$
8. $51 = 16d + 3$
9. $8x - 5 = 27$
10. $15x - 44 = 16$
11. $9x - 36 = 0$
12. $29 = 6n - 25$
13. $10b - 17 = 13$
14. $45 = 18a - 45$
15. $4x - 23 = 19$

16. $0 = 7x - 10$
17. $21 = 15T - 39$
18. $1 - 9a = 10$
19. $n + n = 38$
20. $5x + 9x = 84$
21. $7y - 3y = 28$
22. $8b = 19 + 29$
23. $x + 2x + 3x = 96$
24. $11c - c = 95$
25. $5n = 52 - 27$
26. $68 = 6a + 11a$
27. $9d - 4d + 2d = 63$
28. $p + .12p = 560$

29. $a - .04a = 384$
30. $1.8c + 32 = 68$
31. $30 + .6T = 45$
32. $l - .25l = \$.81$
33. $\frac{2}{5}n = 8$
34. $\frac{7}{8}b = 21$
35. $\frac{3}{4}y = \$8.46$
36. $m + \frac{1}{2}m = 4\frac{1}{2}$
37. $\frac{5}{8}n - 8 = 22$
38. $6x + 53 = 11$
39. $15a - 34 = 41$
40. $c + .18c = 236$
41. $\frac{2}{3}u = 74$

42. $9 + 8n = 13$
43. $15 = 12y - 57$
44. $3x - x = -10$
45. $6n + 15 = 9$
46. $6x - 11x = 35$
47. $9y - 15y = -24$
48. $5x + 6 = 2x + 27$
49. $11y - 16 = 40 - 3y$
50. $6n + 2 = 4n - 14$
51. $8x + 5x - 9 = 30$
52. $9y - 7y + 8 = 6y - 28$
53. $10x + 5(x + 4) = 230$
54. $9a - 43 = 2(a + 3)$

4–15 SOLVING EQUATIONS WHEN THE REPLACEMENTS FOR THE VARIABLE ARE RESTRICTED

The solution of an equation depends not only on the equation but also on how restricted the replacements for the variable are.

> For the equation $5n = 35$
>
> When the replacements are all the whole numbers, the solution is 7.
>
> When the replacements are all the prime numbers, the solution is 7.
>
> When the replacements are all the even numbers, there is no solution.

> For the equation $3x = 2$
>
> When the replacements are all the real numbers, the solution is $\frac{2}{3}$.
>
> When the replacements are all the whole numbers, there is no solution.

When the replacements for the variables are not given in this book, use all the real numbers as the replacements.

EXERCISES

1. Find the solution of each of the following equations:

   a. When the replacements for the variable are all the natural numbers:

   $n + 12 = 9$ $n - 6 = 6$ $5d = 20$ $\frac{b}{4} = 7$

   b. When the replacements for the variable are 0, 1, 2, 3, 4, 5, 6, 7, 8, and 9:

   $6y = 42$ $n - 4 = 1$ $2x + 5 = 11$ $\frac{n}{3} = 6$

   c. When the replacements for the variable are all the integers:

   $-5y = -20$ $c + 8 = 8$ $6b - 4 = 15$ $\frac{c}{2} = -4$

   d. When the replacements for the variable are all the prime numbers:

   $b + 6 = 8$ $n - 4 = 13$ $\frac{t}{3} = 2$ $20s = 60$

   e. When the replacements for the variable are all the non-positive integers:

   $-7x = -21$ $\frac{c}{8} = -5$ $n + 9 = 9$ $8a - 11 = -43$

   f. When the replacements for the variable are $-3, -2, -1, 0, 1, 2,$ and 3:

   $r - 2 = 2$ $-18b = 6$ $\frac{m}{2} = -1$ $5c + 18 = 3$

   g. When the replacements for the variable are all the one-digit even whole numbers:

   $d + 13 = 17$ $6a - 4a = 24$ $-8y = -56$ $\frac{r}{7} = 4$

h. When the replacements for the variable are 0, 1, 2, 3, 4, . . .:

$h - 9 = 16$ $-11t = -79$ $\dfrac{x}{9} = 1$ $3m + 28 = 13$

i. When the replacements for the variable are all the odd whole numbers:

$x + 8 = 13$ $\dfrac{n}{5} = 9$ $15y = 90$ $10b - 7 = 13$

j. When the replacements for the variable are all the two-digit prime numbers less than 25:

$t - 17 = 6$ $2x = 58$ $\dfrac{d}{12} = 8$ $9c + 4c = 169$

2. Do not solve the following equations. Substitute each replacement number to determine which is the root of the open sentence.

Open Sentence	Replacements	Open Sentence	Replacements
a. $n + 5 = 6$	2, −1, and 1	**d.** $6n + 9 = -3$	5, −5, 2 and −2
b. $t - 1 = 8$	−7, 0, and 9	**e.** $17y - y = 32$	−2, −1, 0, 1, and 2
c. $9x = 72$	4, 6, 8, and 10	**f.** $\dfrac{b}{2} = 3$	2, 4, 6, and 8

4–16 EQUATIONS INVOLVING ABSOLUTE VALUES

Since $|+2| = 2$ and $|-2| = 2$, replacement of the variable in the equation $|x| = 2$ by either $+2$ or -2 will make the sentence true. Therefore, the roots of $|x| = 2$ are $+2$ and -2 or its solutions are $+2$ and -2.

> We use the equality axioms to solve equations of the basic types:
>
> $$|y| - 8 = 7$$
> $$|y| - 8 + 8 = 7 + 8$$
> $$|y| = 15$$
> $$y = +15 \text{ or } -15$$
>
> *Answer,* $+15$ and -15

Algebra

EXERCISES

Find the solutions of each of the following equations:

1. $|x| = 8$
2. $|y| = 12$
3. $|s| + 3 = 4$
4. $|t| - 7 = 15$
5. $8 \times |d| = 72$
6. $|n| + 9 = 27$
7. $4 \times |n| = 40$
8. $\frac{|a|}{6} = 5$
9. $16 - |x| = 5$

4–17 EVALUATING FORMULAS

To determine the value of any variable in a formula when the values of the other variables are known, we copy the formula, substitute the given values for the variables, and perform the necessary operations.

We then solve the resulting equation for the value of the required variable.

Find the value of t when $d = 60$ and $r = 15$, using the formula $d = rt$:

$d = rt$
$60 = 15\,t$
$4 = t$
$t = 4$ *Answer,* $t = 4$

EXERCISES

Find the value of:

1. p when $C = 63$ and $n = 9$, using the formula $C = np$.
2. h when $V = 168, l = 7$, and $w = 6$, using the formula $V = lwh$.
3. t when $i = 24, p = 80$, and $r = .06$, using the formula $i = prt$.
4. I when $W = 260$ and $E = 65$, using the formula $W = IE$.
5. a when $v = 105$ and $t = 15$, using the formula $v = at$.
6. p when $i = 80, r = .02$ and $t = 8$, using the formula $i = prt$.

7. A when $F = 720$, $H = 5$, and $D = 24$, using the formula $F = AHD$.
8. r when $i = 36$, $p = 90$, and $t = 10$, using the formula $i = prt$.
9. A when $B = 34$, using the formula $A + B = 90$.
10. D_w when $D_t = 93$ and $D_p = 28$, using the formula $D_t = D_w + D_p$.
11. C when $A = 7{,}500$ and $L = 1{,}800$, using the formula $A = L + C$.
12. N when $M = 109$, using the formula $M + N = 180$.
13. A when $p = 85$ and $i = 6$, using the formula $p = A - i$.
14. K when $C = 47$, using the formula $C = K - 273$.
15. V when $v = 123$, $g = 32$, and $t = 3$, using the formula $v = V + gt$.
16. a when $l = 57$, $n = 13$, and $d = 4$, using the formula $l = a + (n - 1)d$.
17. l when $p = 92$ and $w = 21$, using the formula $p = 2l + 2w$.
18. C when $F = 176$, using the formula $F = 1.8 C + 32$.
19. r when $A = 496$, $p = 400$, and $t = 4$, using the formula $A = p + prt$.
20. h when $A = 2{,}816$, $\pi = \frac{22}{7}$, and $r = 14$, using the formula $A = 2\pi r^2 + 2\pi rh$.
21. W when $H.P. = 3.5$, using the formula $H.P. = \frac{W}{746}$.
22. E when $I = 4$ and $R = 55$, using the formula $I = \frac{E}{R}$.
23. N when $S = 1{,}884$, $\pi = 3.14$, and $D = 36$, using the formula $S = \frac{\pi DN}{12}$.
24. A when $L = 1{,}250$, $C = .5$, $d = .002$, and $V = 100$, using the formula $L = \frac{CdAV^2}{2}$.
25. F when $P = 160$, $d = 8$, and $t = 6$, using the formula $P = \frac{Fd}{t}$.
26. p when $A = 310$, $r = .04$, and $t = 6$, using the formula $A = p + prt$.
27. I when $E = 96$, $r = 7$, and $R = 9$, using the formula $E = Ir + IR$.
28. V when $P = 52$, $P' = 28$, and $V' = 13$, using the formula $PV = P'V'$.
29. C when $F = -13$, using the formula $F = 1.8 C + 32$.
30. T_2 when $P_1 = 18$, $T_1 = 40$, and $P_2 = 9$, using the formula $P_1 T_2 = T_1 P_2$.

Algebra

4–18 ALGEBRAIC REPRESENTATION

We write an algebraic expression using the given variables, numerals, and proper symbols to show the relationship involved.

Observe that when there are two related quantities involved, (1) one quantity may be a given number times the second quantity, or (2) a given number more or less than the second quantity, or (3) the sum of the two quantities may be given.

> (1) There are n degrees in one angle. If a second angle is 6 times as large, how many degrees does it measure?
>
> *Answer,* $6n$
>
> (2) Pierre is x years old. How old will he be in 4 years?
> *Answer,* $x + 4$
>
> (3) Joe and Helen together have 20 books. If Joe has B books, how many books does Helen have?
> *Answer,* $20 - B$

EXERCISES

Write an algebraic expression for each of the following:

1. Erica is n years old. How old was she 8 years ago?
2. How many hours did it take an automobile to travel x kilometers if it averaged 80 km/h?
3. Tom has b dollars in the bank. If he deposits c dollars, how many dollars does he then have in the bank?
4. Rose has s cents. She spends 35 cents for lunch. How many cents does she have left?
5. Ricardo has 3 dollars more than Beulah. If Beulah has d dollars, how many dollars does Ricardo have?
6. Cora is x years old. Liza is 8 times as old as Cora. How old is Liza?

7. If Mrs. Carter saves D dollars each week, how many weeks will it take her to save $500?
8. How many pencils can be bought for n cents if one pencil costs c cents?
9. A streamliner travels R miles per hour. How many miles can it go in h hours?
10. The length of a rectangle is 4 times the width. If the width is m centimeters, what is the length in centimeters?
11. The sum of two numbers is 25. If one number is n, what is the other number?
12. The base of a triangle is t centimeters long. The altitude is 4 centimeters less than the base. How many centimeters is the altitude?
13. If the radius of a circle is y inches, how long is the diameter?
14. A baker has 48 kg of $1.99 and $2.59 cookies. If she has z kg of $1.99 cookies, how many kg of $2.59 cookies does she have?
15. Harry has x $5 bills but 6 times as many $1 bills. How many $5 and $1 bills does he have in all?
16. Ingrid has n dollars. Arnold has $7 less than 3 times as many dollars as Ingrid. How many dollars does Arnold have?
17. One of two angles has d degrees and their sum is 75°. How many degrees are in the second angle?
18. How many miles can an automobile travel on 15 gallons of gasoline if it averages b miles per gallon?
19. There are n pupils in a class. If there are x boys, how many girls are in the class?
20. A board 16 feet long is divided into 2 pieces. If one piece is f feet long, how long is the other piece?
21. At the rate of r km/h, how many hours will it take an airplane to fly s kilometers?
22. Angle A is 5 times as large as angle B. Angle B has h degrees. How many degrees are in angle A?
23. Nathan is 18 years old. How old will he be in y years?
24. Nancy and Maureen together sold 240 magazines. If Nancy sold n magazines, how many did Maureen sell?
25. Ms. Wilson has r 13¢ stamps. She has 9 more 1¢ stamps than 13¢ stamps and 4 times as many 10¢ as 13¢ stamps. How many stamps does she have in all?

Algebra

4–19 PROBLEMS

To solve problems by the use of algebraic techniques, we first read the problem carefully to determine what is required and any facts which are related to this unknown value.

When there is only one required value, we represent it by a variable, usually a letter. When there are two required values, we represent the smaller unknown value by some letter and express the other unknown value in terms of that letter.

We then find two facts, at least one involving the unknown value, which are equal to each other. We translate them into algebraic expressions and form an equation by writing one algebraic expression equal to the other.

We solve the equation to find the required value. We check the answer directly with the given problem.

How long should each piece be if a ribbon 105 centimeters long is to be cut into two pieces, one twice as long as the other?

Solution:
  Let x = length of smaller piece in centimeters
  and $2x$ = length of larger piece in centimeters

$$x + 2x = 105$$
$$3x = 105$$
$$\frac{3x}{3} = \frac{105}{3}$$

Check:
  35 cm
  +70 cm
  ―――
  105 cm

  $35 \overline{)70 \text{ cm}}^{\,2}$

$x = 35$ centimeters, length of smaller piece
$2x = 70$ centimeters, length of larger piece

Answer, 35 centimeters and 70 centimeters

EXERCISES

Solve the following problems:

1. Olga asked Carlos to guess her age. She said, "If you add 16 to six times my age, you get 100." How old is Olga?

2. $\frac{7}{8}$ of what number is 42?
3. 4% of what number is 59?
4. How much money must be invested at 6% to earn $1,500 per year?
5. There are 11 more girls than boys in the graduating class of 167 students. How many girls are in the class?
6. What mark must Renee get in a test to bring her average up to 85 if her other test marks are 79 and 87?
7. The net price of a radio is $54 when a 10% discount is allowed. What is its list price?
8. How much must a salesclerk sell each week on a 15% commission basis to earn $72 weekly?
9. Two girls together sold 60 tickets for the school show. If one girl sold 4 times as many as the other, how many tickets did each girl sell?
10. A dealer sold a rug for $108, making a profit of 35% on the cost. How much did it cost?
11. A board 12 feet long is to be cut into two pieces, one 6 feet longer than the other. Find the length of each piece.
12. A man is 3 times as old as his son. Together the sum of their ages is 52. What are their ages?
13. What is the selling price of a television set if it cost $175 and the profit is 30% of the selling price?
14. Find the regular price of a suit that sold for $41.25 at a 25% reduction sale.
15. The perimeter of a rectangle is 42 meters. Its length is 6 times its width. Find the dimensions.
16. Find the measures of the angles of a triangle if the first angle measures 3 times the second angle and the third angle measures twice the second angle.
17. Joe has 8 more dimes than nickels. If he has 26 coins in all, how many of each does he have?
18. Find two consecutive integers whose sum is 71.
19. How much money was borrowed at the annual rate of $8\frac{1}{4}$% if the interest charged was $536.25 per year?
20. A merchant purchased CB radios at $48 each. What price should each radio be marked so that a 40% profit on the selling price may be realized?

Algebra

4–20 GRAPHING AN EQUATION IN ONE VARIABLE ON THE NUMBER LINE

The *graph of an equation* is the graph of its solutions.

To draw the graph of an equation in one variable on the number line, we first solve the given equation to find its solutions. We then locate the point or points on the number line whose coordinate or coordinates are numbers forming all the solutions to the given equation. We indicate these points by solid or colored dots.

Where the solution is one number, the graph will consist of one point. Where there are two solutions, the graph will consist of two points; see illustration (4). Observe in illustration (5) that the solution of the given equation is an infinite collection of numbers which is shown as the entire number line.

(1) The solution of the equation $n + 4 = 7$ is 3. The graph of $n + 4 = 7$ is the point whose coordinate is 3.

$$-1 \ 0 \ 1 \ 2 \ 3 \ 4 \ 5 \ 6$$

(2) The solution of the equation $3n + 2 = 14$ is 4. The graph of $3n + 2 = 14$ is the point whose coordinate is 4.

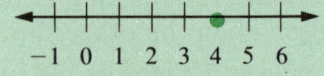

$$-1 \ 0 \ 1 \ 2 \ 3 \ 4 \ 5 \ 6$$

(3) There is no solution of the equation $x = x - 5$. We have no points to draw the graph.

(4) The solutions of $|x| = 1$ are $+1$ and -1. The graph of $|x| = 1$ is the set of points whose coordinates are $+1$ and -1.

$$-3 \ -2 \ -1 \ 0 \ 1 \ 2 \ 3$$

(5) The solutions of $5x - 20 = 5(x - 4)$ are all the real numbers since the equation (identity) is satisfied by every real number. The graph of $5x - 20 = 5(x - 4)$ is the entire number line indicated by a heavy or colored line with an arrowhead in each direction to show that it is endless.

$$-4 \ -3 \ -2 \ -1 \ 0 \ 1 \ 2 \ 3 \ 4$$

(6) The graph

$$-2 \ -1 \ 0 \ 1 \ 2 \ 3 \ 4$$

is a picture of the equation $x = 2$ or any equivalent equation such as $4x = 8$, $x - 6 = -4$, $x + 3 = 5$, etc.

EXERCISES

1. On the number line below, what is the coordinate of point: D? J? P? B? I? L? A? K? N? E?

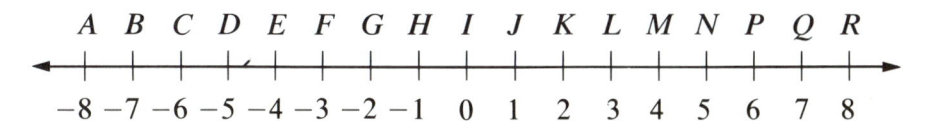

   A B C D E F G H I J K L M N P Q R
   ←──┼──┼──┼──┼──┼──┼──┼──┼──┼──┼──┼──┼──┼──┼──┼──┼──→
   −8 −7 −6 −5 −4 −3 −2 −1 0 1 2 3 4 5 6 7 8

2. On the number line above, what point is the graph of:
   −3? 8? −6? 0? 4? −2? 7? −1? 1? −4?

3. For each of the following equations, draw an appropriate number line, then graph its solutions.

   a. $y = 3$
   b. $n = -2$
   c. $x + 8 = 15$
   d. $7a = 42$
   e. $b - 4 = 1$
   f. $2c + 7 = 5$
   g. $8x - x = 63$
   h. $4x + 3 = 3$
   i. $\frac{x}{5} = 1$
   j. $|x| = 6$
   k. $y = y + 5$
   l. $9x + 27 = 9(x + 3)$

4. On a number line (see section 4–15) draw the graph of:

   a. $6x = 12$ when the replacements for the variable are all prime numbers.
   b. $\frac{x}{2} = 4$ when the replacements for the variable are all natural numbers.
   c. $b + 7 = 2$ when the replacements for the variable are all integers.
   d. $8y - 4 = 44$ when the replacements for the variable are all multiples of 3.
   e. $4x - x = -12$ when the replacements for the variable are −6, −4, −2, 0, 2, 4, and 6.

5. Write a corresponding equation which is pictured by each of the following graphs:

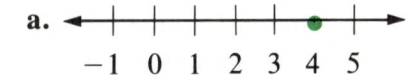

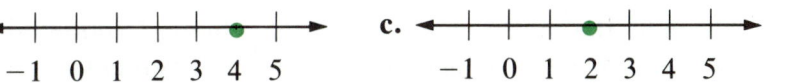

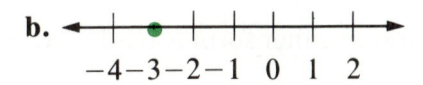

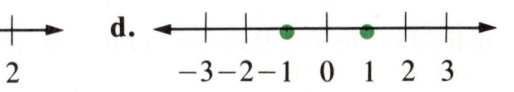

   a. ←──┼──┼──┼──┼──●──┼──→
        −1 0 1 2 3 4 5

   b. ←──┼──●──┼──┼──┼──┼──→
        −4 −3 −2 −1 0 1 2

   c. ←──┼──┼──●──┼──┼──┼──→
        −1 0 1 2 3 4 5

   d. ←──┼──┼──●──┼──●──┼──┼──→
        −3 −2 −1 0 1 2 3

Algebra

4–21 RATIO

We may compare two numbers by subtraction or by division. When we compare 24 to 4, we find by subtraction that 24 is twenty more than 4 and by division that 24 is six times as large as 4. When we compare 4 with 24, we find by subtraction that 4 is twenty less than 24 and by division that 4 is one-sixth as large as 24.

The answer we obtain when we compare two quantities *by division* is called the *ratio* of two quantities.

When 24 is compared to 4, 24 is divided by 4 as $\frac{24}{4}$ which equals $\frac{6}{1}$. The ratio is $\frac{6}{1}$ or 6 to 1.

When 4 is compared to 24, 4 is divided by 24 as $\frac{4}{24}$ which equals $\frac{1}{6}$. The ratio is $\frac{1}{6}$ or 1 to 6.

A ratio has two terms, the number that is being compared (found in the numerator) and the number to which the first number is being compared (found in the denominator). The ratio in fraction form is usually expressed in lowest terms.

The ratio of 7 miles to 8 miles is $\frac{7}{8}$.

The fraction $\frac{7}{8}$ is an indicated division meaning $7 \div 8$.

Usually the ratio is expressed as a common fraction but it may also be expressed as a decimal fraction or a percent.

The ratio $\frac{7}{8}$ is sometimes written as 7:8. The colon may be used instead of the fraction bar. This ratio in either form is read "7 to 8" ("seven to eight").

If the quantities compared are denominate numbers, they must first be expressed in the same units. The ratio is an abstract number; it contains no unit of measurement.

A ratio may be used to express a rate. The average rate of speed of an automobile when it travels 252 kilometers in 3 hours may be expressed by the ratio which is $\frac{252}{3}$ or 84 km/h. If apples sell at the rate of 3 for 35¢, the ratio expressing this rate is $\frac{3}{35}$ and is read "3 for 35."

When a ratio is used to express a rate, the two quantities have different names like kilometers and hours, apples and cents, etc.

Ratios such as $\frac{9}{12}, \frac{18}{24}, \frac{12}{16}, \frac{6}{8}$, although written with different number pairs, express the same comparison or rate. These ratios are called *equivalent ratios*. In simplest form they are the same, the ratio $\frac{3}{4}$.

EXERCISES

1. Find the ratio of:

   a. 4 to 12
   b. 8 to 10
   c. 3 to 5
   d. 15 to 60
   e. 21 to 28
   f. 12 to 6
   g. 7 to 4
   h. 18 to 8
   i. 25 to 5
   j. 40 to 24
   k. 15 to 3
   l. 6 to 24
   m. 16 to 18
   n. 30 to 12
   o. 36 to 16
   p. 48 to 30

2. Find the ratio of:

   a. 6 in. to 9 in.
   b. 8 cm to 1 m
   c. 30 cm to 1 m
   d. 3 mm to 2 cm
   e. 1 ft. 6 in. to 4 yd.
   f. 10 lb. to 35 lb.
   g. 3 kg to 8 g
   h. 2 cL to 1 L
   i. 3 gal. to 3 qt.
   j. 1 gal. to 1 pt.
   k. 20 min. to 45 min.
   l. 30 sec. to 2 min.
   m. 2 hr. to 40 min.
   n. 1 hr. 15 min. to 25 min.
   o. 18 mo. to 2 yr.
   p. 1 min. to 80 sec.

3. What is the ratio of:
   a. A dime to a dollar?
   b. A quarter to a nickel?
   c. A half dollar to a dime?
   d. 4 cents to 2 nickels?
   e. 6 dollars to 3 dimes?
   f. 3 things to a dozen?
   g. 10 things to a dozen?
   h. 1 dozen to 8 things?
   i. 3 dozen to 9 dozen?
   j. $1\frac{1}{2}$ dozen to 4 things?

4. Express each of the following rates as a ratio:
   a. 5 oranges for 48 cents
   b. $7 for 3 shirts
   c. 850 miles per hour
   d. 81¢ for 27 kW·h
   e. 91 miles on 7 gal.
   f. 2,400 liters in 30 minutes
   g. 325 kilometers in 5 hours
   h. 7,200 revolutions in 6 minutes
   i. 128 meters in 4 seconds
   j. 7 books for $10

5. There are 16 girls and 12 boys in a mathematics class. What is the ratio of: girls to boys? boys to girls? girls to entire class? boys to entire class?

6. The distribution of final marks in Science showed 4 A's, 6 B's, 12 C's, 8 D's, and 2 E's. What is the ratio of the number of pupils receiving A to those receiving E? B's to C's? D's to A's? C's to E's? A's to B's?

7. In a school where there are 988 students and 38 teachers, what is the ratio of the number of pupils to the number of teachers?

8. The pitch of a roof is the ratio of the rise to the span. What is the pitch of a roof with a rise of 6 feet and a span of 18 feet?

9. Find the ratio of a 24-tooth gear to a 36-tooth gear.

10. The aspect ratio is the ratio of the length of an airplane wing to its width. What is the aspect ratio of a wing 28 m long and 4 m wide?

11. The lift-drag ratio is the ratio of the life of an airplane to its drag. Find the lift-drag ratio if the lift is 900 kg and the drag is 75 kg.

12. The compression ratio is the ratio of the total volume of a cylinder to its clearance volume. Find the compression ratio of a cylinder of an engine if its total volume is 1,300 cm³ and its clearance volume is 26 cm³.

4–22 PROPORTIONS

A *proportion* is a mathematical sentence which states that two ratios are equivalent.

> Using the equivalent ratios $\frac{2}{5}$ and $\frac{6}{15}$ we may write the proportion $\frac{2}{5} = \frac{6}{15}$. This proportion may also be expressed as $2:5 = 6:15$. The proportion in both forms is read "2 is to 5 as 6 is to 15."

There are four terms in a proportion as shown:

$$\begin{array}{l}\text{first} \rightarrow 2 \\ \text{second} \rightarrow 5\end{array} = \begin{array}{l}6 \leftarrow \text{third} \\ 15 \leftarrow \text{fourth}\end{array} \qquad \begin{array}{cccc}\text{1st} & \text{2nd} & \text{3rd} & \text{4th} \\ \downarrow & \downarrow & \downarrow & \downarrow \\ 2 : & 5 & = 6 : & 15\end{array}$$

In the proportion $2:5 = 6:15$, the first term (2) and the fourth term (15) are called the *extremes*; the second term (5) and the third term (6) are called the *means*. Observe that the product of the extremes (2×15) is equal to the product of the means (5×6). In the form $\frac{2}{5} = \frac{6}{15}$ observe that the cross products are equal ($2 \times 15 = 5 \times 6$). These products are equal *only* when the ratios are equivalent. We can check whether two ratios are equivalent by either expressing each ratio in lowest terms or by determining whether the cross products are equal.

If any three of the four terms of the proportion are known quantities, the fourth may be determined. We use the idea that the product of the extremes is equal to the product of the means to transform the given proportion to a simpler equation.

Thus we find the cross products by multiplying the numerator of the first fraction by the denominator of the second fraction and the numerator of the second fraction by the denominator of the first fraction. We write one product equal to the other, then we solve the resulting equation.

Algebra

> Solve and check: $\frac{n}{28} = \frac{4}{7}$
>
> *Solution:*
>
> $$\frac{n}{28} = \frac{4}{7}$$
> $$7 \times n = 28 \times 4$$
> $$7n = 112$$
> $$\frac{7n}{7} = \frac{112}{7}$$
> $$n = 16$$
>
> *Check:*
>
> $$\frac{16}{28} = \frac{4}{7}$$
> $$\frac{4}{7} = \frac{4}{7} \checkmark$$
>
> *Answer,* 16

EXERCISES

Solve and check:

1. $\frac{x}{12} = \frac{3}{4}$
2. $\frac{n}{36} = \frac{5}{9}$
3. $\frac{c}{27} = \frac{2}{3}$
4. $\frac{b}{6} = \frac{11}{15}$
5. $\frac{a}{5} = \frac{7}{8}$
6. $\frac{2}{5} = \frac{b}{20}$
7. $\frac{3}{8} = \frac{x}{56}$
8. $\frac{5}{6} = \frac{y}{42}$
9. $\frac{7}{12} = \frac{c}{16}$
10. $\frac{14}{18} = \frac{w}{27}$
11. $\frac{9}{c} = \frac{3}{53}$
12. $\frac{2}{y} = \frac{1}{12}$
13. $\frac{10}{d} = \frac{5}{16}$
14. $\frac{18}{x} = \frac{96}{64}$
15. $\frac{6}{b} = \frac{4}{8}$
16. $\frac{16}{7} = \frac{96}{a}$
17. $\frac{7}{9} = \frac{28}{n}$
18. $\frac{72}{180} = \frac{8}{x}$
19. $\frac{36}{54} = \frac{12}{d}$
20. $\frac{4}{13} = \frac{15}{m}$

21. Jacques saved $25 in 9 weeks. At that rate how long will it take him to save $175?

22. A motorist travels 480 kilometers in 6 hours. How long will it take her at that rate to travel 640 kilometers?

23. A tree casts a shadow of 26 feet, while a 5-foot post nearby casts a shadow of 2 feet. Find the height of the tree.

24. Miss Wilson pays $900 taxes on a house assessed at $15,000. Using the same tax rate, find the taxes on a house assessed at $18,400.

25. A picture 6 cm wide and 9 cm high is to be enlarged so that the height will be 18 cm. How wide will it be?

26. Find the value of V if $V' = 9$, $P' = 4$, and $P = 12$, using the formula, $\dfrac{V}{V'} = \dfrac{P'}{P}$

27. Find the value of r if $c = 132$, $c' = 88$, and $r' = 14$, using the formula, $\dfrac{c}{c'} = \dfrac{r}{r'}$

28. Find the value of R_2 if $R_1 = 450$, $R_3 = 600$, and $R_4 = 720$, using the formula, $\dfrac{R_1}{R_2} = \dfrac{R_3}{R_4}$

29. Find the value of L if $W = 120$, $w = 80$, and $l = 6$, using the formula, $\dfrac{W}{w} = \dfrac{l}{L}$

30. Find the value of h when $F = 27$, $W = 15$, and $d = 35$, using the formula, $\dfrac{F}{W} = \dfrac{h}{d}$

4–23 SOLVING INEQUALITIES IN ONE VARIABLE

To solve an inequality in one variable means to find all the numbers each of which, when substituted for the variable, will make the inequality a true sentence.

In order to be able to complete the solutions on inequalities, study the solutions of the basic types of inequalities in their simplest forms, given below. We generally list the solutions where convenient, especially when the replacements for the variable are restricted. See illustrations (1), (2), (4), and (6).

---

(1) The solutions of $x < 5$ in the system of natural numbers consists of every natural number less than five.
Therefore $x = 1, 2, 3,$ or 4.

(2) The solutions of $a > 10$ in the system of whole numbers consists of every whole number greater than 10.
Therefore $a = 11, 12, 13, 14, \ldots$

(3) The solutions of $n \neq 8$ in the system of real numbers consists of every real number except 8.
Therefore $n = $ every real number except 8.

Algebra

(4) The solutions of $d \geq -2$ in the system of integers consists of -2 and all integers greater than -2.

Therefore $d = -2, -1, 0, 1, 2, 3, \ldots$

(5) The solutions of $x \leq 4$ when the replacements for the variable are 5, 6, 7, 8, 9, and 10 do not exist.

Therefore x has no solution.

(6) The solutions of $n \not\geq 1$ in the system of integers is the same as the solutions of $n \leq 1$ which consists of 1 and all integers less than 1.

Therefore $n = \ldots, -2, -1, 0, 1$.

(7) The solutions of $x \not< 2$ in the system of real numbers is the same as the solutions of $x \geq 2$ which consists of 2 and all real numbers greater than 2.

Therefore $x = 2$ or any real number greater than 2.

We solve inequalities just as we do equations. We use transformations based on the axioms and on the additive and multiplicative inverses to get a simple equivalent inequality in each case which has the same solutions as the given inequality. This simple equivalent inequality is expressed with only the variable itself as one member of the inequality and a numeral for the other member.

The axioms used to solve inequalities are separated into the following three parts:

I. When the same number is added to or subtracted from both members of an inequality, another inequality of the *same order* results.

Solve $n + 3 > 5$ in the system of natural numbers:

$$n + 3 > 5$$
$$n + 3 - 3 > 5 - 3$$
$$n > 2$$
$$n = 3, 4, 5, \ldots$$

Answer, 3, 4, 5, . . .

Solve $b - 6 < 4$ in the system of whole numbers:

$$b - 6 < 4$$
$$b - 6 + 6 < 4 + 6$$
$$b < 10$$
$$b = 0, 1, 2, \ldots, 9$$

Answer, 0, 1, 2, . . . , 9

EXERCISES

Find the solutions of each of the following inequalities when the replacements for the variable are all the real numbers:

1.
   a. $x + 6 > 9$
   b. $n + 7 > 7$
   c. $t + .4 > 1.3$
   d. $y + 2 > 1$
   e. $21 + b > 54$
   f. $n - 4 > 7$
   g. $r - 10 > 2$
   g. $a - \frac{2}{3} > 1\frac{1}{3}$
   i. $b - 4 > -6$
   j. $x - 18 > 35$

2.
   a. $c + 2 < 11$
   b. $m + 5 < 2$
   c. $y + 1\frac{1}{2} < 4$
   d. $n + 9 < 19$
   e. $x + 30 < 45$
   f. $d - 6 < 15$
   g. $g - 7 < 3$
   h. $n - .8 < .25$
   i. $y - 1 < -4$
   j. $x - 20 < 12$

3.
   a. $y + 7 \geq 10$
   b. $r + 1 \geq 9$
   c. $a + 2\frac{1}{4} \geq 4\frac{3}{4}$
   d. $x + 24 \not< 0$
   e. $s + 8 \geq 8$
   f. $z - 4 \geq 6$
   g. $t - 13 \geq 5$
   h. $y - 8.2 \geq 6.4$
   i. $m - 17 \not< 21$
   j. $p - 5 \geq -5$

4.
   a. $c + 1 \leq 11$
   b. $n + 15 \leq 14$
   c. $w + .6 \leq -.3$
   d. $x + 9 \not> 9$
   e. $r + 32 \leq 50$
   f. $x - 5 \leq 7$
   g. $n - 2 \leq -8$
   h. $t - 3\frac{1}{3} \leq -2\frac{5}{6}$
   i. $y - 6 \not> 4$
   j. $a - 10 \leq 6$

5.
   a. $g + 3 \neq 15$
   b. $x + 8 \neq 22$
   c. $d + 1.5 \neq 1.1$
   d. $y + 18 \neq -7$
   e. $n + 9 \neq 16$
   f. $b - 2 \neq 2$
   g. $h - 1 \neq 10$
   h. $n - \frac{1}{2} \neq \frac{3}{4}$
   i. $x - 14 \not= 6$
   j. $y - 7 \neq 21$

6.
   a. $s + 4 < 11$
   b. $d - 5 > 8$
   c. $x - 3 \neq 2$
   d. $b + 9 \leq 5$
   e. $r - 13 \geq 1$
   f. $y + 4 \neq 4$
   g. $b - 12 < 0$
   h. $w + 20 > 15$
   i. $x \quad 8 \not< 17$
   j. $q + 7 \not> 16$

Find the solutions of each of the following inequalities:

7. When the replacements for the variable are all the one-digit prime numbers:

   a. $x + 2 < 5$
   b. $n - 4 > 2$
   c. $y + 8 \leq 15$
   d. $c - 7 \not< 1$

8. When the replacements for the variable are $-5, -3, -1, 1, 3,$ and 5:

   a. $y + 7 > 3$
   b. $t - 5 < 1$
   c. $m + 2 \neq 7$
   d. $x - 1 \geq 3$

9. When the replacements for the variable are all the natural numbers:

   a. $n - 4 \neq 4$
   b. $w + 8 > 8$
   c. $y - 8 \not> 2$
   d. $t + 6 < 7$

10. When the replacements for the variable are all the integers:

    a. $t - 9 < 1$
    b. $x + 17 > 20$
    c. $a - 1 \leq -1$
    d. $n + 5 \neq 5$

11. When the replacements for the variable are all the non-positive integers:

    a. $n + 4 > 3$
    b. $x + 6 \not< 2$
    c. $a - 7 \leq 2$
    d. $b - 4 < -4$

Algebra 291

12. When the replacements for the variable are all the multiples of 3 greater than 5 and less than 25:
   a. $y - 4 \neq 8$ b. $c + 5 > 19$ c. $d - 9 < 12$ d. $n + 7 \not> 11$

13. When the replacements for the variable are $-4, -2, 0, 2,$ and 4:
   a. $x - 1 < 3$ b. $m + 5 \neq 1$ c. $r + 3 \geq 5$ d. $y - 2 > 2$

14. When the replacements for the variable are all the negative integers greater than -6:
   a. $n - 3 > -6$ b. $y + 8 < 3$ c. $t - 10 \neq -14$ d. $x + 9 \leq 0$

II. When both members of an inequality are either multiplied or divided by the same *positive* number, another inequality of the *same order* results.

Solve $4b > 20$ in the system of real numbers:

$4b > 20$

$\dfrac{4b}{4} > \dfrac{20}{4}$

$b > 5$

$b =$ all real numbers greater than 5

Answer, all real numbers greater than 5

Solve $6a - 5 \leq 13$ when the replacements for the variable are 1, 3, 5, 7, and 9:

$6a - 5 \leq 13$

$6a - 5 + 5 \leq 13 + 5$

$6a \leq 18$

$a \leq 3$

$a = 1$ or 3

Answer, 1 and 3

EXERCISES

Find the solutions of each of the following inequalities when the replacements for the variable are all the real numbers:

1. a. $3a > 21$
   b. $8y > 56$
   c. $6a > 15$
   d. $.4n > 5$
   e. $10t > 8$
   f. $4y > -14$
   g. $\frac{1}{2}v > 7$
   h. $9z > -54$
   i. $12b > 60$
   j. $7x > 42$
   k. $\dfrac{x}{6} > 2$
   l. $\dfrac{n}{8} > 5$
   m. $\dfrac{y}{9} > 12$
   n. $\dfrac{w}{3} > -3$
   o. $\dfrac{b}{10} > 6$

2. a. $9n < 72$
   b. $5r < 40$
   c. $3w < 11$
   d. $18x < 15$
   e. $.05s < 1$
   f. $6b < -30$
   g. $\frac{2}{3}y < 12$
   h. $4m < -28$
   i. $20t < 100$
   j. $32g < 96$
   k. $\dfrac{m}{3} < 15$
   l. $\dfrac{x}{12} < 1$
   m. $\dfrac{t}{20} < 0$
   n. $\dfrac{a}{8} < -2$
   o. $\dfrac{n}{9} < 4$

3. a. $8m \geq 48$
   b. $6r \geq 114$
   c. $21x \geq 14$
   d. $7b \not< 91$
   e. $1.05y \geq 10.5$
   f. $\frac{3}{4}c \geq 36$
   g. $11a \geq 99$
   h. $5y \geq 42$
   i. $18t \not< 90$
   j. $25x \geq 300$
   k. $4y \not< -24$
   l. $3n \geq -81$
   m. $\frac{n}{4} \geq 1$
   n. $\frac{c}{12} \geq 2$
   o. $\frac{r}{2} \geq -4$
   p. $\frac{v}{5} \not< 7$
   q. $\frac{x}{10} \geq 5$
   r. $\frac{t}{4} \not< -3$

4. a. $4b \leq 64$
   b. $12y \leq 56$
   c. $2d \leq -10$
   d. $5m \not> 100$
   e. $9x \leq -63$
   f. $54n \leq 18$
   g. $.01w \leq 1$
   h. $14c \leq 84$
   i. $24y \not> 72$
   j. $30z \leq 120$
   k. $.7b \not> 1.4$
   l. $8a \leq -56$
   m. $\frac{x}{11} \leq 2$
   n. $\frac{d}{4} \leq 8$
   o. $\frac{g}{2} \leq 19$
   p. $\frac{n}{5} \not> -6$
   q. $\frac{x}{16} \leq 3$
   r. $\frac{r}{6} \leq -9$

5. a. $3x \neq 57$
   b. $12c \neq -12$
   c. $7y \neq 49$
   d. $8m \neq 144$
   e. $48b \neq 36$
   f. $11z \neq -132$
   g. $.2a \neq 5$
   h. $\frac{5}{8}t \neq 30$
   i. $9g \neq 51$
   j. $4s \neq 104$
   k. $\frac{c}{5} \neq 2$
   l. $\frac{m}{3} \neq -4$
   m. $\frac{r}{10} \neq 0$
   n. $\frac{x}{6} \neq 8$
   o. $\frac{w}{2} \neq -10$

6. a. $8a > 96$
   b. $10g < 400$
   c. $13x \neq 39$
   d. $\frac{h}{6} > 9$
   e. $16n \geq 12$
   f. $5b \leq -40$
   g. $\frac{r}{8} < 8$
   h. $\frac{7}{10}a > 14$
   i. $\frac{c}{9} \leq -2$
   j. $17m \not< -51$
   k. $\frac{s}{8} \neq 0$
   l. $\frac{x}{14} \geq 2$
   m. $3d \not> 26$
   n. $17y < 0$

Find the solutions of each of the following inequalities:

7. When the replacements for the variable are all the natural numbers:

   a. $2x > 14$
   b. $4y < -4$
   c. $10n \leq 80$
   d. $\frac{x}{6} > 7$

8. When the replacements for the variable are all the integers:

   a. $9d < -45$
   b. $16n \not> 80$
   c. $\frac{5}{6}y \neq 20$
   d. $\frac{x}{4} \geq -1$

9. When the replacements for the variable are $-8, -6, -4, -2, 0, 2, 4, 6,$ and 8:

   a. $6b > -12$
   b. $8x \not< 48$
   c. $\frac{n}{2} < -3$
   d. $3t \leq 0$

10. When the replacements for the variable are all the odd prime numbers less than 15:

    a. $25n < 25$
    b. $\frac{s}{5} \not> 3$
    c. $7x \leq 63$
    d. $24y \neq 120$

Algebra

III. When both members of an inequality are either multiplied or divided by the same *negative* number, an inequality of the *reverse order* results.

> Solve $-3c > 21$ in the system of integers:
> $$-3c > 21$$
> $$\frac{-3c}{-3} < \frac{21}{-3}$$
> $$c < -7$$
> $$c = \ldots, -10, -9, -8$$
>
> Observe that -6 which is greater than -7 makes $-3c > 21$ false ($18 > 21$) but -8 makes it true ($24 > 21$).
>
> *Answer*, $\ldots, -10, -9, -8$
>
> Solve $-5x < -20$ in the system of integers:
> $$-5x < -20$$
> $$\frac{-5x}{-5} > \frac{-20}{-5}$$
> $$x > 4$$
> $$x = 5, 6, 7, \ldots$$
>
> Observe that 3 which is less than 4 makes $-5x < -20$ false ($-15 < -20$) but 5 makes it true ($-25 < -20$).
>
> *Answer*, $5, 6, 7, \ldots$

EXERCISES

Find the solutions of each of the following inequalities when the replacements for the variable are all the real numbers:

1. **a.** $-3x > 15$
   b. $-8y > 80$
   c. $-.7d > 21$
   d. $-12a > 108$
   e. $-25n > 100$
   f. $-2y > -14$
   g. $-5d > -23$
   h. $-\frac{1}{3}t > -60$
   i. $-14n > -84$

2. **a.** $-4m < 48$
   b. $-6c < 0$
   c. $-\frac{3}{5}x < 12$
   d. $-10z < 4$
   e. $-16s < 64$
   f. $-5b < -10$
   g. $-9m < -63$
   h. $-1.2y < -6$
   i. $-6n < -15$

3. **a.** $-4n \geq 52$
   b. $-x \geq 1$
   c. $-.9b \geq 5.4$
   d. $-3a \not< 19$
   e. $-18y \geq 90$
   f. $-9z \geq -36$
   g. $-15d \geq -105$
   h. $-\frac{5}{7}x \geq -20$
   i. $-n \not< -1$

4. **a.** $-6s \leq 120$
   b. $-24c \leq 16$
   c. $-.05n \leq 2.5$
   d. $-17m \not> 51$
   e. $-5x \leq 0$
   f. $-9b \leq -117$
   g. $-10y \leq -35$
   h. $-\frac{3}{4}a \leq -24$
   i. $-7n \not> -147$

5. a. $-8x > -16$
   b. $-3y < 81$
   c. $-4z \geq -72$
   d. $-9t \leq -90$
   e. $-7m < 23$
   f. $-6p > 0$
   g. $-10n \not< 30$
   h. $-5a \not> 18$
   i. $-g < -1$
   j. $-12h \leq 48$
   k. $-2w > -1$
   l. $-24c \geq -144$

6. a. $\dfrac{b}{-2} < 6$
   b. $\dfrac{n}{-5} > 4$
   c. $\dfrac{c}{-3} \leq 7$
   d. $\dfrac{a}{-4} > -2$
   e. $\dfrac{x}{-6} < -8$
   f. $\dfrac{d}{-8} \geq 9$
   g. $\dfrac{n}{-10} < -2$
   h. $\dfrac{y}{-7} \not< -10$
   i. $\dfrac{a}{-9} > 5$
   j. $\dfrac{x}{-12} < 7$
   k. $\dfrac{w}{-3} \not> -1$
   l. $\dfrac{y}{-15} > 12$

Find the solutions of each of the following inequalities:

7. When the replacements for the variable are all the integers:
   a. $-7x > 42$
   b. $-12c < -24$
   c. $-4t \not< 20$
   d. $\dfrac{n}{-9} \not> 2$

8. When the replacements for the variable are all the natural numbers:
   a. $-15b < 45$
   b. $-y > -1$
   c. $-3a \not> 0$
   d. $\dfrac{w}{-3} \not< -5$

9. When the replacements for the variable are $-2, -1, 0, 1,$ and 2:
   a. $-4c > -4$
   b. $-6t \not> 12$
   c. $-10n \geq 30$
   d. $\dfrac{x}{-1} < 1$

10. When the replacements for the variable are all the non-negative integers:
    a. $-2y < 8$
    b. $-6x > -12$
    c. $-x \leq 0$
    d. $\dfrac{v}{-9} \not< 1$

11. When the replacements for the variable are the integers greater than -5 and less than 2:
    a. $-5x \geq -5$
    b. $-3a < 0$
    c. $-4y \not> -8$
    d. $\dfrac{b}{-4} > -1$

MISCELLANEOUS EXERCISES

1. Find the solutions of the following inequalities when the replacements for the variable are all the real numbers:
   a. $x + 6 < 17$
   b. $w - 11 > 13$
   c. $9y \neq 90$
   d. $-6s > 84$
   e. $\dfrac{n}{4} < -3$
   f. $3x + 9 \leq 18$
   g. $-8y > 56$
   h. $11a - 4a < 14$
   i. $4m - 5 \geq -21$
   j. $\dfrac{d}{-5} \not> 2$
   k. $-5c < -10$
   l. $2x - x > -1$
   m. $-7x \not< 0$
   n. $10y + 3 \neq 8$
   o. $\dfrac{b}{9} > 6$
   p. $d + 7d \geq -64$
   q. $24n - 3 < 93$
   r. $8h + 6 \neq 6$
   s. $4x + 7 > -11$
   t. $\dfrac{x}{-2} \not< -9$

Algebra

2. Find the solutions of the following inequalities when the replacements for the variable are all the real numbers:

   a. $12b + 7 < 43$
   b. $5c - 3 > 22$
   c. $8m - 4 \leq 60$
   d. $18y - 8 \neq 16$
   e. $3n + 17 \geq 5$
   f. $7c + 9c > 32$
   g. $6n + 12n < -9$
   h. $2b - 8b \neq -54$
   i. $10x - x \geq 63$
   j. $14c + 11c \leq 10$
   k. $4m - 7 < 35$
   l. $x + x > -8$
   m. $16y + 5 \not> 37$
   n. $6t - 11t \neq 72$
   o. $3c - 12 \not< 0$

3. Find the solutions of each of the following inequalities when the replacements for the variable are all the natural numbers:

   a. $6x + 3 > 3$
   b. $9n - 5 < 13$
   c. $12a + 2a > 42$
   d. $11b - b < 50$

4. Find the solutions of each of the following inequalities when the replacements for the variable are all the integers:

   a. $7n - n < -48$
   b. $4y + 10 > -2$
   c. $d + d \leq 0$
   d. $3x - 1 \geq 2$

5. Find the solutions of each of the following inequalities when the replacements for the variable are all the non-positive integers:

   a. $2x - 7x > 10$
   b. $8n - 5 < -19$
   c. $3b + 6 > 15$
   d. $13w - 4w < 0$

4–24 GRAPHING ON THE NUMBER LINE AN INEQUALITY IN ONE VARIABLE IN SIMPLEST FORM

The graph of an inequality in one variable is the collection of all points on the number line whose coordinates are the solutions of the given inequality.

We shall see below that the basic graph on a number line of an inequality in one variable in simplest form may be a half-line, a ray, a line segment, or an interval.

A *line* is a collection of points. A point on the line separates the line into two *half-lines*. The half-line extends indefinitely in one direction only and does not include the endpoint separating it from the other half-line. An *open dot* indicates the exclusion of this endpoint and the arrowhead indicates the direction and that the half-line is endless.

A *ray* is a half-line which includes one endpoint. A heavy solid or colored dot is used to indicate the inclusion of this endpoint.

A definite part of a line which includes its endpoints is called a *line segment* or *segment*. Heavy solid or colored dots are used to indicate the inclusion of the two endpoints.

A definite part of a line which excludes its endpoints is called an *interval*. Open dots are used to indicate that the two endpoints are excluded from the graph.

Study each of the following basic graphs of inequalities in simplest form. The real numbers are used as the replacements for the variables.

> The graph of all real numbers greater than a given number is a half-line extending to the right along the number line.
>
> Graph of $x > 3$ 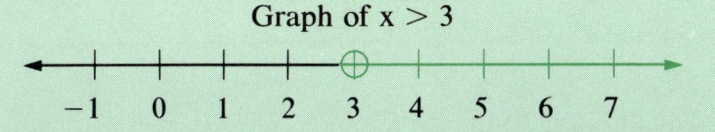
>
> The graph of all real numbers less than a given number is a half-line extending to the left along the number line.
>
> Graph of $x < 3$
>
> The graph of all real numbers greater than or equal to a given number is a ray extending to the right along the number line. A *ray* is a half-line which includes one endpoint. A solid dot indicates the inclusion of this endpoint.
>
> Graph of $x \geq 3$ 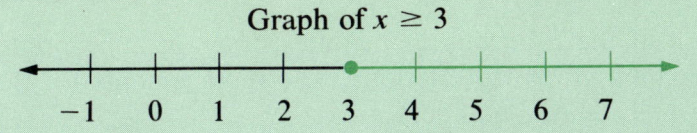
>
> The graph of all real numbers less than or equal to a given number is a ray extending to the left along the number line.
>
> Graph of $x \leq 3$

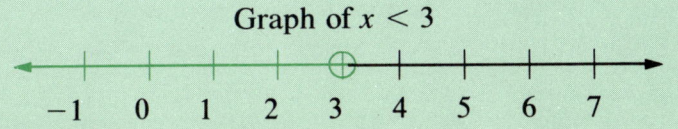

Algebra

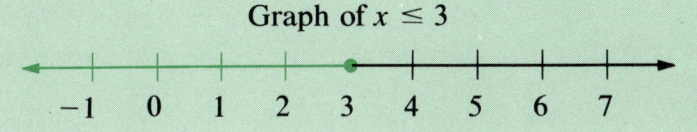

The graph of all real numbers excluding one number is the entire real number line except the point corresponding to the excluded number—or two half-lines.

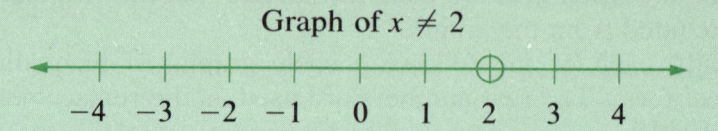

Graph of $x \neq 2$

The graph of all real numbers between two given numbers and including the two given numbers is a line segment which is a definite part of a line including the two endpoints.

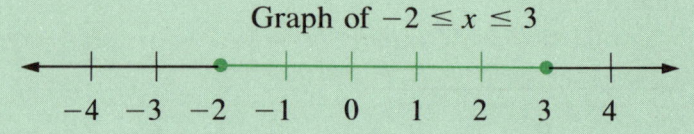

Graph of $-2 \leq x \leq 3$

The graph of all real numbers between the two given numbers but not including the two given numbers is an interval which is a definite part of a line excluding the two endpoints.

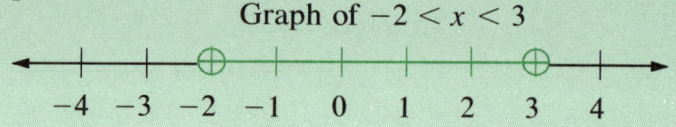

Graph of $-2 < x < 3$

EXERCISES

For each of the following inequalities draw an appropriate number line, then graph its solutions. The replacements for the variables are all the real numbers.

1. $x > 1$
2. $n < -3$
3. $x \neq 0$
4. $y \geq 2$
5. $-1 < x < 1$
6. $b \not> 2$
7. $-4 \leq n \leq 3$
8. $a \not< -4$
9. $0 \leq x < 5$
10. $z \leq 0$
11. $x > -4$
12. $-3 < n \leq 3$
13. $y \neq -2$
14. $|x| < 1$
15. $n \geq -1$
16. $-2 < y < 4$
17. $|y| > 2$
18. $-5 \leq x \leq -1$
19. $x < -2$
20. $y > 0$
21. $n \neq -1$
22. $x \leq 2$
23. $x \geq -4$
24. $1 < a < 4$

4–25 GRAPHING ON THE NUMBER LINE AN INEQUALITY IN ONE VARIABLE NOT IN SIMPLEST FORM

We have studied that the graph on the number line of an inequality in one variable is the collection of all points on the number line whose coordinates are the solutions of the given inequality.

To draw the graph on a number line of an inequality in one variable not in simplest form, we first find the solutions of the inequality and then draw the graph of these solutions. Or generally when the given inequality is simplified, a basic type may be identified and its graph easily drawn.

Draw the graph of $5x - 6 > 4$ when all the real numbers are used as replacements for the variable.

Solving the inequality $5x - 6 > 4$, we find $x > 2$.

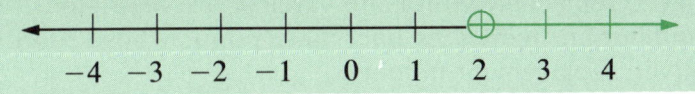

The graph of $5x - 6 > 4$ is the graph of $x > 2$.

EXERCISES

1. For each of the following inequalities draw an appropriate number line, then graph its solutions. The replacements for the variables are all the real numbers.

a. $x + 3 < 6$
b. $n - 1 > 4$
c. $5b \leq 20$
d. $a + 1 \geq 5$
e. $\frac{x}{2} < 3$
f. $-3y > 12$
g. $6x - x \not< 15$
h. $y + y \not> 6$
i. $4x + 5 \neq 9$
j. $8n - 12 < 4$

k. $c - 5 \leq 7$
l. $-2x > -4$
m. $11x + 9 < -13$
n. $7z - 14 \neq 0$
o. $\frac{n}{5} \geq -1$
p. $3y - 8y < 10$
q. $t - 2 \not> 4$
r. $9n + 6 > 6$
s. $4w - 10 \not< -2$
t. $12y - 7 < 5$

Algebra 299

2. Write the corresponding inequality which is pictured by each of these graphs:

   a. ←—+—+—+—⊕—+—+—+—→
      −1 0 1 2 3 4 5

   b. ←—+—+—⊕—+—+—+—+—→
      −3 −2 −1 0 1 2 3

   c. ←—+—⊕—+—+—+—+—⊕—→
      −5 −4 −3 −2 −1 0 1

   d. ←—+—●—+—+—+—+—+—→
      −4 −3 −2 −1 0 1 2

   e. ←—+—+—+—+—●—+—+—→
      −2 −1 0 1 2 3 4

   f. ←—+—+—●—+—+—●—+—→
      −3 −2 −1 0 1 2 3

3. Draw the graph of each of the following:

   a. $n + 3 > n$ b. $x - 2 < x$ c. $y + 2 < y$ d. $n - 5 > n$

4–26 GRAPHING AN INEQUALITY ON THE NUMBER LINE WHEN THE REPLACEMENTS FOR THE VARIABLE ARE RESTRICTED

When the replacements for the variable are restricted, we first draw a number line indicating only those points whose coordinates are the given replacement numbers.

To draw the graph of the given inequality on this number line, we locate and indicate by heavy solid or colored dots all the points found on this line whose coordinates are numbers forming the solutions to the given inequality.

> The graph of $x > 3$ when the replacements for the variable are 0, 1, 2, 3, 4, 5, 6, 7, and 8 is:
>
> ←—+—+—+—+—●—●—●—●—●—→
>   0 1 2 3 4 5 6 7 8

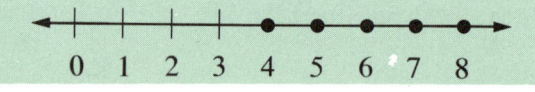

EXERCISES

For each of the following inequalities draw an appropriate number line, then graph its solutions:

1. When the replacements for the variable are −5, −4, −3, −2, −1, 0, 1, 2, 3, 4, and 5:

   a. $x > -2$ b. $y + 3 < 4$ c. $a - 1 \geq -4$ d. $2n + 5 \neq 7$

2. When the replacements for the variable are −4, −2, 0, 2, and 4:

   a. $-x < 0$ **b.** $n - 2 > -1$ **c.** $4d + d \leq 5$ **d.** $7n - 1 \not< -15$

3. When the replacements for the variable are −1, 0, and 1:

   a. $-9w > 9$ **b.** $x - 5 < -5$ **c.** $3y - y \not> -2$ **d.** $6x + 9 \geq 15$

4. When the replacements for the variable are all one-digit prime numbers:

   a. $y + y < 10$ **b.** $9x - 2 > 16$ **c.** $8c \neq 56$ **d.** $12n + 5 \leq 41$

5. When the replacements for the variable are all integers greater than −3 but less than 4:

   a. $4x - 6 > -10$ **b.** $n - 5n < -8$ **c.** $9s - 3s \leq -12$ **d.** $7y + 11 \geq 11$

POINTS ON A NUMBER PLANE

4–27 INTRODUCTION

Number Plane

When a vertical number line and a horizontal number line are drawn perpendicular* to each other, a number plane is determined. If the number lines are real number lines (see page 167), the number plane consists of an infinite number of points and is called the *real number plane*.

These two perpendicular number lines form a pair of *coordinate axes* or guide lines so that points on the plane may be located or plotted.

The horizontal number line is the horizontal axis or *x-axis*. The vertical number line is the vertical axis or *y-axis*.

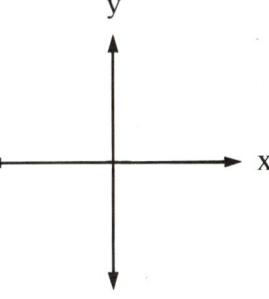

Ordered Pair of Numbers

Numbers are used in pairs in many ways. We use number pairs when we work with fractions and ratios and when we compare quantities or show relationships between quantities by tables, graphs, and some formulas.

---

* Each angle formed measures 90°.

An *ordered pair of numbers* is a pair of numbers expressed in a definite order so that one number is first (*first component*) and the other number is second (*second component*).

To express an ordered pair of numbers, we write within parentheses the numerals for the numbers with the numeral for the first component written first, followed by a comma, and then the numeral for the second component like (2,5).

> The ordered pair of numbers (2,5) and (5,2) are different. (2,5) has 2 as the first component and 5 as the second component. (5,2) has 5 as the first component and 2 as the second component.

EXERCISES

1. Write the ordered pair of numbers which has:
   a. 3 as the first component and 4 as the second component.
   b. 6 as the first component and 0 as the second component.
   c. 8 as the first component and -2 as the second component.
   d. -5 as the first component and 3 as the second component.
   e. -4 as the first component and -1 as the second component.

2. Write all the ordered pairs of numbers which have:
   a. 1, 2, or 3 as the first component and 2 as the second component.
   b. 1 or 2 as the first component and 1 or 2 as the second component.
   c. 1, 2, or 3 as the first component and 0 or 1 as the second component.
   d. 1, 2, 3, or 4 as the first component and 1, 2, 3, or 4 as the second component.
   e. -2, 0, or 2 as the first component and -3, -1, 1, or 3 as the second component.

4-28 LOCATING POINTS ON A NUMBER PLANE

We have found on page 167 that we required only a single number (coordinate) to locate a point on a number line. However, we shall see that an ordered pair of numbers is needed to locate a point on a number plane because the point is located with respect to both the x-axis and y-axis.

To locate a point on a number plane,
(1) We first draw a perpendicular line from the point to the x-axis. The number corresponding to the point on the x-axis where this line intersects the x-axis is called the *x-coordinate* or *abscissa* of the point. The x-coordinate indicates the horizontal distance, measured parallel to the x-axis, that a point is located to the left or to the right of the y-axis. This distance and direction is represented to the left of the y-axis by a negative number and to the right by a positive number.

> The x-coordinate of the point in the illustration below is +5 or 5.

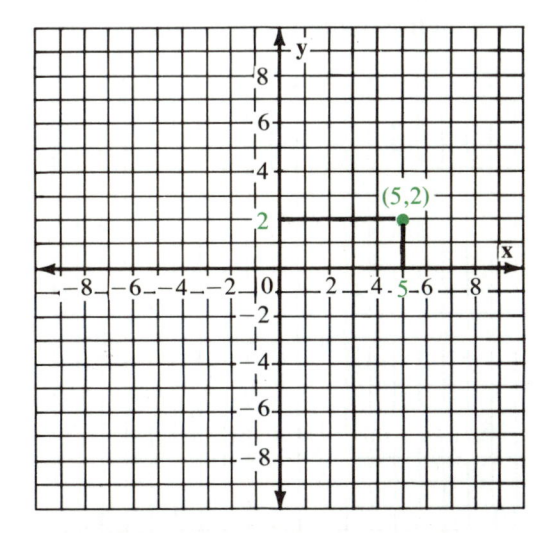

(2) We then draw a perpendicular line from the point to the y-axis. The number corresponding to the point on the y-axis where this line intersects the y-axis is called the *y-coordinate* or *ordinate*

Algebra 303

of the point. The y-coordinate indicates the vertical distance, measured parallel to the y-axis, that a point is located up or down from the x-axis.

> The y-coordinate of the point in the previous illustration is +2 or 2.

The coordinates of a point are represented by an ordered pair of numbers in which the x-coordinate is expressed first and the y-coordinate is second.

> The coordinates of the point in the previous illustration is the ordered pair of numbers named by (5,2).

Observe below that the real number plane is divided into four regions, each of which is called a *quadrant*.

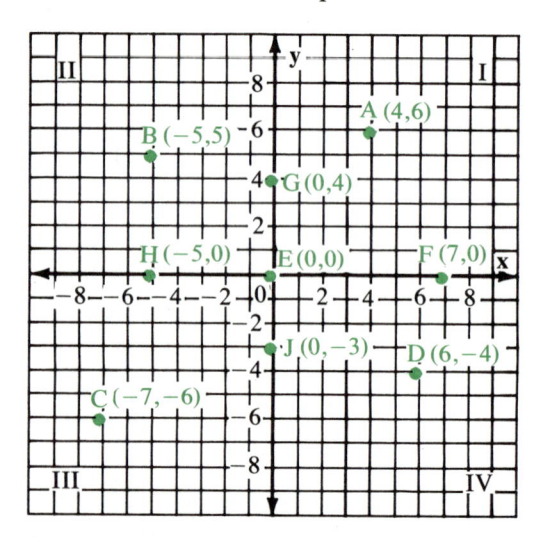

In the quadrant marked I, each point has a positive x-coordinate and a positive y-coordinate.

> The coordinates of point A are (4,6).

In the quadrant marked II, each point has a negative x-coordinate and a positive y-coordinate.

> The coordinates of point B are (−5,5).

In the quadrant marked III, each point has a negative x-coordinate and a negative y-coordinate.

> The coordinates of point C are (−7,−6).

In the quadrant marked IV, each point has a positive x-coordinate and a negative y-coordinate.

> The coordinates of point D are (6,−4).

The x-axis and the y-axis intersect at a point called the *origin,* see point E, where the coordinates are (0,0).

Any point on the x-axis has 0 for its y-coordinate.

> The coordinates of point H are (−5,0), and of point F are (7,0).

Any point on the y-axis has 0 for its x-coordinate.

> The coordinates of point G are (0,4) and of point J are (0,−3).

Algebra

EXERCISES

What are the coordinates of each of the following indicated points?

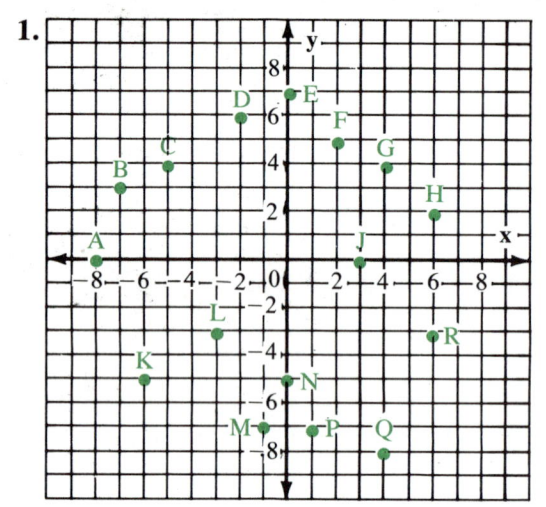

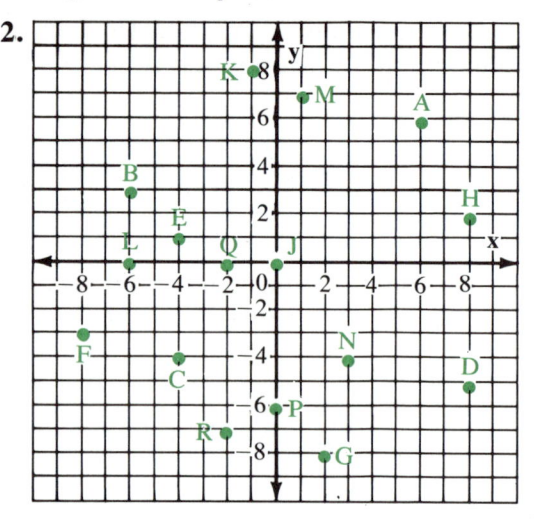

1.

2.

4–29 PLOTTING POINTS ON A NUMBER PLANE

To *plot a point* that corresponds to a given ordered pair of numbers means to locate this point on the number plane and to indicate its position by a dot.

(1) To locate this point, we use the first component of the given ordered pair of numbers as the x-coordinate and the second component as the y-coordinate.

(2) We use either a solid or a colored dot to indicate the position of the point.

> (1) Plot point K whose coordinates are (3,4). The x-coordinate is 3 and the y-coordinate is 4.
>
> (2) Plot point L whose coordinates are (−4,5). The x-coordinate is −4 and the y-coordinate is 5.

(3) Plot point M whose coordinates are (−3,−7). The x-coordinate is −3 and the y-coordinate is −7.

(4) Plot point N whose coordinates are (6,0). The x-coordinate is 6 and the y-coordinate is 0.

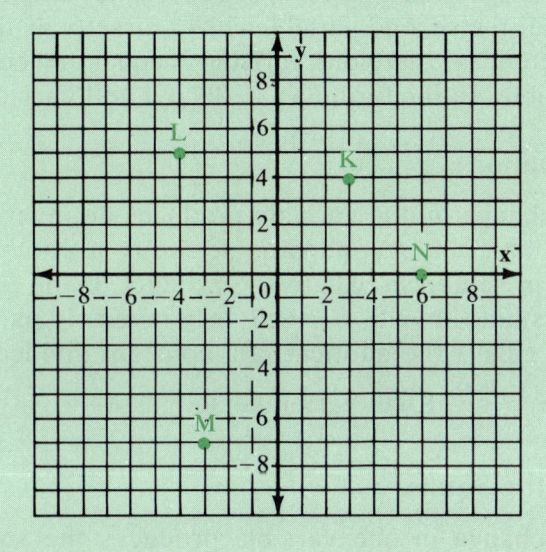

EXERCISES

On graph paper draw the axes and plot the points that have the following coordinates:

1. (1,6)
2. (4,−8)
3. (−3,5)
4. (−2,−2)
5. (0,−5)
6. (−7,−3)
7. (9,4)
8. (5,0)
9. (−6,1)
10. (−4,−5)
11. (3,7)
12. (−8,8)
13. (−3,−6)
14. (0,0)
15. (2,−3)
16. (−1,0)
17. (−5,−5)
18. (0,4)
19. (6,−7)
20. (−9,6)

Algebra

4–30 VARIATION

In a formula or an equation having two variables, any change in one variable will produce a change in the other.

Direct Variation

When a change in one variable produces the same change in the other variable, the second variable *varies directly* as the first. This occurs in sentences where one variable is equal to a constant times another variable. In the formula $i = .06p$, i will vary directly as p.

Inverse Variation

When a change in one variable produces an inverse change in the other variable, the second variable *varies inversely* as the first. This occurs in sentences where the product of the two variables equals a constant (in $lw = 6$, l will vary inversely as w) or where one variable equals the quotient of a constant divided by another variable (in $I = \dfrac{100}{R}$, I will vary inversely as R).

Directly as the Square

When a change in one variable produces the square of this change in the other variable, the second variable *varies directly as the square* of the first. This occurs in sentences where one variable is equal to the square of the other variable or to the product of a constant and the square of the other variable. In the formula $A = s^2$, A will vary directly as the square of s.

EXERCISES

In each of the following examples find what effect a change in one quantity has upon a related quantity:

1. Using the formula $p = 4s$, substitute the values of s given in the table below to determine the corresponding values of p, then complete the following statements:

s	2	3	4	6	12
p					

a. Doubling the value of s (from 3 to 6) will ? the value of p.
b. Halving the value of s (from 4 to 2) will ? the value of p.
c. Trebling the value of s (from 2 to 6) will ? the value of p.
d. A change in the value of s produces the ? change in the value of p. Or, p varies ? as s.

2. Using the formula $a = \dfrac{360}{n}$, substitute the values of n given in the table below to determine the corresponding values of a, then complete the following statements:

n	5	10	15	20	40
a					

a. Doubling the value of n (from 10 to 20) will ? the value of a.
b. Halving the value of n (from 10 to 5) will ? the value of a.
c. A change in the value of n produces an ? change in the value of a. Or, a varies ? as n.

3. Using the formula $A = 6s^2$, substitute the values of s given in the table below to determine the corresponding values of A, then complete the following statements:

s	2	3	4	6	12
A					

a. Doubling (or 2 times) the value of s (from 2 to 4) makes A equal ? its original value.
b. Halving the value of s (from 6 to 3) makes A equal ? its original value.
c. Trebling (or 3 times) the value of s (from 4 to 12) makes A equal ? its original value.
d. A change in the value of s produces the ? of this change in the value of A. Or, A varies ? of s.

4. In the formula $A = lw$, if w is doubled and l remains the same, the value of A is ?.

5. In the formula $r = \dfrac{d}{t}$, if t is halved and d remains the same, the value of r is ?.

6. In the formula $A = \pi r^2$, if r is doubled, the value of A is ?.

Find the required values, using the principles of variation.

7. In the formula $c = \pi d$, $c = 44$ when $d = 14$. Find the value of c when $d = 28$.

Algebra   309

8. In the formula $s = 16\,t^2$, $s = 64$ when $t = 2$. Find the value of s when $t = 8$.

9. In the formula $I = \dfrac{E}{R}$, $I = 6$ when $R = 20$. Find the value of I when $R = 60$ and E is constant.

10. The area of a circle varies directly as the square of the diameter. Compare the area of a circle having an 18-centimeter diameter to the area of a circle with a 3-centimeter diameter.

4–31 TRANSFORMING FORMULAS

Transformation

To transform a formula means to solve for a specific variable in terms of the other related variables of the formula.

To do this, we copy the given formula and solve for the required variable using the principles developed for the solution of numerical equations. See below left.

Derivation

To derive a formula from two given formulas by eliminating a common variable, we replace the variable to be eliminated in one formula by an equivalent expression found in the second formula. We then transform and simplify as necessary. See below right.

Transform formula $C = np$, solving for p.

Solution:
$$C = np$$
$$\frac{C}{n} = \frac{np}{n}$$
$$\frac{C}{n} = p$$
$$p = \frac{C}{n}$$

Answer, $p = \dfrac{C}{n}$

Derive a formula for A in terms of π and d, using the formulas $A = \pi r^2$ and $r = \dfrac{d}{2}$.

$$A = \pi r^2$$
$$A = \pi \times \left(\frac{d}{2}\right)^2$$
$$A = \pi\,\frac{d^2}{4}$$
or
$$A = \tfrac{1}{4}\,\pi d^2$$

Answer, $A = \tfrac{1}{4}\,\pi d^2$

EXERCISES

1. Transform formula $A = lw$, solving for w.
2. Transform formula $d = rt$, solving for r.
3. Transform formula $i = prt$, solving for t.
4. Transform formula $V = lwh$, solving for w.
5. Transform formula $A = p + i$, solving for i.
6. Transform formula $C = A - L$, solving for A.
7. Transform formula $I = \dfrac{E}{R}$, solving for E.
8. Transform formula $d = \dfrac{m}{v}$, solving for v.
9. Transform formula $F = 1.8\,C + 32$, solving for C.
10. Transform formula $\dfrac{D}{d} = \dfrac{r}{R}$, solving for R.
11. Using formulas $c = \pi d$ and $d = 2\,r$, derive a formula for c in terms of π and r.
12. Using formulas $A = p + i$ and $i = prt$, derive a formula for A in terms of p, r, and t.
13. Derive a formula for A in terms of p, using the formulas $A = s^2$ and $p = 4\,s$.
14. Using the formulas $W = IE$ and $E = IR$, derive a formula for W in terms of I and R.
15. Derive a formula for P in terms of F and v, using the formulas $P = \dfrac{W}{t}$, $W = Fd$, and $v = \dfrac{d}{t}$.

4-32 QUADRATIC EQUATIONS

A *quadratic equation* is an equation which, after its terms have been collected and simplified, contains the second power (but no higher power) of the variable. Every quadratic equation has two roots. In the evaluation of formulas sometimes a quadratic equation develops and its solution is required.

To solve a quadratic equation containing the second power but not the first power of the variable (an *incomplete quadratic equation*), we divide both sides by the numerical coefficient of the variable. We take the square root of both members of the equation and prefix a $\pm$ sign to the answer. The two answers are numerically equal but opposite in sign.

If the given equation is a proportion, we first cross multiply, then we solve the resulting equation.

Algebra

> Solve and check: $6x^2 = 24$
>
> *Solution:*   *Check:*
> $6x^2 = 24$ $6 \times (+2)^2 = 24$
> $x^2 = 4$ $6 \times 4 = 24$
> $x = \pm 2$ $24 = 24$ ✓
> $6 \times (-2)^2 = 24$
> $6 \times 4 = 24$
> $24 = 24$ ✓
>
> *Answer,* $+2$ and -2

EXERCISES

Solve and check each of the following equations. Whenever necessary, find answer correct to nearest hundredth.

1. $x^2 = 25$
2. $x^2 = 9$
3. $x^2 = 64$
4. $x^2 = 81$
5. $x^2 = 169$
6. $8x^2 = 32$
7. $2x^2 = 98$
8. $5x^2 = 180$
9. $7x^2 = 252$
10. $16x^2 = 400$
11. $x^2 = 19$
12. $x^2 = 7$
13. $x^2 = 30$
14. $x^2 = 48$
15. $x^2 = 54$
16. $5x^2 = 60$
17. $6x^2 = 90$
18. $14x^2 = 84$
19. $9x^2 = 72$
20. $7x^2 = 140$

21. $\dfrac{x}{3} = \dfrac{27}{x}$
22. $\dfrac{x}{5} = \dfrac{20}{x}$
23. $\dfrac{x}{4} = \dfrac{9}{x}$
24. $\dfrac{x}{125} = \dfrac{5}{x}$
25. $\dfrac{x}{32} = \dfrac{8}{x}$

Find the value of:

26. s when $A = 196$, using the formula $A = s^2$.
27. t when $s = 1{,}024$, using the formula $s = 16t^2$.
28. r when $A = 314$ and $\pi = 3.14$, using the formula $A = \pi r^2$.
29. s when $A = 150$, using the formula $A = 6s^2$.
30. r when $A = 616$ and $\pi = \frac{22}{7}$, using the formula $A = 4\pi r^2$.
31. d when $A = 1{,}963.5$, using the formula $A = .7854\, d^2$.
32. I when $W = 720$ and $R = 20$, using the formula $W = I^2 R$.
33. r when $V = 6{,}280$, $h = 5$, and $\pi = 3.14$, using the formula $V = \pi r^2 h$.
34. v when $K = 324$ and $m = 8$, using the formula $K = \frac{1}{2} mv^2$.
35. b when $h = 91$ and $a = 35$, using the formula $h^2 = a^2 + b^2$.
36. E when $P = 720$ and $R = 20$, using the formula $P = \dfrac{E^2}{R}$.

CHAPTER REVIEW

1. Write each of the following as an algebraic expression: (4–2)
   a. The sum of angles A, B, and C.
   b. The product of the acceleration of gravity (g) and time (t) added to the velocity (V).
   c. The difference between the selling price (s) and the margin (m).
   d. The annual depreciation (d) divided by the original cost (c).
   e. The product of the sum of m and n and the difference between x and y.

2. Write each of the following as an open sentence symbolically: (4–3)
   a. Some number x increased by nine is equal to forty-one.
   b. Each number y decreased by four is less than sixty.
   c. Seven times each number b plus five is greater than twenty-nine.
   d. Each number r divided by three is greater than or equal to sixteen.

3. Express as a formula: The average (A) of two numbers m and n equals the sum of the two numbers divided by two. (4–4)

4. Find the value of: (4–5)
   a. $d + 5y$ when $d = 9$ and $y = -2$.
   b. $n + x(n - x)$ when $n = 10$ and $x = 7$.
   c. $a^2 + 2ab - 3b^2$ when $a = 5$ and $b = 3$.
   d. $\dfrac{(x - y)^2}{x^2 - y^2}$ when $x = 6$ and $y = -3$.
   e. F when $C = 30$, using the formula $F = 1.8C + 32$. (4–6)

5. Add: (4–7)

   $+7a$
   $-8a$
   $+5a$

   $4c^2 - 3cd - d^2$
   $5c^2 - 7cd + 2d^2$
   $-6c^2 - cd - d^2$

   $(2a - b) + (4b - a) + (-a - 3b)$

6. Add $9c$ to $-5b$. (4–7)
7. Simplify: $4c - 5d - 3 + 8d - 9d - 7c + c - 5$ (4–7)

Algebra 313

8. Subtract (4–8)

$$\begin{array}{r}-3\,m^2\\ -11\,m^2\end{array} \qquad \begin{array}{r}8\,x^3 - 5\,x^2 + 6\\ 2\,x^3 - x^2 + 3\,x\end{array}$$

$(2\,x^2 - 3\,y^2) - (y^2 - 3\,x^2)$

9. Subtract $-6\,y$ from $4\,x$. (4–8)
10. What must be added to $2\,n^2 - 5$ to get $8\,n$? (4–8)
11. Multiply: (4–9)

$$\begin{array}{l}-9\,a^4b^3y^2\\ -4\,a^2b^9x^7\end{array} \qquad (8\,c^3n^2)(-c^4n^3) \qquad (-3\,ac)(4\,bc)(-2\,ab)(3\,a^2bc^3)$$

$(-3\,x^3)^3 \qquad -4\,d^4x^2(3\,d^4 - 5\,d^3x^2 + dx^5)$

12. Divide as indicated: (4–10)

$$\dfrac{-30\,m^9x^8y^4}{6\,m^6x^8y} \qquad (-8\,c^{12}t) \div (-8\,c^4) \qquad \dfrac{56\,x^8y^4 - 64\,x^5y^5 + 16\,x^4y^7}{-8\,x^4y^3}$$

13. Name each of the following, using negative exponents: $\dfrac{1}{10^4}$; $\dfrac{1}{2^5}$ (4–11)

14. Find the numerical value of each of the following: 10^{-5}; 3^{-3} (4–11)

15. Which of these equations have -2 as their solution? (4–12)
    a. $7\,x - 10 = 4$
    b. $4\,x + 9 = 1$
    c. $3\,x = x - 4$
    d. $x = -2$

    Which of the above equations are equivalent? Which equation of the equivalent equations is in simplest form?

16. What operation with what number do you use on both sides of the equation: (4–13)
    a. $n + 9 = 0$ to get the equation $n = -9$
    b. $r - \tfrac{3}{4} = 3\tfrac{1}{2}$ to get the equation $r = 4\tfrac{1}{4}$
    c. $-6\,y = 60$ to get the equation $y = -10$
    d. $\tfrac{x}{8} = 2$ to get the equation $x = 16$

17. Solve and check, using all real numbers: (4–14)

   a. $n + 17 = 36$
   b. $28 = b - 42$
   c. $-7x = 56$
   d. $\dfrac{a}{5} = 15$
   e. $4y + 5 = 33$
   f. $6x - x = 19$
   g. $9n + 27 = 0$
   h. $\dfrac{4}{9}x = 36$

18. Find the solution of: (4–15)

   a. $\dfrac{x}{2} = 2$ when the replacements for the variable are all the natural numbers.
   b. $4c + 7 = 8$ when the replacements for the variable are all the integers.
   c. $9n = 27$ when the replacements for the variable are all the odd prime numbers.
   d. $2x - 5 = 3$ when the replacements for the variable are all the integers greater than -6 but less than 9.
   e. $11b - 5b = 30$ when the replacements for the variable are 0, 2, 4, 6, 8, and 10.

19. Find the solution of: (4–16)

   a. $|x| = 12$
   b. $|n| + 4 = 11$

20. Find the value of w when $p = 96$ and $l = 29$, using the formula $p = 2l + 2w$. (4–17)

21. How many miles can an airplane fly on n gallons of fuel if it averages r miles per gallon? (4–18)

22. At what price must a dealer sell a lamp which costs $28 to make 20% on the selling price? (4–19)

23. On a number line draw the graph of: (4–20)

   a. $3x + 7 = 16$ when the replacements for the variable are all the real numbers.
   b. $-11b = 55$ when the replacements for the variable are all the integers.

24. Find the ratio of 72 to 8. Of 6 to 48. Of 18 minutes to 2 hours. (4–21)

25. Solve and check: $\dfrac{n}{90} = \dfrac{13}{15}$ (4–22)

Algebra

26. Find the value of c' when $c = 66$, $d = 21$, and $d' = 35$, using the formula $\dfrac{c}{c'} = \dfrac{d}{d'}$. (4–22)

27. Find the solutions of each of the following inequalities when the replacements for the variable are all the real numbers: (4–23)

    a. $x + 3 \neq 9$ e. $3x + 8 \neq 2$ i. $\dfrac{b}{-2} \geq 1$

    b. $y - 4 < 3$ f. $5y - y < 0$ j. $-2x < -10$

    c. $6b \not> 12$ g. $-7y > 42$ k. $-\dfrac{2}{3}y \not> 4$

    d. $\dfrac{n}{3} \not< 1$ h. $4n + 5 \leq -11$ l. $2a - 6a > -12$

28. Find the solution of: (4–23)

    a. $4n - 9 < 7$ when the replacements for the variable are all the natural numbers.

    b. $5x - 2x > 21$ when the replacements for the variable are all the one-digit prime numbers.

    c. $-6y \leq 6$ when the replacements for the variable are -3, -2, -1, 0, 1, 2, and 3.

    d. $2a + 4 \not> 16$ when the replacements for the variable are all the multiples of 2.

    e. $\dfrac{n}{-1} \geq 0$ when the replacements for the variable are all the non-positive integers.

29. On a number line draw the graph of each of the following inequalities when the replacements for the variable are all the real numbers: (4–24, 4–25)

    a. $x < 5$ e. $-5 \leq n \leq -1$

    b. $x - 2 \geq -2$ f. $3x + 7 \not> 1$

    c. $-4b \leq 16$ g. $x > -1$

    d. $-3 < x < 6$ h. $x + x \neq 8$

30. On a number line draw the graph of: (4–26)

    a. $x + 6 > 3$ when the replacements for the variable are -4, -3, -2, -1, 0, 1, 2, and 3.

    b. $5x - 9 < 11$ when the replacements for the variable are all the one-digit odd natural numbers.

    c. $-7y \geq 7$ when the replacements for the variable are -1, 0, and 1.

31. a. Write the ordered pair of numbers which has 10 as its first component and −3 as its second component. (4–27)

b. Write all the ordered pairs of numbers which have 0, 1, 2, 3 as the first component and 0, 1, 2 as the second component. (4–27)

32. a. What are the coordinates of each of the following indicated points: (4–28)

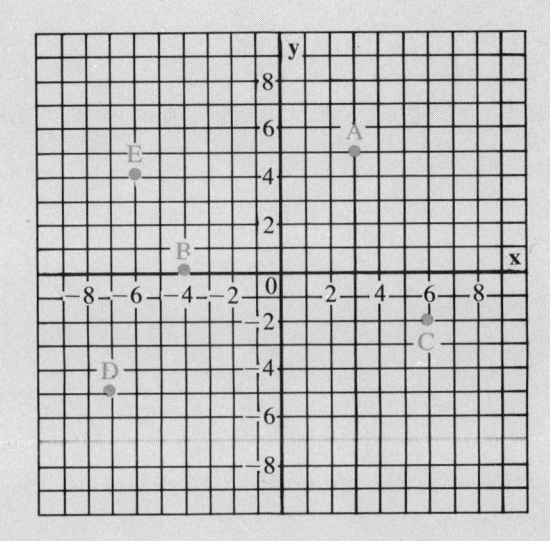

b. On graph paper draw the axes and plot the points which have the following coordinates: (4–29)

(1) (−4,6) (2) (0,−7) (3) (9,5) (4) (−5,−3) (5) (1,−6)

33. Which of the following are true? (4–30)

a. In $n = \frac{360}{a}$, n varies inversely as a.

b. In $p = 5s$, p varies directly as s.

c. In $A = \pi r^2$, A varies directly as the square of r.

34. a. Transform formula $A = bh$, solving for h. (4–31)

b. Derive a formula for V in terms of B and h, using the formulas $V = lwh$ and $B = lw$. (4–31)

35. a. Solve and check: $6x^2 = 150$. (4–32)

b. Find the value of t when $s = 1{,}296$ and $g = 32$, using the formula $s = \frac{1}{2}gt^2$. (4–32)

Algebra

ACHIEVEMENT TEST

1. Add:
   62,968 + 7,809 + 35,674
   + 21,988 + 75,347 (1–2)

2. Subtract:
   526,043
   395,967 (1–3)

3. Multiply:
   4,906
     718 (1–4)

4. Divide:
   508)354,584 (1–5)

5. Add:
   $2\frac{7}{8} + 3\frac{3}{4} + 1\frac{9}{16}$ (1–12)

6. Subtract:
   $8\frac{1}{2} - 6\frac{2}{3}$ (1–13)

7. Multiply:
   $5\frac{5}{8} \times 1\frac{3}{5}$ (1–14)

8. Divide:
   $2\frac{1}{4} \div 1\frac{1}{2}$ (1–15)

9. Add:
   6.54 + .984 + 26.1 (1–18)

10. Subtract:
    80 − .54 (1–19)

11. Multiply:
    .7854
      2.5 (1–20)

12. Divide:
    .9).0639 (1–22)

13. Find 19% of $426. (1–30)
14. What percent of 48 is 42? (1–31)
15. 5% of what number is 26? (1–32)
16. Find the square root of 751,689. (1–36)
17. Which measurement is greater: 5,000 mm or 8 m? (2–1)
18. Which weight is lighter: 6 kg or 4,500 g? (2–2)
19. Change 5.6 liters to milliliters. (2–3)
20. How many m^2 are in 7.4 hectares? (2–4)
21. How many cm^3 are in 9 m^3? (2–5)

22. 35.8 L occupies a space of _____ cm³. This volume of water weighs _____ kg. (2–5)
23. How many yards are in 9 miles? (2–6)
24. Change 2 hours 45 minutes to minutes. (2–13)
25. Is a temperature of 6°C colder than a temperature of 40°F? (2–16)
26. If it is 1 A.M. in Detroit, what time is it in St. Louis? In Seattle? (2–18)
27. Which of the following name integers? Which name rational numbers? Which name irrational numbers? Which name real numbers? (3–1, 3–2)

   $^{-}\frac{4}{5}$ $^{+}\sqrt{91}$ $^{-}63$ $^{+}.38$ $^{-}2\frac{7}{10}$ $^{+}\frac{54}{9}$

28. Draw the graph of $^-5$, $^-3$, 0, $^+2$, $^+4$, and $^+7$ in a number line. (3–4)
29. Use the number line as a scale to draw a vector that illustrates each of the following movements. Write the numeral that is represented by each of these vectors. (3–7)

   a. From $^-2$ to $^+5$ b. From $^+3$ to $^-1$

30. a. What is the opposite of $^-\frac{3}{10}$? (3–8)
    b. What is the additive inverse of $^+11$? (3–8)

31. a. Write each of the following as an algebraic expression: (4–2)
       (1) The discount (d) subtracted from the list price (l).
       (2) The sum of the product of twice pi (π) and the radius (r) and the product of pi, the radius, and the height (h).
    b. Write each of the following as an open sentence symbolically: (4–3)
       (1) Some number n decreased by five is equal to forty.
       (2) Each number y increased by twleve is less than four.
       (3) Six times each number x minus ten is greater than negative nine.

32. The rate of commission (r) is equal to the commission (c) divided by the sales (s). (4–4)

33. a. When $x = 3$, what is the value of $9x - 4$? $9(x - 4)$? (4–5)
    b. Find the value of v when $V = 18$, $g = 32$, and $t = 5$, using the formula $v = V + gt$. (4–6)

Algebra

34. Add: (4–7)

   a. $+6x$
        $-9x$
        $-4x$

   b. $-5n^2 + 7n - 6$
        $3n^2 - 7n + 8$

   c. $(x - 3y) + (2x + 7y) + (5y - 6x)$

   d. Simplify: $4s^2 - 6 - 3s + s^2 - 5s + 4 - 2s^2 - s + 2$

35. Subtract: (4–8)

   a. $-6r^4$
        $9r^4$

   b. $2t^3 \quad\quad -4t + 3$
        $3t^3 - 6t^2 \quad\quad - 8$

   c. $(3a^2 - 5b^2) - (4b^2 - 7a^2)$

   d. Subtract $6mn - 4n^2$ from $5m^2 - mn$

36. Multiply: (4–9)

   a. $-7x^3yz^4$
        $5x^5y^3z^3$

   b. $(-a^5d^2)(4ad^4)$

   c. $(-2y^2)^3$

   d. $(2st)(-5rt^2)(-3rs)(4r^3s^2t)$

   e. $-6m^3y^4(3m^6 - m^2y^5 + 2my^9)$

37. Divide as indicated: (4–10)

   a. $\dfrac{42c^8d^4x^3}{-7c^3d^4x}$

   c. $\dfrac{54m^6n^2 - 30m^4n^4 + 48m^2n^6}{-6m^2n}$

   b. $(-18a^2y^5) \div (-2y^5)$

38. Solve and check, using all the real numbers: (4–14)

   a. $x + 16 = 71$
   b. $9c = 54$
   c. $n - 8 = -12$
   d. $\dfrac{a}{15} = 5$
   e. $11y - y = 50$
   f. $7n - 49 = 0$

39. Find the value of: (4–17)

   a. d when $W = 825$ and $F = 75$, using the formula $W = Fd$.

   b. R when $E = 60$, $I = 4$, and $r = 10$, using the formula $E = Ir + IR$.

40. How much money must be invested at the annual rate of 8% to earn $10,000 per year? (4–19)

41. Using all the real numbers as replacements for the variable, draw on a number line the graph of $5x - 2 = 17$. (4–20)

42. Find the ratio of 6 to 27. Of 63 to 21. Of 250 cm to 2 m. (4–21)

43. a. Solve and check: $\dfrac{12}{x} - \dfrac{3}{20}$ (4–22)

  b. Find the value of P when $V = 8$, $V' = 10$, and $P' = 12$, using the formula $\dfrac{V}{V'} = \dfrac{P'}{P}$. (4–22)

44. a. Find the solutions of each of the following inequalities when the replacements for the variable are all the real numbers: (4–23)

  (1) $n - 5 < 14$ (4) $x - 7x \not> -12$

  (2) $-3x > 9$ (5) $16n - 9 \leq 39$

  (3) $5y + 18 \neq 3$ (6) $\dfrac{a}{3} \geq -2$

  b. Find the solutions of: (4–23)

  (1) $5n - 3 > 7$ when the replacements for the variable are all the one-digit even whole numbers.

  (2) $2x + 7x < 81$ when the replacements for the variable are all the prime numbers less than 20.

45. On a number line draw the graph of each of the following inequalities when the replacements for the variable are all the real numbers: (4–25)

  a. $y + 3 > 5$ **d.** $5x \geq -20$

  b. $\dfrac{n}{2} \neq -1$ **e.** $15m + 6 \leq 51$

  c. $r - 1 < 4$ **f.** $8t - 11t \not> -12$

46. Write all the ordered pairs of numbers which have $-1, 0, 1$ as the first component and $-2, 0, 2$ as the second component. (4–27)

Algebra

47. What are the coordinates of each of the following indicated points: (4–28)

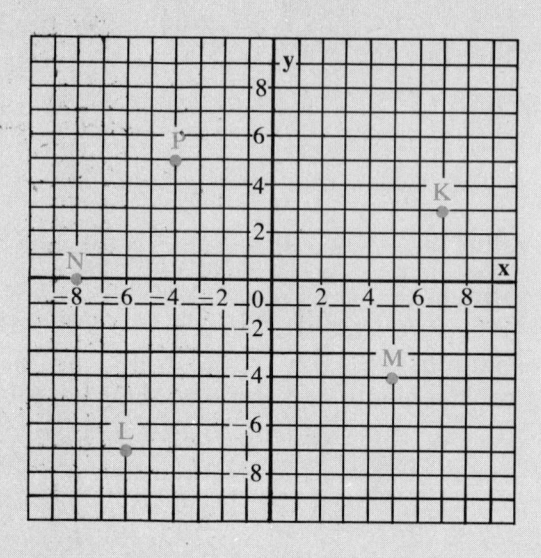

48. On graph paper draw the axes and plot the points which have the following coordinates: (4–29)

   a. $(-3,-1)$ **b.** $(2,8)$ **c.** $(-5,0)$ **d.** $(-2,4)$ **e.** $(6,-3)$

49. Solve and check: $7x^2 = 63$ (4–32)

50. Find the value of r when $V = 594$, $\pi = \frac{22}{7}$, and $h = 21$, using the formula $V = \pi r^2 h$. (4–31)

ELECTION RETURNS

POPULAR VOTE
2 086 772
4 076 357

5

ELECTORAL VOTE

1960	1964	1968	1972	1976
303 / 219 / 15	486 / 52	307 / 191 / 46	520 / 17	297 / 240 / 1

1976 ELECTION RESULTS
- REPUBLICAN
- DEMOCRAT

POPULATION OF VOTING AGE

DID NOT VOTE

VOTED

MILLIONS
- 150
- 125
- 100
- 75
- 50
- 25

REGISTER TO VOTE TODAY

PRESIDENTIAL

CONGRESSIONAL

% VOTING

CHAPTER 5
Graphs and Statistics

Statistics is the study of collecting, organizing, analyzing, and interpreting data or facts. In this chapter we review the bar, line, and circle graphs and develop an understanding of a frequency distribution, of averages (measures of central tendency), of percentiles and quartiles, and of the variability of data (measures of dispersion).

5–1 GRAPHS

Bar Graphs

The *bar graph* is used to compare the size of quantities in statistics.

To construct a bar graph, we draw a horizontal guide line at the bottom of the squared paper and a vertical guide line at the left. We select a convenient scale for the numbers that are being compared, first rounding large numbers. For a vertical bar graph, we place and label the number scale along the vertical guide line; for a horizontal bar graph, we use the horizontal guide line.

We print the items being compared in alternate squares along the second guide line, labeling these items. We mark off for each item the height corresponding to the given number and draw lines to complete bars. All bars should have the same width. We select and print an appropriate title describing the graph.

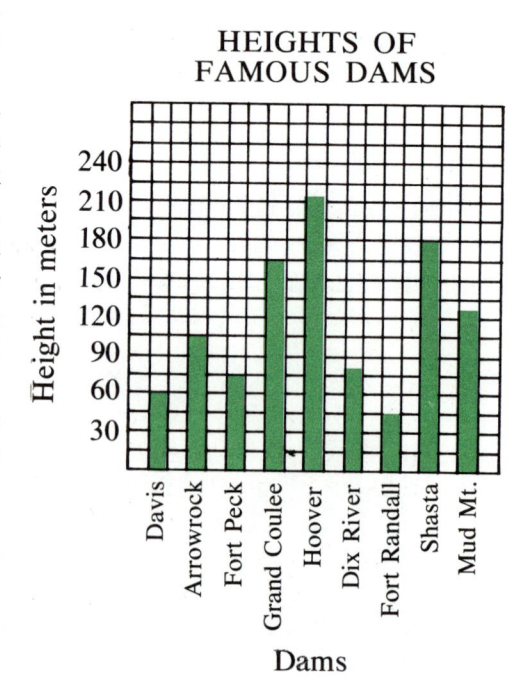

HEIGHTS OF FAMOUS DAMS

Line Graphs

The *line graph* is used to show changes and the relationship between quantities.

To construct a line graph, we draw a horizontal guide line at the bottom of the squared paper and a vertical guide line at the left. We select a convenient scale for the related numbers, first rounding large numbers. We then place and label this scale along one of the guide lines.

We print and label items below the other guide line, using a separate line for each item. On each of these lines we mark with a dot the location of the value corresponding to the given number. We then draw straight lines to connect successive dots. We select and print an appropriate title describing the graph.

NORMAL PRECIPITATION IN CINCINNATI (SOLID LINE) AND PHOENIX (DOTTED LINE)

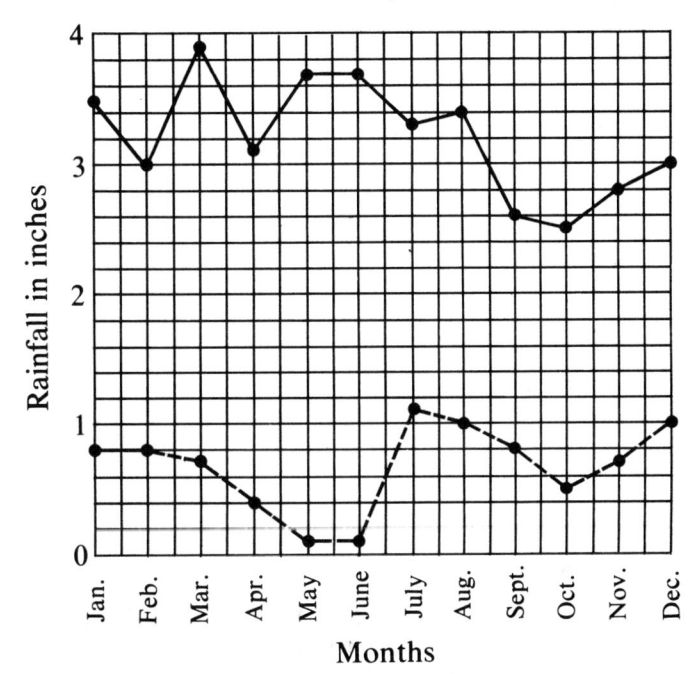

Circle Graphs

The *circle graph* is used to show the relation of the parts to the whole and to each other.

To construct a circle graph, we make a table showing: (a) given facts; (b) fractional part or percent each quantity is of the whole; (c) the number of degrees representing each fractional part or percent, obtained by multiplying 360° by the fraction or percent.

We draw a convenient circle and with a protractor construct successive central angles, using the number of degrees representing each part. We label each sector and select and print an appropriate title describing the graph.

MARKS IN A SOCIAL STUDIES TEST

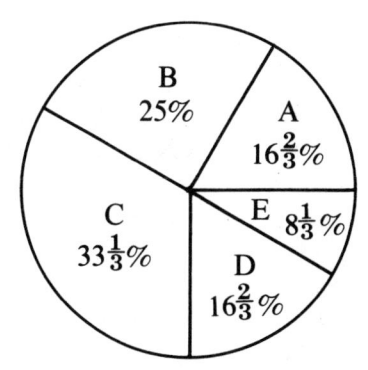

A $16\frac{2}{3}\%$
B 25%
C $33\frac{1}{3}\%$
D $16\frac{2}{3}\%$
E $8\frac{1}{3}\%$

Graphs and Statistics

EXERCISES

1. Construct a bar graph showing the area of the continents: Africa, 11,500,000 sq. mi.; Antarctica, 5,500,000 sq. mi.; Asia, 16,900,000 sq. mi.; Australia, 2,945,000 sq. mi.; Europe, 3,750,000 sq. mi.; North America, 8,440,000 sq. mi.; South America, 6,800,000 sq. mi.

2. Construct a line graph showing the population of the United States, 1800–1970:

Year	Population	Year	Population
1800	5,308,483	1890	62,947,714
1810	7,239,881	1900	75,994,575
1820	9,638,453	1910	91,972,266
1830	12,866,020	1920	105,710,620
1840	17,069,453	1930	122,775,046
1850	23,191,876	1940	131,669,275
1860	31,443,321	1950	150,697,361
1870	38,558,371	1960	179,323,175
1880	50,155,783	1970	203,235,298

3. Construct a circle graph showing the enrollment at the Township Junior-Senior High School: seventh grade, 440 students; eighth grade, 480 students; ninth grade, 420 students; tenth grade, 380 students; eleventh grade, 320 students; twelfth grade, 360 students.

5–2 FREQUENCY DISTRIBUTION

Data may be arranged in tabular form. For example, we arrange the scores made by a class on a test: 80, 85, 65, 60, 90, 80, 85, 95, 100, 85, 60, 75, 95, 85, 80, 75, 80, 65, 90, 80, 75, 85, 70, 65, 70, 85, 80, 90, 85, 75 by first tallying the scores and then summarizing the results as shown in the following table. The number of times each score occurs is called its *frequency*. Arrangement of data in this form is called a *frequency distribution*.

Score	Tally	Frequency
100	|	1
95	||	2
90	|||	3
85	⧸⧸⧸⧸ ||	7
80	⧸⧸⧸⧸ |	6
75	||||	4
70	||	2
65	|||	3
60	||	2

A frequency distribution can be pictured by a special bar graph called a *histogram* or by a special line graph called a *frequency polygon*.

In a histogram each score or group of scores is represented by a bar, its base representing the score and its height representing the frequency of the score. Unlike the conventional bar graph, there is no space left between bars in a histogram.

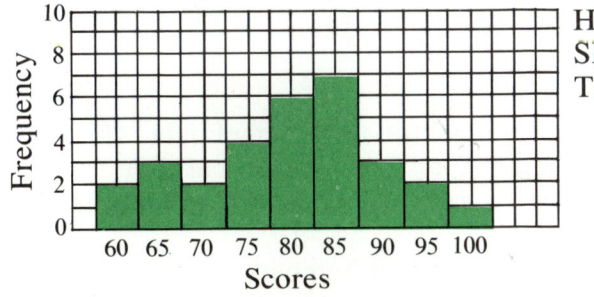

HISTOGRAM SHOWING TEST SCORES

In a frequency polygon, like the histogram, the horizontal scale indicates the score and the vertical scale indicates the frequency of these scores. The line segments and the axis of the horizontal scale form a polygon.

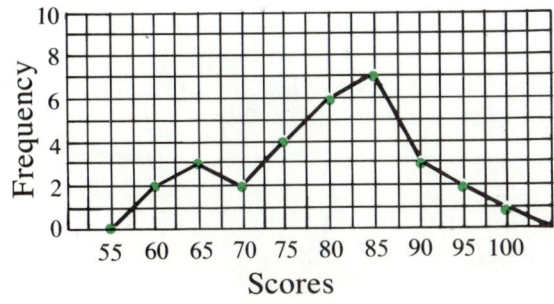

FREQUENCY POLYGON SHOWING TEST SCORES

Graphs and Statistics

EXERCISES

1. Make a frequency distribution table for each of the following lists of scores:
   a. 7, 9, 5, 2, 8, 9, 6, 7, 5, 8, 7, 9, 10, 6, 4, 8, 6, 7, 9, 8, 10, 5, 7, 9
   b. 75, 90, 65, 80, 100, 45, 65, 60, 85, 75, 60, 95, 80, 70, 75, 65, 95, 100, 45, 80
   c. 23, 17, 20, 19, 20, 24, 18, 23, 25, 22, 19, 20, 18, 21, 24, 23, 17, 21, 19, 24, 22, 21, 20, 22, 23, 21, 24, 22, 19, 21, 20
   d. 39, 47, 43, 45, 43, 40, 38, 45, 44, 46, 48, 47, 45, 42, 43, 40, 44, 45, 47, 50, 44, 36, 47, 45, 42, 46, 41, 49, 46, 42, 41, 48, 47, 45, 42, 44, 43, 45, 42, 46
   e. 4, 6, 7, 8, 0, 5, 9, 1, 6, 7, 4, 6, 5, 3, 9, 1, 10, 5, 4, 6, 9, 0, 7, 8, 3, 8, 6, 2, 7, 8, 5, 4, 9, 6, 7, 10, 3, 6, 6, 5, 7, 3, 5, 8, 6

2. Make a histogram for each list of scores given in problem 1.
3. Make a frequency polygon for each list of scores given in problem 1.

5-3 AVERAGES—MEASURES OF CENTRAL TENDENCY

An *average* is a measure of central tendency of data. This data may be scores, measurements, or other numerical facts. Three commonly used averages are the arithmetic mean (or simply mean), the median, and the mode.

The *arithmetic mean* of a group of numbers is determined by dividing the sum of the numbers by the number of items in the group. High numbers and low numbers are balanced off to make all numbers the same size. An extremely high number or an extremely low number distorts the picture of central tendency if the arithmetic mean is used.

> Find the arithmetic mean of the following set of scores: 70, 86, 83, and 77.
> $$\frac{70 + 86 + 83 + 77}{4} = \frac{316}{4} = 79$$
> *Answer,* 79

The *median* of a group of numbers is the middle number when the numbers are arranged in order of size. To find this number we may count from either end, smallest to largest or vice-versa. If the number of members is even, the median is determined by dividing the sum of the two middle numbers by 2. Since the median is a positional average, an extremely high or low number does not affect it.

The *mode* of a group of numbers is the number that occurs most frequently in the group. There may be more than one mode.

A class of 25 pupils made the following scores on a test:
19, 18, 14, 16, 18, 19, 17, 17, 16, 20, 19, 20, 16, 17, 20, 17, 19, 19, 19, 18, 18, 19, 18, 19, and 18.

Find the arithmetic mean, median, and mode.

Score	Tally	Frequency	Arithmetic Mean
20	|||	3	20 × 3 = 60
19	ⵌ |||	8	19 × 8 = 152
18	ⵌ |	6	18 × 6 = 108
17	||||	4	17 × 4 = 68
16	|||	3	16 × 3 = 48
14	|	1	14 × 1 = 14
		Total = 25	Sum = 450

$$\frac{450}{25} = 18$$

Arithmetic mean = 18

Median or Middle Score is the 13th score from top or bottom = 18

Mode or most frequent score = 19

The Greek letter Σ, capital sigma, is the symbol that is generally used to indicate in a concise way the sum of a group of numbers.

Graphs and Statistics

The symbol $\sum_{i=1}^{6} x_i$ is read "the summation of x sub i for i equal to 1 through 6" and represents the sum "$x_1 + x_2 + x_3 + x_4 + x_5 + x_6$" where each addend (represented by x with a different subscript) is a specific number, score, or measurement.

> If $x_1 = 18$, $x_2 = 25$, $x_3 = 16$, $x_4 = 27$, $x_5 = 30$, and $x_6 = 24$,
>
> then $\sum_{i=1}^{6} x_i = x_1 + x_2 + x_3 + x_4 + x_5 + x_6$
> $= 18 + 25 + 16 + 27 + 30 + 24 = 140$
>
> and $\sum_{i=3}^{5} x_i = x_3 + x_4 + x_5 = 16 + 27 + 30 = 73$

Observe that the numeral below the $\sum$ is the first subscript and the numeral above the $\sum$ is the last subscript of the required addends. The sigma notation is used in the following formula for finding the arithmetic mean:

$M = \dfrac{\sum x_i}{n}$ where M represents the arithmetic mean, $\sum x_i$, the sum of the scores, and n, the number of scores.

EXERCISES

1. Find the mean for each of the following groups of scores:
   a. 90, 85, 94, 78, 80, 89, 93
   b. 18, 16, 19, 20, 15, 12, 18, 17, 14, 17, 16, 19
   c. 2.7, 2.6, 2.4, 2.9, 2.4, 2.7
   d. 6, 8, 4, 5, 9, 9, 7, 8, 4, 10, 9, 10, 8, 7, 10, 8, 9, 8
   e. $3\tfrac{1}{2}$, $4\tfrac{3}{8}$, $3\tfrac{7}{8}$, $4\tfrac{3}{4}$

2. Find the median for each of the following groups of scores:
   a. 5, 6, 8, 7, 8, 10, 9
   b. 75, 90, 85, 93, 79, 82
   c. 3.6, 4.9, 2.8, 5.7, 6.3
   d. $1\tfrac{1}{2}$, $2\tfrac{1}{4}$, $1\tfrac{3}{4}$, $2\tfrac{3}{8}$, $2\tfrac{1}{2}$, $2\tfrac{1}{8}$, $1\tfrac{7}{8}$, $2\tfrac{1}{4}$, $2\tfrac{5}{8}$
   e. 15, 17, 13, 10, 19, 14, 18, 13, 16, 18, 19, 17, 14, 15

3. Find the mode for each of the following groups of scores:
   a. 3, 9, 8, 7, 6, 7, 5
   b. 90, 55, 75, 85, 80, 85, 65, 80, 70
   c. 7.4, 8.1, 7.5, 6.9, 7.7, 7.8
   d. 16, 19, 25, 14, 18, 32, 29, 33, 27, 21, 19, 25, 30, 24
   e. 4, 10, 5, 9, 6, 13, 8, 15, 9, 5, 13, 8, 12, 7, 8, 14, 6, 11, 8, 4, 5

4. Find the mean, median, and mode for each of the following groups of scores:
   a. 5, 8, 9, 6, 2, 4, 8, 3, 6, 7, 5, 9, 6
   b. 80, 75, 60, 90, 95, 80, 70, 85, 75, 80, 70, 65, 95, 85, 80
   c. 11, 10, 13, 12, 15, 16, 14, 12, 13, 10, 14, 16, 13, 15, 11, 13, 16, 10, 13, 15
   d. 63, 72, 65, 68, 74, 69, 73, 68, 67, 69, 74, 68, 69, 70, 68, 66, 75, 69, 71
   e. 15, 21, 16, 24, 25, 14, 10, 22, 21, 18, 21, 20, 19, 16, 24, 20, 23, 19, 19, 16, 23, 21, 25, 18, 21

5. Tally and arrange each of the following groups of scores in a frequency distribution table, then find the mean, median, and mode:
   a. 12, 16, 19, 13, 20, 11, 14, 17, 15, 12, 13, 18, 16, 17, 18, 20, 11, 15, 18, 19, 12, 19, 15, 18, 17, 16, 17, 16, 14, 15, 18, 12
   b. 6, 4, 9, 3, 8, 2, 5, 8, 7, 4, 2, 3, 9, 7, 1, 6, 5, 5, 2, 6, 4, 8, 9, 3, 1, 5, 9, 8, 6, 8, 6, 9, 7, 4, 9, 8, 6, 5, 7, 3, 5, 6, 8, 7, 9, 8, 5, 8, 7, 6

6. Find the mean and median of the following group of scores: 65, 69, 77, 800, 75, 73, 68, 71, 66, 70, 72. If the score of 800 were dropped, which measure would be more affected, the mean or the median?

7. If $x_1 = 8$, $x_2 = 15$, $x_3 = 11$, $x_4 = 16$, and $x_5 = 9$, find:

   a. $\sum_{i=1}^{5} x_i$
   b. $\sum_{i=1}^{3} x_i$
   c. $\sum_{i=1}^{4} x_i$
   d. $\sum_{i=2}^{5} x_i$
   e. $\sum_{i=3}^{4} x_i$
   f. $\sum_{i=4}^{5} x_i$

Graphs and Statistics

8. If $x_1 = 14$, $x_2 = 6$, $x_3 = 5$, $x_4 = 10$, $x_5 = 12$, $x_6 = 2$, $x_7 = 17$, and $x_8 = 9$, find:

   a. $\sum_{i=1}^{8} x_i$ c. $\sum_{i=3}^{7} x_i$ e. $\sum_{i=4}^{6} x_i$

   b. $\sum_{i=1}^{6} x_i$ d. $\sum_{i=2}^{8} x_i$ f. $\sum_{i=5}^{8} x_i$

9. Use the formula $M = \dfrac{\sum x_i}{n}$ to find the arithmetic mean, M, of the following six scores:

   $x_1 = 10$, $x_2 = 9$, $x_3 = 8$, $x_4 = 5$, $x_5 = 4$, and $x_6 = 12$.

10. The individual weights of the seven linemen on the school football team starting with the left end are: 88.6, 99.0, 101.8, 95.5, 106.8, 100.0, and 87.3 kilograms respectively. The individual weights of the four backfield men are: 79.5, 82.7, 76.8, and 81.8 kilograms. Find the arithmetic mean of the weights of the:

   a. linemen   b. backfield men   c. entire team

5–4 PERCENTILES AND QUARTILES

We may compare an individual score with all the other scores by giving the score a positional standing or a *rank* from the top. However, unless we know how many scores there are, this comparison is unsatisfactory. For example, Steve ranks ninth in his class. In a class of 12 students, he would have a low rank. In a class of 180 students, he would have a high rank. Thus it is better to make a comparison by using percentile rank.

The *percentile rank* of a score tells us the percent of all the scores that are below this given score. If the rank of a particular score is the 60th percentile, it means that 60% of all the scores are lower than this score.

For example, if Steve ranked ninth in a class of 12 students, there are 3 students of 12 students, or 25%, with a lower rank. He would have a percentile rank of 25, or a rank of the 25th percentile. If Steve ranked ninth in a class of 180 students, there are 171 students of 180 students, or 95%, with a lower rank. He would then have a percentile rank of 95, or a rank of the 95th percentile.

$$\frac{3}{12} = \frac{1}{4} = 25\%$$

$$\frac{171}{180} = \frac{19}{20} = 95\%$$

If an arranged group of scores (or other data) is divided into four equal parts, the score at each point of division is called a *quartile*. A percentile also refers to a score.

The *upper quartile* (or third quartile) is the score at the point below which 75% of all the scores fall. It corresponds to the 75th percentile.

The *median* (or second quartile) corresponds to the 50th percentile.

The *lower quartile* (or first quartile) is the score below which 25% of all the scores fall. It corresponds to the 25th percentile.

EXERCISES

1. Find the percentile rank in each of the following:
   a. Marilyn ranked tenth in a class of 100 students.
   b. Charlotte ranked sixth in a class of 40 students.
   c. Scott ranked third in a class of 50 students.
   d. Anita ranked eighteenth in a class of 72 students.
   e. Ronald ranked twenty-seventh in a class of 90 students.
2. In a class of 50 students Elaine has a percentile rank of 80. What is her rank in the class? Does she have a higher rank than Pierre who ranks fourteenth in this class?
3. If the top 15 students of a graduating class of 120 students are to receive awards, who among the following would get an award? A student with a percentile rank of 75? 90? 85? 80? 89?
4. If all students above the upper quartile are to be exempt from taking the final examination, who among the following would be exempt? A student with a percentile rank of 81? 76? 73? 66? 92?
5. Find the 50th percentile, upper quartile, and lower quartile for the following group of scores: 55, 60, 60, 70, 75, 80, 80, 85, 85, 95, 95, 100.

Graphs and Statistics

5-5 MEASURES OF DISPERSION

Measures of dispersion show the spread or scatter or variability of the distribution of scores (or other statistical facts) around the average.

The *range* is the difference between the highest and the lowest scores.

The *deviation from the mean* is the difference between the score and the mean. Positive and negative numbers are used to indicate whether a score is above the mean (+ deviation) or below the mean (− deviation).

The *mean deviation from the mean* is the mean of the absolute values of all the deviations from the mean.

Variance is the mean of all the squares of all the deviations from the mean.

Standard deviation is the principal square root of the variance and is denoted by the Greek letter sigma, σ.

To find the variance of a group of scores, we first find the deviation from the mean for each score and square each of these deviations. We then find the sum of the squares of these deviations and divide this sum by the number of scores in the group. The quotient, thus found, is the variance.

To find the standard deviation, we take the positive square root of the variance.

Find the range, mean deviation from the mean, variance, and standard deviation for the following group of scores: 80, 95, 90, 85, 100.

Scores	Deviation from Mean	Square of Deviation
100	+10	100
95	+ 5	25
90	0	0
85	− 5	25
80	−10	100
450	0	250

Sum of absolute values
30

Range: $100 - 80 = 20$

Mean: $\frac{450}{5} = 90$

Mean Deviation from Mean: $\frac{30}{5} = 6$

Variance: $\frac{250}{5} = 50$

Standard Deviation: $\sqrt{50} = 7.07$

EXERCISES

1. Find the range for each of the following groups of scores:
   a. 82, 53, 75, 94, 67, 46, 85, 62, 90, 79
   b. 21, 18, 25, 13, 29, 34, 17, 25, 21, 18, 28
   c. 9, 16, 25, 6, 39, 51, 18, 42, 27, 56, 33, 19
   d. 73, 91, 56, 67, 84, 39, 75, 93, 80, 57, 89, 61, 72
   e. 6, 5, 8, 3, 10, 9, 15, 7, 14, 21, 4, 19, 25, 18, 20

2. Find the mean and mean deviation from the mean for each of the following groups of scores:
   a. 100, 85, 70, 90, 80
   b. 60, 75, 90, 85, 90, 75, 80, 85
   c. 6, 4, 8, 2, 5, 7, 8, 7, 4, 6, 9
   d. 20, 15, 16, 25, 17, 22, 18, 15, 20, 22, 17, 21
   e. 11, 16, 24, 9, 7, 10, 22, 13, 8, 15, 18, 16, 20, 9, 12

3. During the semester Peter's test marks in social studies were: 85, 91, 89, 87, 92, and 84 while John's test marks were: 94, 65, 100, 92, 79, and 98. Find the mean and range for each group of marks. Whose marks were more consistent with the mean?

4. Find the range, mean deviation from the mean, variance, and standard deviation for each of the following groups of scores:
   a. 5, 6, 7, 8, 9
   b. 12, 8, 16, 4, 20
   c. 75, 86, 64, 72, 94, 79, 83
   d. 48, 56, 67, 69, 70
   e. 7, 8, 10, 11, 14, 16, 18
   f. 1, 3, 5, 7, 9, 11, 13, 15, 17, 19

5. The arithmetic inventory test scores of a certain class at the beginning of the school year were:

   9, 14, 6, 11, 15, 4, 7, 13, 5, 10, 16.

   The midyear arithmetic achievement test scores were:

   11, 16, 8, 14, 15, 10, 9, 12, 7, 13, 17.

   The final arithmetic achievement test scores were:

   12, 18, 16, 10, 14, 9, 13, 15, 8, 19, 20.

   For each group of scores find the range, mean, mean deviation from the mean, variance, and standard deviation.

Graphs and Statistics

CHAPTER REVIEW

1. Construct a circle graph showing the enrollment at the Community Junior-Senior High School: seventh grade, 340 students; eighth grade, 300 students; ninth grade, 325 students; tenth grade, 280 students; eleventh grade, 270 students; twelfth grade, 285 students. (5–1)

2. A science class made the following scores in a test:
   90, 75, 80, 65, 80, 70, 95, 80, 90, 70, 85, 90, 50, 85, 75, 90, 60, 75, 100, 70, 65, 75, 65, 90, 80.

   a. Make a frequency distribution table, a histogram, and a frequency polygon. (5–2)
   b. Find the mean, median, and mode. (5–3)

3. Find the range, mean deviation from the mean, variance, and standard deviation for the following group of scores:
   60, 65, 70, 75, 80, 85, 90. (5–5)

4. Find the median, upper quartile, and lower quartile for the following group of scores: 5, 6, 8, 9, 11, 12, 14, 15. (5–4)

5. If Felipe ranks twenty-first in his class of 75 students, what is his percentile rank? (5–4)

MORTALITY TABLES
United States

Age	Deaths Per 1,000	Expectation of Life (Years)
0	25.93	69.89
1	1.70	70.75
2	1.04	69.87
3	.80	68.94
4	.67	67.99
5	.59	67.04
6	.52	66.08
7	.47	65.11
8	.43	64.14
9	.39	63.17
10	.37	62.19
11	.37	61.22
12	.40	60.24
13	.48	59.26
14	.59	58.29
15	.71	57.33
16	.82	56.37
17	.93	55.41
18	1.02	54.46
19	1.08	53.52
20	1.15	52.58
21	1.22	51.64
22	1.27	50.70
23	1.28	49.76
24	1.27	48.83
25	1.26	47.89
26	1.25	46.95
27	1.26	46.00
28	1.30	45.06
29	1.36	44.12
30	1.43	43.18
31	1.51	42.24
32	1.60	41.30
33	1.70	40.37
34	1.81	39.44
35	1.94	38.51
36	2.09	37.58
37	2.28	36.66
38	2.49	35.74
39	2.73	34.83
40	3.02	33.92
41		

6

Increasing the Chances of a Heart Attack

Official studies show that cigarette smoking, high blood pressure or high fat content in blood raises the risk that a man will suffer a heart attack — and that any combination of these three "risk factors" increases the dangers even more.

heart attacks per 1,000 men

no risks	20
smoking	45
high blood fat	52
high blood pressure	52
high blood fat & high blood pressure	85
smoking & high blood fat	92
smoking & high blood pressure	92
all three risks	171

BILLIONS OF DOLLARS

$1.5 — 1954
$2.8 — 1959
$4.7 — 1964
$7.6 — 1969
$13.6 — 1974

HEALTH INSURANCE BENEFIT PAYMENTS

EXPECTATION OF LIFE AT VARIOUS AGES IN THE UNITED STATES

male / female — age — years

CHAPTER 6
Probability

6–1 PROBABILITY; SAMPLE SPACES; COMPLEMENTARY EVENTS; ODDS

Probability

When the weather service predicts that the probability of precipitation (rain or snow) for the next day is 70%, it is informing us that the likelihood for this precipitation to occur is 7 chances out of 10.

Probability is a numerical measure indicating the chance or likelihood for a particular event to occur. It is usually expressed as a ratio named by a common fraction or a decimal numeral or a percent.

> For example, when we toss a coin in the air, it is equally likely that it will land heads up as tails up. Thus the chance that this coin will land heads up is one out of two and the probability ratio is $\frac{1}{2}$.

This may be indicated by the notation:

$P(H) = \frac{1}{2}$ which reads "the probability of heads is one-half."

> The standard deck of 52 playing cards has 13 diamond cards. The probability of drawing at random (without looking or without preference) a diamond card from this full well-shuffled deck is 13 out of 52 or the ratio $\frac{13}{52}$ which is equivalent to $\frac{1}{4}$. The chance to draw a diamond card therefore is 1 out of 4. This may be indicated as $P(D) = \frac{1}{4}$.

In the study of probability, activities like coin-tossing, drawing cards, etc. are used as experiments. Tossing a coin so that it lands heads up is an example of an *event* or an *outcome,* a particular way in which something occurs.

When a coin is tossed, there are 2 outcomes, heads and tails. However, if the desired result is heads, then there is only 1 favorable outcome, heads.

> In a bag containing 10 marbles there are 10 possible outcomes when drawing a marble at random. However, if the event is to draw a green marble from a bag of 10 marbles of which 2 are green marbles, there are only 2 favorable outcomes, the two green marbles. The probability of drawing a green marble is $\frac{2}{10}$ which is equivalent to $\frac{1}{5}$ and may be written as $P(\text{green marble}) = \frac{1}{5}$.

Probability is the ratio of favorable outcomes to the total number of possible outcomes. This is indicated by $P = \frac{f}{n}$ where P is the probability, f is the number of favorable outcomes, and n is the total number of possible outcomes. If there are no favorable outcomes, the probability is 0. If all are favorable outcomes, the probability is 1.

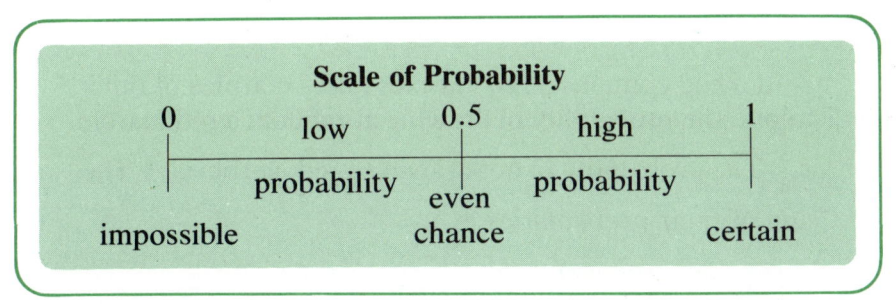

Scale of Probability

```
0           low          0.5        high          1
|———————————|————————————|———————————|
        probability    even    probability
   impossible         chance              certain
```

Sample Space

All the possible outcomes of a given experiment make up the *sample space*.

The following are examples of sample spaces:

(1) When a coin is tossed, the sample space *H,T* indicates the 2 possible outcomes: *heads* or *tails*.

Probability

(2) When two coins are tossed, the sample space *HH, HT, TH, TT* indicates the 4 outcomes: *heads heads, heads tails, tails heads, tails tails.*

(3) When drawing a marble from a bag containing 3 red marbles and 4 white marbles, the sample space $R_1, R_2, R_3, W_1, W_2, W_3, W_4$ indicates the 7 possible outcomes since each *R* represents one of the red marbles and each *W* represents one of the white marbles.

A *die* is a cube each of whose six faces is marked with a different number of dots, from one through six.

When a die is rolled, the sample space *1, 2, 3, 4, 5, 6* indicates the 6 possible outcomes, each representing a different face of the die that may turn up. The plural of "die" is "dice."

An *event* is any outcome or group of outcomes of the sample space.

A *simple event* consists of one outcome of the sample space.

When a coin is tossed, *T* is a simple event. When a die is rolled, *5* is a simple event.

Complementary Events

The event that something happens and the event that this something does not happen are called *complementary events*.

If an event can happen in *m* ways and fail to happen in *n* ways and each is equally likely, the probability of its happening is $\dfrac{m}{m+n}$; the probability of its failing to happen is $\dfrac{n}{m+n}$.

> If a bag contains 3 red marbles and 5 marbles of other colors, the probability of drawing at random a red marble is $\dfrac{3}{8}$. The probability of not drawing a red marble is $\dfrac{5}{8}$. The sum of their probabilities is 1.

If *event A* represents the event of drawing a red marble and *event not A* represents the event of not drawing a red marble, then *event A* and *event not A* are complementary events and

$$P(A) + P(\text{not } A) = 1$$

or

$$P(A) = 1 - P(\text{not } A)$$

Odds

When a coin is tossed, the odds of getting heads are 1 to 1 since there are 1 head and 1 tail. When there are 2 green marbles and 8 other marbles in a bag, the odds of selecting a green marble are 2 to 8 or 1 to 4. The odds against selecting a green marble are 8 to 2 or 4 to 1.

The *odds for an event to happen* against its failure to happen are the ratio of its favorable outcomes to unfavorable outcomes.

The *odds against an event to happen* are the ratio of its unfavorable outcomes to favorable outcomes.

EXERCISES

1. Experiment—To draw at random a ball from a box that contains 1 red ball and 1 white ball.
   a. Write the sample space listing all the possible outcomes (colors) of drawing a ball from the box.
   b. List the favorable outcomes for the event of drawing a red ball.
   c. What is the probability of drawing a red ball on the first draw?
   d. What is the probability of not drawing a red ball on the first draw?
   e. What are the odds of drawing a red ball on the first draw? Of not drawing a red ball on the first draw?

2. Experiment—To draw at random a card from a hat containing three cards of the same size numbered 1 to 3 inclusive.
   a. Write the sample space listing all the possible outcomes (numerals) of drawing a card from the hat.
   b. List the favorable outcomes for the event of drawing a card with numeral 2 on it.
   c. What is the probability of drawing the card with numeral 2 on it on the first draw?
   d. What is the probability of not drawing the card with numeral 2 on it on the first draw?
   e. What are the odds of drawing the card with numeral 2 on it on the first draw? Of not drawing this card on the first draw?

Probability

3. Experiment—To pick at random a crayon from a box of ten crayons: 1 blue, 1 white, 1 brown, 1 yellow, 1 orange, 1 green, 1 purple, 1 pink, 1 black, 1 red.

   a. Write the sample space listing all the possible outcomes (colors) of picking a crayon from the box.
   b. List the favorable outcomes for the event of picking a yellow crayon from the box.
   c. What is the probability of picking a yellow crayon from the box on the first draw?
   d. What is the probability of not picking a yellow crayon on the first draw?
   e. What are the odds of picking a yellow crayon on the first draw? Of not picking a yellow crayon on the first draw?

4. Experiment—To spin the pointer on a spinner whose circle has eight sectors each of the same size numbered 1 through 8.

   a. Write the sample space listing all the possible outcomes (numerals on sectors) where the pointer may stop.
   b. List the favorable outcomes for the event that the pointer stops on the sector marked 6.
   c. What is the probability that the pointer stops on the sector marked 6?
   d. What is the probability that the pointer does not stop on the sector marked 6?
   e. What are the odds that the pointer stops on the sector marked 6? Does not stop on the sector marked 6?

5. Experiment—To draw at random a slip of paper from twelve identical slips of paper with three slips of paper marked with an X.

   a. How many outcomes are in the sample space?
   b. If the event is to select a slip of paper with an X marked on it, how many favorable outcomes are there?
   c. What is the probability that a slip with an X marked on it will be selected on the first draw?
   d. What is the probability that a slip without an X marked on it will be selected on the first draw?
   e. What are the odds of selecting a slip with an X marked on it on the first draw? Of not selecting a slip with an X marked on it on the first draw?

6. Experiment—To draw at random a marble from a bag containing 6 blue marbles and 8 yellow marbles.

   a. How many outcomes are in the sample space? List them.
   b. If the event is to select a blue marble, how many favorable outcomes are there? List them.
   c. What is the probability of drawing a blue marble from the bag on the first draw?
   d. What is the probability of not drawing a blue marble from the bag on the first draw?
   e. What are the odds of drawing a blue marble on the first draw? What are the odds of not drawing a blue marble on the first draw?

7. Experiment—To draw at random two beads from a bowl containing two blue beads and two gold beads, replacing the first bead before drawing the second bead.

   a. Write the sample space listing all the outcomes of drawing two beads such as (*BG*), etc.
   b. If the event is to select at random two blue beads, list the favorable outcomes.
   c. What is the probability of drawing two blue beads on two draws?
   d. What is the probability of not drawing two blue beads on two draws?
   e. What are the odds of drawing two blue beads on the first two draws? What are the odds of not drawing two blue beads on the first two draws?

8. Experiment—Tossing three coins.

   a. Write the sample space listing all the outcomes when the three coins are tossed, such as (*HHH*), (*HHT*), etc.
   b. List the favorable outcomes if the event is to have the three coins fall either all heads or all tails.
   c. What is the probability that the three coins will fall either all heads or all tails?
   d. What is the probability that the three coins will fall neither all heads nor all tails?
   e. What are the odds that the three coins will fall either all heads or all tails? What are the odds that the three coins will fall neither all heads nor all tails?

Probability

9. A standard deck of 52 playing cards consists of an ace, king, queen, jack, 10, 9, 8, 7, 6, 5, 4, 3, and 2 in four different suits: clubs, diamonds, hearts, and spades. Club and spade suits are in black; diamond and heart suits are in red. The king, queen, and jack are called picture cards.

   Experiment—To draw at random a card from a well-shuffled deck.

   a. What is the probability that the first draw will be:
      (1) a black card?
      (2) an ace?
      (3) a picture card?
      (4) a red 2?
      (5) a black 7?
      (6) a red picture card?
      (7) the queen of spades?
      (8) any jack?

   b. What is the probability that the first draw will not be:
      (1) the ace of spades?
      (2) any king?
      (3) a black 9?
      (4) a red card?
      (5) a picture card?
      (6) a black picture card?

   c. What are the odds in favor of the first draw being:
      (1) the king of clubs?
      (2) a queen?
      (3) a red card?
      (4) a picture card?
      (5) a black ace?
      (6) a red picture card?

   d. What are the odds against the first draw being:
      (1) an ace?
      (2) the jack of diamonds?
      (3) a red 5?
      (4) a black card?
      (5) a black picture card?
      (6) a picture card?

10. Experiment—Tossing a pair of dice.

    a. Write the sample space listing all the possible outcomes when a pair of dice is rolled. Observe that each outcome is designated by a number pair such as (5, 3) in which the first component names the number of dots in the upper face of the first die and the second component names the number of dots in the upper face of the second die. How many possible outcomes are there in the sample space?

    b. Write all the outcomes for the event that the sum of the dots on the two dice is 7. What is the probability that the sum will be 7 when the two dice are rolled? What is the probability that the sum will not be 7? What are the odds that the sum of 7 will occur on the first roll? What are the odds that the sum of 7 will not occur on the first roll?

c. Write the outcomes for each of the following events. Then find the probability for each event to happen and the probability for it not to happen. Also find the odds in favor of the events to happen and the odds against the event to happen.

Event: Sum of the dots on the upper faces of the two dice is:

(1) 5   (4) 12   (7) 4
(2) 11   (5) 9   (8) 2
(3) 6   (6) 3   (9) 8

11. What is the probability value of:

   a. An event having as much chance to happen as it has to fail to happen?
   b. An event that is impossible to happen?
   c. An event that is absolutely certain to happen?

12. Which has the higher probability:

   Drawing at random on the first draw a blue marble from a bag containing 5 blue marbles and 8 green marbles or drawing on the first draw a blue marble from a bag containing 15 blue marbles and 24 white marbles?

13. You have 6 nickels, 9 dimes, and 5 quarters in your pocket. What is the probability of picking a nickel at random from your pocket?

14. A total of 500 raffle tickets were sold at a charity affair. If you hold 15 raffle tickets, what is the probability of your winning the prize?

15. One hundred fifty tickets for the door prize were sold at a charity affair. If you hold 3 of these tickets, what is the probability of your winning the door prize?

16. In a class of 12 boys and 16 girls, a student is to be selected to be the school council member by drawing a name of one of these students. What is the probability that a boy will be selected? What is the probability that a girl will be selected? What are the odds that a boy will be selected? What are the odds that a girl will be selected?

6-2 PROBABILITY OF A OR B; PROBABILITY OF A AND B

Probability of A or B

Mutually exclusive events are events which cannot occur at the same time. A coin falls heads or tails but not both heads and tails at the same time.

When a box contains 4 white marbles, 5 red marbles, and 3 yellow marbles, drawing a white marble or drawing a red marble are mutually exclusive events.

Since there are 12 marbles in the box and 4 are white, the probability of drawing at random a white marble is $\frac{4}{12}$ or $\frac{1}{3}$.

$$\text{That is: } P(W) = \frac{1}{3}$$

Since there are 12 marbles in the box and 5 are red, the probability of drawing at random a red marble is $\frac{5}{12}$.

$$\text{That is: } P(R) = \frac{5}{12}$$

Since there are 12 marbles in the box and the total of white and red marbles is 9, the probability of drawing a white marble or a red marble at random is $\frac{9}{12}$ or $\frac{3}{4}$.

$$\text{That is: } P(W \text{ or } R) = \frac{3}{4}$$

Observe that since $P(W) = \frac{1}{3}$ and $P(R) = \frac{5}{12}$, it follows that $P(W) + P(R) = \frac{1}{3} + \frac{5}{12} = \frac{9}{12} = \frac{3}{4}$, the same answer as that found for $P(W \text{ or } R)$.

$$\text{That is: } P(W \text{ or } R) = P(W) + P(R)$$

When two events are mutually exclusive, the probability of one event or the other event to happen is equal to the sum of their separate probabilities.

Thus if A and B are mutually exclusive events, then the *probability of A or B* is the sum of the probability of A and the probability of B.

This may be expressed as: $P(A \text{ or } B) = P(A) + P(B)$

Probability of A and B

Two events are *independent* when the outcome of one event does not affect the outcome of the other.

When we toss two coins, the event of the first coin falling heads up and the event of the second coin falling heads up are independent of each other.

The probability of the first coin falling heads up is $\frac{1}{2}$.

$$\text{That is: } P(H) = \frac{1}{2}$$

The probability of the second coin falling heads up is $\frac{1}{2}$.

$$\text{That is: } P(H) = \frac{1}{2}$$

The sample space for the experiment of tossing two coins may be expressed as $(HH), (HT), (TH), (TT)$

or in tabular form:

		Second Coin	
		H	T
First Coin	H	HH	HT
	T	TH	TT

In this sample space the first letter indicates the outcome of the toss of the first coin and the second letter indicates the outcome of the toss of the second coin. Observe there are 4 outcomes: HH, HT, TH, and TT but there is only 1 favorable outcome: HH.

Thus the probability of both coins falling heads up is $\frac{1}{4}$.

$$\text{That is: } P(H \text{ and } H) \text{ or } P(HH) = \frac{1}{4}$$

We have found that on the first toss $P(H) = \frac{1}{2}$ and on the second toss $P(H) = \frac{1}{2}$. If we multiply these probabilities $\left(\frac{1}{2} \times \frac{1}{2}\right)$, the product $\left(\frac{1}{4}\right)$ is the same answer as that found for $P(H \text{ and } H)$ or $P(HH)$.

The probability that two independent events will both occur is equal to the product of the separate probabilities.

Thus if A and B are independent events, then the *probability of A and B* is the product of the probability of A and the probability of B.

This may be expressed as: $P(A \text{ and } B) = P(A) \cdot P(B)$

Probability

EXERCISES

1. A bag contains 5 white marbles, 6 green marbles, and 9 red marbles. What is the probability of selecting at random from this bag on the first draw:
   a. a green marble or a red marble?
   b. a white marble or a red marble?
   c. a green marble or a white marble?
   d. a white, green, or red marble?
   e. a blue marble?

2. When you roll a die, what is the probability that on the first roll:
   a. a 5 or 6 will turn up?
   b. a 1, 2, or 3 will turn up?
   c. a 3 or 4 will not turn up?

3. There are nine cards of the same size in a box, each card bearing a different numeral from 1 through 9. What is the probability of selecting at random from this box on the first draw a card marked with:
   a. the numeral 4 or 9?
   b. a numeral naming a number greater than 1 but less than 8?
   c. a numeral naming an odd number?
   d. a numeral naming a prime number?
   e. a numeral naming a number divisible by 3?

4. If a spinner has twelve sectors, each of the same size, numbered 1 through 12, what is the probability that the pointer will stop on:
   a. the sector marked 7 or 11?
   b. the sector marked 1, 2 or 3?
   c. the sectors marked by numerals naming a number greater than 8?
   d. the sectors marked by numerals naming a prime number?
   e. the sectors marked by numerals naming an even number?

5. From a well-shuffled deck of 52 playing cards, what is the probability of selecting at random on the first draw:
   a. a king of hearts or a queen of spades?
   b. a 3 of clubs or 3 of diamonds or 3 of spades?
   c. any ace or any jack?
   d. any diamond or any club?
   e. a picture card or an ace?

6. When a red die and a white die are rolled, what is the probability of getting:

   a. a 6 on the white die and a 4 on the red die?
   b. a 2 on the white die and a 2 on the red die?

7. One spinner has four sectors, each of the same size, numbered 1 through 4. A second spinner has six sectors, each of the same size, numbered 1 through 6. What is the probability:

   a. that the pointer on the first spinner will stop at 3 and on the second spinner at 5?
   b. that the pointers on both spinners will stop at 4?

8. One bag contains 6 green balls and 4 yellow balls. A second bag contains 8 green balls and 2 yellow balls. If one ball is selected at random from each bag, what is the probability that:

   a. a green ball is selected from the first bag and a yellow ball from the second bag?
   b. a yellow ball is selected from the first bag and a green ball from the second bag?
   c. both balls are yellow?
   d. both balls are green?

9. A bowl contains 9 white marbles and 3 blue marbles. If two marbles are drawn at random from this bowl, replacing the first marble before drawing the second marble, what is the probability that:

   a. a white marble is selected on the first draw and a blue marble on the second draw?
   b. a blue marble is selected on the first draw and a white marble on the second draw?
   c. two blue marbles are drawn?
   d. two white marbles are drawn?

10. If two cards are selected at random from a well-shuffled deck of 52 playing cards, replacing the first card before the second card is drawn, what is the probability that:

    a. the first card is a 10 and the second card is a jack?
    b. the first card is a red card and the second card is a black card?
    c. the first card is a black 9 and the second card is a red 9?
    d. the first card is a picture card and the second card is an ace?

Probability

CHAPTER REVIEW

1. What does the weather service mean when it predicts that the probability of rain for tonight is 20%, for tomorrow morning is 50%, and for tomorrow afternoon is 100%? (6–1)
2. What is the probability of drawing at random on the first draw a marked card from a hat containing 60 cards of which 36 are marked? What are the odds of selecting an unmarked card on the first draw? (6–1)
3. Experiment—To draw at random a checker from a box that contains 9 black checkers and 15 red checkers. (6–1)

   a. How many outcomes are in the sample space? List them.
   b. If the event is to select a black checker, how many favorable outcomes are there?
   c. What is the probability of drawing a black checker from the box on the first draw? Of not drawing a black checker on the first draw?
   d. If the event is to select a red checker, how many favorable outcomes are there?
   e. What is the probability of drawing a red checker from the box on the first draw? Of not drawing a red checker on the first draw?
   f. What are the odds of drawing a black checker on the first draw? What are the odds of drawing a red checker on the first draw?

4. From a well-shuffled deck of 52 playing cards, what is the probability of selecting at random on the first draw a card that is: (6–2)

   a. any jack or a red 10?   b. any spade or any heart?

5. One bag contains 8 yellow marbles and 4 blue marbles. A second bag contains 10 yellow marbles and 15 blue marbles. If one marble is selected at random from each bag, what is the probability that: (6–2)

   a. a yellow marble is selected from the first bag and a blue marble from the second bag?
   b. two yellow marbles are drawn?
   c. two blue marbles are drawn?

7

CHAPTER 7
Flow Charts and Logic

7–1 FLOW CHARTS

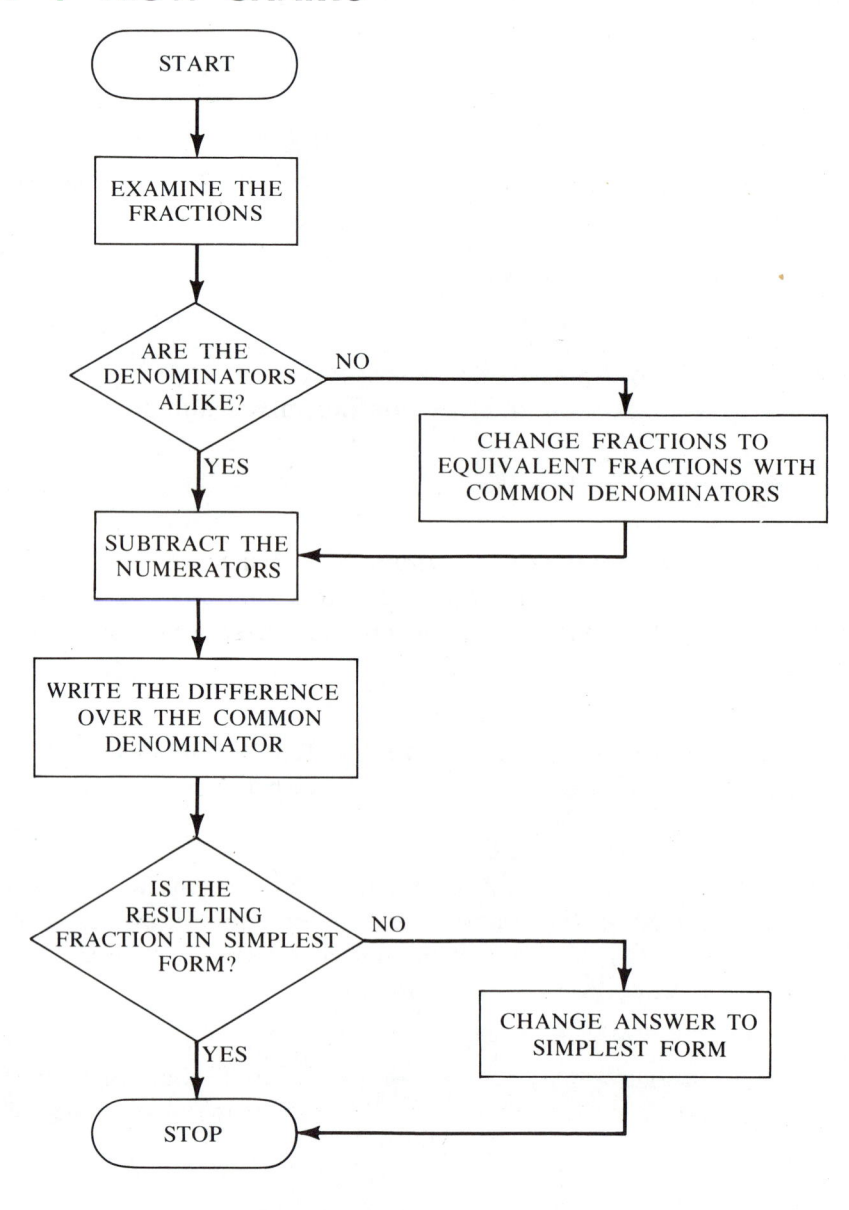

START

↓

EXAMINE THE FRACTIONS

↓

ARE THE DENOMINATORS ALIKE? —NO→ CHANGE FRACTIONS TO EQUIVALENT FRACTIONS WITH COMMON DENOMINATORS

↓ YES

SUBTRACT THE NUMERATORS

↓

WRITE THE DIFFERENCE OVER THE COMMON DENOMINATOR

↓

IS THE RESULTING FRACTION IN SIMPLEST FORM? —NO→ CHANGE ANSWER TO SIMPLEST FORM

↓ YES

STOP

Flow charts are used to indicate and picture the order of the steps to follow in a program.

Various geometric shapes are used. Ovals indicate the beginning and the end of the program. We use rectangles to enclose instructions and diamonds to enclose questions with a two-choice answer nearby.

For example, the flow chart on the opposite page indicates the order of the steps in subtracting a common fraction from a larger common fraction.

EXERCISES

Construct flow charts indicating the steps in each of the following programs:

1. Adding two common fractions.
2. Rounding a decimal.
3. Adding two mixed decimals.
4. Multiplying two 2-place whole numbers.
5. Dividing one mixed number by another mixed number.

7–2 LOGIC

A *compound sentence* (or *statement*) consists of two simple sentences joined by a connective. If the connective is "and," represented by the symbol $\wedge$, the compound sentence is called a *conjunction*.

> The number sentence $8 < n$ and $n < 12$ is a conjunction.
> It could be written as $8 < n \wedge n < 12$ or as $8 < n < 12$.

Usually in logic the letter p represents the first sentence and the letter q the second sentence so that $p \wedge q$ represents a conjuction.

If the connective is "or," represented by the symbol $\vee$, the compound sentence is called a *disjunction*.

> The number sentence $n > 8$ or $n = 8$ is a disjunction.
> It could be written as $n > 8 \vee n = 8$ or as $n \geq 8$.

Flow Charts and Logic

In logic $p \vee q$ symbolizes a disjunction.

A conjunction is true if and only if *both* simple sentences are true. If either sentence is false or both sentences are false, the conjunction is false.

A disjunction is true if *either one* of the simple sentences is true or *both* sentences are true. If both sentences are false, the disjunction is false. See the summary in the truth table at the right.

p	q	$p \wedge q$	$p \vee q$
T	T	T	T
T	F	F	T
F	T	F	T
F	F	F	F

The *negation* of a sentence is the opposite of the sentence. The negation of $b = 4$ is $b \neq 4$; the negation of $x \not> 10$ is $x > 10$. If the sentence is true, the negation is false; if the sentence is false, the negation is true. Negation is indicated by the prefix symbol $\sim$. If p is the sentence, then $\sim p$ (read "not p") is the negation. The negation of a negation is the original sentence.

$$\sim(\sim p) = p$$

p	$\sim p$
T	F
F	T

An *implication* is a conditional sentence (sometimes called a *conditional*) consisting of an *antecedent* (sentence following "if") and a *consequent* (sentence following "then").

> The sentences: "If a triangle has 3 equal sides, then it has 3 equal angles.," and "If $4x = 24$, then $x = 6$," are implications.

The antededent is sometimes called the *hypothesis* and the consequent is called the *conclusion*. Using p as the antecedent, q for the consequent, and an arrow between them, we express an implication as $p \rightarrow q$ which means "if p, then q" or "p implies q."

If we interchange the antecedent and consequent of a conditional sentence, another implication results which is the *converse* of the first. The converse of $p \rightarrow q$ is $q \rightarrow p$. The expression $p \leftrightarrow q$ symbolizes that an implication and its converse are equivalent since each implies the other.

If we negate both the antecedent and consequent of an implication, the resulting implication is the *inverse* of the first. The inverse of $p \rightarrow q$ is $\sim p \rightarrow \sim q$ which is read "not p implies not q."

If we negate both the antecedent and consequent of an implication and then interchange them, the resulting implication is the *contrapositive* of the first. The contrapositive of $p \to q$ is $\sim q \to \sim p$ which is read "not q implies not p."

> An example of each of the above implications is:
> **Implication:** If a triangle is equilateral, then it is equiangular.
> **Converse:** If a triangle is equiangular, then it is equilateral.
> **Inverse:** If a triangle is not equilateral,
>   then it is not equiangular.
> **Contrapositive:** If a triangle is not equiangular,
>   then it is not equilateral.

EXERCISES

1. Read, or write in words, each of the following:
   a. $6 > 5 \land 4 < 7$ c. $\sim a$ e. $r \leftrightarrow s$
   b. $12 < 9 \lor 12 > 9$ d. $m \to n$ f. $\sim c \to \sim d$

2. Write a conjunction for each of the following pairs of sentences and indicate which are true:
   a. $2 > 3$;   b. $8 > 6$;   c. $9 > 5$;   d. $3 = 7$;
      $7 < 10$ $4 < 5$ $1 < 0$ $2 > 5$
   e. Express $9 < n \land n < 15$ in a more concise form.

3. Write a disjunction for each of the following pairs of sentences and indicate which are true:
   a. $4 < 7$;   b. $3 > 10$;   c. $5 < 9$;   d. $6 > 8$;
      $5 > 6$ $2 = 1$ $3 > 0$ $4 < 5$
   e. Express $n < 4 \lor n = 4$ in a more concise form.

4. Write the negation of each of the following:
   a. $9y = 27$ c. $2 + 3 > 7$ e. All students are boys.
   b. $8x \neq 16$ d. $5 - 2 \not< 4$ f. Every line is straight.

5. a. What do we call a sentence like "If $3x = 18$, then $x = 6$"?
   b. What is the antecedent or hypothesis of the above sentence? What is the consequent or conclusion of the above sentence?

6. Write the converse of "If $2a = 10$, then $a = 5$."

Flow Charts and Logic

7. Which of the following indicates that the implication and its converse are equivalent?
   a. $4x = 8 \leftarrow x = 2$
   b. $4x = 8 \rightarrow x = 2$
   c. $4x = 8 \leftrightarrow x = 2$

8. Write the inverse of "If line m is perpendicular to line n, then line n is perpendicular to line m."

9. Write the contrapositive of "If the skies will be clear, the sun will shine."

10. Write the converse, inverse, and contrapositive of each of the following implications:
    a. If b is greater than c, then c is less than b.
    b. If two angles are right angles, then the two angles have the same measure.
    c. If lines a and b are parallel lines, then lines a and b do not intersect.
    d. If $2x + 3 = 11$, then $x = 4$.
    e. If I will go to the ball game, then I will miss the television show.

11. If the original implication or statement is true, is the converse always true? Is the inverse always true? Is the contrapositive always true?

12. If the converse is true, is the inverse always true?

CHAPTER REVIEW

1. Construct a flow chart indicating the steps in changing a common fraction to a percent. (7–1)
2. Express each of the following in a more concise form: (7–2)
   a. $8 < n \wedge n < 20$ b. $n > 5 \vee n = 5$
3. Write the negation of: a. $7x = 42$ b. $6 - 4 \not> 1$ (7–2)
4. State the antecedent and the consequent in the implication: "If $5n = 40$, then $n = 8$." (7–2)
5. Write the converse, inverse, and contrapositive of the implication: "If $3x + 5 = 11$, then $x = 2$." (7–2)

8

CHAPTER 8
Geometry

INVENTORY TEST

The numeral at the end of each problem indicates the section where explanatory material may be found.

1. Read, or write in words, each of the following: (8–1)
   a. $\overleftrightarrow{MN}$ b. $\overline{DE}$ c. $\vec{SR}$

2. Name each of the following: $\binom{8-1}{8-6}$

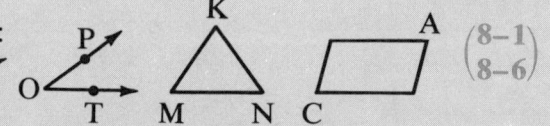

3. Indicate by writing the corresponding numeral which line is: (8–2)
   a. A straight line.
   b. A curved line.
   c. A broken line.

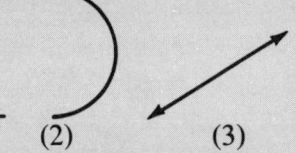

4. Indicate by writing the corresponding numeral which line is in: (8–3)
   a. A horizontal position.
   b. A vertical position.
   c. A slanting position.

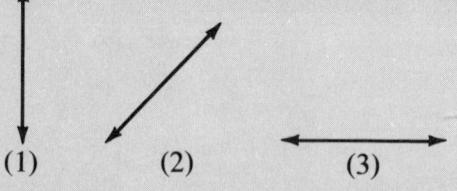

5. Indicate by writing the corresponding numeral which pair of lines are: (8–4)
   a. Perpendicular
   b. Parallel

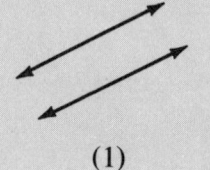

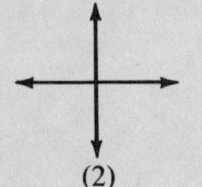

358   CHAPTER 8

6. a. Using a ruler, draw line segments having the dimensions: (8–9)
   (1) 62 mm   (2) 8 cm   (3) $3\frac{5}{8}$ in.   (4) $2\frac{11}{16}$ in.   (5) 108 mm
   b. Measure the following line segments: (8–9)
      (1) With metric ruler: _____
      (2) With customary ruler: _____
7. If the scale is 1 cm = 20 m, what actual distance is represented by 6.7 cm? (8–11)
8. a. With protractor draw an angle of 125°. (8–13)
   b. Draw an acute angle; a right angle; an obtuse angle.
9. a. Construct a triangle with sides measuring 46 mm, 61 mm, and 53 mm. (8–15)
   b. Construct a triangle with two sides measuring $2\frac{3}{4}$ in. and $3\frac{1}{8}$ in. and an included angle of 65°. (8–15)
   c. Construct a triangle with two angles measuring 41° and 78° and an included side measuring 5.7 cm. (8–15)
10. Draw any line. Using a compass, construct a perpendicular:
    a. To your line at a point on your line. (8–16)
    b. To your line from a point not on your line. (8–17)
11. Draw any line segment. Bisect this line segment, using a compass. Check with a ruler. (8–18)
12. Draw an angle of 59°, using a protractor. Then copy this angle, using a compass. Check with a protractor. (8–19)
13. Draw any angle. Bisect this angle, using a compass. Check with a protractor. (8–20)
14. Draw any line. Locate a point not on this line. Through this point construct a line parallel to the line you have drawn. (8–21)
15. Draw a circle with a diameter of 48 mm. (8–35)
16. Draw a regular hexagon, each side measuring 6.3 cm. (8–22)
17. a. What is the complement of an angle measuring 29°? (8–24)
    b. What is the supplement of an angle measuring 103°?
    c. Find the measure of the third angle of a triangle when the other two angles measure 81° and 35°. (8–23)
18. Find the perimeter of a rectangle 45 m long and 19 m wide. (8–32)
19. What is the perimeter of a square whose side measures 37 feet? (8–33)
20. Find the circumference of a circle whose: (8–35)
    a. diameter is 16 centimeters   b. radius is 21 inches

Geometry

Geometry is the study of points, lines, planes, and space, of measurement and construction of geometric figures, and of geometric facts and relationships. The word "geometry" means "earth measure." The Egyptians developed many geometric ideas because the flooding of the Nile River destroyed land boundaries which had to be restored. Euclid, a Greek mathematician, in the third century B.C. collected and organized the geometric principles known at that time and wrote a book on geometry.

POINTS, LINES, PLANES, AND SPACE

A geometric *point* is an exact location in space. It has no size nor can it be seen. Any dot we generally use to indicate it is only a representation of a geometric point.

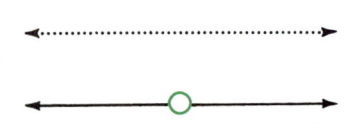

A geometric *line* is a collection of points. The pencil or chalk lines we draw are only representations of geometric lines. A line may be extended indefinitely in both directions because it is endless; it has an infinite number of points but no endpoints. A definite part of a line has length but no width or thickness. We cannot see a geometric line.

A point separates a line into two *half-lines*. Each half-line extends indefinitely in one direction only and does not include the point that separates the line into two half-lines.

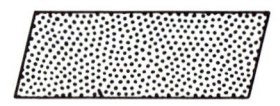

A geometric *plane* or flat surface is a collection of points. It is endless in extent and it has no boundaries but it has length and width which can be measured when the plane is limited. A desk top, wall, floor, and sheet of paper are common representations of a limited plane. We cannot see a geometric plane. It extends beyond any line boundaries we use to represent it.

A line separates a plane into two half-planes.

Space is the infinite collection of all points. Its length, width, and height are endless. A limited space can be measured. A plane separates space into two half-spaces.

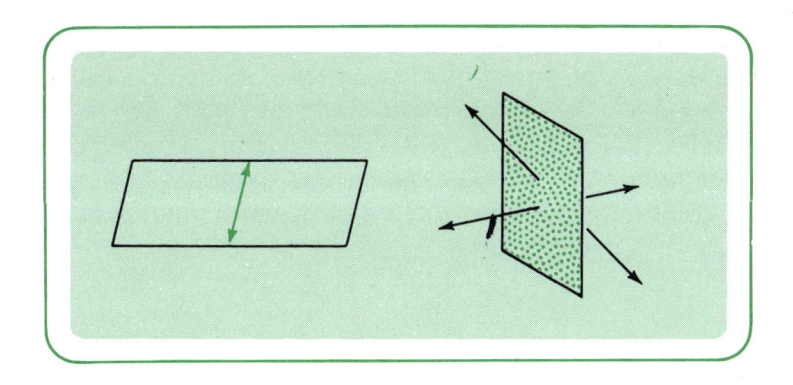

8–1 NAMING POINTS, LINES, AND PLANES

A capital letter is used to label and name a point.

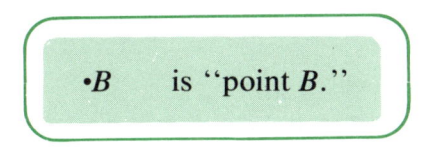

•B   is "point B."

A line is represented as: ⟷ The arrowheads are used to show that a line is endless in both directions.

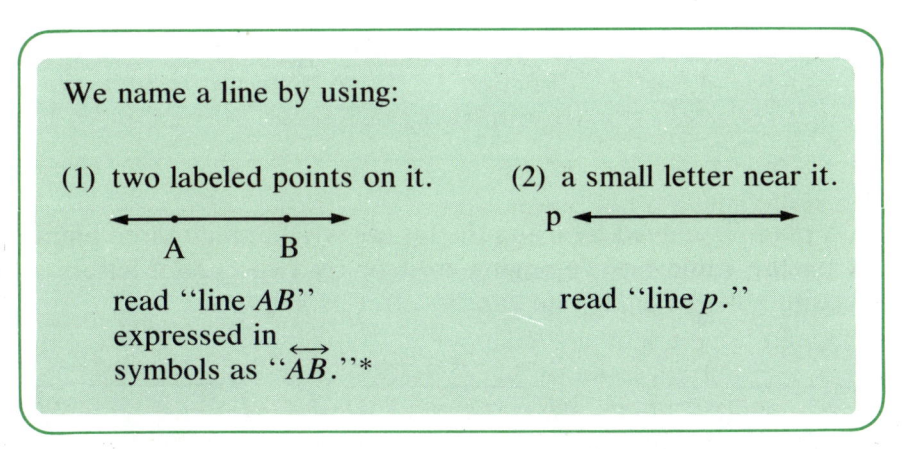

We name a line by using:

(1) two labeled points on it.

⟵•—————•⟶
   A   B

read "line AB"
expressed in symbols as "$\overleftrightarrow{AB}$."*

(2) a small letter near it.

p ⟵————⟶

read "line p."

A definite part of a line including both of its endpoints is called a *line segment* or *segment*. It consists of two endpoints and all the points between. We name a line segment by its endpoints.

———
* Also read "line BA" and expressed as $\overleftrightarrow{BA}$.

Geometry

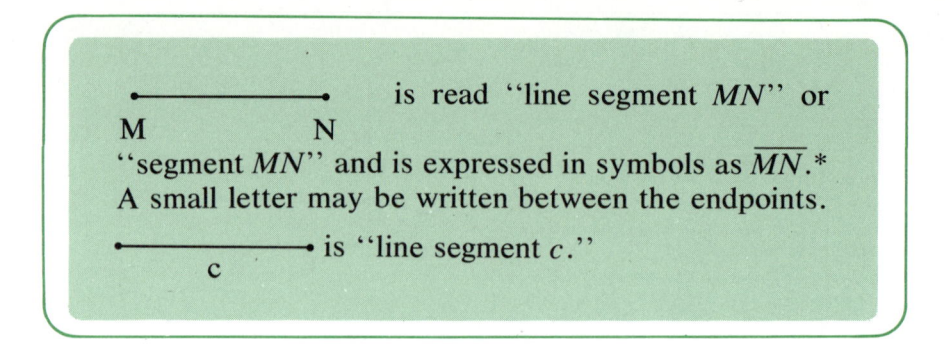

> •────────•   is read "line segment *MN*" or
> M   N
> "segment *MN*" and is expressed in symbols as $\overline{MN}$.*
> A small letter may be written between the endpoints.
> •────────• is "line segment *c*."
>   c

A definite part of a line excluding its endpoints is called an *interval*.

A half-line which includes one endpoint is called a *ray*. This endpoint is the one that separates the line into two half-lines. To name a ray we use the letter first which names the endpoint and then the letter which names one other point on the ray.

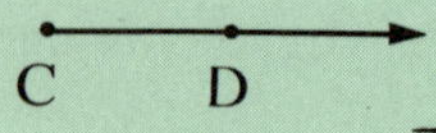

> • •─────▶   is named "ray *CD*," expressed in
> C D
> symbols as $\overrightarrow{CD}$. However, ◀─────• • is named
>   C D
> "ray *DC*," expressed in symbols as $\overrightarrow{DC}$.

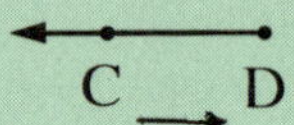

A plane is named by using the letters which name three points not on the same line belonging to it or by two capital letters at opposite corners or by one capital letter as shown.

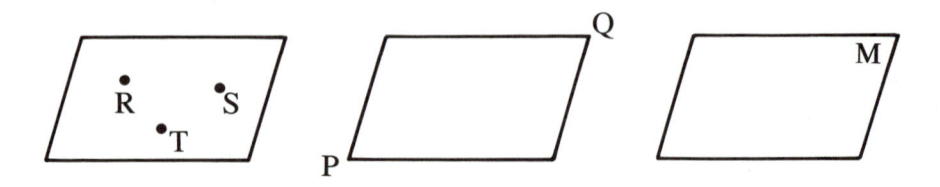

---
* Also read "line segment *NM*" and expressed as $\overline{NM}$.

EXERCISES

1. Name the endpoints of the following line segment:

2. Name the following lines:

3. Name the following line segments:

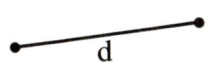

4. Express the name of each of the following symbolically:

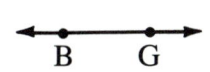

5. Name the following rays and express them symbolically:

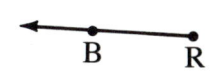

6. Name [M R S T] in three ways. Write the names symbolically.

7. Read, or write in words, each of the following:
   $\overleftrightarrow{BC}$, $\overrightarrow{RT}$, $\overline{NO}$, $\overrightarrow{GF}$, $\overleftrightarrow{KL}$, $\overline{EF}$, $\overleftrightarrow{NY}$, $\overrightarrow{PA}$, $\overleftrightarrow{OC}$, $\overleftrightarrow{DR}$

8. Name the point where $\overleftrightarrow{PQ}$ and $\overleftrightarrow{RS}$ cross.

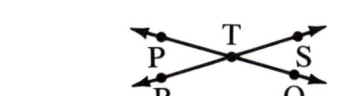

9. $\overline{AC}$ is divided into 2 parts by point B. Name the two segments.

10. Points O and P separate $\overline{MN}$ into three parts. Name the three segments.

11. Point G bisects $\overline{EF}$. Name the two equal segments.

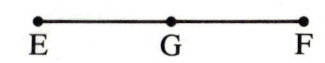

Geometry 363

12. Name each of the following planes:

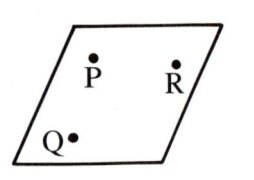

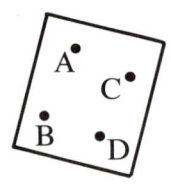

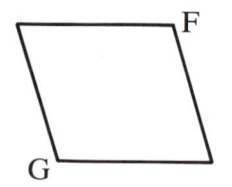

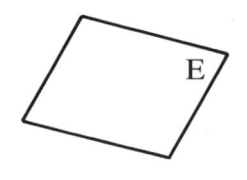

Name the line segments in each of the following figures:

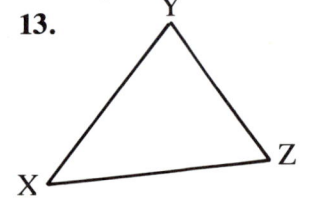

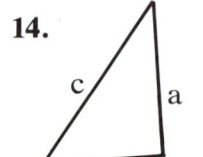

13.

14.

15.

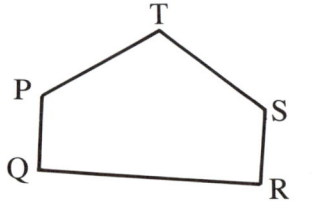

16. $\overline{LM}$ is the sum of what segments?

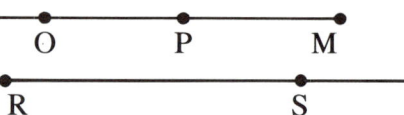

17. $\overline{RS}$ is the difference of what segments?

8–2 KINDS OF LINES

Lines may be *straight, curved,* or *broken.* Usually a straight line is simply called a line. All these lines are sometimes called curves.

Straight Line   Curved Line   Broken Line

EXERCISES

What kind of line is shown by figure *a? b? c? d? e? f? g? h? i? j?*

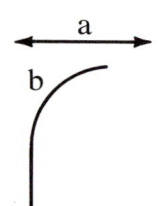

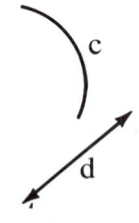

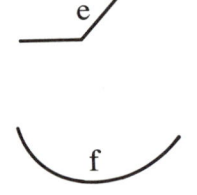

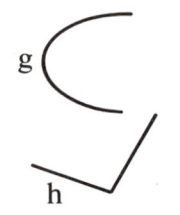

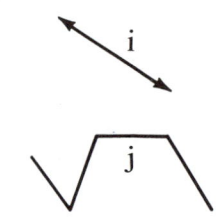

364   CHAPTER 8

8–3 POSITION OF LINES

Lines may be in *vertical, horizontal,* or *slanting* (sometimes called *oblique*) positions.

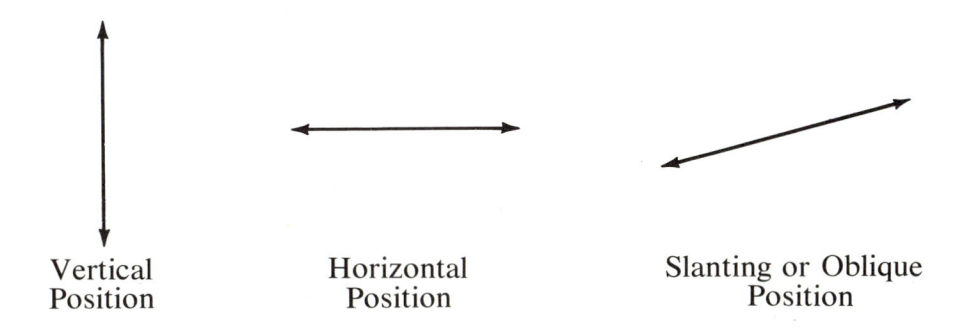

Vertical Position

Horizontal Position

Slanting or Oblique Position

EXERCISES

What position is shown by the line in figure *a? b? c? d? e? f? g? h? i? j?*

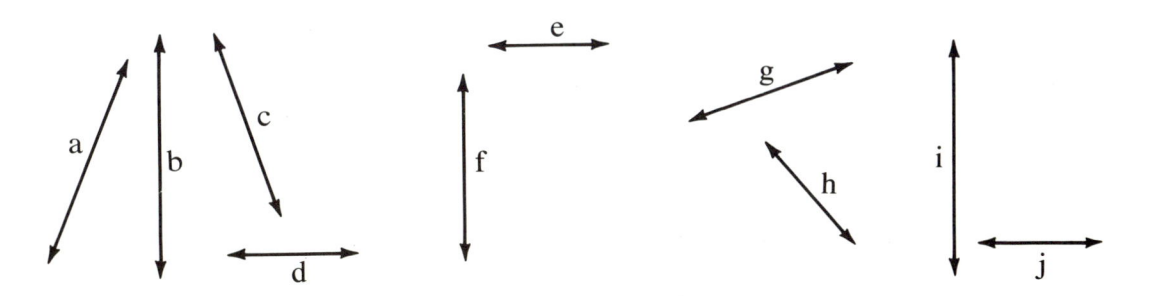

8–4 INTERSECTING, PARALLEL, AND PERPENDICULAR LINES

Lines that meet are *intersecting lines*. Since they have a common point, intersecting lines are sometimes called *concurrent lines*. Two lines in the same plane that do not meet are called *parallel lines*. Two lines not in the same plane that do not meet are called *skew lines*. Two intersecting lines or rays or segments or a line and ray or a line and segment or a ray and segment that form a right angle (see page 371), are said to be *perpendicular* to each other.

Geometry

Intersecting Lines   Parallel Lines   Perpendicular Lines and Rays

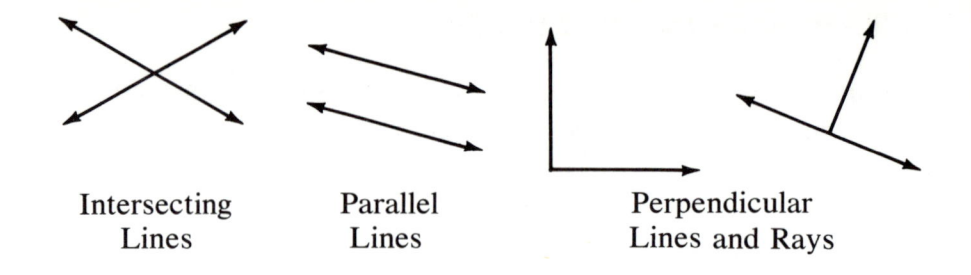

EXERCISES

1. Which of the following are intersecting lines or rays? Which are parallel lines? Which are perpendicular lines or rays?

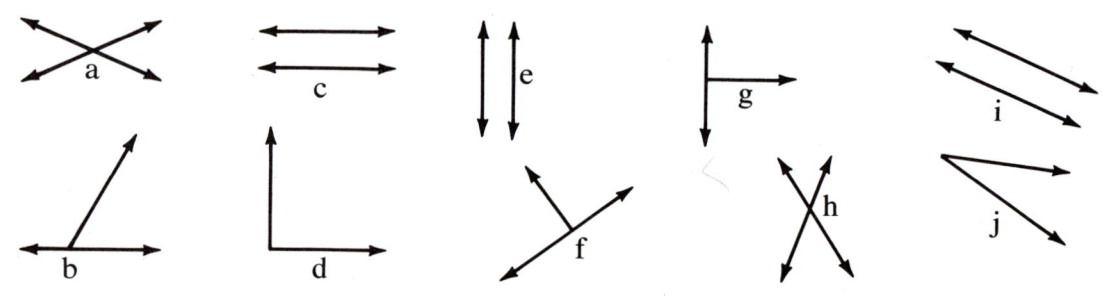

2. Which line is parallel to line *d*?

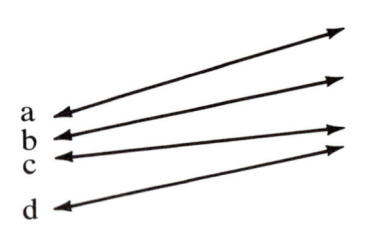

3. a. Which ray is perpendicular to $\overleftrightarrow{CD}$?
   b. Which ray is perpendicular to $\overrightarrow{OE}$?

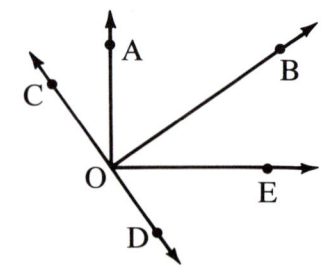

4. At what point do $\overleftrightarrow{GH}$ and $\overleftrightarrow{MN}$ intersect?

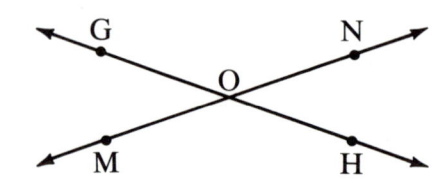

366   CHAPTER 8

5. Name the point where each of the following pairs of segments intersect.

   a. $\overline{EF}$ and $\overline{FG}$ c. $\overline{EG}$ and $\overline{GF}$
   b. $\overline{FE}$ and $\overline{EG}$

6. Name the point where each of the following pairs of segments intersect.

   a. $\overline{BC}$ and $\overline{CD}$ c. $\overline{BA}$ and $\overline{AD}$
   b. $\overline{AD}$ and $\overline{DC}$ d. $\overline{AB}$ and $\overline{BC}$

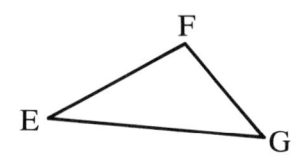

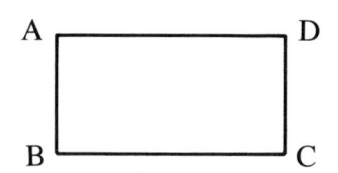

8–5 MORE FACTS ABOUT POINTS, LINES, AND PLANES

(1) Points that lie on the same straight line are called *collinear points*.

   (a) Are points A, B, and C collinear?

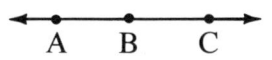

   (b) Are points D, E, and F collinear?
   Are points D and E collinear?
   Are points E and F collinear?
   Are points D and F collinear?

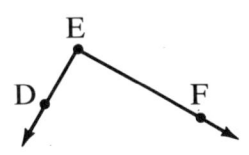

   (c) Are points M, O, and N collinear?
   Are points P, O, and N collinear?
   Are points Q, O, and M collinear?
   Are points P, O, and Q collinear?

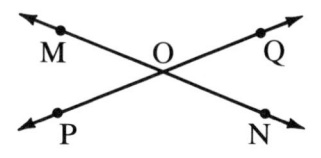

(2) Label a point on your paper as R. Draw a line through point R. Draw a different line through point R. Draw a third line through point R. Can more than one straight line be drawn through a point? How many lines can be drawn through a point? Can we say that *an infinite number of straight lines can be drawn through a point*? Since these lines have a common point, can we call them concurrent lines?

Geometry   367

(3) Label two points on your paper as *C* and *D*. Draw a straight line through points *C* and *D*. Draw another line through points *C* and *D*. How many straight lines can be drawn through two points? Can we say that *two points determine a straight line and that one and only one straight line can pass through any two points*? Are any two points collinear?

(4) Draw a pair of intersecting lines. Draw another pair of intersecting lines. At how many points can two straight lines intersect? How many points in common does a pair of intersecting lines have? Can we say that *two straight lines can intersect in only one point*?

(5) Which of the three kinds of lines is the shortest path between points *A* and *B*? Can we say that *the shortest path between two points is along a straight line*?

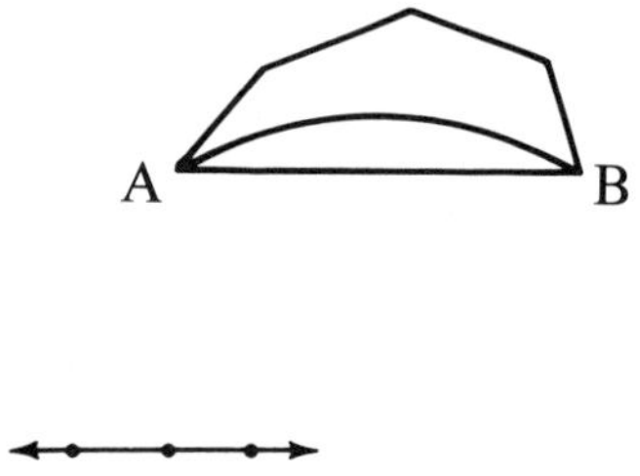

(6) Select 3 points on a straight line. Can more than one plane pass through these 3 points? Select 3 points not on the line. Can more than one plane pass through these 3 points? Can we say that *through 3 points not on the same straight line one and only one plane can pass*? Points in the same plane are called *coplanar points*.

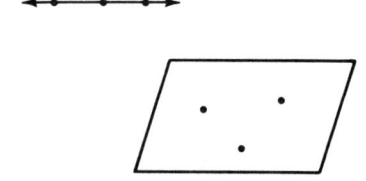

(7) How many points does a geometric plane have? How many lines can be drawn through these points? How many lines does a geometric plane contain? Can we say that *a geometric plane contains an infinite number of points and lines*? Lines in the same plane are sometimes called *coplanar lines*. Can we say that *two intersecting lines determine one and only one plane*?

(8) Draw two different planes that intersect as shown. What geometric figure is their intersection? Can we say that *when two different planes intersect, their intersection is a straight line*? Planes that do not intersect are called *parallel planes*. They have no point in common.

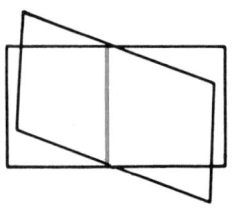

368   CHAPTER 8

(9) Draw a plane and a line that is not in this plane but which intersects the plane. Can we say that their intersection is one and only one point? A plane and a line not in this plane are parallel when they have no point in common.

(10) A *line perpendicular to a plane* is a line that is perpendicular to every line in the plane that passes through its foot (the point of intersection). Can we say that *a line that is perpendicular to each of two intersecting lines at their point of intersection is perpendicular to the plane in which these lines lie?*

ANGLES

An *angle* is the figure formed by two different rays having the same endpoint. It is the union of two rays. This common endpoint is called the *vertex* of the angle and the two rays are called the *sides* of the angle. An angle may be considered as the rotation of a ray about a fixed endpoint, the angle being formed as the ray turns from one position to another. An angle is sometimes used to show direction. The symbol ∠ designates the word "angle."

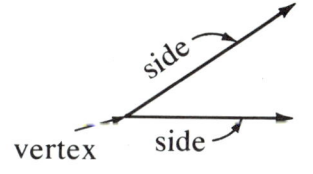

8-6 NAMING ANGLES

Angles are identified or named in the following ways:

(1) By reading the capital letter at the vertex:

(2) By reading the inside letter or numeral:

(3) By reading the three letters associated with the vertex and one point on each of the sides. The middle letter always indicates the vertex.

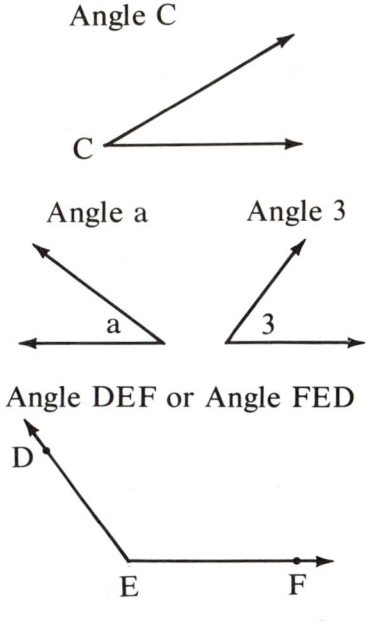

Geometry

369

EXERCISES

1. Name the sides and vertex of the following angle:

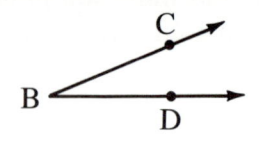

2. What is the symbol used to denote the word "angle"?

3. Name the following angle in three ways:

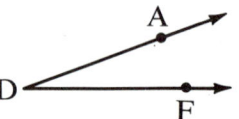

4. Name the following angle in two ways:

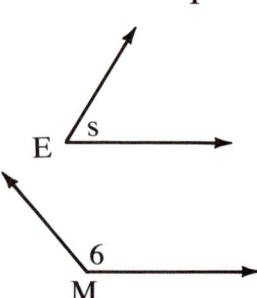

5. Name the following angle in two ways:

6. Name each of the following angles:

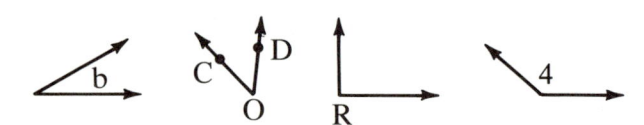

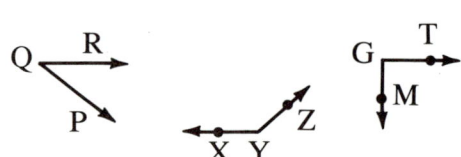

7. Name ∠1, ∠2, and ∠3 in the following figure, using three letters.

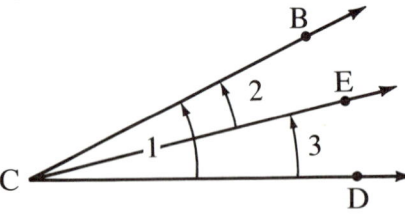

8. Name each angle of the following triangle in four ways.

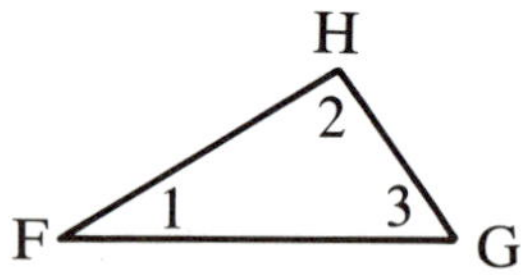

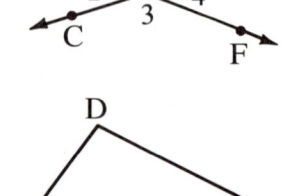

9. Name the four angles formed when $\overleftrightarrow{CD}$ intersects $\overrightarrow{EF}$ at point G, using 3 letters for each angle.

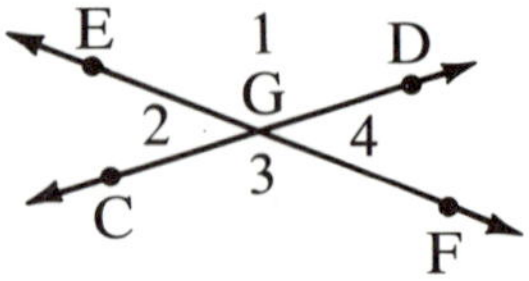

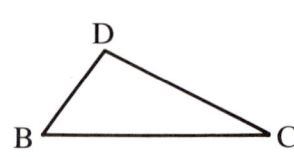

10. In the following triangle, find the angle opposite to:
    a. Side DC b. Side BD c. Side BC

8–7 KINDS OF ANGLES

The *degree*, indicated by the symbol (°) is the unit of measure of angles and arcs. A degree is $\frac{1}{360}$ part of the entire angular measure about a point in a plane. If a circle is divided into 360 equal parts and lines are drawn from the center to these points of division, 360 equal central angles are formed each measuring 1 degree. Each of the corresponding 360 equal arcs also measures 1 degree.

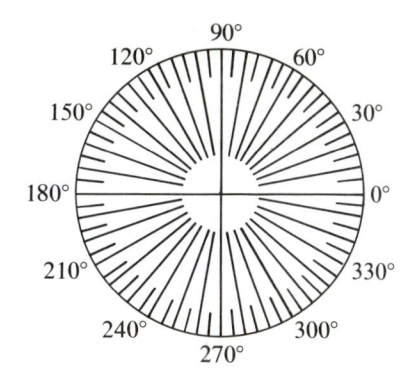

The degree (°) is divided into 60 equal parts called minutes (′). The minute (′) is divided into 60 equal parts called seconds (″).

> To use these symbols:
>
> 360 degrees is written 360°,
> 60 minutes is written 60′,
> 60 seconds is written 60″.

When a ray turns from one position to another about its fixed endpoint, one complete rotation is equal to 360°.

A *right angle* is one-fourth of a complete rotation; it is an angle whose measure is 90°.

An *acute angle* is an angle whose measure is greater than 0° but less than 90°.

An *obtuse angle* is an angle whose measure is greater than 90° but less than 180°.

A *straight angle* is one half of a complete rotation; it is an angle whose measure is 180°. The two rays that form a straight angle extend in opposite directions along a straight line that passes through the vertex.

A *reflex angle* is an angle whose measure is greater than 180° but less than 360°.

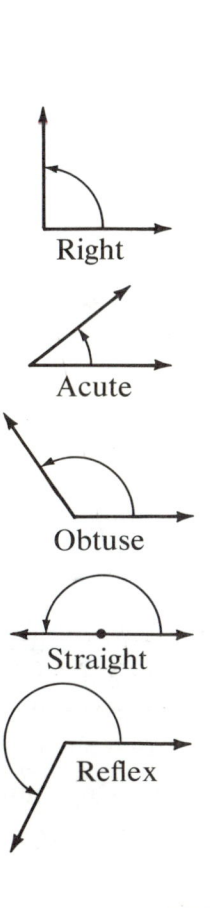

Geometry 371

EXERCISES

1. What is the unit called that is used to measure angles? What symbol designates it?

2. The measure of an acute angle is greater than ?° and less than ?°.

3. One complete rotation measures ?°.

4. A right angle measures ?° and is ? of a rotation.

5. The measure of an obtuse angle is greater than ?° and less than ?°.

6. A straight angle is ? of a rotation and measures ?°.

7. Which of the following are measures of an acute angle:
   90°? 26°? 173°? 89°? 212°?

8. Which of the following are measures of an obtuse angle:
   49°? 101°? 305°? 93°? 183°?

9. Indicate by writing corresponding letters which of the following angles are:

   a. acute angles   **b.** right angles   **c.** obtuse angles   **d.** straight angles

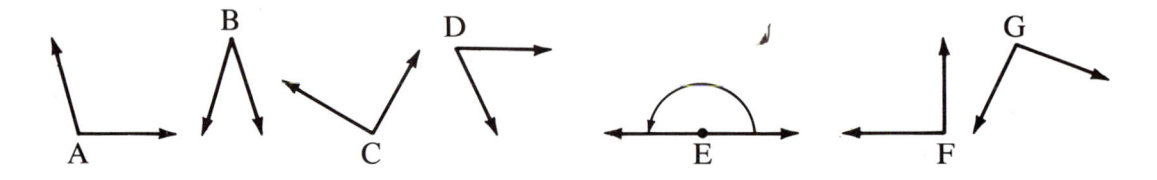

10. What kind of angle is formed by the hands of a clock at 2 o'clock? Describe only the smaller of the two angles formed.

11. Draw five clock faces, then use arrows to indicate the position of the hands at 3 o'clock; at 10 o'clock; at 5 o'clock; at 6 o'clock; at 4 o'clock. What kind of angle is formed by the hands in each case? Describe only the smaller of the two angles formed.

12. **a.** Draw any obtuse angle.

    b. Draw any acute angle.

8–8 GEOMETRIC FIGURES

Simple Closed Plane Figures or Simple Closed Curves

Geometric figures consist of collections of points. Plane geometric figures are figures with all of their points in the same plane. A simple closed plane figure (or curve) begins at a point and returns to this point without crossing itself. It divides a plane into three collections of points, those in the interior, those in the exterior, and those on the figure.

A *polygon* is a simple closed plane figure made up of line segments (called sides). It is the union of three or more line segments. Each pair of intersecting sides meets in a point called a *vertex*. In a polygon the union of two adjacent sides (line segments) forms an angle, since these line segments are parts of rays. A polygon with all sides of equal length and all angles of equal measure is called a *regular polygon*. It should be noted that the collection of points contained in the sides is the figure and not the region enclosed by the figure. A polygon is named by reading the letters at the vertices. A line segment connecting two nonadjacent vertices of a polygon is called a *diagonal*. Some common polygons are: *triangle,* 3 sides; *quadrilateral,* 4 sides; *pentagon,* 5 sides; *hexagon,* 6 sides; *octagon,* 8 sides; *decagon,* 10 sides; and *dodecagon,* 12 sides.

Triangles

When all three sides of a triangle are equal in length, the triangle is called an *equilateral triangle*; when two sides are equal, an *isosceles triangle*; when no sides are equal, a *scalene triangle*. When all three angles of a triangle are equal in size, the triangle is called an *equiangular triangle*; a triangle with a right angle, a *right triangle*; with an obtuse angle, an *obtuse triangle*; and with three acute angles, an *acute triangle*. The *altitude* of a triangle is the perpendicular segment from any vertex of a triangle to the opposite side or extension of that side. The *median* of a triangle is the line segment connecting any vertex of a triangle to the midpoint of the opposite side.

In a right triangle the side opposite the right angle is called the *hypotenuse*. The other two sides or legs are the *altitude* and *base* of the triangle. The *base* is generally the side on which the triangle rests. In an isosceles triangle the angle formed by the two equal sides is called the *vertex angle*. It is opposite to the base. The angles opposite the equal sides are called the base angles.

Congruent triangles are triangles which have exactly the same shape and the same size. See page 409.

Similar triangles are triangles which have the same shape but differ in size. See page 410.

Quadrilaterals

The following properties describe special quadrilaterals:

The *rectangle* has two pairs of opposite sides which are equal and parallel and four angles which are right angles.

The *square* has four equal sides with the opposite sides parallel and four angles which are right angles.

The *parallelogram* has two pairs of opposite sides which are parallel and equal.

The *trapezoid* has only one pair of opposite sides that are parallel.

The square is a special rectangle, and the rectangle and square are special parallelograms.

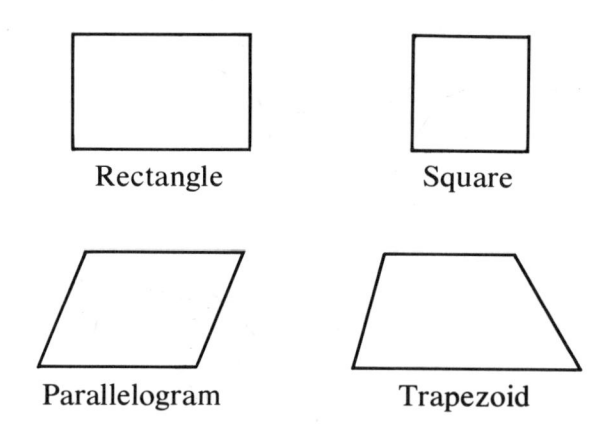

Rectangle   Square

Parallelogram   Trapezoid

Circles

A *circle* is the collection of points in a plane which are equidistant from a fixed point in the plane called the *center*. It is a simple closed curve. The *radius* of a circle is a line segment which has one endpoint at the center of the circle and the other endpoint on the circle. The *diameter* of a circle is a line segment which has both of its endpoints on the circle but passes through the center. A *chord* of a circle is a line segment which has both of its endpoints on the circle. An *arc* is a part of the circle. If the endpoints of an arc are the endpoints of a diameter, the arc is a *semi-circle*. An angle whose vertex is at the center of a circle is called a *central angle*. The *circumference* is the distance around the circle. A *tangent to a circle* is a line that has one and only one point in common with the circle.

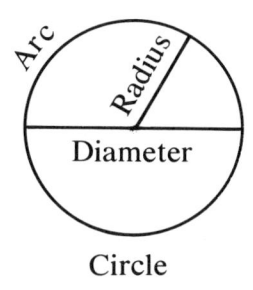

Circle

Concentric circles are circles in the same plane which have the same center but different radii.

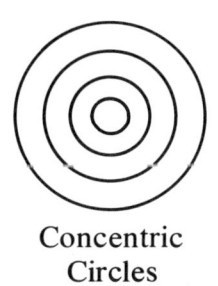

Concentric Circles

Solid or Space Figures

A closed geometric figure consisting of four or more polygons and their interiors, all in different planes, is called a *polyhedron*. The polygons and their interiors are called *faces*. These faces intersect in line segments called *edges*. These edges intersect in points called *vertices*.

Common polyhedra are the rectangular solid (right rectangular prism), the cube, and the pyramid.

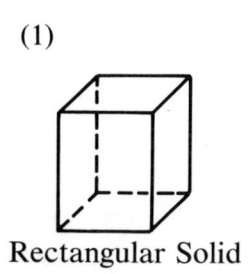

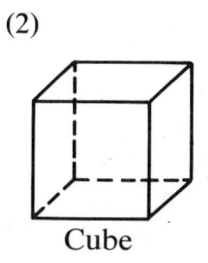

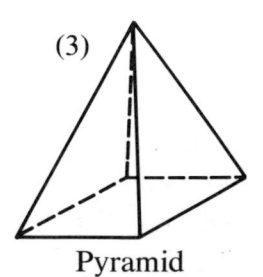

(1) Rectangular Solid

(2) Cube

(3) Pyramid

Geometry

(1) The *rectangular solid* has six rectangular faces.
(2) The *cube* has six squares for its faces. All the edges are equal in length.
(3) The *pyramid* has any polygon as its base and triangular faces that meet in a common vertex.

Other common solid geometric figures are the cylinder, sphere, and cone.

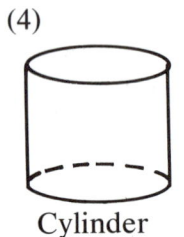

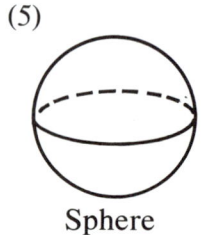

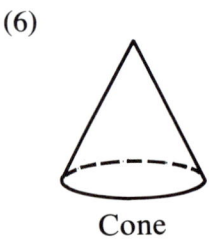

(4) Cylinder   (5) Sphere   (6) Cone

(4) The *cylinder* has two equal and parallel circles as bases and a lateral curved surface.
(5) The *sphere* has a curved surface on which every point is the same distance from the center within.
(6) The *cone* has a circle for the base and a curved surface that comes to a point called the vertex.

Euler Formula

The Euler formula expresses the relationship of the faces, edges and vertices of a polyhedron. The formula $F + V - E = 2$ tells us that "the number of faces plus the number of vertices minus the number of edges is equal to two."

EXERCISES

1. What is the name of each of the following figures?

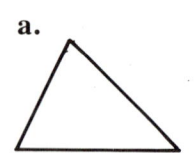

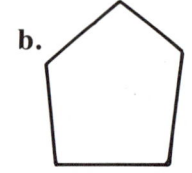

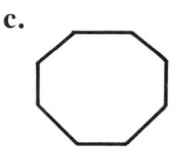

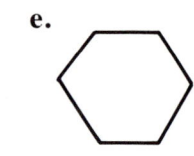

a.   b.   c.   d.   e.

f. g. 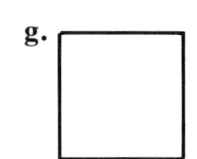 h. 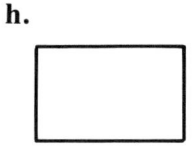 i. 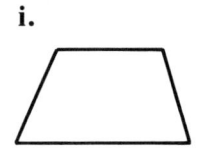 j.

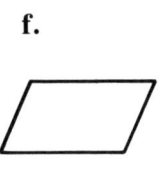

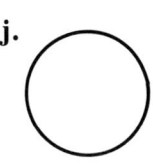

k. 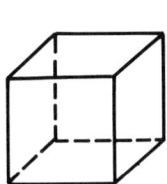 l.  m. 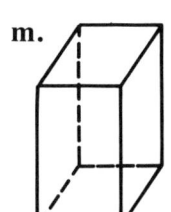 n.  o. 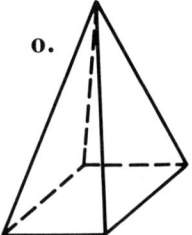 p.

2. What kind of triangle is each of the following?

a. 12 mm, 12 mm, 17 mm
b. 5", 8", 9"
c. 4 cm, 4 cm, 4 cm
d. 90°
e. 60°, 60°, 60°
f. 110°

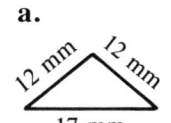

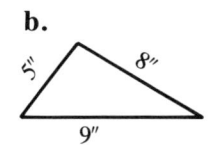

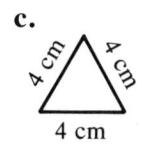

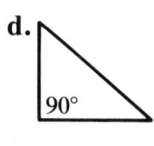

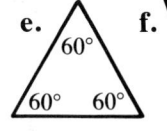

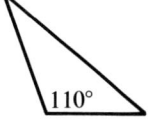

3. Which of the following figures are parallelograms?

a. b.  c. 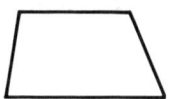 d. 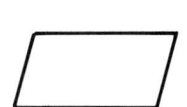 e.

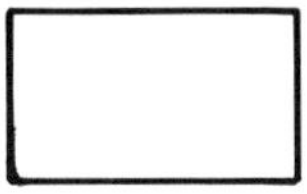

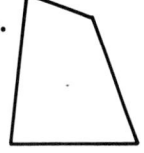

4. How many diagonals can be drawn from any one vertex in each of the following figures?

a. b. 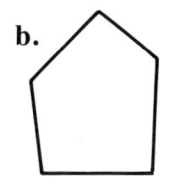 c. 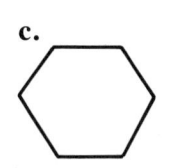 d. 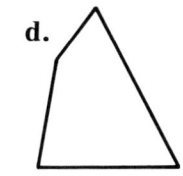 e.

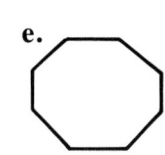

5. a. How many faces (F) does a rectangular solid have? How many vertices (V)? How many edges (E)?
   Does $F + V - E = 2$?

   b. How many faces (F) does a pyramid with a square base have? How many vertices (V)? How many edges (E)? Does $F + V - E = 2$?

Geometry

MEASUREMENT AND CONSTRUCTIONS

8–9 DRAWING AND MEASURING LINE SEGMENTS

A *straightedge* is used to draw line segments. The *ruler,* which is a straightedge with calibrated measurements, is used both to draw and to measure line segments of varying lengths.

Sometimes an instrument called a *compass* is used along with the ruler to measure line segments or to draw line segments of specified lengths. Since the compass is used to draw circles and arcs of circles, the distance between the metal point and the pencil point of the compass corresponds to the radius of the circle or the radius of the arc of the circle that may be drawn.

Generally we measure the length of a line segment by using only the ruler. However when we use both a compass and a ruler to do this, we open the compass so that the metal point and the pencil point fit exactly on the endpoints of the line segment. We then transfer the compass to a ruler to determine the measurement between the metal point and the pencil point.

We reverse this operation to draw a line segment of a given length by using both a compass and a ruler. We first set the compass so that the distance between the metal point and pencil point corresponds to the given measurement on a ruler. We then apply this compass setting to a light pencil working line, using the metal point as one endpoint and the pencil point as the second endpoint.

EXERCISES

1. Using a customary ruler, draw line segments having the following dimensions:

    a. $2\frac{1}{2}$ in.   **b.** $3\frac{3}{4}$ in.   **c.** $2\frac{5}{8}$ in.   **d.** $3\frac{11}{16}$ in.

2. Write your estimate, then measure the length of each of the following line segments, using the ruler only for segments in *a, b,* and *c,* and a compass and ruler for segments in *d* and *e.*

    a. _____
    b. _____
    c. _____
    d. _____
    e. _____

3. Draw a line segment 6 inches long. Using a compass, lay off in succession on this segment lengths of $1\frac{7}{8}$ inches, $2\frac{3}{16}$ inches, and $1\frac{3}{4}$ inches. How long is the remaining segment?

4. Using a metric ruler, draw line segments having the following dimensions:

   a. 12 cm   b. 105 mm   c. 9 cm   d. 86 mm

5. The symbol $m\overline{DC}$ is read "the measure of line segment DC" and represents the length of the segment.

   a. Find $m\overline{DC}$; $m\overline{EF}$; $m\overline{AG}$; $m\overline{RS}$.

   b. Does $m\overline{DC} = m\overline{AG}$?

   c. Does $m\overline{RS} = m\overline{EF}$?

6. Draw any line segment BC. Label a point not on $\overline{BC}$ as M. Draw a line segment MN equal in length to $\overline{BC}$.

7. a. Draw a line segment that is twice as long as $\overline{AG}$ shown in problem 5.
   b. Draw a line segment that is three times as long as $\overline{EF}$ shown in problem 5.

8–10 COPYING A LINE SEGMENT

To copy a line segment means to draw a line segment equal in length to the given line segment.

(1) *Using a ruler,* we first measure the given line segment, then we draw another line segment of the same length.

(2) *Using a compass,* we first draw a pencil working line. Then we select a point on this line where the copy of the given line segment is to begin, indicating it by a dot. We open the compass, placing the metal point on one endpoint of the given line segment and the pencil point on the other endpoint. We transfer this fixed compass

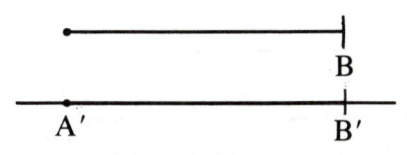

Geometry

setting to the working line, placing the metal point on the marked dot. This point forms one endpoint of the copy of the given line segment. We draw an arc, cutting the working line. This intersection forms the second endpoint of the copy of the given line segment. We then draw the segment of the working line between these endpoints heavier to indicate the required line segment.

EXERCISES

1. Make a copy of each of the following line segments, using a ruler:

2. Make a copy of each of the following line segments, using a compass and a straightedge. Check with ruler.

3. Draw any line segment and label the endpoints E and F. Label a point not on $\overline{EF}$ as G. Using a compass and a straightedge, draw a line segment GH equal in length to $\overline{EF}$.

8–11 SCALE

The *scale* shows the relationship between the dimensions of a drawing, plan, or map and the actual dimensions.

A scale like 1 inch = 4 feet means that 1 scale inch represents 4 actual feet. This scale may also be written as $\frac{1}{4}$ inch = 1 foot or as the representative fraction $\frac{1}{48}$ or as the ratio 1:48. These all indicate that each scale inch represents 48 actual inches or 4 feet.

The scale ratio 1:3,000,000 indicates that 1 scale millimeter represents 3,000,000 actual millimeters which equal 3,000 meters or 3 kilometers. Therefore, the ratio 1:3,000,000 could also be written as 1 mm = 3 km.

On maps we usually find a scale of miles like that shown at the right. This scale indicates that each scale inch represents 80 actual miles and is equivalent to the scale 1 inch = 80 miles.

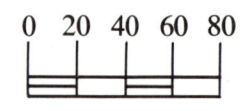

(1) *To find the actual distance when the scale and scale distance are known,* we multiply the scale distance by the scale value of a unit (mm in the following):

> If the scale is 1 mm = 20 km, what actual distance is represented by 19 mm?
>
> Scale: 1 mm = 20 km   Scale distance 19 mm = ? km
>
> $$19 \times 20 = 380 \text{ km}$$
>
> *Answer,* 380 km

(2) *To find the scale distance when the scale and the actual distance are known,* we divide the actual distance by the scale value of a unit (inch in the following):

> If the scale is 1 inch = 8 feet, how many inches represent 30 feet?
>
> Scale: 1 inch = 8 feet   ? scale distance = 30 feet
>
> $$30 \div 8 = 3\frac{3}{4}$$
>
> *Answer,* $3\frac{3}{4}$ inches

(3) *To find the scale when the actual and scale distances are known,* we divide the actual distance by the scale distance.

> Find the scale when the actual distance is 24 kilometers and the scale distance is 6 millimeters.
>
> $$24 \div 6 = 4$$
>
> Each scale millimeter represents 4 actual kilometers or 4,000,000 actual millimeters.
>
> Thus, the scale is 1 mm = 4 km which may be written as the ratio 1:4,000,000.
>
> *Answer,* 1 mm = 4 km or 1:4,000,000

Geometry

EXERCISES

1. If the scale is 1 mm = 25 km, what actual distance is represented by:

   a. 6 mm?   b. 19 mm?   c. 78 mm?   d. 53 mm?   e. 114 mm?

2. If the scale is 1 cm = 10 m, what actual distance is represented by:

   a. 9 cm?   b. 25 cm?   c. 6.5 cm?   d. 0.4 cm?   e. 12.3 cm?

3. If the scale is 1 inch = 64 miles, what actual distance is represented by:

   a. 8 in.?   b. $3\frac{1}{2}$ in.?   c. $9\frac{3}{4}$ in.?   d. $1\frac{7}{8}$ in.?   e. $5\frac{13}{16}$ in.?

4. If the scale is $\frac{1}{8}$ inch = 1 foot, what actual distance is represented by:

   a. 2 in.?   b. $9\frac{1}{2}$ in.?   c. $8\frac{3}{8}$ in.?   d. $6\frac{1}{4}$ in.?   e. $7\frac{11}{16}$ in.?

5. If the scale is $\frac{1}{48}$, what actual distance is represented by:

   a. 1 in.?   b. $5\frac{1}{2}$ in.?   c. $10\frac{1}{4}$ in.?   d. $4\frac{5}{8}$ in.?   e. $8\frac{7}{16}$ in.?

6. The scale ratio 1:500 means:
   1 scale cm = _____ actual cm = _____ actual m

7. The scale ratio 1:2,000,000 means:
   1 scale mm = _____ actual mm = _____ m = _____ km

8. The scale ratio 1:1,250,000 means:
   1 scale mm = _____ actual mm = _____ m = _____ km

9. The scale 1 mm = 15 mm means:
   1 scale mm = _____ actual mm = scale ratio 1:_____

10. The scale 1 cm = 40 km means:
    1 scale cm = _____ actual cm = scale ratio 1:_____

11. Write the scale ratio for each of the following scales:

    a. 1 mm = 3 km   b. 1 cm = 25 km   c. 1 cm = 75 m   d. 1 mm = 50 m

12. Write in another way the scale for each of the following scale ratios:

    a. 1:1,000,000   b. 1:600   c. 1:300,000   d. 1:2,500,000

13. If the scale is 1:200, what actual distance is represented by:

   a. 6 m? b. 37 cm? c. 8.9 cm? d. 58 mm? e. 142 mm?

14. If the scale is 1:1,500,000, what actual distance is represented by:

   a. 15 mm? b. 9 mm? c. 51 mm? d. 2 cm? e. 18.4 cm?

15. If the scale is 1 inch = 32 miles, how many inches represent:

   a. 160 mi.? b. 240 mi.? c. 216 mi.? d. 300 mi.? e. 374 mi.?

16. If the scale is $\frac{1}{16}$ inch = 1 foot, how many inches represent:

   a. 96 ft.? b. 84 ft.? c. 152 ft.? d. 14 ft.? e. 333 ft.?

17. If the scale is $\frac{1}{96}$, how many inches represent:

   a. 24 ft.? b. 62 ft.? c. 7 ft.? d. 588 ft.? e. $121\frac{1}{2}$ ft.?

18. If the scale is 1 cm = 5 m, how many centimeters represent:

   a. 20 m? b. 4 m? c. 115 m? d. 81.5 m? e. 31 m?

19. If the scale is 1:2,500,000, how many millimeters represent:

   a. 5 km? b. 200 km? c. 32.5 km? d. 160 km? e. 113 km?

20. Find the scale when:

   a. The scale length of 8 mm represents an actual distance of 560 km.

   b. The scale length of $4\frac{3}{4}$ in. represents an actual distance of 76 ft.

   c. The actual distance of 265 km is represented by the scale length of 5.3 cm.

   d. The actual distance of 345 mi. is represented by the scale length of $4\frac{5}{16}$ in.

   e. The scale length of 12.6 cm represents an actual distance of 504 m.

   f. The scale length of $5\frac{3}{8}$ in. represents an actual distance of 172 ft.

21. Using the scale 1 cm = 20 m, draw line segments representing:

   a. 60 m b. 110 m c. 18 m d. 32 m e. 94 m

Geometry

22. Using the scale $\frac{1}{2}$ inch = 1 foot, draw line segments representing:

   a. 8 ft.   b. 14 ft.   c. 11 ft.   d. $18\frac{1}{2}$ ft.   e. $3\frac{1}{2}$ ft.

23. Using the scale 1:4,000,000, draw line segments representing:

   a. 12 km   b. 100 km   c. 30 km   d. 86 km   e. 212 km

24. What are the actual dimensions of a gymnasium floor if plans drawn to the scale of 1:500 show dimensions of 8 cm by 5 cm?

25. Draw a floor plan of a family room 26 feet long and 15 feet wide, using the scale 1 inch = 4 feet.

26. Using the scale of *miles* find the distance represented by each of the following line segments:

   a. _____   c. _____
   b. _____   d. _____
   e. _____

27. Using the scale of *kilometers*, find the distance represented by each of the following line segments:

   a. _____   c. _____
   b. _____   d. _____
   e. _____

28. Find the scales used to draw the following line segments representing the given distances:

   a. _____52 ft._____   c. _____108 km_____
      Scale: 1 inch = ?   Scale: 1:?
   b. _____110 mi._____   d. _____35 m_____
      Scale: 1 inch = ?   Scale: 1:?

29. If the distance from C to D is 70 miles, what is the distance from E to F?

   C_____D   E_____F

30. If the distance from M to N is 35 km, what is the distance from P to Q?

   M_____N   P_____Q

8–12 MEASURING ANGLES

Measuring an angle means to determine how many units of angular measure are contained in it. The protractor is an instrument used to measure an angle.

The size of an angle does not depend on the length of its sides. The symbol "$m \angle ABC$" is read "the measure of angle ABC."

To measure an angle, we place the straight edge of the protractor on one side of the angle with its center mark at the vertex of the angle. We read the number of degrees at the point where the other side of the angle cuts the protractor, using the scale which has its zero on one side of the angle.

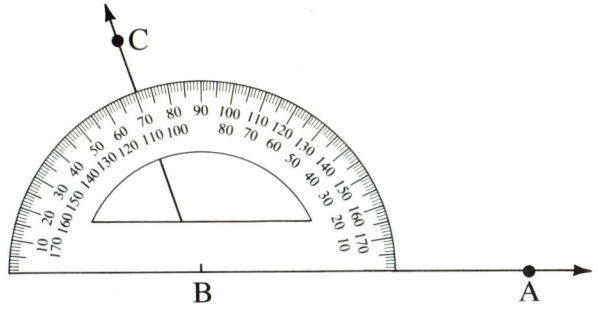

This angle measures 110°, written either as $m \angle ABC = 110°$ or briefly as $\angle ABC = 110°$.

EXERCISES

1. Estimate the size of each of the following angles. Then measure each angle with a protractor.

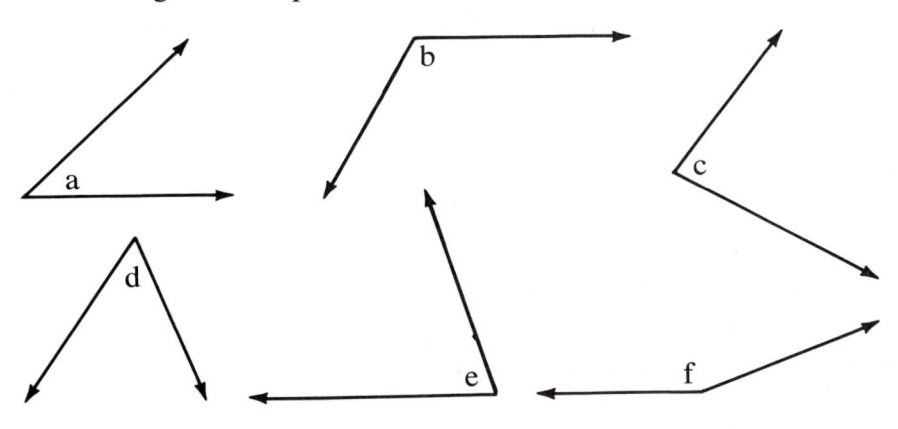

Geometry

385

2. Write the amount of error and percent of error of each estimate you made in Exercise 1.
3. a. Which of the following angles is smaller?
   b. The size of an angle does not depend upon the length of the ?. 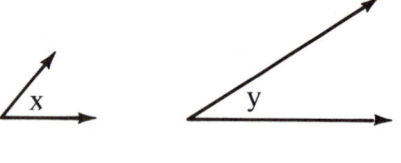
4. How many degrees are in the angle formed by the hands of a clock:
   a. at 4 o'clock?   c. at 3 o'clock?
   b. at 2 o'clock?   d. at 6 o'clock? 
5. Through how many degrees does the minute hand of a clock turn in:
   a. 15 minutes?   c. 10 minutes?   e. 40 minutes?
   b. 30 minutes?   d. 25 minutes?   f. 1 hour?

8–13 DRAWING ANGLES

To draw an angle with a given measure, we draw a ray to represent one side of the angle. Sometimes this ray (or line segment which is part of a ray) is already drawn. We place the protractor so that its straight edge falls on this ray, and its center mark is on the endpoint which becomes the vertex of the angle. This vertex may also be any point on the line.

Draw an angle of 150° using *E* as the vertex.

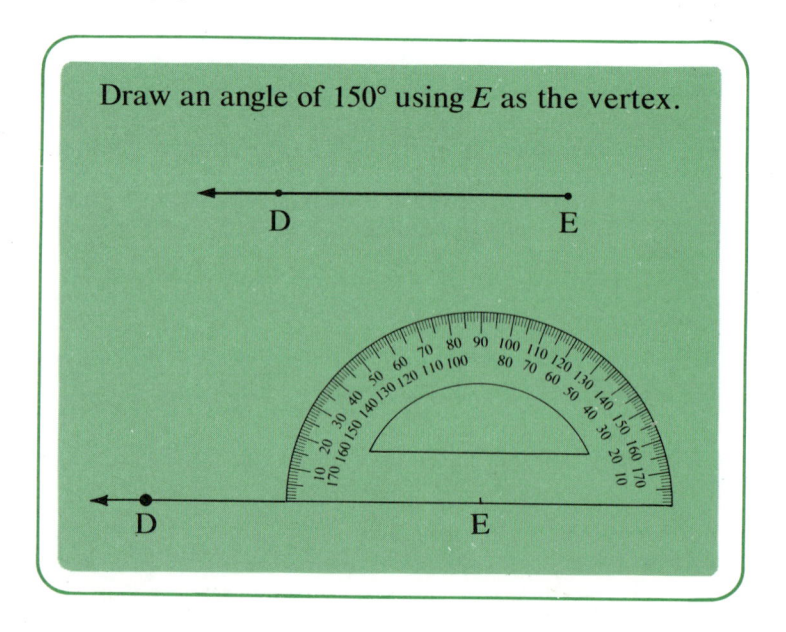

Counting on the scale which has its zero on the ray, we locate the required number of degrees and indicate its position by a dot. We remove the protractor, then we draw a ray from the vertex through this dot.

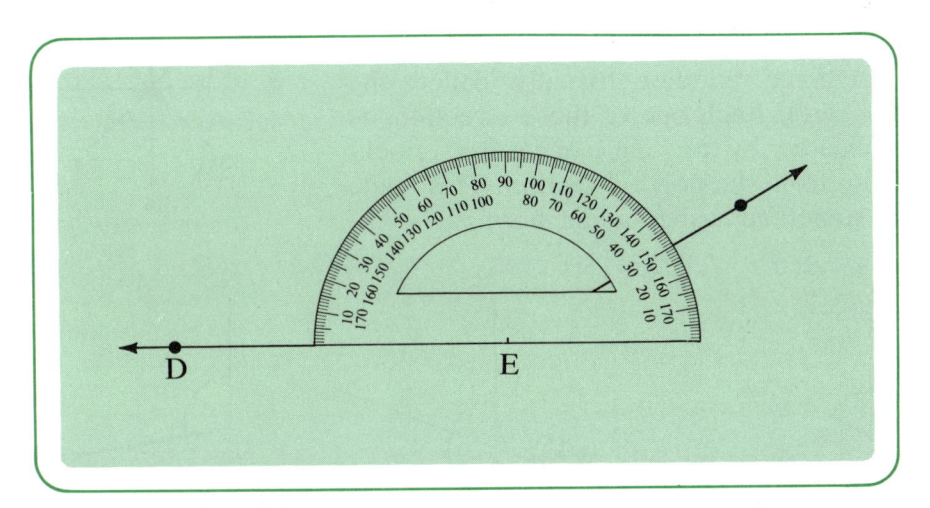

EXERCISES

1. For each of the following first draw a ray with the endpoint either on the left or on the right as required, then with protractor draw the angle of the given measure.

    a. Left endpoint as the vertex:
    40°,   65°,   130°,   155°,   97°,   24°,   109°,   200°,   270°,   345°

    b. Right endpoint as the vertex:
    60°,   75°,   100°,   90°,   84°,   225°,   300°,   340°,   136°,   16°

2. With protractor measure each of the following angles, then draw an angle equal to it.

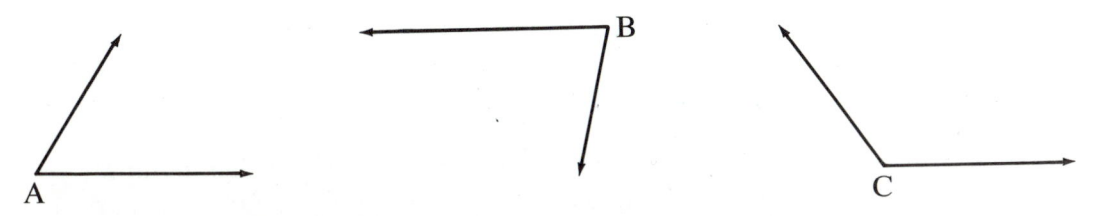

3. With protractor draw: **a.** a right angle   **b.** a straight angle

Geometry

387

8-14 ANGLES IN NAVIGATION

An angle is often used to show the direction (usually called the *course*) in which an airplane flies or the direction from which the wind blows or the direction of one object from another (usually called the *bearing*). Each one of these directions is indicated by an angle measured clockwise from the north direction. This angle is sometimes called an *azimuth*.

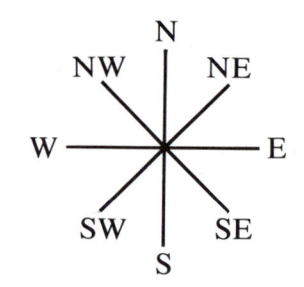

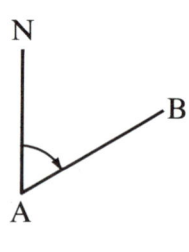

Course: 60°

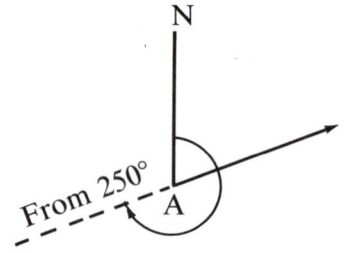

Wind from 250°

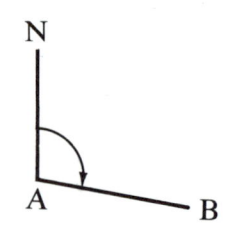

Bearing of B 100° from A

Winds decrease or increase the speed developed by an airplane in flight through the air. *Air speed* is the speed of the airplane in still air. *Ground speed* is the actual speed of the airplane measured by land markings. *Heading* is the direction (expressed by an angle) in which the airplane points. A *vector* is an arrow which represents a speed or force and also indicates its direction. The vector is drawn to scale as one side of the angle which represents the direction measured clockwise from the north.

Find the position of an airplane at the end of 1 hour when flying a heading of 80° at a speed of 200 m.p.h.

Scale 1 inch = 160 m.p.h.

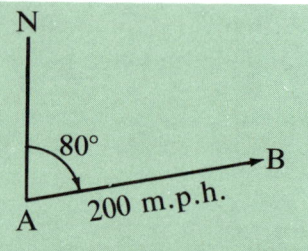

The airplane takes off from point *A*. It reaches point *B* at the end of 1 hour.

---

388   CHAPTER 8

EXERCISES

1. If you face north, then turn clockwise to face east, how many degrees do you turn?
2. If you face northeast, then turn clockwise to face northwest, how many degrees do you turn?
3. If you face north, then turn clockwise to face southwest, how many degrees do you turn?
4. First copy north lines, then with protractor draw angles indicating the required direction, measuring clockwise from the north line with point O as the vertex:

   Due East   Due South   Southeast   Northwest   Due West

5. Find the course from point A to point B in each of the following:

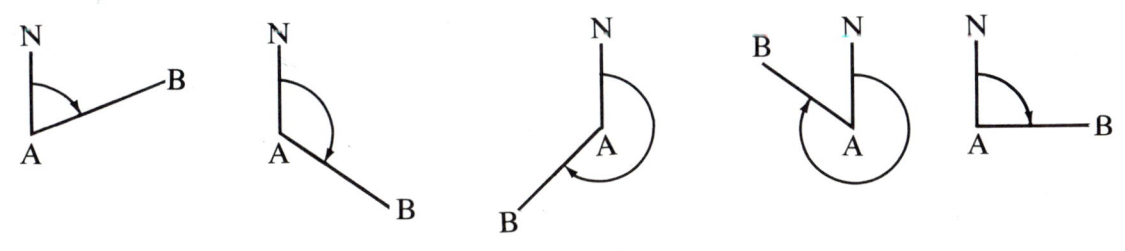

6. Express in degrees each of the following directions from which winds are blowing:

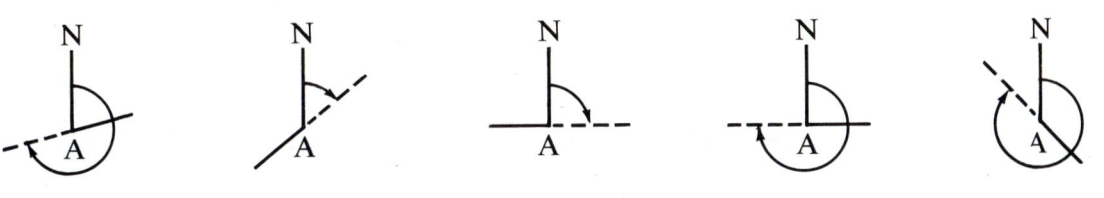

7. Find the bearing of point D from point C in each of the following:

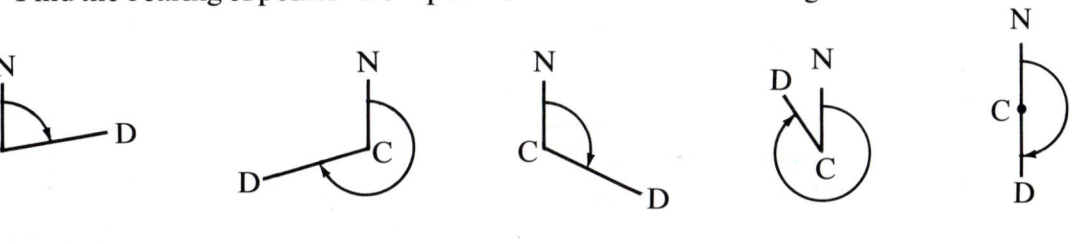

Geometry 389

8. a. Draw an angle showing a course of 35°. Of 140°. Of 230°.
   b. Draw an angle showing that the wind is blowing from 55°. From 110°. From 345°.
9. Make scale drawings showing the positions of airplanes at the end of one hour flying:
   a. Course, 70°; ground speed, 950 km/h; scale 1 cm = 100 km
   b. Course, 120°; ground speed, 300 m.p.h.; scale 1 in. = 160 mi.
   c. Course 200°; ground speed, 640 km/h; scale 1 cm = 200 km
   d. Heading, 60°; air speed, 720 km/h; scale 1 cm = 120 km
   e. Heading, 150°; air speed, 410 m.p.h.; scale 1 in. = 160 mi.
   f. Heading, 280°; air speed, 816 km/h; scale 1 cm = 240 km
10. Draw vectors representing the following wind velocities:
    a. 35 km/h wind from 60°, scale 1 cm = 40 km/h
    b. 25 m.p.h. wind from 135°, scale 1 in. = 20 m.p.h.
    c. 45 km/h wind from 230°, scale 1 cm = 30 km/h

8–15 CONSTRUCTING TRIANGLES

A triangle contains three sides and three angles. A triangle may be constructed when any of the following combinations of three parts are known:
(1) Three sides
(2) Two sides and an included angle
(3) Two angles and an included side.
These are discussed below and on pages 391 and 392.

(1) Three Sides

To construct a triangle whose sides measure 5 cm, 4.4 cm, and 3.8 cm, we first use a compass to lay off a line segment 5 cm long. With one of the endpoints as the center and setting the compass so that the radius is 4.4 cm, we draw an arc. With the other endpoint as the center and a radius of 3.8 cm, we draw an arc crossing the first arc. From this point of intersection we draw line segments to the endpoints of the base line to form the required triangle.

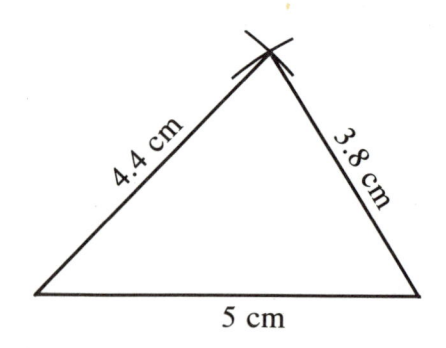

EXERCISES

1. Construct triangles having sides that measure:
   a. 42 mm, 28 mm, 31 mm
   b. $2\frac{1}{4}$ in., 3 in., $1\frac{13}{16}$ in.
   c. 6.1 cm, 4.3 cm, 5.6 cm
2. Construct an equilateral triangle whose sides are each 2.6 cm long. Measure the three angles. Are their measures equal?
3. Construct an isosceles triangle whose base is $2\frac{5}{8}$ in. long and each of whose two equal sides is $1\frac{3}{4}$ in. long.
4. Using the scale 1 in. = 160 mi., construct triangles with sides:
   a. 240 mi., 180 mi., 260 mi.   c. 220 mi., 150 mi., 190 mi.
   b. 300 mi., 190 mi., 240 mi.   d. 340 mi., 280 mi., 370 mi.
5. Using the scale 1:4,000,000 construct triangles with sides:
   a. 80 km, 100 km, 88 km   c. 140 km, 120 km, 112 km
   b. 60 km, 52 km, 72 km   d. 56 km, 64 km, 48 km
6. Construct a triangle with sides equal in length to the following line segments:

   _____   _____   _____

7. Construct a scalene triangle. Check whether:
   a. Of any two sides, the side opposite the greater angle is greater.
   b. Of any two angles, the angle opposite the greater side is greater.

(2) Two Sides and an Included Angle

To construct a triangle with sides measuring $1\frac{1}{2}$ in. and $1\frac{1}{8}$ in. and an included angle of 90°, we draw a line segment $1\frac{1}{2}$ in. long. Using the left endpoint as the vertex, we draw an angle of 90°. Along the ray just drawn we measure $1\frac{1}{8}$ in. from the vertex. We then draw a line segment connecting endpoints to form the required triangle.

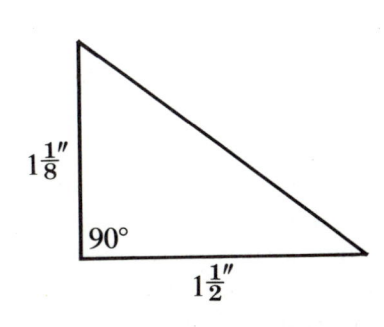

Geometry

EXERCISES

1. Construct triangles having the following sides and included angles that measure:

   a. 58 mm, 41 mm, 65°

   b. 6.7 cm, 4.5 cm, 40°

   c. $3\frac{1}{4}$ in., $3\frac{1}{4}$ in., 90°

2. Construct a right triangle having:

   a. A base of 6.8 cm and an altitude of 5.4 cm

   b. A base of $2\frac{11}{16}$ in. and an altitude of $4\frac{1}{4}$ in.

3. Construct an isosceles triangle in which the equal sides each measure $2\frac{5}{8}$ in. and the vertex angle formed by these sides measures 56°. Measure the angles opposite the equal sides. Are their measures equal?

4. Construct a triangle having two sides and an included angle equal to the following two line segments and angle. Also see sections 8–9, 8–10, and 8–19.

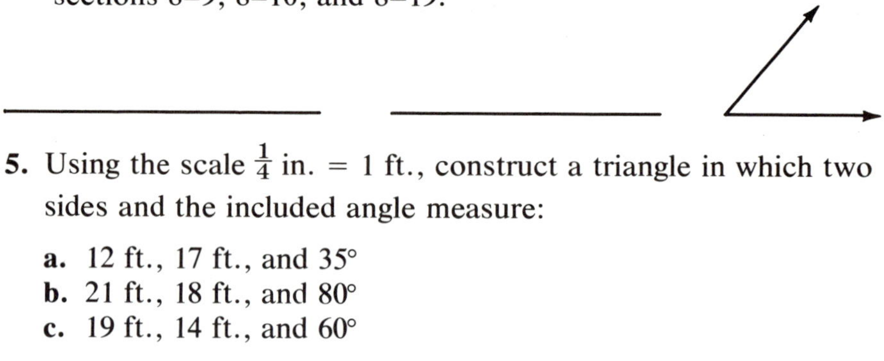

5. Using the scale $\frac{1}{4}$ in. = 1 ft., construct a triangle in which two sides and the included angle measure:

   a. 12 ft., 17 ft., and 35°
   b. 21 ft., 18 ft., and 80°
   c. 19 ft., 14 ft., and 60°

(3) Two Angles and an Included Side

To construct a triangle with angles measuring 50° and 36° and the included side measuring 3 cm, we first draw a line segment 3 cm long. Using the left endpoint as the vertex, we draw an angle of 50°. Using the right endpoint as the vertex, we draw an angle of 36°. We extend the sides until they meet to form the required triangle.

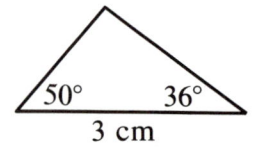

EXERCISES

1. Construct triangles having the following angles and included sides that measure:
   a. 80°, 35°, 6 cm
   b. 110°, 20°, $1\frac{11}{16}$ in.
   c. 45°, 90°, 68 mm
2. Construct a triangle having two equal angles each measuring 40° and the included side measuring 55 mm. Measure the sides opposite the equal angles. Are their measures equal?
3. Construct a triangle having two equal angles each measuring 60° and the included side measuring $3\frac{3}{8}$ in. Measure the third angle. Measure the other two sides. Is the triangle equiangular? Equilateral?
4. Construct a triangle having two angles and an included side equal to the following angles and line segment. Also see sections 8–9, 8–10, and 8–19.

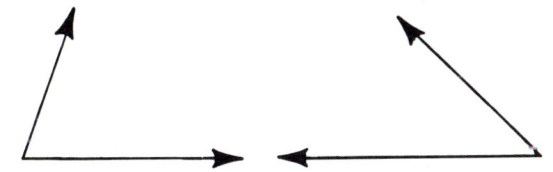

5. Using the scale $\frac{1}{8}$ in. = 1 ft., construct a triangle in which two angles and an included side measure:
   a. 45°, 45°, 20 ft.
   b. 30°, 70°, 25 ft.
   c. 105°, 40°, 31 ft.
   d. 60°, 75°, 15 ft.
6. Construct right triangles having the following angles and included sides that measure:
   a. 90°, 30°, $2\frac{5}{8}$ in.
   b. 90°, 30°, 7.2 cm
   c. 90°, 30°, 49 mm
   d. 90°, 75°, 6.1 cm

   Also select your own measurement for the included side and, using the measures of 90° and 30° for the angles, construct a right triangle. In each of these triangles measure the hypotenuse and the side opposite the 30° angle. Then compare these measurements. In each case is the hypotenuse twice as long as the side opposite the 30° angle?
7. Construct two triangles, each containing two angles measuring 50° and 75°. The included side in the first triangle measures 2 inches and in the second triangle it measures $2\frac{1}{2}$ inches. Are the triangles congruent? Are they similar? See pages 374, 409, and 410.

Geometry

8-16 CONSTRUCTING A PERPENDICULAR TO A GIVEN LINE AT OR THROUGH A GIVEN POINT ON THE GIVEN LINE

Two lines (or rays or segments) that meet to form right angles are called *perpendicular lines* (or rays or segments). Each line is said to be perpendicular to the other. (See page 365.) The symbol ⊥ means "is perpendicular to."

(1) *Using a protractor,* we draw a 90° angle with the given point on the line as the vertex. The ray drawn to form the angle is perpendicular to the given line.

(2) *Using a compass* (see figure): To draw a line (or ray or segment) perpendicular to $\overleftrightarrow{AB}$ at C, we use point C as the center and with any radius we draw an arc cutting $\overleftrightarrow{AB}$ at D and E. With D and E as centers and with a radius greater than $\overline{CD}$, we draw arcs crossing at F. We draw $\overrightarrow{CF}$ which is perpendicular to $\overleftrightarrow{AB}$ at point C. Or we may draw the line FC passing through point C.

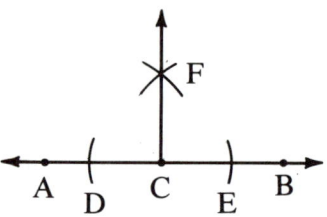

EXERCISES

1. Draw any line. Select a point on this line. Construct a line perpendicular to the line you have drawn at your selected point.
2. Construct a rectangle $2\frac{1}{4}$ inches long and $1\frac{13}{16}$ inches wide.
3. Construct a square with each side 46 mm long.
4. Make a plan of a room 42 ft. by 31 ft., using the scale 1 in. = 8 ft.
5. Draw a circle. Draw any diameter of this circle. At the center construct another diameter perpendicular to the first diameter, dividing the circle into four arcs. Are these four arcs equal in length? Use a compass to check.
6. Draw a circle. Select any point on this circle and label it A. Construct a line perpendicular to the radius at point A. A *tangent* to a circle is a line that has one and only one point in common with the circle. Observe that a line that is perpendicular to the radius at a point on the circle is tangent to the circle.

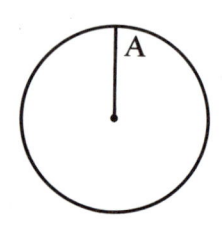

8–17 CONSTRUCTING A PERPENDICULAR TO A GIVEN LINE FROM OR THROUGH A GIVEN POINT NOT ON THE GIVEN LINE

Using a compass (see figure): To draw a ray from point C perpendicular to $\overleftrightarrow{AB}$, we use point C as the center and draw an arc cutting $\overleftrightarrow{AB}$ at D and E. With D and E as centers and a radius of more than one-half the distance from D to E, we draw arcs crossing at F. We draw $\overrightarrow{CF}$ which is perpendicular to $\overleftrightarrow{AB}$. Or we may draw the line FC passing through point C.

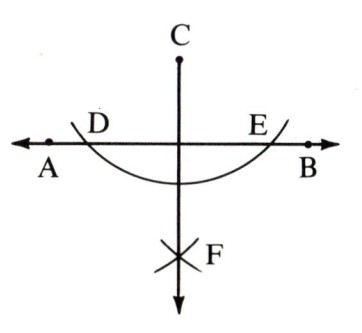

EXERCISES

1. Draw any line. Select a point not on this line. Construct a perpendicular to the line you have drawn from your selected point.
2. Construct an equilateral triangle with each side 7.5 cm long. From each vertex construct a perpendicular to the opposite side. Is each angle bisected? Is each side bisected?
3. Draw any acute triangle. From each vertex construct a perpendicular to the opposite side. What do these perpendicular line segments represent in a triangle? Are they concurrent?
4. Draw any right triangle. Construct the altitude to each side. Are they concurrent? If so, what is the common point?
5. Draw a circle. Draw any chord in this circle except the diameter. Construct a perpendicular from the center of the circle to this chord. Check whether the chord is bisected. Extend the perpendicular line segment so that it intersects the circle. Check with a compass whether the arc corresponding to the chord is also bisected. Observe that a radius of a circle that is perpendicular to a chord bisects the chord and its corresponding arc.

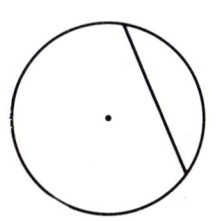

Geometry

8–18 BISECTING A LINE SEGMENT

To *bisect a line segment* means to divide it into two equal parts. The point on a line segment that separates the line segment into equal parts is called the *midpoint* of the line segment.

(1) *Using a ruler,* we first measure the line segment, then we mark off half the measurement.

(2) *Using a compass* (see figure): To bisect $\overline{AB}$, we set the compass so that the radius is more than half the length of $\overline{AB}$. With A and B as centers we draw arcs which cross above and below the segment at C and D. Then we draw $\overleftrightarrow{CD}$ bisecting $\overline{AB}$ at E. Observe that $\overleftrightarrow{CD}$ is also perpendicular to $\overline{AB}$. Thus a line like $\overleftrightarrow{CD}$, which both bisects a line segment and is perpendicular to it, is called the *perpendicular bisector* of the given segment.

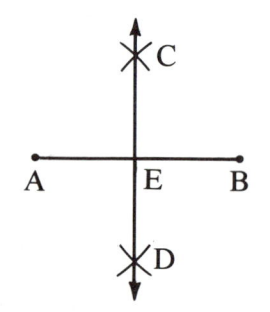

EXERCISES

1. Draw line segments which measure: **a.** $3\frac{3}{4}''$ **b.** 26 mm **c.** 5.8 cm
   Bisect each line segment, using a compass. Check with a ruler.
2. Copy, then using a compass, bisect each of the following segments. Check each with a ruler.

   ─────────────────   ─────────────────────────────   ─────

3. Copy, then using a compass, divide the following segment into four equal parts. Check with a ruler.

   ──

4. Draw any line segment. Bisect it, using a compass. Then use a protractor to check whether each of the four angles formed is a right angle. What name do we give to a line that both bisects and is perpendicular to a line segment?
5. Draw any triangle. Bisect each side by constructing the perpendicular bisector of that side. Are these perpendicular bisectors concurrent? Do they meet in a point equidistant from the

CHAPTER 8

vertices of the triangle? Check by measuring. Using this common point as the center and the distance from this point to any vertex of the triangle as the radius, draw a circle through the three vertices of the triangle.

When each side of the triangle is a chord of the circle or each vertex is a point on the circle, we say that the *circle is circumscribed about the triangle* or that the *triangle is inscribed in a circle*.

6. Draw any triangle. Find the midpoint of each side by constructing the perpendicular bisector of that side. Draw the median from each vertex to the midpoint of the opposite side. Are these medians concurrent? Along each median check whether the distance from this common point to the vertex is twice the distance from this common point to the midpoint of the opposite side.
7. Draw a right triangle. Bisect the hypotenuse of the right triangle. Draw the median from the vertex of the right angle to the hypotenuse. Check whether this median is one-half as long as the hypotenuse.

8–19 COPYING A GIVEN ANGLE

To copy a given angle means to construct an angle equal in size to the given angle.

(1) *Using a protractor,* we measure the given angle and draw another angle of the same size.

(2) *Using a compass* (see figure): To construct an angle at point C on $\overleftrightarrow{AB}$ equal to $\angle MNO$, we take point N as center and draw an arc cutting side MN at P and side NO at Q. With the same radius and point C as center, we draw an arc cutting $\overleftrightarrow{AB}$ at D. With a radius equal to PQ and point D as center, we draw an arc crossing the first arc at E. We draw CE. $m\angle BCE$ is equal to $m\angle MNO$.

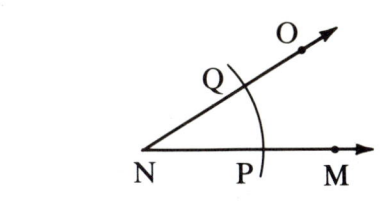

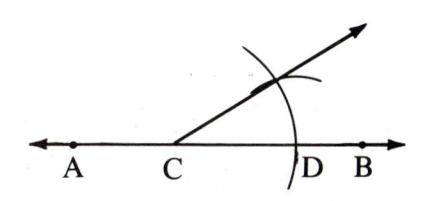

EXERCISES

1. Draw angles having the following measures, using a protractor. For each angle construct with a compass an angle of equal size. Check your copy of the angle with a protractor.

   a. 80°   **b.** 130°   **c.** 45°   **d.** 23°   **e.** 157°   **f.** 90°

2. Draw an acute angle. Construct with a compass an angle of equal size. Check both angles with a protractor.
3. Draw a right angle, using a protractor. Construct with a compass an angle of equal size. Check with a protractor.
4. Draw any obtuse angle. Construct with a compass an angle of equal size. Check both angles with a protractor.
5. Draw any angle. Construct with a compass an angle having the same measure. Check both angles with a protractor.

8–20 BISECTING AN ANGLE

To *bisect an angle* means to divide it into two equal angles.

(1) *Using a protractor,* we measure the given angle and mark off one-half the measurement. We then draw a ray from the vertex.

(2) *Using a compass* (see figure): To bisect ∠ABC with B as the center and any radius, we draw an arc cutting side AB at D and side BC at E. With D and E as centers and a radius of more than half the distance from D to E, we draw arcs crossing at F. We then draw $\overrightarrow{BF}$ bisecting ∠ABC.

EXERCISES

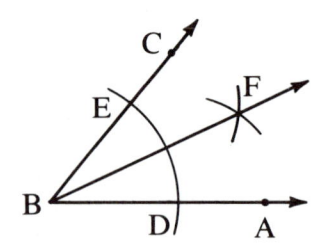

1. Draw angles having the following measures. Bisect each angle using a compass. Check with a protractor.

   a. 60°   **b.** 150°   **c.** 48°   **d.** 165°   **e.** 75°   **f.** 121°

2. Draw any angle. Bisect it by using a compass only. Check by measuring your angle and each bisected angle.
3. Copy, then bisect each of the following angles:

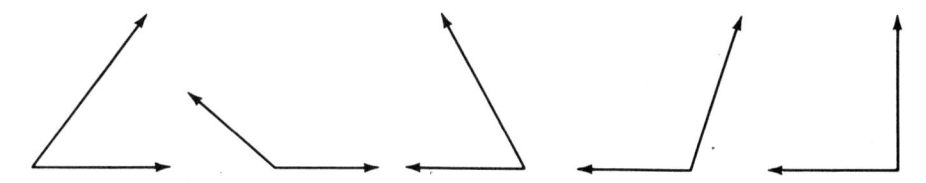

4. Draw any angle. Using a compass only, divide the angle into four equal angles. Check with protractor.
5. Draw any triangle. Bisect each angle. Do the bisectors meet at a common point? Are these bisectors concurrent? From this common point draw a perpendicular to any side. With this common point as center and with the perpendicular distance from the common point to any side as the radius, draw a circle. Observe that each side of the triangle is tangent to the circle.

    When each side of the triangle is tangent to the circle, we say that the *triangle is circumscribed about the circle* or that the *circle is inscribed in the triangle.*
6. Construct with a compass an angle measuring: **a.** 45° **b.** 135°
7. Draw any equilateral triangle. Select an angle and bisect it. Construct from the vertex of this selected angle the altitude (line segment that is perpendicular to the opposite side). Does this perpendicular line bisect this opposite side? From this same vertex also draw the median to the opposite side. In an equilateral triangle, are all three of these lines (angle bisector, altitude, and median) one and the same line? If this is so, will it be true if each of the other angles is selected? Check.
8. Draw any isosceles triangle. Bisect the vertex angle (angle formed by the two equal sides and opposite to the base). Then construct from the vertex of this angle the altitude (line segment that is perpendicular to the base). Does this perpendicular line bisect the base of the isosceles triangle? From this same vertex also draw the median to the base. In an isosceles triangle, are all three of these lines (bisector of the vertex angle, altitude drawn to the base, and the median drawn to the base) one and the same line?

    In an isosceles triangle the angles opposite the equal sides are called *base angles.* Check whether the bisector of a base angle, the altitude and the median both drawn from the vertex of the same base angle are all one and the same line.

9. Draw any scalene triangle and do the following:
   a. Select an angle and bisect it.
   b. Construct from the vertex of this bisected angle the line segment that is perpendicular (altitude) to the opposite side.
   c. Construct the perpendicular bisector of the side opposite to the bisected angle.
   d. Draw the median from the vertex of this bisected angle to the opposite side. Are any of these four lines (angle bisector, altitude, perpendicular bisector, and median) the same line? If so, which?

8–21 CONSTRUCTING A LINE PARALLEL TO A GIVEN LINE THROUGH A GIVEN POINT NOT ON THE GIVEN LINE

Lines in the same plane which do not meet are called *parallel lines*. The symbol ∥ means "is parallel to."

To construct a line parallel to $\overleftrightarrow{AB}$ through point C (see figure), we draw any line $\overleftrightarrow{DE}$ through C meeting $\overleftrightarrow{AB}$ at F.

(1) *Using a protractor*, we measure ∠BFC and draw, at point C on $\overleftrightarrow{DE}$, ∠GCE equal to the corresponding ∠BFC. Then we extend $\overrightarrow{GC}$ through H. $\overleftrightarrow{HG}$ is parallel to $\overleftrightarrow{AB}$.

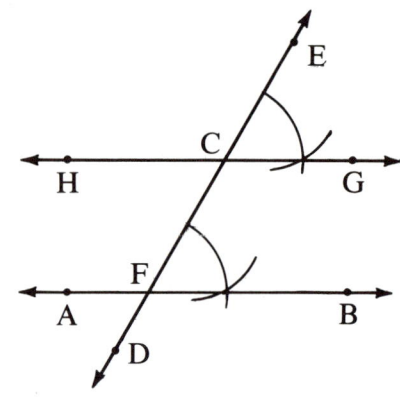

(2) *Using a compass*, we construct, at point C on $\overleftrightarrow{DE}$, ∠GCE equal to the corresponding ∠BFC by following the procedure explained in the preceding construction (8–19). We then extend $\overrightarrow{GC}$ through H. $\overleftrightarrow{HG}$ is parallel to $\overleftrightarrow{AB}$.

EXERCISES

1. Draw any line. Select a point that is not on this line. Through this point construct a line parallel to the line you have drawn.

2. Construct a parallelogram with a base $3\frac{5}{16}$ inches long, a side $2\frac{7}{8}$ inches long, and an included angle of 55°.

3. Draw any acute triangle. Bisect one of the sides. Draw a line segment from this midpoint, parallel to one of the other sides, until it intersects the third side. Does this line segment bisect the third side? Check by measuring. Also check whether this line segment is one-half as long as the side to which it is parallel. Draw other triangles and check whether the above findings are true no matter which side is used first or which side is used as the parallel side.

8–22 CONSTRUCTING REGULAR POLYGONS AND OTHER POLYGONS

A *regular polygon* is a polygon that is both equilateral (all of its sides are of equal length) and equiangular (all of its angles are of equal size). An *inscribed polygon in a circle* is a polygon whose vertices are points on the circle.

Although there are other ways to construct some of the regular polygons, in general we use a method that is based on the geometric fact that equal central angles of a circle intercept equal arcs and equal chords. A regular polygon constructed in this way is inscribed within a circle.

Therefore, *to construct a regular polygon,* we first draw a circle. Then we divide this circle into the same number of equal arcs as there are sides in the required polygon by drawing a corresponding number of equal central angles. We draw line segments (chords) connecting the points of division to form the polygon.

To draw a regular hexagon, which is a polygon of six equal sides, we first determine the measure of each of the six equal central angles by dividing 360° by 6. This measure is 60°. We draw six central angles each measuring 60°, with its sides (radii) intercepting the circle dividing it into six equal arcs. We then draw line segments to connect the points of division to form the regular hexagon.

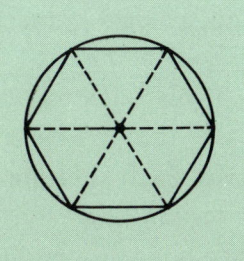

Geometry

A regular hexagon and an equilateral triangle may also be drawn by the following special method:

We divide the circle into six equal arcs by using the radius of the circle as the radius of the arc. We then connect these points of division to get a regular hexagon.

An equilateral triangle may be constructed by drawing line segments to connect alternate points of division after the circle is divided into six equal arcs.

A square may be constructed by drawing two diameters of a circle perpendicular to each other to divide the circle into four equal arcs. We then connect these points of division to get a square.

Rectangles and parallelograms of specific measurement may be constructed provided enough of these measurements are given so that basic constructions may be used.

EXERCISES

1. Construct each of the following regular polygons:
   a. Octagon—8 sides
   b. Pentagon—5 sides
   c. Square
   d. Decagon—10 sides
   e. Dodecagon—12 sides
   f. Equilateral triangle

2. Construct a rectangle:
   a. $2\frac{1}{2}$ inches long and $1\frac{7}{8}$ inches wide
   b. 58 mm long and 32 mm wide
   c. 8.2 cm long and 5.7 cm wide
   d. $3\frac{11}{16}$ inches long and $2\frac{3}{8}$ inches wide

3. Construct a square whose side measures:
   a. $2\frac{1}{2}$ inches
   b. 4 cm
   c. $\frac{3}{4}$ inch
   d. 75 mm

4. Construct a parallelogram with:
   a. A base 67 mm long, a side 46 mm long, and an included angle of 45°
   b. A base $3\frac{3}{4}$ inches long, a side $2\frac{7}{8}$ long, and an included angle of 70°
   c. A base 4 inches long, a side $3\frac{11}{16}$ inches long, and an included angle of 60°
   d. A base 5 cm long, a side 8.6 cm long, and an included angle of 110°

5. Draw a regular hexagon, each side measuring:
   a. 3 cm
   b. 44 mm
   c. $3\frac{5}{16}$ inches.

   In each case draw a circle whose radius has the same measure as each required side.

8–23 SUM OF ANGLES OF POLYGONS

(1) Triangle

Draw any triangle. Measure its three angles. What is the sum of the measure of these three angles? Draw a second triangle. Find the sum of the measures of its three angles. Find the sum of the measures of the three angles of a third triangle.

Do you see that *the sum of the measures of the angles of any triangle is 180°*?

   (a) In each of the following, find the measure of the third angle of a triangle when the other two angles measure:

      1. 56° and 47°
      2. 112° and 39°
      3. 63° and 84°
      4. 18° and 25°
      5. 4° and 141°
      6. 83° and 83°

   (b) What is the measure of each angle of an equiangular triangle?
   (c) If each of the equal angles of an isosceles triangle measures 48°, find the measure of the third angle.
   (d) If the vertex angle of an isosceles triangle measures 80°, find the measure of each of the other two angles.
   (e) Why can there be only one right or one obtuse angle in a triangle? What kind of angle must each of the other two angles be?
   (f) In a right triangle if one of the acute angles measures 51°, what is the measure of the other acute angle?
   (g) If two angles of one triangle are equal respectively to two angles of another triangle, why are the third angles equal?

(2) Quadrilateral

Draw a parallelogram and a trapezoid. Measure the four angles in each. What is the sum of the measures of the angles of the parallelogram? Of the trapezoid?

Draw any other quadrilateral. Find the sum of the measures of the angles of the quadrilateral.

Do you see that *the sum of the measures of the angles of any quadrilateral is 360°*?

   (a) What is the sum of the measures of the angles of a rectangle? Of a square?

Geometry

(b) In each of the following, find the measure of the fourth angle of a quadrilateral when the other three angles measure:

1. 56°, 83°, and 108°
2. 121°, 90°, and 90°
3. 115°, 17°, and 154°

(c) The opposite angles of a parallelogram are equal. If one angle measures 68°, find the measures of the other three angles.

(d) Three angles of a trapezoid measure 90°, 90°, and 117°. What is the measure of the fourth angle?

(3) Other Polygons

Draw a pentagon, a hexagon, an octagon, and a decagon. What is the sum of the measures of the angles of each of these geometric figures? Check whether your sum of measures of the angles of each of these figures matches the angular measure determined by substituting the number of sides in the polygon for n in the expression $180(n - 2)$ and performing the required operations. Is this also true for the triangle and the quadrilateral?

Do you see that *the sum of measures of the angles of a polygon of n sides is $180(n - 2)$*?

8–24 PAIRS OF ANGLES

(1) *Complementary angles* are two angles whose sum of measures is 90°.

(a) Which of the following pairs of angles are complementary?

1. $m\angle B = 60°, m\angle D = 30°$
2. $m\angle A = 22°, m\angle T = 68°$
3. $m\angle 5 = 53°, m\angle 7 = 47°$
4. $m\angle r = 21°, m\angle s = 59°$
5. $m\angle E = 74°, m\angle F = 16°$
6. $m\angle 1 = 100°, m\angle 2 = 80°$

(b) Find the measure of the angle that is the complement of each of the following angles:

1. $m\angle P = 27°$
2. $m\angle 3 = 81°$
3. $m\angle t = 55°$
4. $m\angle A = 8°$
5. $m\angle R = 39°$
6. $m\angle B = 64°$

(c) Angle ABC is a right angle.
  1. If $m\angle 1 = 75°$, find the measure of $\angle 2$.
  2. If $m\angle 2 = 24°$, find the measure of $\angle 1$.
  3. If $m\angle 1 = 63°$, find the measure of $\angle 2$.
  4. If $m\angle 2 = 11°$, find the measure of $\angle 1$.

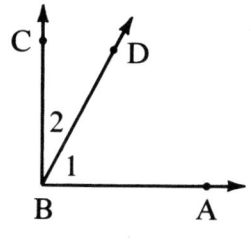

(d) In the figure $\overrightarrow{FB} \perp \overrightarrow{FD}$ and $\overrightarrow{FC} \perp \overleftrightarrow{AE}$.
  1. If $m\angle 1 = 52°$, find the measure of $\angle 2$. Of $\angle 3$. Of $\angle 4$.
  2. If $m\angle 4 = 69°$, find the measure of $\angle 1$. Of $\angle 2$. Of $\angle 3$.
  3. If $m\angle 2 = 33°$, find the measure of $\angle 3$. Of $\angle 1$. Of $\angle 4$.
  4. If $m\angle 3 = 78°$, find the measure of $\angle 1$. Of $\angle 2$. Of $\angle 4$.

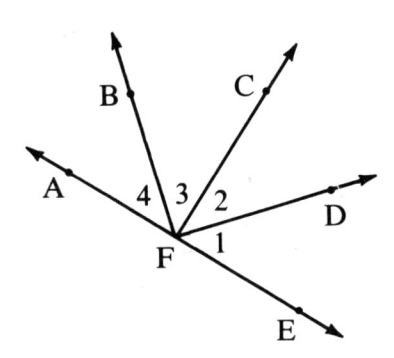

(e) Why is the complement of an acute angle also an acute angle?

(2) *Supplementary angles* are two angles whose sum of measures is 180°.
  (a) Which of the following pairs of angles are supplementary?
    1. $m\angle B = 125°, m\angle R = 55°$
    2. $m\angle 1 = 83°, m\angle 2 = 107°$
    3. $m\angle G = 92°, m\angle H = 78°$
    4. $m\angle a = 47°, m\angle b = 133°$
    5. $m\angle M = 51°, m\angle N = 39°$
    6. $m\angle S = 86°, m\angle T = 94°$

  (b) Find the measure of the angle that is the supplement of each of the following angles:
    1. $m\angle D = 67°$
    2. $m\angle b = 102°$
    3. $m\angle 3 = 5°$
    4. $m\angle T = 90°$
    5. $m\angle L = 148°$
    6. $m\angle E = 39°$

Geometry

(c) Measure ∠CDF and ∠EDF. Is the sum of their measures 180°? Do you see that *when one straight line meets another, the adjacent angles, which have the same vertex and a common side, are supplementary*? 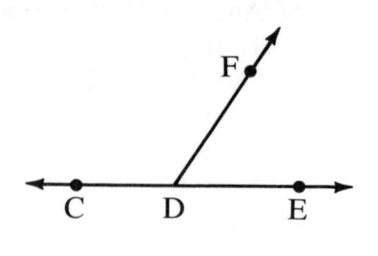
1. Find the measure of ∠EDF when $m\angle CDF = 120°$.
2. What is the measure of ∠CDF when $m\angle EDF = 42°$?
3. What is the measure of ∠EDF when $m\angle CDF = 157°$?
4. Find the measure of ∠CDF when $m\angle EDF = 90°$.

(d) Is the supplement of an obtuse angle also an obtuse angle? Explain your answer.

(3) Opposite or Vertical Angles

Draw two intersecting lines, forming four angles. Measure a pair of angles that are directly opposite to each other. Are their measures equal? Measure the other pair of opposite angles. Are their measures equal?

Do you see that, *when two straight lines intersect, the opposite or vertical angles are equal*?

(a) In the drawing at the right, what angle is opposite to ∠1? Does ∠1 = ∠3? What angle is opposite to ∠4? Does ∠4 = ∠2? 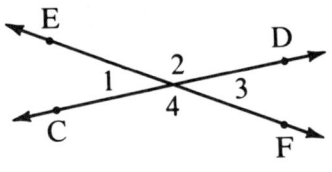
(b) If $m\angle 4 = 106°$, what is the measure of ∠2? Of ∠1? Of ∠3?
(c) If $m\angle 3 = 59°$, what is the measure of ∠1? Of ∠2? Of ∠4?
(d) If $m\angle 2 = 132°$, what is the measure of ∠3? Of ∠4? Of ∠1?
(e) If $m\angle 1 = 27°$, what is the measure of ∠4? Of ∠2? Of ∠3?

(4) Exterior Angle of a Triangle

Draw a triangle and extend one side like the drawing at the right. Label the angles as shown in the drawing.

The angle formed by a side of the triangle and the adjacent side extended (see ∠4 in the drawing) is called the *exterior angle*.

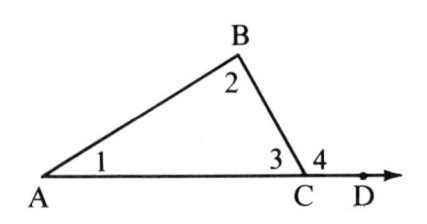

Measure the three angles of the triangle and the indicated exterior angle in your drawing.

What is the sum of the measures of ∠1, ∠2, and ∠3? What is the sum of the measures of ∠4 and ∠3?

Why does the measure of ∠4 equal the sum of the measures of ∠1 and ∠2?

Do you see that *the measure of an exterior angle of a triangle is equal to the sum of the measures of the opposite two interior angles?*

   (a) If $m\angle 1 = 37°$ and $m\angle 2 = 68°$, find the measure of ∠4. Of ∠3.
   (b) If $m\angle 3 = 43°$ and $m\angle 1 = 74°$, find the measure of ∠2. Of ∠4.
   (c) If $m\angle 2 = 81°$ and $m\angle 3 = 79°$, find the measure of ∠1. Of ∠4.
   (d) If $m\angle 4 = 116°$ and $m\angle 1 = 60°$, find the measure of ∠2. Of ∠3.
   (e) If $m\angle 2 = 58°$ and $m\angle 4 = 135°$, find the measure of ∠3. Of ∠1.
   (f) Can an exterior angle of a triangle have the same measure as one of the angles of the triangle? As one of the opposite interior angles of the triangle? Explain your answers.

8–25 PARALLEL LINES AND ANGLE RELATIONSHIPS

In the figure at the right, $\overleftrightarrow{DE}$ and $\overleftrightarrow{FG}$ are parallel and cut by $\overleftrightarrow{MN}$; ∠2 and ∠7 form a pair of *alternate-interior angles* and ∠3 and ∠6 form another pair of alternate-interior angles. These angles are between the parallel lines and the related angles fall on alternate sides.

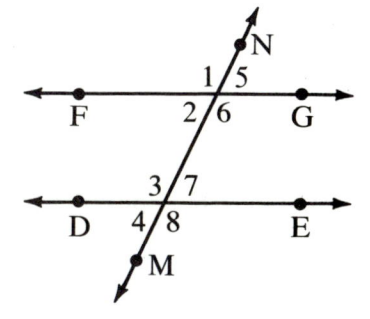

∠1 and ∠3, ∠2 and ∠4, ∠5 and ∠7, and ∠6 and ∠8 are pairs of *corresponding angles*. Each pair of angles is in a corresponding position. Observe that when two parallel lines are cut by a third line (called a *transversal*), both the corresponding angles are equal and the alternate-interior angles are equal.

Also, when two lines are cut by a transversal making a pair of corresponding angles or a pair of alternate-interior angles equal, the lines are parallel.

Geometry 407

EXERCISES

Use the figure below for all problems. Complete each statement as required in problems 1 to 6 inclusive.

1. If lines *AB* and *CD* are cut by a third line *EF*,
    a. the line *EF* is called a ?.
    b. ∠s 4 and 5 are ? angles.
    c. ∠s 3 and 6 are ? angles.
    d. ∠s 4 and 8 are ? angles.
    e. ∠s 5 and 1 are ? angles.
2. If two parallel lines are cut by a transversal, the alternate-interior angles are ?. Therefore, ∠6 = ?, ∠5 = ?.
3. If two parallel lines are cut by a transversal, the corresponding angles are ?. Therefore, ∠2 = ?, ∠7 = ?, ∠1 = ?, ∠8 = ?.
4. What is the sum of the measures of ∠s 6 and 8? They are a pair of ? angles.
5. What is the sum of the measures of ∠s 1 and 3? They are a pair of ? angles.
6. ∠1 and ∠4 are a pair of ? angles.
7. If line *AB* is parallel to line *CD*,
    a. Show that ∠2 = ∠7.
    b. Show that ∠s 1 and 6 are supplementary angles.
    c. Show that ∠8 = ∠1.
    d. Show that ∠s 3 and 5 are supplementary angles.
    e. Show that ∠s 8 and 2 are supplementary angles.
8. If line *AB* is parallel to line *CD*, what is the measure of:
    a. ∠5 if m∠1 = 75°?
    b. ∠6 if m∠3 = 140°?
    c. ∠7 if m∠4 = 61°?
    d. ∠1 if m∠8 = 57°?
    e. ∠8 if m∠2 = 106°?
    f. ∠2 if m∠7 = 132°?
9. If two lines are cut by a transversal making a pair of corresponding angles or a pair of alternate-interior angles equal, the lines are ?.
10. Are lines *AB* and *CD* parallel when:
    a. m∠4 = 39° and m∠8 = 39°?
    b. m∠6 = 97° and m∠3 = 97°?
    c. m∠4 = 54° and m∠6 = 126°?
    d. m∠2 = 135° and m∠8 = 50°?

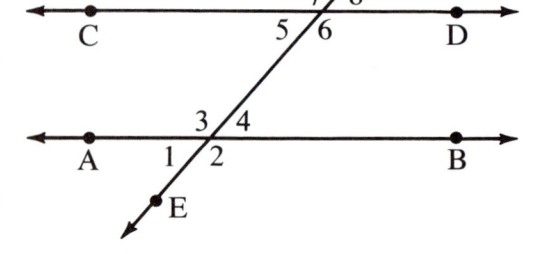

8–26 CONGRUENT TRIANGLES

Congruent triangles are triangles which have exactly the same shape and the same size. The corresponding sides are equal in length and the corresponding angles are equal in size. The symbol ≅ means "is congruent to."

Two triangles are congruent when any of the following combinations of three parts are known:
  (1) Three sides of one triangle are equal to three sides of the second triangle.
  (2) Two sides and an included angle of one triangle are equal respectively to two sides and an included angle of the other.
  (3) Two angles and an included side of one triangle are equal respectively to two angles and an included side of the other.

EXERCISES

Select in each of the following groups two triangles which are congruent and state the reason why.

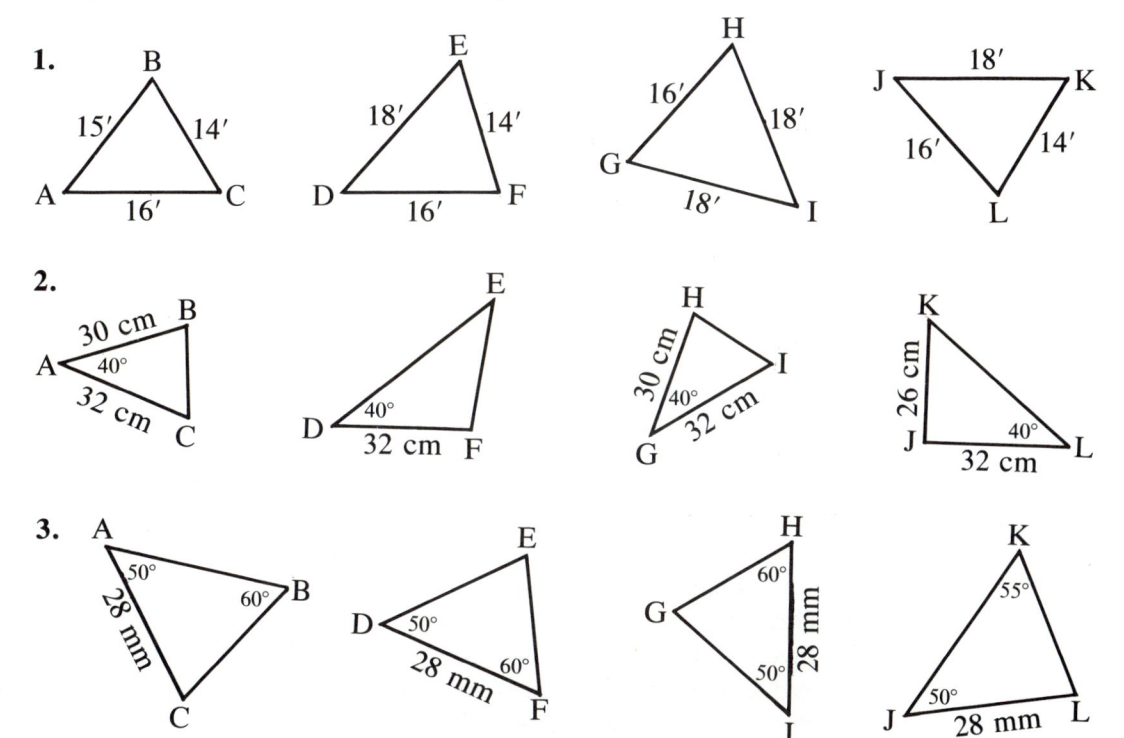

1.

2.

3.

Geometry

Find the indicated missing parts in the following triangles:

4.

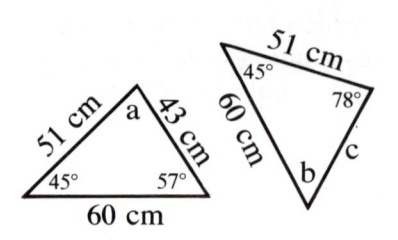

5.

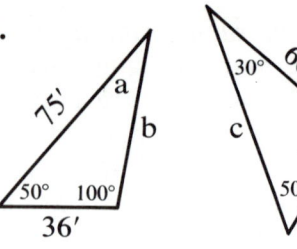

6. In triangle *ABC*, side *AB* = side *BC* and *BD* bisects ∠*ABC* making ∠*ABD* = ∠*DBC*. Prove that triangle *ABD* and triangle *DBC* are congruent.

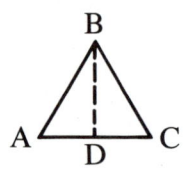

7. In parallelogram *ABCD*, side *AB* = side *DC* and side *AD* = side *BC*. Prove that the diagonal *AC* divides the parallelogram into two congruent triangles.

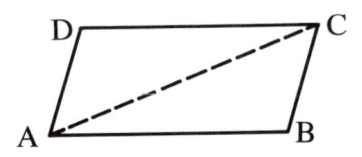

8–27 SIMILAR TRIANGLES

Similar triangles are triangles which have the same shape but differ in size. Two triangles are similar when any of the following conditions are known:
  (1) Two angles of one triangle are equal to two angles of the other triangle.
  (2) The ratios of the corresponding sides are equal.
  (3) Two sides of one triangle are proportional (equal ratios) to two corresponding sides of the other triangle and the included angles are equal.

EXERCISES

Select in each of the following groups two triangles which are similar and state the reason why.

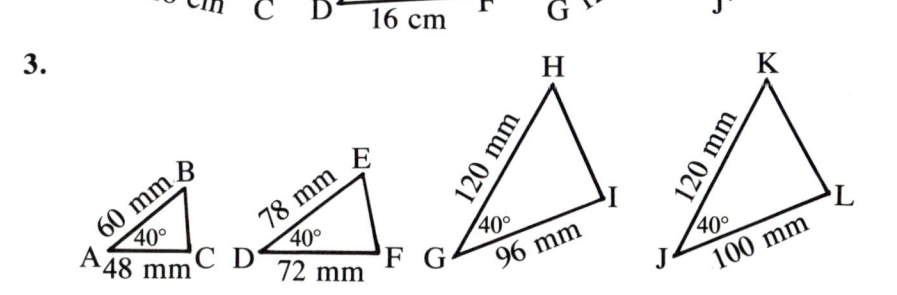

1.

Triangle ABC: A = 70°, C = 50°, B
Triangle DEF: D = 70°, F = 45°, E
Triangle GHI: G = 50°, I = 50°, H
Triangle JKL: K = 50°, L = 70°, J

2.

Triangle ABC: AB = 21 cm, BC = 16 cm, AC = 18 cm
Triangle DEF: DE = 14 cm, EF = 12 cm, DF = 16 cm
Triangle GHI: GH = 11 cm, HI = 9 cm, GI = 12 cm
Triangle JKL: KL = 18 cm, JK = 21 cm, JL = 24 cm

3.

Triangle ABC: AB = 60 mm, A = 40°, AC = 48 mm
Triangle DEF: DE = 78 mm, D = 40°, DF = 72 mm
Triangle GHI: GH = 120 mm, G = 40°, GI = 96 mm
Triangle JKL: JK = 120 mm, J = 40°, JL = 100 mm

Find the indicated missing parts in the following figures:

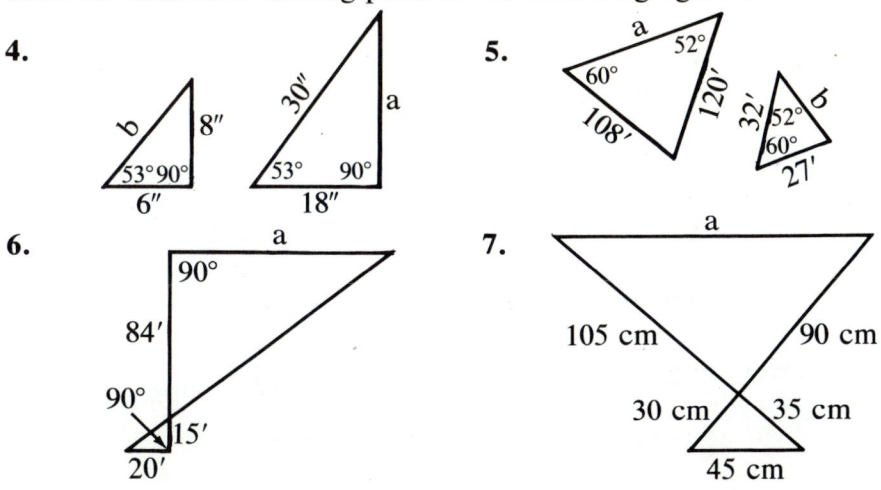

4. Triangle with b, 8", 53°, 90°, 6"; Triangle with 30", a, 53°, 90°, 18"

5. Triangle with a, 52°, 60°, 108', 120'; Triangle with 32', b, 52°, 60°, 27'

6. 90°, 84', a, 90°, 15', 20'

7. a, 105 cm, 90 cm, 30 cm, 35 cm, 45 cm

Geometry 411

INDIRECT MEASUREMENT

8–28 BY RULE OF PYTHAGORAS

The Rule of Pythagoras, who was a Greek mathematician, is used to find distances by indirect means. It expresses the relationship of the sides of a right triangle. The side opposite the right angle is called the *hypotenuse*. The other two sides or legs are the *altitude* and *base* of the triangle.

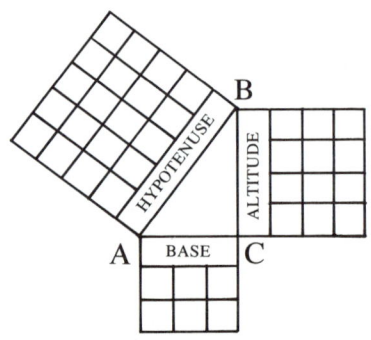

The diagram shows that the area of the square drawn on the hypotenuse is 25 square units and is equal to the sum of the areas of the squares drawn on the altitude and base (16 square units and 9 square units respectively). The rule states: *The square of the hypotenuse is equal to the sum of the squares of the other two sides.* This relationship is expressed by the formula $h^2 = a^2 + b^2$ where h represents the hypotenuse, a the altitude, and b the base.

If any two sides of a right triangle are known, the third side may be determined by the Pythagorean relation expressed in one of the following simplified forms:

$$h = \sqrt{a^2 + b^2} \qquad a = \sqrt{h^2 - b^2} \qquad b = \sqrt{h^2 - a^2}$$

---

(1) Find the hypotenuse of a right triangle if the altitude is 36 centimeters and the base is 27 centimeters:

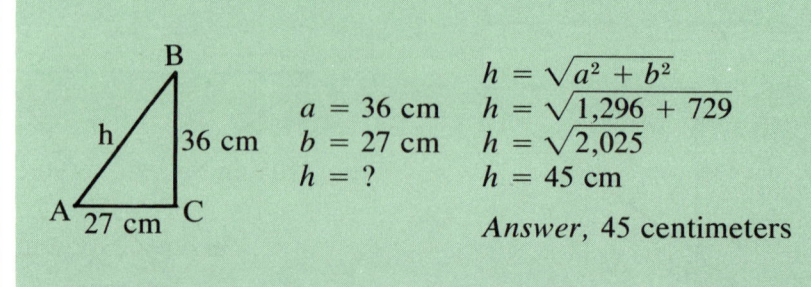

$a = 36$ cm
$b = 27$ cm
$h = ?$

$h = \sqrt{a^2 + b^2}$
$h = \sqrt{1{,}296 + 729}$
$h = \sqrt{2{,}025}$
$h = 45$ cm

Answer, 45 centimeters

(2) Find the altitude of a right triangle if the hypotenuse is 26 feet and the base is 24 feet:

$h = 26$ ft.
$b = 24$ ft.
$a = ?$

$a = \sqrt{h^2 - b^2}$
$a = \sqrt{676 - 576}$
$a = \sqrt{100}$
$a = 10$ ft.

Answer, 10 ft.

(3) Find the base of a right triangle if the hypotenuse is 17 meters and the altitude is 8 meters:

$h = 17$ m
$a = 8$ m
$b = ?$

$b = \sqrt{h^2 - a^2}$
$b = \sqrt{289 - 64}$
$b = \sqrt{225}$
$b = 15$ m

Answer, 15 meters

EXERCISES

1. Find the hypotenuse of each right triangle with the following dimensions:

Altitude	28 in.	192 dm	108 yd.	75 m	2.1 km
Base	45 in.	56 dm	315 yd.	180 m	3.2 km

2. Find the altitude of each right triangle with the following dimensions:

Hypotenuse	73 m	136 cm	455 ft.	117 km	9.5 mi.
Base	48 m	64 cm	112 ft.	108 km	5.7 mi.

3. Find the base of each right triangle with the following dimensions:

Hypotenuse	65 mm	116 yd.	680 ft.	182 m	4.5 km
Altitude	33 mm	84 yd.	104 ft.	168 m	3.6 km

4. How long is a path running diagonally across a rectangular lot 405 meters long and 216 meters wide?

Geometry

5. If the foot of a ladder is 14 feet from the wall, how high up on the wall does a 50-foot ladder reach?
6. Find the length of the diagonal of a square whose side measures 18 cm.
7. What is the distance from the pitcher's box to second base if the distance from the pitcher's box to home plate is 60 feet and the distance between bases is 90 feet?
8. In the installation of an antenna 16 feet high, how long must a wire be to reach from the top of the antenna to a point on the roof 12 feet from the foot of the antenna?
9. A boat left port, sailing 28 kilometers due west and 45 kilometers due south. How far away from the port was the boat?
10. Mr. Harris plans to brace his 4 ft. by 7 ft. garage door by nailing a board diagonally across. How long a board does he need to reach from corner to corner?
11. What distance did an airplane actually fly when, traveling due east 312 kilometers from town X to town Y, it drifted off its course in a straight line and landed at town Z, 91 kilometers due north of town Y?
12. How high is a kite if a girl holds the string 3 ft. from the ground and lets out 200 ft. of string? The distance from a point on the ground directly under the kite to where the girl stands is 56 ft.

8–29 BY SIMILAR TRIANGLES

To measure indirectly a distance or length by similar triangles, we first draw two similar triangles from the given facts. Since the corresponding sides of similar triangles are proportional (have equal ratios), we form and solve a proportion using three given sides and the unknown distance as the fourth side.

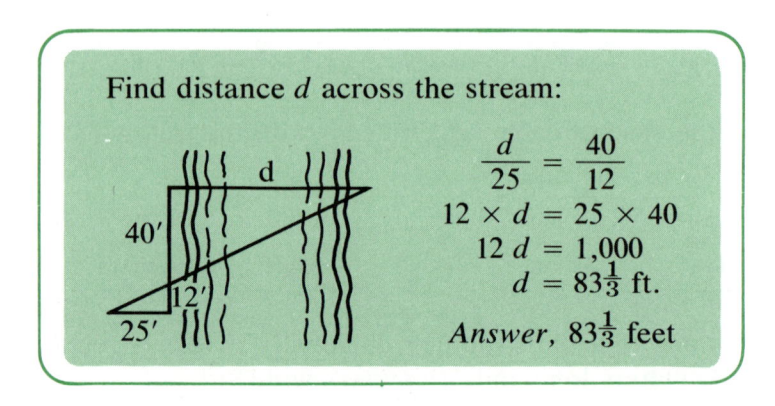

Find distance d across the stream:

$$\frac{d}{25} = \frac{40}{12}$$
$$12 \times d = 25 \times 40$$
$$12\,d = 1{,}000$$
$$d = 83\tfrac{1}{3} \text{ ft.}$$

Answer, $83\tfrac{1}{3}$ feet

EXERCISES

1. What is the height (*BC*) of a tree that casts a shadow (*AB*) of 28 m while a 3-meter post (*EF*) nearby casts a shadow (*DE*) of 2 m? Why are triangles *ABC* and *DEF* similar?

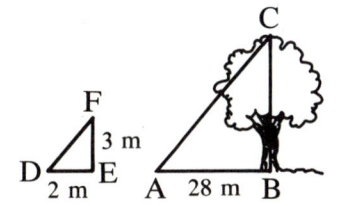

2. Find the height of a flagpole that casts a shadow of 15 ft. at a time when a girl, 5 feet tall, casts a shadow of 3 ft.?
3. A building casts a shadow of 72 ft. At the same time an 8-foot pole casts a shadow of 12 ft. How high is the building?
4. Find the length (*AB*) of the pond shown in the figure below. Why are triangles *ABE* and *CDE* similar?

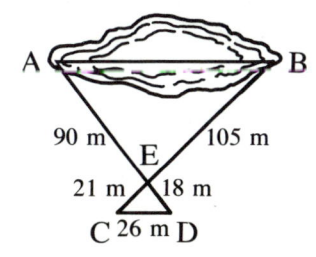

5. What is the distance (*AB*) across the stream shown in the figure below? Why are triangles *ABC* and *CDE* similar?

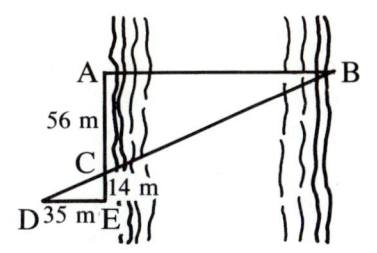

8–30 BY SCALE DRAWING

We determine the required lengths by first finding the scale (see page 380) if it is not known, then measuring the scale lengths and applying the scale to find the actual length.

Geometry

EXERCISES

1. Determine the required lengths indicated in the following triangles which are drawn to scale:

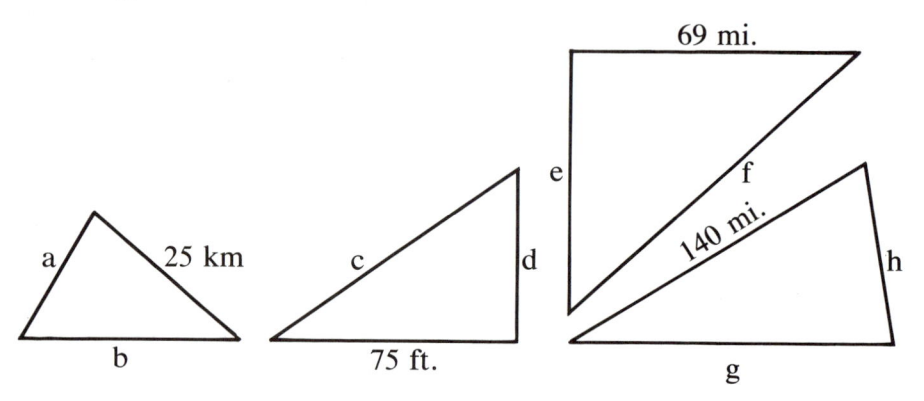

2. An airplane, flying due west (270°) from starting point A at an air speed of 400 km/h was blown off its course by a 50 km/h south wind (180°). Using the scale 1 mm = 10 km, make a drawing showing the position of the airplane at the end of one hour. How far was the airplane from its starting point? What course (or track) was it actually flying?

8–31 NUMERICAL TRIGONOMETRY

In trigonometry (meaning triangle measure) the relationships between the sides and the angles of the right triangle are used to determine certain parts of the triangle when the other parts are known. The ratios of the sides of the right triangle are related to the acute angles as follows:

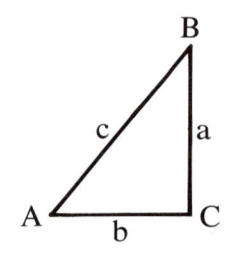

The ratio of the side opposite an acute angle to the adjacent side is called the *tangent* of the angle (abbreviated tan). The expression tan A means tangent of angle A. In right triangle ABC, $\tan A = \frac{a}{b}$ and $\tan B = \frac{b}{a}$.

The ratio of the side opposite an acute angle to the hypotenuse is called the *sine* of the angle (abbreviated sin). In the right triangle ABC, $\sin A = \frac{a}{c}$ and $\sin B = \frac{b}{c}$.

416   CHAPTER 8

The ratio of the adjacent side of an acute angle to the hypotenuse is called the *cosine* of the angle (abbreviated cos). In the right triangle ABC, $\cos A = \dfrac{b}{c}$ and $\cos B = \dfrac{a}{c}$.

Angles are measured both vertically and horizontally. An angle measured vertically between the horizontal line and the observer's line of sight to an object is called the *angle of elevation* when the object is above the observer and the *angle of depression* when the object is below the observer.

To solve problems using trigonometric ratios, we draw a right triangle if one is not given and place the given dimensions on it. We select the proper formula and substitute the given values, using the table of trigonometric values when necessary. Then we solve the resulting equation.

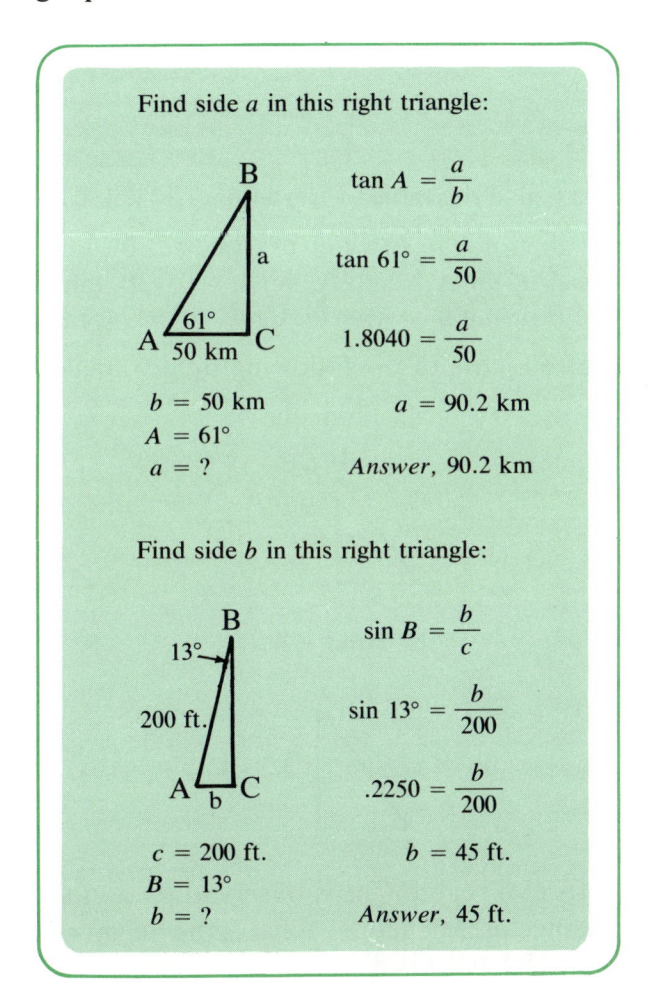

Find side a in this right triangle:

$\tan A = \dfrac{a}{b}$

$\tan 61° = \dfrac{a}{50}$

$1.8040 = \dfrac{a}{50}$

$b = 50$ km
$A = 61°$
$a = ?$

$a = 90.2$ km

Answer, 90.2 km

Find side b in this right triangle:

$\sin B = \dfrac{b}{c}$

$\sin 13° = \dfrac{b}{200}$

$.2250 = \dfrac{b}{200}$

$c = 200$ ft.
$B = 13°$
$b = ?$

$b = 45$ ft.

Answer, 45 ft.

Geometry

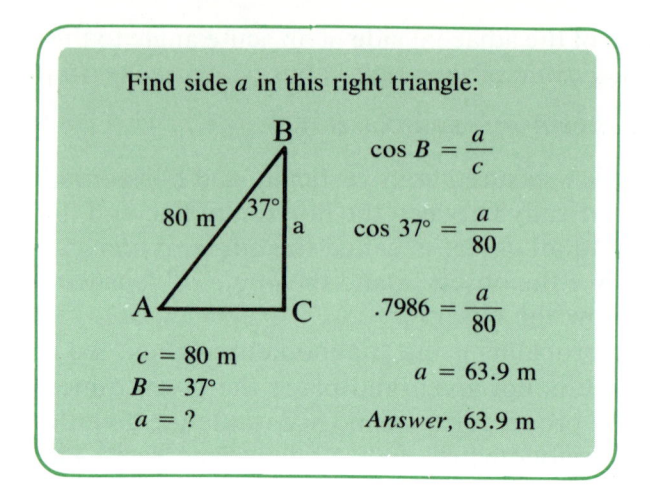

> Find side *a* in this right triangle:
>
> B
> 80 m / 37°
> a
> A ———— C
>
> $c = 80$ m
> $B = 37°$
> $a = ?$
>
> $\cos B = \dfrac{a}{c}$
>
> $\cos 37° = \dfrac{a}{80}$
>
> $.7986 = \dfrac{a}{80}$
>
> $a = 63.9$ m
>
> Answer, 63.9 m

EXERCISES

Tangent Ratio

When necessary, use the table of trigonometric values on page 420.

1. Find the value of: tan 61°, tan 17°, tan 42°30′, tan 9°15′.
2. Find angle A if: tan A = .8693, tan A = .5430, tan A = 2.3559.
3. Find angle B if: tan B = 1.9626, tan B = .9574, tan B = .0699.

Find the indicated parts of the following right triangles:

4. Find side *a*:

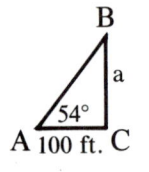

   B
   a
   54°
   A 100 ft. C

5. Find side *b*:

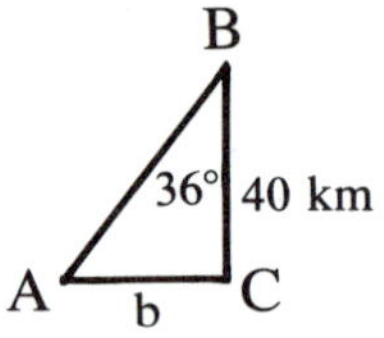

   B
   36° 40 km
   A — b — C

6. Find side *b*:

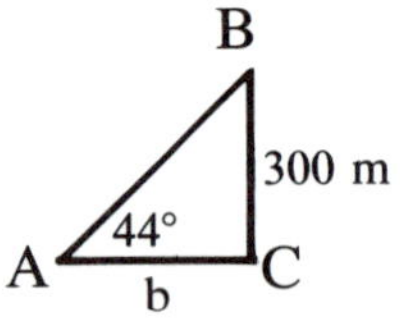

   B
   300 m
   44°
   A — b — C

7. Find side *a*:

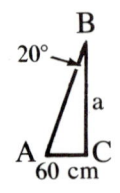

   B
   20°
   a
   A — C
   60 cm

8. Find ∠A:

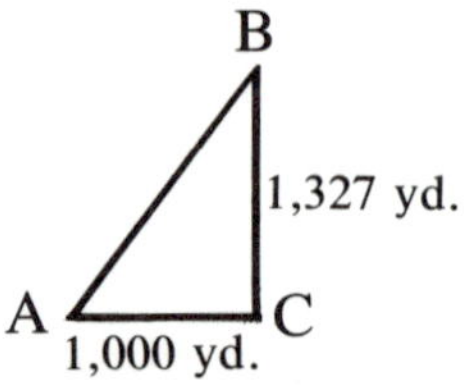

   B
   1,327 yd.
   A — C
   1,000 yd.

9. Find ∠B:

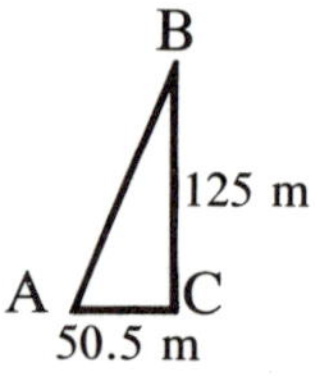

   B
   125 m
   A — C
   50.5 m

10. The light from a searchlight is observed on a cloud at a horizontal distance of 600 meters. What is the height of the cloud if the angle of elevation is 82°?

Sine Ratio

When necessary, use the table of trigonometric values on page 420.

11. Find the value of: sin 26°, sin 83°, sin 54°45′, sin 16°30′.
12. Find angle A if: sin A = .8290, sin A = .6626, sin A = .3090.
13. Find angle B if: sin B = .5446, sin B = .9288, sin B = .8829.

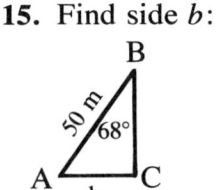

Find the indicated parts of the following right triangles:

14. Find side a: (400 ft., 19°, find a)
15. Find side b: (50 m, 68°, find b)
16. Find side b: (100 yd., 72°, find b)
17. Find side a: (750 km, 40°, find a) 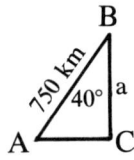
18. Find side c: (391 cm, 9°, find c)
19. Find side c: (120 m, 43°, find c)
20. Find ∠A: (126 mm, 63 mm)
21. Find ∠B: (1,000 yd., 829 yd.)

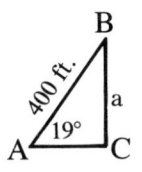

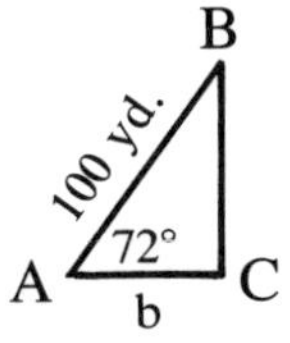

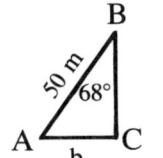

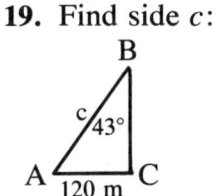

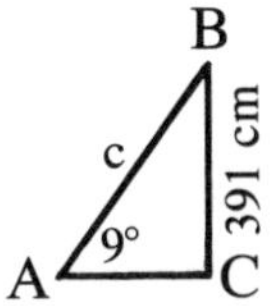

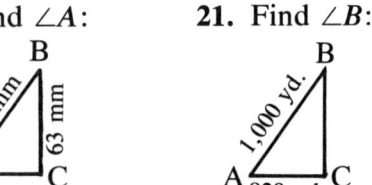

22. How high is a kite if 240 feet of string is let out and the string makes an angle of 52° with the ground?

Cosine Ratio

When necessary, use the table of trigonometric values on page 420.

23. Find the value of: cos 47°, cos 6°, cos 73°30′, cos 19°20′.
24. Find angle A if cos A = .2588, cos A = .8064, cos A = .9994.
25. Find angle B if: cos B = .9744, cos B = .5075, cos B = .0872.

Find the indicated parts of the following right triangles, using the cosine ratio:

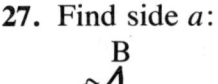

26. Find side b: (70 km, 27°, find b)
27. Find side a: (225 cm, 63°, find a) 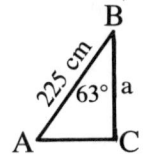
28. Find side a: (50 mi., 58°, find a)
29. Find side b: (850 cm, 20°, find b) 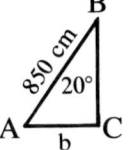
30. Find side c: (341 m, 47°, find c)
31. Find side c: (675 yd., 77°, find c)
32. Find ∠A: (500 mm, 383 mm)
33. Find ∠B: (2,000 yd., 908 yd.)

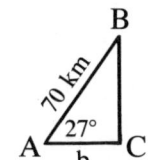

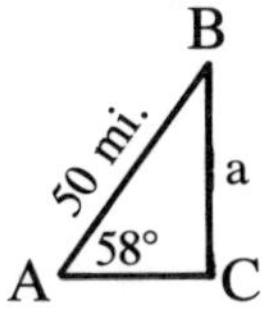

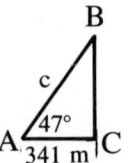

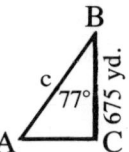

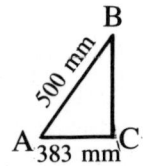

Geometry

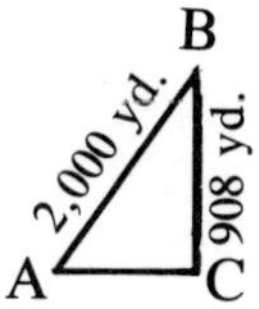

Table of Trigonometric Values

Angle	Sine	Cosine	Tangent	Angle	Sine	Cosine	Tangent
0°	.0000	1.0000	.0000	46°	.7193	.6947	1.0355
1°	.0175	.9998	.0175	47°	.7314	.6820	1.0724
2°	.0349	.9994	.0349	48°	.7431	.6691	1.1106
3°	.0523	.9986	.0524	49°	.7547	.6561	1.1504
4°	.0698	.9976	.0699	50°	.7660	.6428	1.1918
5°	.0872	.9962	.0875	51°	.7771	.6293	1.2349
6°	.1045	.9945	.1051	52°	.7880	.6157	1.2799
7°	.1219	.9925	.1228	53°	.7986	.6018	1.3270
8°	.1392	.9903	.1405	54°	.8090	.5878	1.3764
9°	.1564	.9877	.1584	55°	.8192	.5736	1.4281
10°	.1736	.9848	.1763	56°	.8290	.5592	1.4826
11°	.1908	.9816	.1944	57°	.8387	.5446	1.5399
12°	.2079	.9781	.2126	58°	.8480	.5299	1.6003
13°	.2250	.9744	.2309	59°	.8572	.5150	1.6643
14°	.2419	.9703	.2493	60°	.8660	.5000	1.7321
15°	.2588	.9659	.2679	61°	.8746	.4848	1.8040
16°	.2756	.9613	.2867	62°	.8829	.4695	1.8807
17°	.2924	.9563	.3057	63°	.8910	.4540	1.9626
18°	.3090	.9511	.3249	64°	.8988	.4384	2.0503
19°	.3256	.9455	.3443	65°	.9063	.4226	2.1445
20°	.3420	.9397	.3640	66°	.9135	.4067	2.2460
21°	.3584	.9336	.3839	67°	.9205	.3907	2.3559
22°	.3746	.9272	.4040	68°	.9272	.3746	2.4751
23°	.3907	.9205	.4245	69°	.9336	.3584	2.6051
24°	.4067	.9135	.4452	70°	.9397	.3420	2.7475
25°	.4226	.9063	.4663	71°	.9455	.3256	2.9042
26°	.4384	.8988	.4877	72°	.9511	.3090	3.0777
27°	.4540	.8910	.5095	73°	.9563	.2924	3.2709
28°	.4695	.8829	.5317	74°	.9613	.2756	3.4874
29°	.4848	.8746	.5543	75°	.9659	.2588	3.7321
30°	.5000	.8660	.5774	76°	.9703	.2419	4.0108
31°	.5150	.8572	.6009	77°	.9744	.2250	4.3315
32°	.5299	.8480	.6249	78°	.9781	.2079	4.7046
33°	.5446	.8387	.6494	79°	.9816	.1908	5.1446
34°	.5592	.8290	.6745	80°	.9848	.1736	5.6713
35°	.5736	.8192	.7002	81°	.9877	.1564	6.3138
36°	.5878	.8090	.7265	82°	.9903	.1392	7.1154
37°	.6018	.7986	.7536	83°	.9925	.1219	8.1443
38°	.6157	.7880	.7813	84°	.9945	.1045	9.5144
39°	.6293	.7771	.8098	85°	.9962	.0872	11.4301
40°	.6428	.7660	.8391	86°	.9976	.0698	14.3007
41°	.6561	.7547	.8693	87°	.9986	.0523	19.0811
42°	.6691	.7431	.9004	88°	.9994	.0349	28.6363
43°	.6820	.7314	.9325	89°	.9998	.0175	57.2900
44°	.6947	.7193	.9657	90°	1.0000	.0000	
45°	.7071	.7071	1.0000				

PERIMETER AND CIRCUMFERENCE

The *perimeter* of a polygon is the sum of the lengths of its sides. It is the distance around the polygon.

8–32 PERIMETER OF A RECTANGLE

We see from the figure below that the perimeter of a rectangle is equal to twice its length plus twice its width. Expressed as a formula this relationship is $p = 2l + 2w$ or $p = 2(l + w)$.

Find the perimeter of a rectangle 18 feet long and 16 feet wide:

$l = 18$ ft.
$w = 16$ ft.
$p = ?$

$p = 2l + 2w$
$p = 2 \times 18 + 2 \times 16$
$p = 36 + 32$
$p = 68$ feet

```
 18      16      36
×2      ×2     +32
 36     32      68
```

Answer, 68 feet

EXERCISES

Find the perimeters of rectangles having the following dimensions:

1.
Length	23 dm	145 ft.	216 mm	6.5 cm	8.25 m	5.625 km
Width	17 dm	98 ft.	307 mm	1.2 cm	9.75 m	4.875 km

2.
Length	$5\frac{3}{4}$ ft.	$7\frac{5}{8}$ in.	$4\frac{1}{2}$ yd.	2 ft. 7 in.	6 ft.	9 ft. 11 in.
Width	$3\frac{1}{2}$ ft.	$2\frac{13}{16}$ in.	$3\frac{2}{3}$ yd.	1 ft. 5 in.	7 ft. 4 in.	5 ft. 8 in.

3. How many meters of picture moulding are needed to go around the walls of a room 17 m long and 15 m wide?
4. Deducting 1 meter for a doorway, how many meters of baseboard are needed for a room 6 meters long and 5 meters wide?
5. At $1.70 a yard, how much will it cost to bind a rug 9 ft. by 12 ft.?
6. A rectangular field is 146 ft. long and 121 ft. wide. How many feet of fencing are required to enclose the field? At $4.20 per yard, how much will it cost?

Geometry

8–33 PERIMETER OF A SQUARE

Since the 4 sides of a square are of equal length, the perimeter of a square is 4 times the length of its side. Expressed as a formula this is $p = 4s$.

Find the perimeter of a square whose side is 31 meters:

31 m

$s = 31$ m
$p = ?$

$p = 4s$
$p = 4 \times 31$
$p = 124$ meters

$$\begin{array}{r} 31 \\ \times 4 \\ \hline 124 \end{array}$$

Answer, 124 meters

EXERCISES

Find the perimeters of squares whose sides measure:

1. a. 9 cm
   b. 47 ft.
   c. 6,080 m
   d. 1,760 mm
2. a. 0.75 m
   b. 1.25 dm
   c. 28.5 cm
   d. 5.9 km
3. a. $\frac{5}{8}$ in.
   b. $32\frac{3}{4}$ ft.
   c. $7\frac{13}{16}$ in.
   d. $29\frac{2}{3}$ yd.
4. a. 1 ft. 6 in.
   b. 3 ft. 4 in.
   c. 5 ft. 11 in.
   d. 8 yd. 2 ft.

5. How many yards of linoleum border are needed for a kitchen floor 11 feet by 11 feet?
6. If the distance between bases on a softball diamond is 60 ft., how many yards does a batter run when he hits a home-run?
7. At $.50 per meter, what would be the cost of weatherstripping for a window measuring 1.5 meters by 1.5 meters?
8. How much would it cost, at $4.80 per yard, to fence a 90-foot square garden?

8–34 PERIMETER OF A TRIANGLE

The perimeter of a triangle is equal to the sum of the lengths of its sides. The formula $p = a + b + c$ is sometimes used.

Since the sides of an equilateral triangle are equal, its perimeter is equal to 3 times the length of its side which expressed as a formula is $p = 3\,s$.

EXERCISES

1. Find the perimeters of triangles with sides measuring:
   a. 16 cm, 23 cm, 19 cm
   b. 3.5 m, 4.6 m, 2.75 m
   c. $9\frac{1}{2}$ yd., $8\frac{3}{4}$ yd., $10\frac{2}{3}$ yd.
   d. 1 ft. 7 in., 2 ft., 2 ft. 1 in.

2. Find the perimeters of equilateral triangles with sides measuring:
   a. 26 cm
   b. 9.25 km
   c. $7\frac{9}{16}$ in.
   d. 2 ft. 5 in.

3. Find the perimeters of isosceles triangles with the following dimensions:

Base	13 cm	74 mm	$4\frac{1}{2}$ ft.	7.5 m	2 ft. 4 in.
Each equal side	17 cm	59 mm	$3\frac{3}{4}$ ft.	4.3 m	1 ft. 11 in.

4. How many meters of fencing are needed to enclose a triangular lot with sides measuring 128 m, 106 m, and 117 m?

5. Joan wishes to sew ribbon around her school pennant which measures $29\frac{1}{2}$ in., $29\frac{1}{2}$ in., and 13 in. How many yards of ribbon will she need? At $.60 a yard, how much will the ribbon cost?

8–35 CIRCUMFERENCE OF A CIRCLE

The *circumference* of a circle is the length of the circle or the distance around the circle. To develop the relationship of the circumference, diameter, and radius of a circle, let us do the following:
  (1) Draw a circle having a radius of 2 centimeters. Determine the diameter by measuring it. Is the diameter of a circle twice as long as the radius? Does the formula $d = 2\,r$ express this relationship when d represents the diameter, and r the radius?

Geometry

(2) Draw a circle having a diameter of 5 centimeters. What is its radius? Is the radius of a circle one-half as long as the diameter? Does the radius equal the diameter divided by two? Does the formula $r = \frac{d}{2}$ express this relationship.

(3) Draw a circle with a radius of 3.5 centimeters. Draw three line segments from the center to points on the circle. Measure these segments. Are all radii of a circle of equal length? Extend each of these line segments until it meets the circle, thus forming a diameter. Measure the diameters. Are all diameters of a circle of equal length?

(4) Measure the diameters and circumferences of three circular objects. Copy the chart below and tabulate your measurements. Divide the circumference of each object by its diameter, finding the quotient correct to the nearest hundredth. Find the average of these quotients.

Object	Circumference (c) of object	Diameter (d) of object	Circumference divided by diameter $(c \div d) = \pi$

The circumference is a little more than 3 times as long as the diameter. This constant ratio is represented by the Greek letter π (called pi) which equals 3.14 or $3\frac{1}{7}$ or $\frac{22}{7}$. For greater accuracy 3.1416 is used.

Thus, the circumference of a circle is equal to pi (π) times the diameter.

Expressed as a formula it is: $c = \pi d$.

Since $d = 2r$, we may substitute $2r$ for d in the formula $c = \pi d$ to get $c = \pi\, 2\, r$ or better $c = 2\, \pi r$. Therefore we may say that the circumference of a circle is equal to two times pi (π) times the radius. The diameter of a circle is equal to the circumference divided by pi (π). Expressed as a formula it is $d = \frac{c}{\pi}$.

CHAPTER 8

Find the circumference of a circle whose diameter is 6 centimeters:

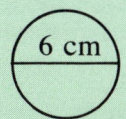

$d = 6$ cm
$\pi = 3.14$
$c = ?$

$c = \pi d$
$c = 3.14 \times 6$
$c = 18.84$ cm

$\begin{array}{r} 3.14 \\ \times 6 \\ \hline 18.84 \end{array}$

Answer, 18.84 centimeters

Find the circumference of a circle whose radius is 21 feet:

$r = 21$ ft.
$\pi = \frac{22}{7}$
$c = ?$

$c = 2\pi r$
$c = 2 \times \frac{22}{7} \times 21$
$c = 132$ ft.

$\frac{2}{1} \cdot \frac{22}{\cancel{7}} \cdot \frac{\cancel{21}^{3}}{1} = 132$

Answer, 132 feet

EXERCISES

Find the circumference of a circle having a diameter of:

1. a. 8 cm
   b. 21 ft.
   c. 37 mm
   d. 63 m

2. a. 2.5 km
   b. $8\frac{3}{4}$ in.
   c. $\frac{7}{8}$ mi.
   d. 1.625 m

3. a. $5\frac{1}{4}$ ft.
   b. 4 ft. 8 in.
   c. 3 ft. 6 in.
   d. 1 ft. 5 in.

Geometry 425

Find the circumference of a circle having a radius of:

4. a. 9 mm   **c.** 85 cm
   b. 42 in.   **d.** 280 dm

5. a. $7\frac{7}{8}$ in.   **c.** $1\frac{5}{16}$ in.
   b. $2\frac{1}{3}$ yd.   **d.** 3.75 km

6. a. 8.6 m   **c.** 1 ft. 3 in.
   b. 3 ft. 1 in.   **d.** 2 ft. 4 in.

7. What distance do you ride in one turn of a Ferris wheel when you sit 6 meters from the center?

8. How many times must you go around a circular track with a diameter of 105 feet in order to run a mile?

9. Over what distance will the tip of the minute hand of a clock move in 1 hour? The minute hand is 12 cm long.

10. How long a metal bar is needed to make a hoop for a barrel with a diameter of 28 inches?

11. What distance does a point on a circular saw travel in one turn if the saw has a diameter of 40 centimeters?

12. The diameter of the earth is 7,918 miles. Find the circumference of the earth.

13. How many yards of wire fencing are needed to enclose a circular flower bed 16 feet in diameter?

14. The wheels on a truck have a diameter of 35 inches. How many times does each wheel revolve when the truck goes a mile?

15. Find the diameter of a tree whose circumference is 110 cm.

AREA

The *area* of the interior (or closed region) of any plane figure is the number of units of square measure it contains. When computing the area of a geometric figure, we express all linear units in the same denomination.

8–36 AREA OF A RECTANGLE

At the right we see that one measurement indicates the number of square units in a row and the other measurement indicates how many rows there are:

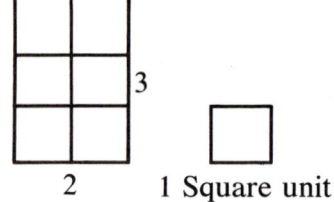

                                      2   1 Square unit

Thus to find the area of the interior of a rectangle we multiply the length by the width. Expressed as a formula this relationship is $A = lw$. Or, the area is equal to the altitude times the base. Formula: $A = ab$.

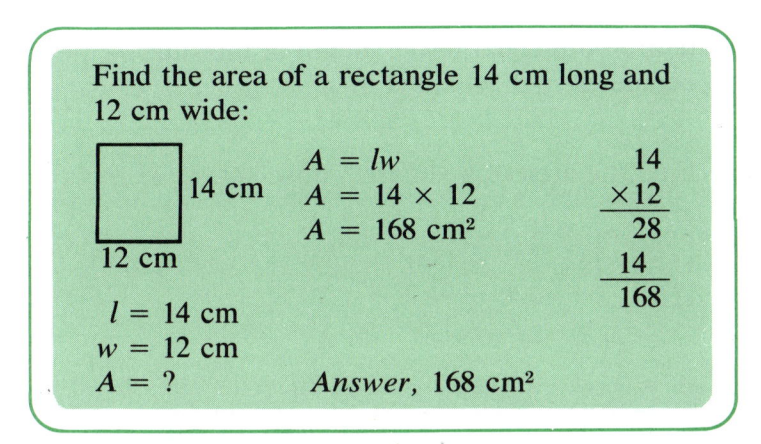

> Find the area of a rectangle 14 cm long and 12 cm wide:
>
> 14 cm
> 12 cm
>
> $A = lw$
> $A = 14 \times 12$
> $A = 168 \text{ cm}^2$
>
> $\begin{array}{r} 14 \\ \times 12 \\ \hline 28 \\ 14 \\ \hline 168 \end{array}$
>
> $l = 14$ cm
> $w = 12$ cm
> $A = ?$
>
> *Answer*, 168 cm²

EXERCISES

Find the area of rectangles having the following dimensions:

1.

Length	16 mm	58 ft.	175 km	3.4 cm	9.5 m	4.6 km
Width	9 mm	65 ft.	128 km	0.8 cm	6.25 m	8.3 km

2.

Length	$7\frac{1}{2}$ in.	$2\frac{3}{4}$ yd.	$6\frac{1}{4}$ in.	5 ft. 9 in.	1 ft. 6 in.	8 ft. 2 in.
Width	5 in.	$8\frac{2}{3}$ yd.	$1\frac{3}{8}$ in.	4 ft.	2 ft. 8 in.	6 ft. 10 in.

3. Find the cost of each of the following:
   a. Cementing a sidewalk 15 m by 2 m at $13.50 per sq. m.
   b. Covering a floor 18 ft. by 12 ft. with broadloom at $14.95 per sq. yd.
   c. Sodding a lawn 12 ft. 6 in. by 9 ft. 8 in. at $.20 per sq. ft.

4. What is the picture area of a television screen 47 cm wide and 36 cm high?
5. The official length of the United States flag is 1.9 times its width. What is the area of a flag 4 m wide?
6. How many square feet of floor space remain uncovered if a 9 ft. by 12 ft. rug is placed on a floor 17 ft. long and 14 ft. wide?

Geometry

8-37 AREA OF A SQUARE

Although we may use the formula $A = lw$ to find the area of the interior of a square, we generally use the formula $A = s^2$ where s is the length of the side of the square. Since the length and width of a square are both equal to the length of the side, s, the formula $A = lw$ becomes $A = s \times s$ or $A = s^2$.

Find the area of a square whose side is 43 feet:

43'

$s = 43$ ft.
$A = ?$

$A = s^2$
$A = (43)^2$
$A = 43 \times 43$
$A = 1,849$ sq. ft.

Answer, 1,849 sq. ft.

```
   43
 ×43
  ───
  129
  172
 ─────
 1,849
```

EXERCISES

Find the areas of squares whose sides measure:

1. **a.** 8 km **b.** 59 mm **c.** 440 dm **d.** 5,280 ft.
2. **a.** 9.5 cm **b.** 1.09 m **c.** 3.28 km **d.** 0.62 m
3. **a.** $5\frac{1}{2}$ yd. **b.** $16\frac{1}{2}$ ft. **c.** $5\frac{3}{4}$ ft. **d.** $\frac{7}{8}$ in.
4. **a.** 1 ft. 3 in. **b.** 2 ft. 5 in. **c.** 7 ft. 8 in. **d.** 2 yd. 1 ft.
5. Find the cost of each of the following:
    a. Covering a floor 10 ft. by 10 ft. with linoleum at $7.75 per sq. yd.
    b. Resilvering a mirror 18 in. by 18 in. at $6.90 per sq. ft.
    c. Refinishing a floor 15 ft. 6 in. by 15 ft. 6 in. at $.28 per sq. ft.
6. How many square yards of tarpaulin are needed to cover the infield of a baseball diamond if the distance between bases is 90 feet?
7. How many times as large as a 2-centimeter square is a 10-centimeter square?
8. How many square meters of ground remain when a building 105 m square is erected on a lot 180 m square?

8-38 AREA OF A PARALLELOGRAM

The *base* of a parallelogram is the side on which it rests. The *altitude* or *height* is the perpendicular segment between the base and its opposite side.

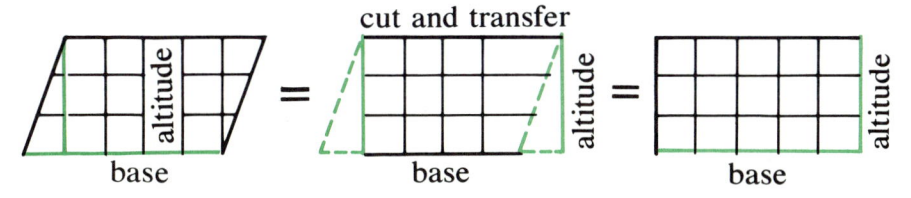

The above diagrams illustrate that the area of the interior of a parallelogram is equal to the product of the altitude and the base, formula $A = ab$, or the product of the base and height; formula: $A = bh$.

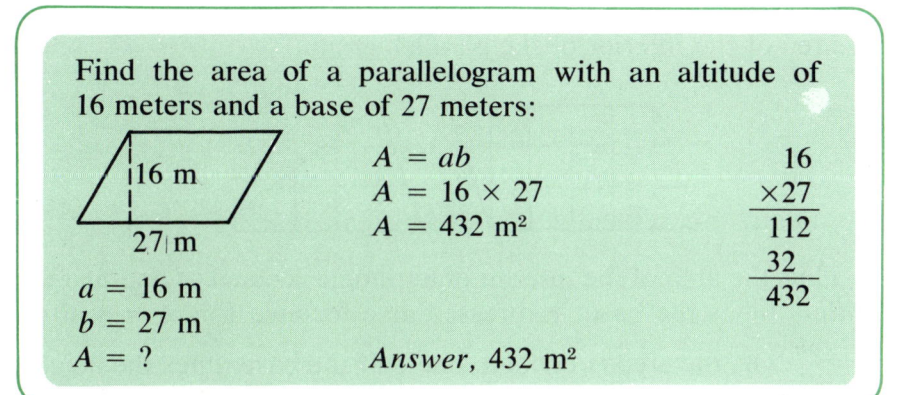

Find the area of a parallelogram with an altitude of 16 meters and a base of 27 meters:

$A = ab$
$A = 16 \times 27$
$A = 432$ m²

$a = 16$ m
$b = 27$ m
$A = ?$

```
   16
  ×27
  ---
  112
   32
  ---
  432
```

Answer, 432 m²

EXERCISES

Find the areas of parallelograms having the following dimensions:

1.
Altitude	19 m	53 in.	165 mm	3.5 cm	6.8 km	9.5 m
Base	21 m	37 in.	86 mm	6 cm	1.4 km	6.25 m

2.
Altitude	$1\frac{7}{8}$ in.	$4\frac{2}{3}$ ft.	$2\frac{1}{2}$ yd.	1 ft. 3 in.	5 yd. 1 ft.	1 yd. 18 in.
Base	$1\frac{3}{4}$ in.	$2\frac{1}{4}$ ft.	$3\frac{3}{4}$ yd.	2 ft.	3 yd. 2 ft.	2 yd. 6 in.

Geometry

3. What is the area of a lot shaped like a parallelogram with a base of 145 m and an altitude of 124 m?
4. A parking lot has 46 spaces, each a parallelogram with a base of 7 ft. and an altitude of 15 ft. What is the total parking area of the lot?
5. Which parallelogram has the greater area, one with an altitude of 8 ft. 6 in. and a base of 6 ft. 6 in. or one with an altitude of 3 yd. and a base of 2 yd.? How much greater?
6. How many square meters of sod are needed for a lawn shaped like a parallelogram with a base of 32 m and an altitude of 15 m? At $1.98 per m² what is the cost of the sod?

8-39 AREA OF A TRIANGLE

A diagonal separates a parallelogram into two congruent triangles. The area of the interior of each triangle is equal to one-half the area of the interior of the parallelogram.

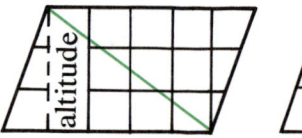

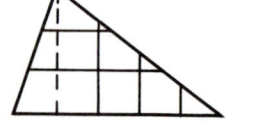

Thus the area of the interior of a triangle is equal to one-half the altitude times the base. Expressed as a formula it is $A = \frac{1}{2} ab$ or $A = \frac{ab}{2}$. Or, the area is equal to one-half the base times the height. Formula: $A = \frac{1}{2} bh$.

Find the area of a triangle with an altitude of 12 millimeters and a base of 9 millimeters:

$a = 12$ mm
$b = 9$ mm
$A = ?$

$A = \frac{1}{2} ab$
$A = \frac{1}{2} \times 12 \times 9$
$A = 54$ mm

$\frac{1}{2} \times \frac{\cancel{12}^{6}}{1} \times \frac{9}{1} = 54$

Answer, 54 mm²

---

430   CHAPTER 8

EXERCISES

Find the area of a triangle having the following dimensions:

1.
Altitude	20 cm	8 mm	53 km	$2\frac{1}{2}$ in.	$8\frac{3}{4}$ ft.	7.4 m
Base	32 cm	13 mm	47 km	3 in.	$4\frac{1}{4}$ ft.	5.9 m

2.
Base	$6\frac{2}{3}$ yd.	9.3 km	$1\frac{5}{8}$ in.	4 ft.	1 ft. 4 in.	5 yd. 9 in.
Height	$1\frac{7}{8}$ yd.	4.5 km	$2\frac{3}{4}$ in.	2 ft. 9 in.	1 ft. 11 in.	3 yd. 6 in.

3. What is the area of a triangular plot having a base of 69 m and an altitude of 86 m?
4. How many square feet of felt are used in a pennant with a base of 12 inches and an altitude of 30 inches?
5. A triangular sail has a base of 4.7 m and an altitude of 5.8 m. How many m² of surface does each side of the sail expose?
6. Allowing 3 sq. ft. per person, find the number of persons a triangular safety traffic island can hold if it has a base of 16 ft. and an altitude of 4 ft. 6 in.

8–40 AREA OF A TRAPEZOID

The figure below shows that the diagonal separates the trapezoid into two triangles which have a common height but different bases. The area of one triangle is $\frac{1}{2}b_1h$ and of the other is $\frac{1}{2}b_2h$. The area of the interior of the trapezoid is equal to the sum of the areas of the interiors of the two triangles or $\frac{1}{2}b_1h + \frac{1}{2}b_2h$. Using the distributive principle we find that the area of the interior of a trapezoid is equal to the height times the average of the two parallel sides (bases). Expressed as a formula it is $A = h \times \frac{b_1 + b_2}{2}$ or sometimes as $A = \frac{h}{2}(b_1 + b_2)$.

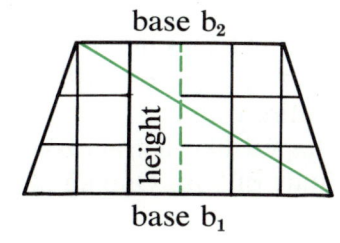

Geometry

> Find the area of a trapezoid with bases of 18 meters and 14 meters and a height of 15 meters:
>
> $A = h \times \dfrac{b_1 + b_2}{2}$
>
> $A = 15 \times \dfrac{18 + 14}{2}$
>
> $A = 15 \times 16$
>
> $A = 240 \text{ m}^2$
>
> $h = 15$ m
> $b_1 = 18$ m
> $b_2 = 14$ m
> $A = ?$
>
> ```
>   18          15
>  +14         ×16
>   32          90
>              15
>             ---
>             240
>
>    16
>  2)32
> ```
>
> Answer, 240 m²

EXERCISES

Find the areas of trapezoids having the following dimensions:

1.
Height	9 dm	21 km	13 yd.	15 cm	85 mm	37 m
Upper Base	8 dm	17 km	46 yd.	19 cm	102 mm	26 m
Lower Base	16 dm	29 km	53 yd.	26 cm	97 mm	42 m

2.
Height	$5\frac{3}{4}$ in.	8 in.	6 yd.	8 in.	1 ft.	1 ft. 7 in.
Upper Base	$4\frac{1}{2}$ in.	$2\frac{3}{4}$ in.	$1\frac{2}{3}$ yd.	2 ft.	1 ft. 11 in.	2 ft. 4 in.
Lower Base	$11\frac{1}{2}$ in.	$6\frac{7}{8}$ in.	$3\frac{1}{4}$ yd.	1 ft. 6 in.	2 ft. 9 in.	1 yd. 8 in.

3. Find the area of a section of a tapered airplane wing which has the shape of a trapezoid. The two parallel sides, measuring 4 ft. 3 in. and 5 ft. 9 in., are 16 ft. apart.
4. A porch roof consists of 2 sections each shaped like a trapezoid. The bases of one section are 11 ft. and 13 ft. and of the other 17 ft. and 19 ft. The distance between the bases in both sections is 8 ft. At the spreading rate of 250 sq. ft. per gallon, will a gallon of paint be sufficient to cover the entire roof?
5. Find the cost of seeding a lawn shaped like a trapezoid with bases of 23 m and 17 m and an altitude of 9 m. One kg of grass seed covers 30 m² and costs $6.49.

8-41 AREA OF A CIRCLE

If we divide the interior of a circle into sectors and arrange them in a shape approximating a parallelogram, we see that the area of the interior of a circle equals the area of the interior of the parallelogram whose base is one-half the circumference ($\frac{1}{2} c$) and the altitude is the radius (r). The formula for this is $A = \frac{1}{2} cr$. Since $c = 2 \pi r$, we may substitute $2 \pi r$ for c in the formula of $A = \frac{1}{2} cr$ to obtain $A = \frac{1}{2} \cdot 2 \pi r \cdot r$ which simplified is $A = \pi r^2$. The area of a circle is equal to pi (π) times the radius squared.

Or by substituting $\frac{d}{2}$ for r in the formula $A = \pi r^2$ we obtain the formula $A = \frac{1}{4} \pi d^2$ which may also be used to find the area of the interior of a circle.

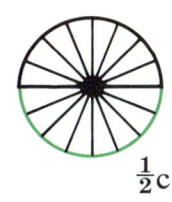

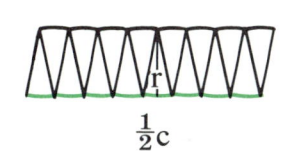

Find the area of a circle having a radius of 8 inches:

$r = 8$ in.
$\pi = 3.14$
$A = ?$

$A = \pi r^2$
$A = 3.14 \times (8)^2$
$A = 3.14 \times 64$
$A = 200.96$ sq. in.

```
    8          3.14
   ×8          ×64
   ──         ────
   64         1256
              1884
             ──────
             200.96
```

Answer, 200.96 sq. in.

Find the area of a circle having a diameter of 28 millimeters:

$d = 28$ mm
$\pi = \frac{22}{7}$
$A = ?$

$A = \frac{1}{4} \pi d^2$
$A = \frac{1}{4} \times \frac{22}{7} \times (28)^2$
$A = 616$ mm²

Answer, 616 mm²

Geometry

EXERCISES

Find the area of a circle having a radius of:

1. **a.** 6 m   **b.** 35 in.   **c.** 78 cm   **d.** 112 mm
2. **a.** 7.3 km   **b.** $1\frac{3}{4}$ in.   **c.** $\frac{5}{16}$ mi.   **d.** 4.625 m
3. **a.** $11\frac{3}{8}$ in.   **b.** 2 ft. 11 in.   **c.** 9 yd. 1 ft.   **d.** 8 ft. 9 in.

Find the area of a circle having a diameter of:

4. **a.** 5 km   **b.** 42 cm   **c.** 110 mm   **d.** 56 m
5. **a.** 8.6 cm   **b.** $4\frac{1}{2}$ ft.   **c.** $3\frac{15}{16}$ in.   **d.** 0.18 km
6. **a.** $\frac{7}{8}$ in.   **b.** 2 ft. 6 in.   **c.** 8 ft. 2 in.   **d.** 1 yd. 1 ft.

7. Over how many square kilometers can a program be received if a TV station can televise programs for a distance of 63 kilometers?
8. What is the area of a circular skating rink 50 meters in diameter?
9. The viewing area of a 21-inch circular television screen is how many times as large as the area of a 14-inch circular screen?
10. If the lookout in the "crow's nest" of a ship can see for a distance of 12 kilometers, over how many square kilometers can he observe?
11. What is the area of the largest circle that can be cut out of paper 21 cm by 36 cm?
12. If the pressure is the same, how many times as much water will flow through a pipe 24 inches in diameter as through a 2-inch pipe?
13. A circular fountain 10 m in diameter has a circular walk 3 m wide paved around it. What is the area of the walk? At $15.75 per m², what was the cost of the walk?
14. How many times as large is the area of the 200-inch mirror of the world's largest telescope compared to the area of the previous record 100-inch telescope?
15. Find the area of the head of the piston if the diameter is $4\frac{3}{8}$ inches.
16. What is the cross-sectional area of the metal of a steel tube with a wall thickness of 0.25 centimeter and an outside diameter of 3 centimeters?

8-42 TOTAL AREA OF A RECTANGULAR SOLID AND A CUBE

The total area of the outside surface of a rectangular solid is the area of its six faces. Expressed as a formula it is:

$$A = 2\,lw + 2\,lh + 2\,wh$$

The total area of the outside surface of a cube is the area of its six congruent square faces. Expressed as a formula it is:

$$A = 6\,e^2 \text{ or } A = 6\,s^2$$

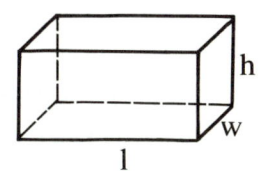

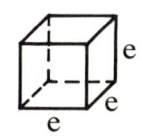

EXERCISES

1. Find the total areas of rectangular solids with the following dimensions:

Length	18 m	23 in.	54 mm	9.2 cm	$5\frac{3}{4}$ in.	2 ft. 6 in.
Width	6 m	17 in.	25 mm	5.4 cm	4 in.	1 ft. 9 in.
Height	15 m	39 in.	40 mm	7.5 cm	$6\frac{1}{2}$ in.	3 ft.

2. Find the total areas of cubes whose sides measure:

   a. 9 m
   b. 25 ft.
   c. 40 mm
   d. 17 m
   e. $3\frac{1}{2}$ ft.
   f. 5.8 cm
   g. $4\frac{3}{4}$ in.
   h. 2 ft. 8 in.

3. How many square centimeters of cardboard are needed to make 5 dozen boxes with lids? Each box is to be 12 cm long, 9 cm wide, and 6 cm deep. Each lid is to have an overlapping band 1 cm wide all around. Allow 10% for waste.

4. At $5.40 per sq. yd., find the cost of plastering the walls and ceiling of a room 15 ft. long, 13 ft. wide, and 9 ft. high. Allow 33 sq. ft. for openings.

Geometry

5. Allowing 82 sq. ft. for windows and doorway, how many gallons of paint are needed to cover the walls and ceiling of a room 17 ft. long, 16 ft. wide, and 10 ft. high with two coats of paint? One gallon will cover 425 sq. ft. in one coat. At $9.25 a gallon, how much will the paint cost?

6. How many double rolls of wallpaper are needed to cover the walls of a room 16 ft. long, 12 ft. wide, and 8 ft. high if each double roll covers 72 sq. ft.? Deduct 52 sq. ft. for openings and allow $\frac{1}{2}$ roll extra paper for matching. Find the cost of the wallpaper at $9.95 per roll.

8–43 LATERAL AREA AND TOTAL AREA OF A RIGHT CIRCULAR CYLINDER

If we cut the label on a can vertically and flatten it out, we see that the lateral area or the area of the curved surface is equal to the area of a rectangle whose dimensions are those of the circumference and the height of the can.

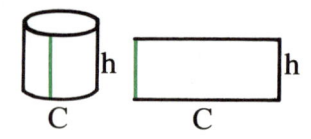

The lateral area is equal to the product of the circumference and height or $A = ch$. Since $c = \pi d$, we may substitute πd for c in $A = ch$ which then becomes $A = \pi dh$. Since $c = 2\pi r$, we may substitute $2\pi r$ for c in $A = ch$ which then becomes $A = 2\pi rh$. Thus the lateral area may be computed by using any one of the following formulas:

$$A = ch \quad \text{or} \quad A = \pi dh \quad \text{or} \quad A = 2\pi rh$$

Since the lower and upper bases are congruent circles (area of each is πr^2), the total area of a right circular cylinder equals the lateral area plus the area of the two bases. Any one of the following formulas may be used:

$$A = 2\pi rh + 2\pi r^2 \quad \text{or} \quad A = 2\pi r(r + h) \quad \text{or} \quad A = \pi dh + \tfrac{1}{2}\pi d^2$$

Find the total area of a cylinder when its radius is 6 feet and its height is 11 feet:

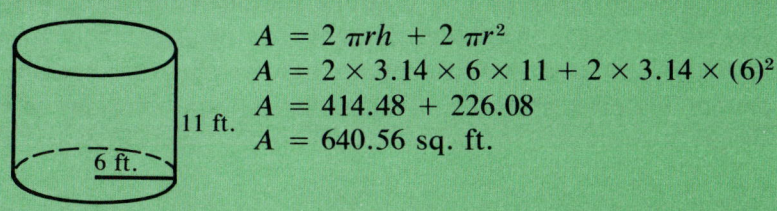

$A = 2\pi rh + 2\pi r^2$
$A = 2 \times 3.14 \times 6 \times 11 + 2 \times 3.14 \times (6)^2$
$A = 414.48 + 226.08$
$A = 640.56$ sq. ft.

$r = 6$ ft.
$h = 11$ ft.
$\pi = 3.14$
$A = ?$

Answer, 640.56 sq. ft.

EXERCISES

1. Find the lateral area of cylinders with the following dimensions:

a.
Radius	9 m	35 in.	16 mm	4.5 cm	$19\frac{1}{4}$ ft.	5 ft. 8 in.
Height	14 m	40 in.	13 mm	8 cm	$24\frac{2}{3}$ ft.	10 ft.

b.
Diameter	5 m	13 ft.	42 mm	$3\frac{15}{16}$ in.	10.5 cm	3 ft. 6 in.
Height	6 m	29 ft.	25 mm	4 in.	15.7 cm	8 ft. 4 in.

2. Find the total areas of cylinders with the following dimensions:

a.
Radius	6 m	14 cm	$17\frac{1}{2}$ ft.
Height	8 m	30 cm	$27\frac{1}{2}$ ft.

b.
Diameter	4 cm	49 mm	$6\frac{1}{2}$ in.
Height	7 cm	38 mm	$3\frac{1}{8}$ in.

3. A tennis court roller is 18 inches in diameter and 3 feet long. How much surface is rolled in one rotation?

4. How many square meters of paper are needed to make 7,200 labels for cans, each 7 centimeters high and 4 centimeters in diameter?

5. How many square meters of asbestos covering are used to enclose the curved surface and the two ends of a water storage tank 35 centimeters in diameter and 2 meters high?

VOLUME—MEASURE OF SPACE

Space has three dimensions: length, width, and height (or depth or thickness). We live in a three-dimensional world. The volume, also called capacity or cubical contents, is the number of units of cubic measure contained in a given space. When computing the volume of a geometric solid, we express all linear units in the same denomination.

8–44 VOLUME OF A RECTANGULAR SOLID

A one-centimeter cube contains a volume of 1 cubic centimeter. The rectangular solid below has 4 cubes in each row, 3 rows of cubes and 2 layers of cubes. In one layer there are 4×3 or 12 cubes; in two layers there are $4 \times 3 \times 2$ or 24 cubes which contain a total volume of 24 cubic centimeters. The volume of a rectangular solid is equal to the length times the width times the height. Expressed as a formula it is $V = lwh$. Sometimes the formula $V = Bh$ is used where B is the area of the base (lw) of the rectangular solid.

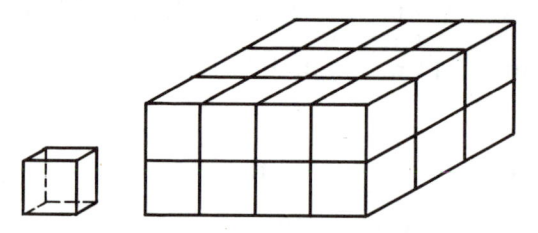

Find the volume of a rectangular solid 9 meters long, 4 meters wide, and 8 meters high:

$l = 9$ m
$w = 4$ m
$h = 8$ m
$V = ?$

$V = lwh$
$V = 9 \times 4 \times 8$
$V = 288$ m³

$$\begin{array}{r} 9 \\ \times 4 \\ \hline 36 \end{array} \qquad \begin{array}{r} 36 \\ \times 8 \\ \hline 288 \end{array}$$

Answer, 288 m³

---

438 CHAPTER 8

EXERCISES

Find the volume of rectangular solids having the following dimensions:

1.

Length	15 km	27 in.	59 mm	6.5 cm	9.25 m
Width	9 km	16 in.	32 mm	4.8 cm	6.2 m
Height	13 km	14 in.	70 mm	5.3 cm	4.75 m

2.

Length	$4\frac{1}{2}$ in.	$2\frac{1}{4}$ ft.	1 yd.	3 ft. 8 in.	8 ft.
Width	$5\frac{3}{4}$ in.	$1\frac{1}{3}$ ft.	2 ft. 6 in.	4 ft. 2 in.	6 ft. 6 in.
Height	8 in.	$3\frac{1}{2}$ ft.	2 ft.	5 ft. 6 in.	3 ft. 9 in.

3. A classroom is 10 m long, 9 m wide, and 5 m high. If 30 pupils are assigned to the room, how many cubic meters of air space does this allow for each pupil?
4. How many cubic decimeters of dirt will fill a box 4 meters long, 25 centimeters wide, and 16 centimeters deep?
5. At $8.75 per m³, how much will it cost to dig a basement 14 m long, 7 m wide, and 3 m deep?
6. A swimming pool 100 ft. long and 32 ft. wide is filled to an average depth of 5 ft. What is the weight of the water if a cu. ft. of water weighs $62\frac{1}{2}$ lb.? How many gallons of water does the swimming pool contain if a cubic foot holds $7\frac{1}{2}$ gallons? At the rate of 400 gallons per minute, how long will it take to fill the pool?
7. How many cartons, 2 ft. by 2 ft. 6 in. by 1 ft., can be stored on a freight car with inside dimensions of 40 ft. by 8 ft. by 8 ft.?
8. What is the weight of a steel bar 6 meters long, 7.5 centimeters wide, and 2 centimeters thick if 1 cm³ of steel weighs 7.75 g?

8–45 VOLUME OF A CUBE

Since the length, width, and height of a cube are all equal to the length of the edge of the cube, the formula $V = lwh$ becomes $V = e \times e \times e$ or $V = e^3$. The volume of a cube is equal to the length of its edge cubed. Sometimes the formula $V = s^3$ is used.

Geometry

Find the volume of a cube whose edge is 14 centimeters:

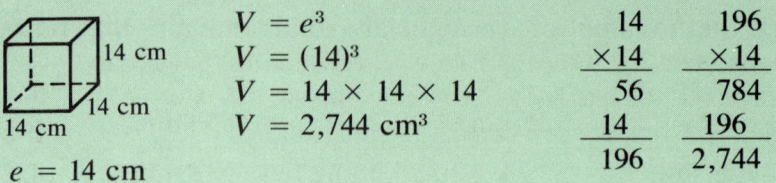

$V = e^3$
$V = (14)^3$
$V = 14 \times 14 \times 14$
$V = 2{,}744 \text{ cm}^3$

$e = 14$ cm
$V = ?$

Answer, 2,744 cm³

```
   14        196
 ×14        ×14
  56        784
  14        196
 196      2,744
```

EXERCISES

Find the volumes of cubes whose sides measure:

1. **a.** 8 m **c.** 50 mm
   b. 34 ft. **d.** 26 cm

2. **a.** .04 m **c.** 3.28 m
   b. 39.3 cm **d.** 21.75 cm

3. **a.** $16\frac{1}{2}$ ft. **c.** $1\frac{7}{8}$ in.
   b. $2\frac{3}{4}$ in. **d.** $8\frac{2}{3}$ ft.

4. **a.** 1 ft. 6 in. **c.** 3 yd. 1 ft.
   b. 4 ft. 2 in. **d.** 2 yd. 10 in.

5. Find the weight of a cubical column of water, 6 meters long, wide, and deep. A cubic decimeter of water weighs 1 kilogram.

6. How many bushels will a storage bin hold if it measures $7\frac{1}{2}$ ft. on each side? $1\frac{1}{4}$ cu. ft. = 1 bushel.

7. What is the weight of a cake of ice, $1\frac{1}{2}$ ft. long, wide, and thick? Ice weighs 57 lb. per cu. ft.

8. If 1 ton of coal occupies 35 cu. ft., how many tons of coal could be stored in a bin measuring $10\frac{1}{2}$ ft. by $10\frac{1}{2}$ ft. by $10\frac{1}{2}$ ft.?

9. Which has a greater volume and how much greater, a group of three 4-centimeter cubes or four 3-centimeter cubes?

10. The volume of a 12-millimeter cube is how many times as large as the volume of a 2-millimeter cube?

8–46 VOLUME OF A RIGHT CIRCULAR CYLINDER

To determine the formula for the volume of a right circular cylinder we apply the principle that the volume is equal to the area of the base of the cylinder times the height ($V = Bh$). The same principle is used in determining the volume of a rectangular solid. Since the area of the base of the cylinder is the area of a circle, the formula $V = Bh$ becomes $V = \pi r^2 h$. Thus the volume of a cylinder is equal to pi (π) times the square of the radius of the base times the height.

Find the volume of a cylinder 45 feet high with its base having a radius of 20 feet:

$r = 20$ ft.
$h = 45$ ft.
$\pi = 3.14$
$V = ?$

$V = \pi r^2 h$
$V = 3.14 \times (20)^2 \times 45$
$V = 3.14 \times 400 \times 45$
$V = 56{,}520$ cu. ft.

$\begin{array}{r} 20 \\ \times 20 \\ \hline 400 \end{array}$ $\quad$ $\begin{array}{r} 3.14 \\ \times 400 \\ \hline 1{,}256.00 \end{array}$

$\begin{array}{r} 1{,}256 \\ \times 45 \\ \hline 6280 \\ 5024 \\ \hline 56{,}520 \end{array}$

Answer, 56,520 cu. ft.

When the diameter is known, the formula $V = \frac{1}{4}\pi d^2 h$ may be used.

EXERCISES

Find the volumes of cylinders having the following dimensions:

1.

Radius	6 m	7 in.	42 mm	9.8 cm	3.5 m
Height	9 m	20 in.	35 mm	6.5 cm	10 m

2.

Radius	$3\frac{1}{2}$ in.	$4\frac{3}{8}$ in.	$5\frac{3}{4}$ ft.	2 ft. 4 in.	1 ft. 3 in.
Height	10 in.	$9\frac{1}{4}$ in.	$4\frac{2}{3}$ ft.	8 ft.	4 ft. 2 in.

Geometry

3.

Diameter	5 m	28 cm	39 mm	4.1 m	2.8 cm
Height	8 m	25 cm	56 mm	3.7 m	8.25 cm

4.

Diameter	$3\frac{1}{2}$ in.	$5\frac{1}{4}$ ft.	$6\frac{3}{4}$ in.	1 yd. 6 in.	5 ft. 8 in.
Height	4 in.	$10\frac{2}{3}$ ft.	$2\frac{5}{8}$ in.	10 ft.	2 ft. 10 in.

5. How many cubic meters of dirt must be removed in digging a well 2 meters in diameter and 18 meters deep? At $16.50 per m³, how much will the excavation cost?
6. An oil drum is 63 centimeters in diameter and 1.2 meters high. How many liters of oil will it hold?
7. How many gallons will a gasoline tank hold if its diameter is 35 ft. and it is 30 ft. high?
8. How many bushels of silage will a silo hold if it has an inside diameter of 24 ft. and a height of 42 ft.?
9. A water storage tank has a diameter of 49 cm and a height of 2 m. How many liters of water will it hold?

VOLUME OF A SPHERE, RIGHT CIRCULAR CONE, AND PYRAMID

Sphere

The volume of a sphere is equal to $\frac{4}{3}$ times pi (π) times the cube of the radius. Expressed as a formula it is $V = \frac{4}{3} \pi r^3$. Sometimes the formula $V = \frac{\pi d^3}{6}$ is used when the diameter is known.

EXERCISES

1. Find the volumes of spheres having the following radii:
   a. 5 m   b. 21 cm   c. $8\frac{3}{4}$ in.   d. 2 ft. 6 in.
2. Find the volumes of spheres having the following diameters:
   a. 18 cm   b. 49 mm   c. $2\frac{5}{8}$ in.   d. 6 ft. 5 in.

Right Circular Cone

The volume of a right circular cone is equal to $\frac{1}{3}$ times pi (π) times the square of the radius of the base times the height. Expressed as a formula it is $V = \frac{1}{3}\pi r^2 h$.

EXERCISES

Find the volumes of right circular cones having the following dimensions:

a. Radius: 6 mm, height: 10 mm
b. Radius: $3\frac{1}{2}$ ft., height: 9 ft.
c. Diameter: 20 cm, height: 35 cm
d. Diameter: 8 ft. 9 in., height: 15 ft.

Pyramid

The volume of a pyramid is equal to $\frac{1}{3}$ times the area of the base times the height. Expressed as a formula it is $V = \frac{1}{3} Bh$.

EXERCISES

Find the volumes of square pyramids having the following dimensions:

Side of base	8 mm	200 cm	7 ft. 6 in.	$5\frac{1}{2}$ in.	6.4 m
Height	13 mm	50 cm	9 ft.	18 in.	5.1 m

Problems

1. Find the volume of each of the following figures:

a. 4 m (sphere)

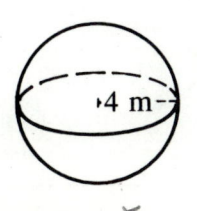

b. 6 cm, 7 cm (cone)

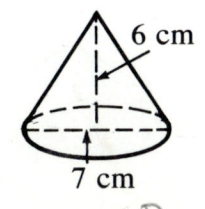

c. 16 ft., 12 ft., 12 ft., 12 ft. (pyramid)

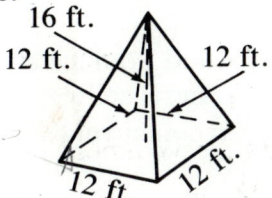

d. 14 mm, 10 mm (cone)

Geometry 443

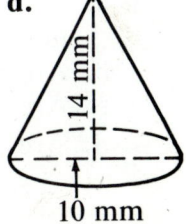

2. What is the weight of a steel ball 28 centimeters in diameter? A cubic centimeter of steel weighs 7.75 g.
3. How many tons of coal are in a conical pile of coal 21 ft. in diameter and 10 ft. high? 1 ton of coal occupies 35 cu. ft.
4. How many m³ of space are inside a tent in the shape of a square pyramid 5 m on each side of the base and 4.8 m high?
5. How many cu. ft. of space does the earth satellite, Vanguard I, occupy if its diameter is 6.4 inches?
6. How many m³ of sand are in a conical pile of sand 4.9 m in diameter and 3 m high?

CHAPTER REVIEW

Part 1

1. a. Read, or write in words, each of the following: $\overrightarrow{AN}, \overrightarrow{FG}, \overleftrightarrow{ED}$. (8–1)
   b. Explain the difference between: line and ray; ray and segment; line and segment.
2. Name each of the following: (8–1) (8–6)

   a. C N
   b. R X
   c. F B
   d. M R T
   e. E F D
   f. B C

3. a. At what point do $\overleftrightarrow{GL}$ and $\overleftrightarrow{NT}$ intersect?
   b. Are points G, S, and T collinear?
   c. Are points G, S, and L collinear?
   d. Are $\overleftrightarrow{GL}$ and $\overleftrightarrow{NT}$ concurrent?

   (8–4)
   (8–5)

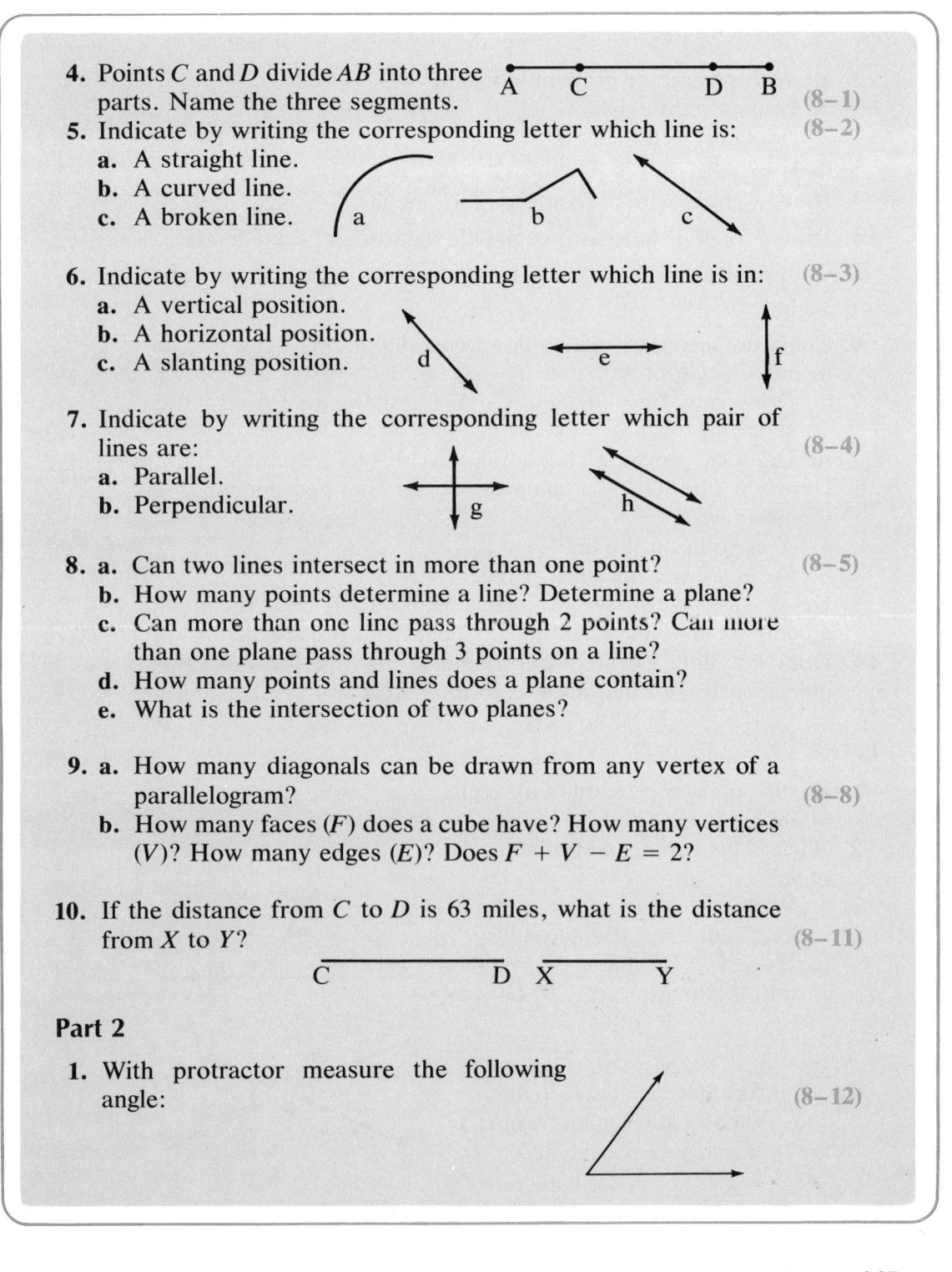

4. Points *C* and *D* divide *AB* into three parts. Name the three segments. (8–1)

5. Indicate by writing the corresponding letter which line is: (8–2)
   a. A straight line.
   b. A curved line.
   c. A broken line.

6. Indicate by writing the corresponding letter which line is in: (8–3)
   a. A vertical position.
   b. A horizontal position.
   c. A slanting position.

7. Indicate by writing the corresponding letter which pair of lines are: (8–4)
   a. Parallel.
   b. Perpendicular.

8. a. Can two lines intersect in more than one point? (8–5)
   b. How many points determine a line? Determine a plane?
   c. Can more than one line pass through 2 points? Can more than one plane pass through 3 points on a line?
   d. How many points and lines does a plane contain?
   e. What is the intersection of two planes?

9. a. How many diagonals can be drawn from any vertex of a parallelogram? (8–8)
   b. How many faces (*F*) does a cube have? How many vertices (*V*)? How many edges (*E*)? Does $F + V - E = 2$?

10. If the distance from *C* to *D* is 63 miles, what is the distance from *X* to *Y*? (8–11)

Part 2

1. With protractor measure the following angle: (8–12)

Geometry 445

2. a. With protractor draw an angle of 145°. (8–13)
   b. Draw a right angle.
   c. Draw any acute angle.
   d. Draw any obtuse angle. (8–7)
3. Draw a circle with a diameter of $1\frac{3}{8}$ inches. (8–8)
4. Draw a regular hexagon, each side measuring 38 millimeters. (8–22)
5. Construct a triangle with sides measuring $1\frac{1}{2}$ in., $1\frac{3}{4}$ in., and $1\frac{3}{8}$ in. (8–15)
6. Construct a triangle with sides measuring 3 cm, 4 cm, and an included angle of 40°. (8–15)
7. a. Draw any line segment. Using a compass, bisect this segment. (8–18)
   b. Draw any angle. Using a compass, bisect this angle. (8–20)
8. Draw any line. Using a compass, construct a perpendicular to this line:
   a. At a point on the line. (8–16)
   b. From a point not on the line. (8–17)
9. Draw with protractor an angle of 65°. Then construct with a compass an angle equal to it. Check with protractor. (8–19)
10. Draw any line. Locate a point outside this line. Through this point construct a line parallel to the first line. (8–21)

Part 3

1. What is the complement of an angle of 38°? (8–24)
2. What is the supplement of an angle of 96°?
3. a. What is the sum of measures of ∠s 1, 2, and 3 of triangle *ABC*? (8–23)
   b. If $m\angle 1 = 56°$ and $m\angle 4 = 118°$, find the measure of ∠3. Of ∠2. (8–24)

4. $\overleftrightarrow{AB}$ and $\overleftrightarrow{CD}$ intersect at *E*. If $m\angle 4 = 115°$, what is the measure of ∠1? ∠2? ∠3? (8–24)

5. Parallel lines *AB* and *CD* are cut by transversal *EF*. If $m\angle 7 = 104°$, what is the measure of $\angle 5$? $\angle 6$? $\angle 4$? $\angle 1$? $\angle 2$? $\angle 8$? $\angle 3$? (8–25)

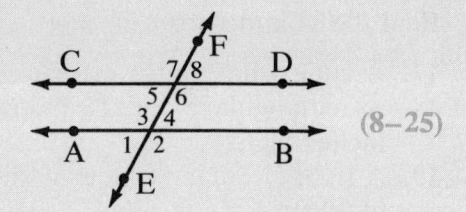

Part 4

Find the perimeter of:

1. A rectangle 23 centimeters long and 19 centimeters wide. (8–32)
2. A triangle with sides measuring $6\frac{3}{4}$ in., $5\frac{7}{8}$ in., and $7\frac{3}{8}$ in. (8–34)
3. A square whose side measures 34 meters. (8–33)

Find the circumference of:

4. A circle whose diameter is 42 feet. (8–35)
5. A circle whose radius is 57 millimeters. (8–35)
6. How many feet of fencing are required to enclose a rectangular garden 73 ft. long and 59 ft. wide. At $4.30 a yard, how much will the fencing cost? (8–32)
7. How many meters do you ride in one turn of a merry-go-round if you sit 7 meters from the center? (8–35)

Find the area of:

8. A square whose side measures 27 meters. (8–37)
9. A circle whose radius is 49 millimeters. (8–41)
10. A rectangle 104 centimeters long and 95 centimeters wide. (8–36)
11. A parallelogram with an altitude of 76 ft. and a base of 68 ft. (8–38)
12. A circle whose diameter is 60 meters. (8–41)
13. A triangle whose altitude is 17 inches and base is 26 inches. (8–39)
14. A trapezoid with bases of 91 mm and 57 mm and a height of 48 mm. (8–40)
15. A house 42 ft. square and a garage 16 ft. long and 10 ft. wide are built on a lot 108 ft. long and 61 ft. wide. How many sq. ft. of the lot remains? (8–36, 8–37)
16. At $6.00 per sq. ft., how much will it cost to resilver a circular mirror 35 inches in diameter? (8–41)

Geometry

Find the volume of:

17. A cube whose edge measures 39 millimeters. (8–45)
18. A rectangular solid 17 inches long, 13 inches wide, and 25 inches high. (8–44)
19. A right circular cylinder with a diameter of 28 ft. and a height of 30 ft. (8–46)
20. A sphere whose diameter is 6 centimeters. (8–47)
21. A right circular cylinder with a radius of 4 m and a height of 9 m. (8–46)
22. A square pyramid 20 ft. on each side and 18 ft. high. (8–47)
23. A right circular cone 14 dm in diameter and 17 dm high. (8–47)
24. A right circular cylinder 70 cm in diameter and 45 cm high. (8–43)
25. A circular wading pool 40 ft. in diameter is filled to an average depth of 2 ft. How many gallons of water does the pool contain? (8–46)
26. How many bushels will a storage bin hold if it is 15 ft. long, 12 ft. wide, and 8 ft. deep? (8–44)
27. A cube 15 centimeters on a side. (8–42)
28. A rectangular solid 28 m long, 23 m wide, and 14 m high. (8–42)

Find the total area of:

29. At $1.30 per sq. yd., find the cost of painting the walls and ceiling of a room 18 ft. long, 14 ft. wide, and 10 ft. high? Allow 55 sq. ft. for openings. (8–42)
30. Allowing 2 centimeters for the seam, how many square meters of tin are needed to make a pipe 14 meters long and 10 centimeters in diameter? (8–43)

Part 5

1. Find the hypotenuse of a right triangle if the altitude is 105 cm and the base is 56 cm. (8–28)
2. Find the altitude of a right triangle if the hypotenuse is 195 ft. and the base is 48 ft. (8–28)
3. Find the base of a right triangle if the altitude is 270 mm and the hypotenuse is 318 mm. (8–28)
4. How many feet does a person save by walking diagonally across a field 385 m long and 336 m wide instead of around two sides? (8–28)

Find the indicated missing parts in the following:

5. Congruent triangles (8–26)

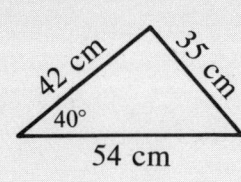

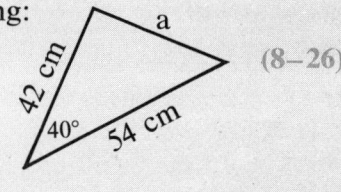

6. Similar triangles (8–27)

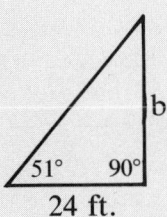

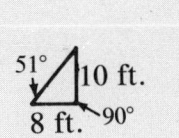

7. Find the height of a telegraph pole that casts a shadow of 10 feet at a time when a boy, 5 feet tall, casts a shadow of 2 feet. (8–29)

8. Find the distance across the stream (MN) shown in the figure below. (8–29)

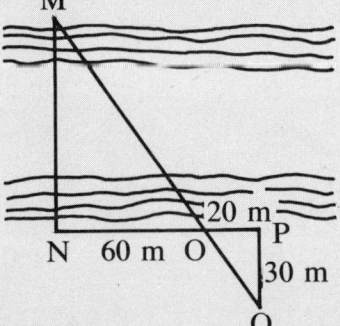

9. Two telegraph poles, 26 ft. and 39 ft. respectively, are 84 ft. apart. Find the length of the wire that is attached to each pole, 2 ft. below the top. (8–28)

10. Find the indicated parts of the following right triangles: (8–31)
  a. Find side a. b. Find side b. c. Find side b.

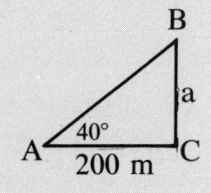

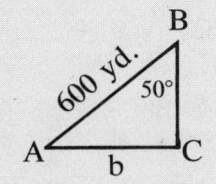

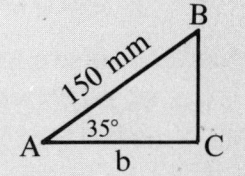

Geometry

ACHIEVEMENT TEST

1. Add:
   63,847
   58,696
   27,388
   45,172
   29,887   (1–2)

2. Subtract:
   753,000
      8,706   (1–3)

3. Multiply:
   9,068
   8,507   (1–4)

4. Divide:
   5,218)3,543,022   (1–5)

5. Add:
   $4\frac{5}{6} + 3\frac{3}{8} + \frac{7}{12}$ (1–12)

6. Subtract:
   $8\frac{2}{3} - 7\frac{4}{5}$ (1–13)

7. Multiply:
   $9\frac{1}{6} \times 4\frac{1}{5}$ (1–14)

8. Divide:
   $18\frac{3}{4} \div 10$ (1–15)

9. Add:
   .65 + 3.8 + .574   (1–18)

10. Subtract:
    $300 − $159.87   (1–19)

11. Multiply:
    2.5 × .006   (1–20)

12. Divide:
    .05)4   (1–22)

13. Find 7% of $63.49 (to nearest cent).   (1–30, 1–33, 1–34)
14. $3.20 is what percent of $2.40?   (1–31, 1–33, 1–34)
15. $33\frac{1}{3}$% of what amount is $180?   (1–32, 1–33, 1–34)
16. Find the square root of 37,136,836.   (1–36)
17. Name the prime numbers that are greater than 90 and less than 100.   (1–39)

450   CHAPTER 8

18. **a.** Find the greatest common factor of 16, 36, and 48. (1–38)
    b. What is the least common multiple of 18, 27, and 45? (1–40)

19. **a.** Which distance is the shortest? .75 km; 76.3 cm; 746 m (2–1)
    b. Which weight is the heaviest? 40.9 mg; 4,100 g; 5.85 kg (2–2)
    c. Which capacity is the largest? 1,050 cL; 10.5 L; 105 dL (2–3)
    d. Which area is the smallest? 3 hectares; 450 m²; 40,000 cm² (2–4)
    e. Which volume is the largest? 496.8 dm³; 300,000 cm³; 2.6 m³ (2–5)
    f. Which is the faster speed? 100 km/h; or 50 m/s. (2–15)
    g. Which is the warmer temperature? 122°F; or 60°C (2–16)
    h. 52.6 liters of water weighs _____ kg and occupies _____ cm³. (2–5)

20. **a.** Which is the longest period of time? 9 hr; 2,000 sec.; $\frac{2}{3}$ da; 600 min. (2–13)
    b. Write in A.M. or P.M. time: (1) 0005 (2) 2346 (2–17)
    c. When it is 11 P.M. in Dallas, what time is it in Milwaukee? Louisville? Spokane? Providence? (2–18)

21. Which of the following numerals name whole numbers? Which name integers? Which name rational numbers? Which name irrational numbers? Which name real numbers?

    $^-\frac{7}{10}$ $\frac{45}{5}$ $^-\sqrt{57}$ $^+6.1$ $^-83$ 12% $^+7\frac{1}{3}$ (3–2)

22. On a number line, draw the graph of: $^-6, ^-4, ^-1, 0, 2, 5$ (3–4)

23. Which of the following sentences are true? (3–5)
    a. $^-9 > ^-2$
    b. $^-4 < ^+3$
    c. $^+1 \not> ^+6$
    d. $0 \not< ^-1$
    e. $^-6 > ^+5$

24. **a.** Find the opposite of: (1) $^-36$ (2) $^+.7$ (3) $^-\frac{2}{5}$ (3–8)
    b. Find the value of: (1) $|^-2.5|$ (2) $|^+76|$ (3) $|^-1\frac{2}{3}|$ (3–3)
    c. Find the multiplicative inverse of: (1) $\frac{1}{12}$ (2) -28 (3) $2\frac{3}{4}$ (3–16)
    d. Find the additive inverse of: (1) -54 (2) $\frac{13}{16}$ (3) 9.8   (3–8)

25. Compute as indicated:
    a. (1) $^{+}11 + {}^{-}15$; (2) $^{-}4 + {}^{-}10$; (3) $(+6) + (-5) + (-8) + (+7)$ (3–10)
    b. Find the value of: $4 - 7 - 3 + 9 - 8 + 5$ (3–19)
    c. (1) $^{-}7 - {}^{-}12$ (2) $^{-}6 - {}^{+}6$ (3) $0 - {}^{-}8$ (4) $(+5) - (-5)$ (3–13)
    d. (1) $^{-}1 \times {}^{+}5$ (2) $^{-}3 \times {}^{-}8$ (3) $4(-9)$ (4) $(-6)(-2)(-1)$ (3–15)
    e. (1) $^{+}63 \div {}^{-}9$ (3) $(-36) \div (-12)$
       (2) $^{-}42 \div {}^{-}6$ (4) $(-56) \div (+7)$ (3–17)

26. Express as a formula: The perimeter of an isosceles triangle (p) is equal to the sum of the base (b) and twice the length of one of the equal sides (e).

27. a. Add as indicated: (4–4)
       (1) $(-3m^2) + (-5m^2) + (+9m^2)$
       (2) $(2a^2 - b^2) + (3ab - 3b^2) + (-4a^2 - ab)$
    b. Simplify: $5n - n^2 + 2 - 6n^2 - 4n - 3 + n + n^2 + 6 - 2n$ (4–7)

28. Subtract as indicated: (4–8)
    a. $(5a^3b^2x) - (9a^3b^2x)$
    b. $(c^2 - 5cd) - (-2c^2 + 5cd - 3d^2)$

29. Multiply as indicated: (4–9)
    a. $(-3y^2)(-my^5)$ c. $(3a^2b)(-5ac)(-ab^2c)(4bc^3)$
    b. $(-4xz^4)^3$ d. $-2a^5b^3(6a^4 - 3a^2b^2 + b^4)$

30. Divide as indicated: (4–10)
    a. $\dfrac{-60\,r^7sx^9}{5\,r^6x^2}$ b. $(-7a^3b^6) \div (-ab^4)$
    c. $\dfrac{63x^8y^4 - 9x^6y^6 + 27x^5y^8}{-9x^5y^2}$

31. When the replacements for the variable are all the real numbers,
    a. Solve and check: (4–14)
       (1) $6n = 54$ (3) $y + 14 = 31$ (5) $3x - 5x = -24$
       (2) $x - 25 = 17$ (4) $\dfrac{x}{10} = 5$ (6) $7b + 30 = 2$
    b. Find the solutions of: (4–23)
       (1) $x - 12 > 5$ (3) $-9b \not< 72$ (5) $3n + 7 \geq 16$
       (2) $\dfrac{r}{6} < -6$ (4) $t + 18 \neq 18$ (6) $4x - 11x \leq -56$

32. Find the value of: (4–17)
   a. F when $W = 76$, $v = 40$, $g = 32$, and $r = 25$, using the formula $F = \dfrac{Wv^2}{gr}$
   b. a when $l = 34$, $n = 10$, and $d = 3$, using the formula $l = a + (n - 1)d$
   c. m when $d = 8$ and $v = 31$, using the formula $d = \dfrac{m}{v}$
   d. C when $F = -49$, using the formula $F = 1.8\,C + 32$

33. A merchant bought a table, costing him $260. At what price must he sell it to make 35% profit on the selling price? (4–19)

34. Using all the real numbers as the replacements for the variable, draw on a number line the graph of: (4–20, 4–24, 4–25)
   a. (1) $x + 8 = 4$
   b. (1) $b + 4 < 6$
   (2) $\dfrac{m}{3} > 1$
   (3) $6x \neq 20$
   (2) $16n - 9 = 39$
   (4) $y - 2y \not> -2$
   (5) $4m - 12 \leq 0$
   (6) $-3 < x < 4$

35. On graph paper draw the axes and plot the points which have the following coordinates: (4–29)
   a. $(-6, -2)$ b. $(9, -3)$ c. $(5, 8)$ d. $(4, -5)$ e. $(-1, 7)$

36. The students in a mathematics class received the following scores in a test: 95, 85, 65, 70, 100, 80, 70, 85, 60, 75, 90, 70, 85, 70, 90, 85, 95, 85, 65
   a. Tally and make a frequency distribution table. (5–2)
   b. Make a histogram. (5–2)
   c. Make a frequency polygon. (5–2)
   d. Find the mean, median, and mode. (5–3)
   e. Find the range. (5–5)

37. a. What is the probability of drawing at random on the first draw a marked card from a box containing 60 cards of which 24 are marked? (6–1)
   b. What are the odds against selecting a marked card on the first draw? (6–1)

Geometry

38. Construct a flow chart indicating the steps in subtracting a two-place decimal fraction from a larger two-place decimal fraction. (7–1)
39. a. Name each of the following: (8–1)

   (1)   (2)   (3)   (4)   (5)   (6)

   b. Draw line segments measuring: (8–9)
      (1) 46 mm   (2) 8.3 cm   (3) $4\frac{7}{8}$ in.
   c. Draw angles measuring: (1) 24°   (2) 115°   (3) 90° (8–13)
   d. If the scale is 1:2,500,000, what distance is represented by: (8–11)
      (1) 3 mm?   (2) 5.1 cm?

40. Construct triangles with the following measurements: (8–15)
   a. Three sides measuring: $2\frac{3}{4}$ in., $3\frac{5}{8}$ in., and $4\frac{3}{16}$ in.
   b. Angles measuring 67° and 58° with an included side of 73 mm.
   c. Sides measuring 5.4 cm and 7.1 cm with an included angle of 49°.

41. a. Draw a line segment 4.6 cm long. Using a compass, bisect this segment. Check with a ruler. (8–18)
   b. Draw an angle measuring 72°. Using a compass, bisect this angle. Check with a protractor. (8–20)
   c. Draw an angle of 110°. Then construct with a compass an angle equal to it. Check with a protractor. (8–19)

42. a. Draw any line. Using a compass, construct a perpendicular to this line at a point on the line. (8–16)
   b. Draw any line. Using a compass, construct a perpendicular to this line from a point not on the line. (8–17)
   c. Draw any line. Locate a point outside this line. Through this point construct a line parallel to the first line. (8–21)

43. **a.** What is the supplement of an angle measuring 107°? (8–24)
   b. What is the complement of an angle measuring 24°? (8–24)
   c. If two angles of a triangle measure 61° and 38°, what is the measure of the third angle? (8–23)

44. Find the perimeter of:
   a. A square whose side measures 87 centimeters. (8–33)
   b. A rectangle 53 meters long and 48 meters wide. (8–32)
   c. A triangle with sides measuring 28 mm, 19 mm, and 33 mm. (8–34)

45. Find the circumference of: (8–35)
   a. A circle whose radius is 56 feet.
   b. A circle whose diameter is 105 meters.

46. Find the area of:
   a. A parallelogram with an altitude of 93 mm and a base of 76 mm. (8–38)
   b. A square whose side measures $20\frac{1}{2}$ feet. (8–37)
   c. A circle whose radius is 21 meters. (8–41)
   d. A rectangle 42 cm long and 37 cm wide. (8–36)
   e. A triangle whose altitude is 26 inches and base is 19 inches. (8–39)
   f. A circle whose diameter is 61 millimeters. (8–41)
   g. A trapezoid with bases of 103 mm and 85 mm and a height of 52 mm. (8–40)

47. Find the volume of:
   a. A cube whose edge measures 11.6 centimeters. (8–45)
   b. A right circular cylinder with a diameter of 42 ft. and a height of 20 ft. (8–46)
   c. A rectangular solid 3.6 meters long, 2.9 meters wide, and 4.7 meters high. (8–44)
   d. A sphere whose diameter is 98 millimeters. (8–47)

48. Find the total area of:
   a. A cube whose edge measures 22 centimeters. (8–42)
   b. A rectangular solid 30 inches long, 26 inches wide, and 19 inches high. (8–42)

Geometry

c. A right circular cylinder with a radius of 35 meters and a height of 10 meters. (8–43)

49. a. Find the height of a building that casts a shadow of 48 feet at a time when a 10-foot pole casts a shadow of 15 feet. (8–29)
  b. Find the hypotenuse of a right triangle if the altitude is 30 mm and the base is 72 mm. (8–28)
  c. Find the base of a right triangle if the hypotenuse is 89 yd. and the altitude is 80 yd. (8–28)

50. Find the indicated parts of the following right triangles: (8–31)

  a. Find side b: **b.** Find side a: **c.** Find side b:

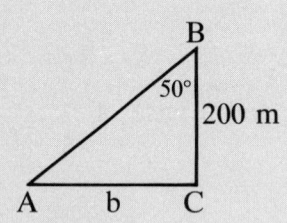

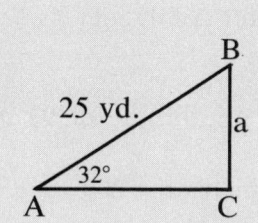

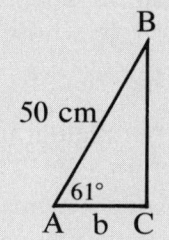

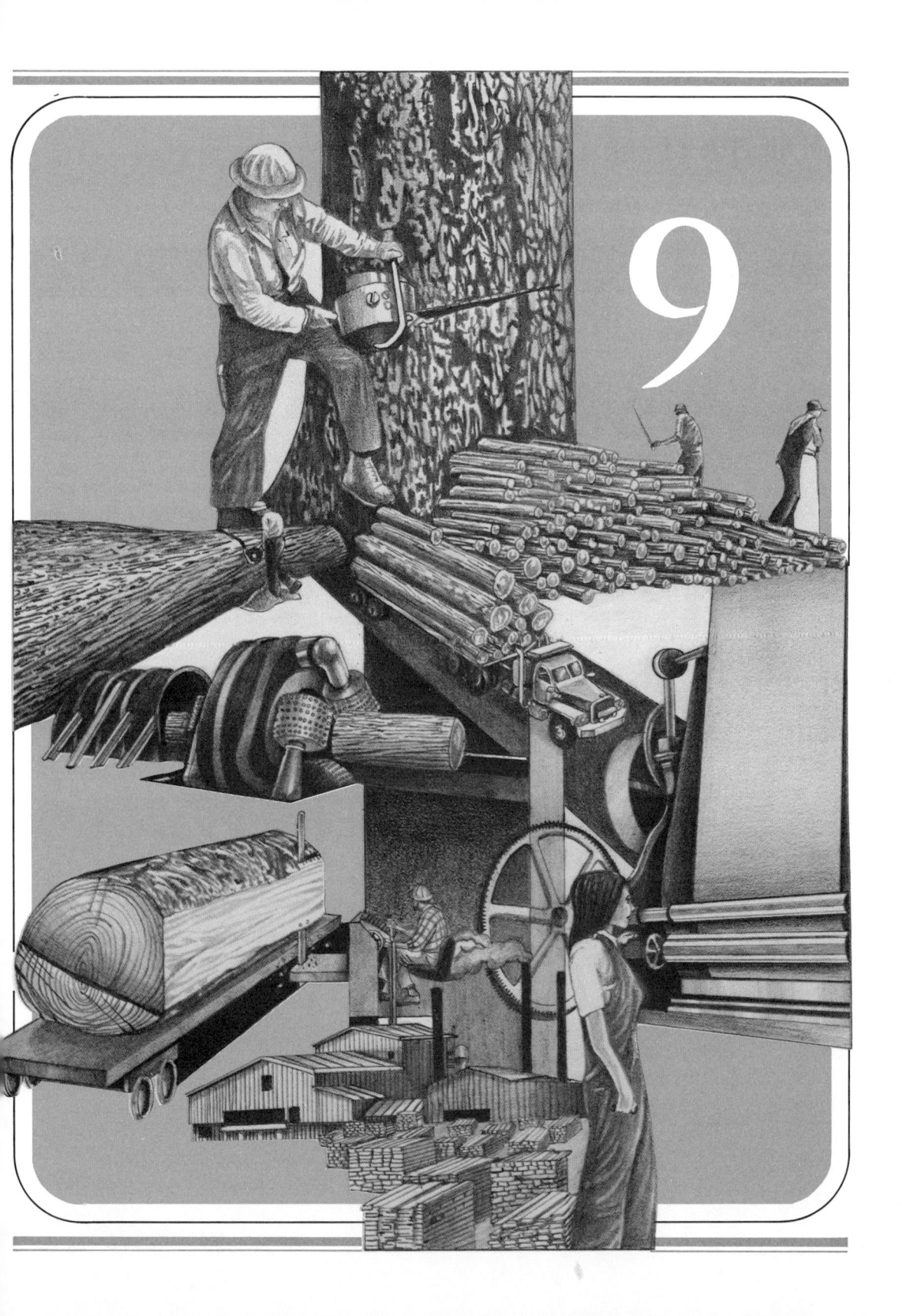

9

CHAPTER 9
Applications of Mathematics

INCOME, TAKE-HOME PAY, INCOME TAX

9–1 COMPUTING INCOME

A person's income usually consists of earnings in the form of a salary or wages per hour, day, week, month, or year. Other types of income include commissions (see page 513), profits (see page 517), bonuses, fees, tips, interest (see page 484), dividends, and pensions.

In some occupations wages are calculated at hourly rates of pay which vary depending on the job. In other occupations the employed persons have fixed salaries on a weekly, semi-monthly, or annual basis. A *salary* is a fixed sum of money paid for working a definite period of time.

Some people work where their earnings are based on the number of pieces or units of work that a person completes at a given pay rate per piece or unit. Salespersons work on commission or salary and commission. Sometimes bonuses are paid.

Repair persons and most professional people usually charge a fee for their work.

When a person works at an hourly rate more than 40 hours per week or more than 8 hours per day, *time and a half* is usually paid for the overtime hours and *double time* for all hours worked on holidays and Sundays.

To find the weekly wages when the hourly rate and the number of working hours are known, we multiply the number of hours of work by the hourly rate, using the overtime rate for the time worked over 40 hours.

EXERCISES

1. a. Name two occupations in which wages or salaries are paid:
   - (1) On an hourly basis
   - (2) On a weekly basis
   - (3) On an annual basis
   - (4) On a piece-work basis

b. Name two occupations in which:
   (1) Fees are paid   (2) Commissions are paid

2. If a person works 40 hours per week, what are the weekly earnings at each of the following hourly rates:
   a. $7.60   **b.** $5.85   **c.** $3.10   **d.** $9.45   **e.** $12.35

3. Find the weekly earnings of a person who works:
   a. 33 hours at $4.80 per hour
   b. 27 hours at $7.30 per hour
   c. 38 hours at $3.90 per hour
   d. $36\frac{1}{2}$ hours at $6.00 per hour
   e. $19\frac{1}{2}$ hours at $8.10 per hour
   f. $28\frac{1}{2}$ hours at $5.75 per hour

4. If the overtime rate is $1\frac{1}{2}$ times the regular rate, what is the hourly overtime rate corresponding to each of the following regular hour rates:
   a. $6.40   **b.** $3.50   **c.** $4.96   **d.** $7.25   **e.** $10.65

5. Find the weekly earnings (time and a half rate over 40 hours) of a person who works:
   a. 42 hours at $5.60 per hour
   b. 50 hours at $8.75 per hour
   c. 43 hours at $3.50 per hour
   d. 45 hours at $6.25 per hour
   e. $46\frac{1}{2}$ hours at $10.58 per hour
   f. $41\frac{1}{2}$ hours at $7.45 per hour

6. For each of the following persons find the earnings for the week when the number of units and the pay rate per unit are as follows:

	Number of Units					Total Units	Unit Rate	Total Wages
	M	T	W	T	F			
Linda Marshall	53	48	57	61	46		$.63	
Richard Tucker	37	42	39	41	40		$.75	
Rita Vincelli	125	131	118	133	128		$.27	
David Segal	89	86	94	92	101		$.32	
Marie Varga	167	173	182	177	169		$.17	

7. Find the hourly rate of pay when a person's weekly earnings are:
   a. $160 for 40 hours of work
   b. $246 for 40 hours of work
   c. $298.80 for 36 hours of work
   d. $163.41 for 39 hours of work
   e. $255.30 for $34\frac{1}{2}$ hours of work
   f. $358.50 for $37\frac{1}{2}$ hours of work

Applications of Mathematics

8. Find the annual salary of a person earning:

   a. $195 per week
   b. $83.75 per week
   c. $1,060 per month
   d. $765.40 per month
   e. $590 semi-monthly
   f. $482.50 semi-monthly
   g. $261.25 per week
   h. $1,137.75 per month
   i. $328.60 per week
   j. $427.50 per week

9. If a woman works 40 hours per week, what is her weekly salary, annual salary, and monthly salary, at each of the following hourly rates:

   a. $2.70?
   b. $6.25?
   c. $4.50?
   d. $8.60?
   e. $3.15?
   f. $5.36?
   g. $9.75?
   h. $11.20?
   i. $7.69?
   j. $12.95?
   k. $8.85?
   l. $15.50?

10. What is your monthly salary, semi-monthly salary, and weekly salary if your annual salary is:

    a. $6,240?
    b. $7,800?
    c. $11,700?
    d. $9,600?
    e. $15,600?
    f. $10,000?
    g. $5,400?
    h. $24,000?
    i. $18,200?
    j. $32,500?
    k. $29,000?
    l. $37,500?

11. Which is a better wage:

    a. $580 per month or $7,000 per year?
    b. $10,000 per year or $195 per week?
    c. $315 per week or $1,260 per month?
    d. $13,000 per year or $1,075 per month?
    e. $960 per month or $230 per week?
    f. $285 per week or $14,500 per year?

12. Miss Sullivan receives $175 per week and 3% commission on sales. Last week her sales totaled $4,108. What were her total earnings for the week?

13. Mr. Hadden receives a salary of $115 per week and 5% bonus on all sales over the quota of $2,500. If his weekly sales are $4,470, what are his total weekly earnings?

14. A lawyer collected 80% of a debt of $1,800. If she charged $33\frac{1}{3}\%$ commission, how much did the lawyer receive?

15. Last year Mr. Cartier received a 6% increase on his weekly salary of $320. Last month his salary was reduced 6%. How does his present salary compare with his original salary before the increase?

9-2 WITHHOLDING TAX

Withholding tax is the amount of income tax that employers are required by the federal government to withhold from their employees' earnings. This money is forwarded to the Collector of Internal Revenue's offices, where it is entered on each employee's account. The tax rates may change from year to year.

The employer may use either the percentage method or the income tax withholding tables to compute the amount of income tax to be deducted and withheld from wages to an employee.

To compute the amount of weekly withholding tax for a married person by the percentage method,

(1) We multiply the number of withholding allowances (exemptions) by $14.40.
(2) We subtract this product from the total wage payment to find the amount of wages subject to withholding.
(3) Then we use the following rates:

If the amount of wages subject to withholding is:
Not over $48

The amount of income tax to be withheld shall be:
0

Over—	But not over—		Of excess over—
$48	—$96	17%	—$48
$96	—$173	$8.16 plus 20%	—$96
$173	—$264	$23.56 plus 17%	—$173
$264	—$346	$39.03 plus 25%	—$264
$346	—$433	$59.53 plus 28%	—$346
$433	—$500	$83.89 plus 32%	—$433
$500		$105.33 plus 36%	—$500

Compute by the percentage method the amount of income tax to be withheld each week when a married person's weekly wages are $260 with 3 allowances.

(1) Total wage payment		$260.00
(2) Amount of one allowance	$14.40	
(3) Number of allowances	3	
(4) Line 2 multiplied by line 3		$43.20
(5) Amount subject to withholding (line 1 minus line 4)		$216.80
(6) Tax to be withheld on $216.80, using above rates:		
Tax on first $173.00		$23.56
Tax on remainder $43.80 @ 17%		$ 7.45
Total to be withheld		$31.01
	Answer,	$31.01

Applications of Mathematics

To compute the amount of weekly withholding tax for a married person, using the income tax withholding tables,
 (1) We locate the row indicating the weekly salary
 (2) Then we select the amount of tax from the column corresponding to the number of withholding allowances.

MARRIED Persons—WEEKLY Payroll Period

And the wages are—		And the number of withholding allowances claimed is—						
At least	But less than	0	1	2	3	4	5	6
		The amount of income tax to be withheld shall be—						
$ 80	82	5.60	3.10	.70	0	0	0	0
82	84	5.90	3.50	1.00	0	0	0	0
84	86	6.30	3.80	1.40	0	0	0	0
86	88	6.60	4.20	1.70	0	0	0	0
88	90	7.00	4.50	2.10	0	0	0	0
90	92	7.30	4.80	2.40	0	0	0	0
92	94	7.60	5.20	2.70	.30	0	0	0
94	96	8.00	5.50	3.10	.60	0	0	0
96	98	8.30	5.90	3.40	1.00	0	0	0
98	100	8.70	6.20	3.80	1.30	0	0	0
100	105	9.40	6.80	4.30	1.90	0	0	0
105	110	10.40	7.70	5.20	2.70	.30	0	0
110	115	11.40	8.60	6.00	3.60	1.10	0	0
115	120	12.40	9.60	6.90	4.40	2.00	0	0
120	125	13.40	10.60	7.70	5.30	2.80	.40	0
125	130	14.40	11.60	8.70	6.10	3.70	1.20	0
130	135	15.40	12.60	9.70	7.00	4.50	2.10	0
135	140	16.40	13.60	10.70	7.80	5.40	2.90	.50
140	145	17.40	14.60	11.70	8.80	6.20	3.80	1.30
145	150	18.40	15.60	12.70	9.80	7.10	4.60	2.20
150	160	19.90	17.10	14.20	11.30	8.40	5.90	3.50
160	170	21.90	19.10	16.20	13.30	10.40	7.60	5.20
170	180	23.90	21.10	18.20	15.30	12.40	9.50	6.90
180	190	25.60	23.10	20.20	17.30	14.40	11.50	8.60
190	200	27.30	24.80	22.20	19.30	16.40	13.50	10.60
200	210	29.00	26.50	24.10	21.30	18.40	15.50	12.60
210	220	30.70	28.20	25.80	23.30	20.40	17.50	14.60
220	230	32.40	29.90	27.50	25.00	22.40	19.50	16.60
230	240	34.10	31.60	29.20	26.70	24.30	21.50	18.60
240	250	35.80	33.30	30.90	28.40	26.00	23.50	20.60
250	260	37.50	35.00	32.60	30.10	27.70	25.20	22.60
260	270	39.20	36.70	34.30	31.80	29.40	26.90	24.50
270	280	41.70	38.40	36.00	33.50	31.10	28.60	26.20
280	290	44.20	40.60	37.70	35.20	32.80	30.30	27.90
290	300	46.70	43.10	39.50	36.90	34.50	32.00	29.60
300	310	49.20	45.60	42.00	38.60	36.20	33.70	31.30

Use the table of withholding tax to compute the amount of income tax to be withheld each week when a married person's weekly wages are $195 and 3 allowances are claimed.

(1) We locate the row that reads "wages at least $190 but less than $200."
(2) Then we move to the column headed "3" meaning 3 allowances claimed to find $19.30.
(3) $19.30 is the weekly amount of withholding tax.

Answer, $19.30

EXERCISES

1. Use the percentage method to compute the amount of income tax to be withheld each week if a married person's weekly wages are:

    a. $225 and 4 allowances are claimed.
    b. $387 and 3 allowances are claimed.
    c. $132 and 1 allowance is claimed.
    d. $93 and 2 allowances are claimed.
    e. $256 and 0 allowance is claimed.
    f. $98.75 and 1 allowance is claimed.
    g. $240.50 and 5 allowances are claimed.
    h. $326.25 and 4 allowances are claimed.
    i. $183.40 and 6 allowances are claimed.
    j. $297.75 and 2 allowances are claimed.

2. Use the table of withholding tax to compute the amount of income tax to be withheld each week if a married person's weekly wages are:

    a. $105 and 2 allowances are claimed.
    b. $230 and 5 allowances are claimed.
    c. $95 and 1 allowance is claimed.
    d. $296 and 3 allowances are claimed.
    e. $158 and 0 allowance is claimed.
    f. $162.75 and 2 allowances are claimed.
    g. $253.25 and 4 allowances are claimed.
    h. $87.50 and 3 allowances are claimed.
    i. $281.75 and 1 allowance is claimed.
    j. $134.50 and 6 allowances are claimed.

3. Use the table of withholding tax to compute the amount of income tax withheld for the year if a married person's weekly wages are:

    a. $234 and 3 allowances are claimed.
    b. $89 and 2 allowances are claimed.
    c. $165.25 and 1 allowance is claimed.
    d. $286.50 and 4 allowances are claimed.

Applications of Mathematics

4. Use the percentage method to compute the amount of income tax withheld for the year if a married person's weekly wages are:

   a. $425 and 3 allowances are claimed.
   b. $178 and 5 allowances are claimed.
   c. $254.50 and 1 allowance is claimed.
   d. $312.60 and 2 allowances are claimed.

9–3 SOCIAL SECURITY

Social security tax is money deducted from employees' earnings to help pay old age pensions. Employees pay a tax at the rate of 5.85% of the first $16,500 of their wages. Employers pay the same amount, matching the amounts paid by the employees. The social security rate and the earnings on which it is applied may change from year to year.

EXERCISES

1. Find the amount deducted from the employee's wages for social security tax if an employee earns each week:

   a. $64
   b. $137
   c. $240
   d. $72
   e. $105
   f. $185
   g. $300
   h. $226
   i. $168
   j. $295
   k. $58.50
   l. $103.25
   m. $177.75
   n. $212.60
   o. $78.20
   p. $253.40
   q. $180.50
   r. $306.75
   s. $231.50
   t. $116.30

2. Find the amount deducted from the employee's wages for social security tax for the given work period if an employee earns:

   a. $960 monthly
   b. $15,000 annually
   c. $375 semi-monthly
   d. $8,450 annually
   e. $687.50 semi-monthly
   f. $863.25 monthly
   g. $18,200 annually
   h. $712.75 monthly

9–4 STATE AND CITY INCOME TAX— WAGE TAX

Most of the states and some cities in the United States have, in addition to the federal income tax, another tax on earnings. This income tax is sometimes called a *wage tax*. It is usually expressed as a fixed rate of percent of the earnings.

EXERCISES

1. If the wage tax is 2% of the wages, find the tax owed for the week when the person's weekly wages are:
   a. $75
   b. $160
   c. $245
   d. $86.50
   e. $302.75

2. If the state income tax rate is 3% of earnings, find the tax owed for the week when the person's weekly earnings are:
   a. $139
   b. $90
   c. $305
   d. $118.25
   e. $293.50

3. If the city income tax rate is $1\frac{1}{2}$% of earnings, find the tax owed for the week when the person's weekly earnings are:
   a. $87
   b. $215
   c. $196
   d. $322.50
   e. $141.90

4. If the wage tax rate is 2.3% of earnings, find the tax owed for the given work period when the person's earnings are:
   a. $820 monthly
   b. $7,400 annually
   c. $395 semi-monthly
   d. $12,000 annually
   e. $482.75 semi-monthly
   f. $637.50 monthly
   g. $591.25 semi-monthly
   h. $17,500 annually

5. If the state income tax rate is 3.6% of earnings and the city income tax rate is 1.8% of earnings, find the state income tax, city income tax, and total tax owed for the given work period when the person's earnings are:
   a. $9,650 annually
   b. $1,000 monthly
   c. $540 semi-monthly
   d. $16,800 annually
   e. $815.50 monthly
   f. $360.75 semi-monthly
   g. $25,000 annually
   h. $693.25 monthly

9–5 TAKE-HOME PAY

Take-home pay is the amount of a person's earnings remaining after federal withholding tax, social security taxes, and state and city income taxes, if any, are deducted from his or her wages. Sometimes union dues and health insurance costs are also deducted.

To determine the take-home pay,
(1) Find the federal withholding tax, the social security tax, and, if any, the state and city income taxes.
(2) Then subtract the sum of these taxes from the amount of the person's earnings.

Applications of Mathematics

> Determine the take-home pay for the week of a married person who earns a weekly wage of $200 and claims 4 allowances.
>
> | Withholding Tax | | | |
> |---|---|---|---|
> | by Table | $18.40 | $200.00 | Total Wages |
> | Social Security Tax | | 30.10 | Deductions |
> | (5.85% of $200) | 11.70 | $169.90 | Take-Home Pay |
> | Sum of Taxes* | $30.10 | | |
> | (deductions) | | | |
>
> * When there is also a wage tax, add to this sum.
>
> *Answer,* $169.90 Take-Home Pay

EXERCISES

In the following problems, use the table of withholding tax (page 462) for all wages between $80 and $300; for all other wages use the rates given on page 461. The social security tax rate is 5.85% on the first $16,500 earned per year.

1. Mrs. Scott earns $275 each week and claims 2 withholding allowances. How much income tax is withheld each week? How much tax is deducted for social security? What is her take-home pay?

2. Find the take-home pay when a married person's weekly wages are:
   a. $185 with 4 withholding allowances claimed
   b. $250 with 3 withholding allowances claimed
   c. $137 with 2 withholding allowances claimed
   d. $98 with 1 withholding allowance claimed
   e. $310 with 5 withholding allowances claimed
   f. $86.50 with 2 withholding allowances claimed
   g. $113.90 with 4 withholding allowances claimed
   h. $261.60 with 3 withholding allowances claimed
   i. $227.25 with 1 withholding allowance claimed
   j. $306.75 with 6 withholding allowances claimed

3. Which take-home pay is greater, a weekly wage of $144 with 4 withholding allowances claimed or $132 with 1 withholding allowance claimed?

4. Mr. Ryan earns $186.40 each week and claims 3 withholding allowances. He works in a state where a 2.4% tax is levied on all income. How much federal income tax is withheld each week? How much is deducted for social security? What is his weekly state income tax? What is his take-home pay?

5. Find the take-home pay for each of the following married person's weekly wages, using the table of withholding tax, a 5.85% deduction for social security, and the indicated state and city income tax rates:
   a. $225 with 2 allowances claimed; 3.2% state tax
   b. $146.90 with 3 allowances claimed; 1.8% city tax
   c. $305.75 with 1 allowance claimed; 4% state tax and 1.75% city tax
   d. $96.45 with 2 allowances claimed; 2.6% state tax and 3% city tax
   e. $274.20 with 4 allowances claimed; 4.5% state tax and 1.2% city tax

9–6 INCOME TAX

An *income tax* is a tax on earnings. Below are parts of tables that are used only when two or four exemptions are claimed by persons whose incomes are under $10,000 and who do not itemize deductions. For other tables see the Internal Revenue Service tax publications. Income tax rates are subject to change.

To find the federal income tax owed, select the tax table which covers the number of exemptions claimed, locate the row indicating the income, and read across to the column headed by the required marital status.

To determine the federal income tax owed by a married person earning $9,860 per year who files a joint return and claims 4 exemptions, use Table 4 (page 468) for tax returns claiming 4 exemptions.
   (1) Locate the row that reads "at least $9,850 but less than $9,900."
   (2) Select the column "Married filing joint return" to find $885 as the amount of income tax.

Answer, $885

Applications of Mathematics

Table 2—Returns claiming TWO exemptions (and not itemizing deductions)

| If line 17 (adjusted gross income) is— || And you are— |||| If line 17 (adjusted gross income) is— || And you are— ||||
At least	But less than	Single, not head of house-hold	Head of house-hold	Married filing joint return	Low income allow-ance	% Standard deduc-tion	At least	But less than	Single, not head of house-hold	Head of house-hold	Married filing joint return	Low income allow-ance	% Standard deduc-tion
				Married filing separate return claiming—								Married filing separate return claiming—	
				Your tax is—								Your tax is—	
$7,900	$ 7,950	$ 926	$ 874	$ 834	$1,081	$1,004	$8,950	$ 9,000	$1,141	$1,068	$1,024	$1,336	$1,249
7,950	8,000	937	883	843	1,092	1,015	9,000	9,050	1,151	1,078	1,033	1,349	1,261
8,000	8,050	947	893	853	1,103	1,026	9,050	9,100	1,161	1,087	1,041	1,361	1,274
8,050	8,100	958	902	862	1,114	1,037	9,100	9,150	1,172	1,096	1,049	1,374	1,286
8,100	8,150	968	912	872	1,125	1,048	9,150	9,200	1,182	1,106	1,057	1,386	1,299
8,150	8,200	979	921	881	1,136	1,059	9,200	9,250	1,192	1,115	1,065	1,399	1,311
8,200	8,250	989	931	891	1,149	1,070	9,250	9,300	1,202	1,124	1,073	1,411	1,324
8,250	8,300	1,000	940	900	1,161	1,081	9,300	9,350	1,212	1,134	1,081	1,424	1,336
8,300	8,350	1,010	950	910	1,174	1,092	9,350	9,400	1,223	1,143	1,089	1,436	1,349
8,350	8,400	1,021	959	919	1,186	1,103	9,400	9,450	1,233	1,152	1,097	1,449	1,361
8,400	8,450	1,031	969	929	1,199	1,114	9,450	9,500	1,243	1,162	1,105	1,461	1,374
8,450	8,500	1,042	978	938	1,211	1,125	9,500	9,550	1,253	1,171	1,113	1,474	1,386
8,500	8,550	1,052	988	948	1,224	1,136	9,550	9,600	1,263	1,181	1,121	1,486	1,399
8,550	8,600	1,063	997	957	1,236	1,149	9,600	9,650	1,274	1,190	1,129	1,499	1,411
8,600	8,650	1,073	1,007	967	1,249	1,161	9,650	9,700	1,284	1,199	1,138	1,511	1,424
8,650	8,700	1,083	1,016	976	1,261	1,174	9,700	9,750	1,294	1,209	1,146	1,524	1,436
8,700	8,750	1,092	1,024	984	1,274	1,186	9,750	9,800	1,304	1,218	1,154	1,536	1,449
8,750	8,800	1,101	1,032	992	1,286	1,199	9,800	9,850	1,314	1,227	1,162	1,549	1,461
8,800	8,850	1,110	1,040	1,000	1,299	1,211	9,850	9,900	1,325	1,237	1,170	1,561	1,474
8,850	8,900	1,121	1,050	1,008	1,311	1,224	9,900	9,950	1,335	1,246	1,178	1,574	1,486
8,900	8,950	1,131	1,059	1,016	1,324	1,236	9,950	10,000	1,345	1,255	1,186	1,586	1,499

Table 4—Returns claiming FOUR exemptions (and not itemizing deductions)

At least	But less than	Single, not head of house-hold	Head of house-hold	Married filing joint return	Low income allow-ance	% Standard deduc-tion	At least	But less than	Single, not head of house-hold	Head of house-hold	Married filing joint return	Low income allow-ance	% Standard deduc-tion
$7,900	$ 7,950	$ 619	$593	$556	$ 751	$ 676	$8,950	$ 9,000	$ 822	$779	$739	$ 982	$ 905
7,950	8,000	628	602	565	762	685	9,000	9,050	831	788	748	993	916
8,000	8,050	638	611	573	773	696	9,050	9,100	840	796	756	1,004	927
8,050	8,100	647	620	582	784	707	9,100	9,150	849	804	764	1,015	938
8,100	8,150	657	629	590	795	718	9,150	9,200	858	812	772	1,026	949
8,150	8,200	666	638	599	806	729	9,200	9,250	867	820	780	1,037	960
8,200	8,250	676	647	607	817	740	9,250	9,300	876	828	788	1,048	971
8,250	8,300	685	656	616	828	751	9,300	9,350	885	836	796	1,059	982
8,300	8,350	695	665	625	839	762	9,350	9,400	893	844	804	1,070	993
8,350	8,400	706	674	634	850	773	9,400	9,450	902	852	812	1,081	1,004
8,400	8,450	716	684	644	861	784	9,450	9,500	911	860	820	1,092	1,015
8,450	8,500	727	693	653	872	795	9,500	9,550	920	868	828	1,103	1,026
8,500	8,550	737	703	663	883	806	9,550	9,600	929	876	836	1,114	1,037
8,550	8,600	748	712	672	894	817	9,600	9,650	938	884	844	1,125	1,048
8,600	8,650	758	722	682	905	828	9,650	9,700	947	893	853	1,136	1,059
8,650	8,700	768	731	691	916	839	9,700	9,750	956	901	861	1,149	1,070
8,700	8,750	777	739	699	927	850	9,750	9,800	965	909	869	1,161	1,081
8,750	8,800	786	747	707	938	861	9,800	9,850	974	917	877	1,174	1,092
8,800	8,850	795	755	715	949	872	9,850	9,900	983	925	885	1,186	1,103
8,850	8,900	804	763	723	960	883	9,900	9,950	992	933	893	1,199	1,114
8,900	8,950	813	771	731	971	894	9,950	10,000	1,001	941	901	1,211	1,125

EXERCISES

Use the income tax tables to answer each of the following:

1. How much federal income tax does a person, single, not head of household, with 2 exemptions, owe for the year if she earns:
   a. $9,300 per year
   b. $8,425 per year?
   c. $163 per week?
   d. $181.70 per week?

2. How much federal income tax does a person, single, head of household, with 4 exemptions, owe for the year if he earns:
   a. $8,850 per year?
   b. $9,775 per year?
   c. $176 per week?
   d. $190.85 per week?

3. How much federal income tax does a married person, filing a joint return, owe for the year when earning annually:
   a. $9,000 and claiming 2 exemptions?
   b. $8,350 and claiming 4 exemptions?
   c. $9,750 and claiming 2 exemptions?
   d. $8,900 and claiming 4 exemptions?
   e. $9,875 and claiming 4 exemptions?
   f. $8,225 and claiming 2 exemptions?
   g. $9,690 and claiming 4 exemptions?
   h. $8,815 and claiming 2 exemptions?
   i. $9,464 and claiming 2 exemptions?
   j. $9,929 and claiming 4 exemptions?

4. How much federal income tax does a married person, filing a joint return, owe for the year when earning weekly:
   a. $160 and claiming 2 exemptions?
   b. $185 and claiming 4 exemptions?
   c. $191 and claiming 2 exemptions?
   d. $173 and claiming 2 exemptions?
   e. $156 and claiming 4 exemptions?
   f. $181.50 and claiming 4 exemptions?
   g. $169.25 and claiming 2 exemptions?
   h. $190.75 and claiming 4 exemptions?
   i. $174.30 and claiming 4 exemptions?
   j. $189.15 and claiming 2 exemptions?

5. The Davis family consists of a husband, a wife, and two young children. Mr. Davis is the only wage earner in the family. How many tax exemptions may he claim? If he earns $186.50 per week, how much federal income tax is withheld from his pay each week?

   How much federal income tax is withheld from his pay for the year?

   How much federal income tax does he owe for the entire year?

   Does he still owe some taxes or did he overpay? How much?

Applications of Mathematics

6. Using the income tax tables and the table of withholding tax, find the balance of the federal income tax due or the amount of overpayment for the year when a married person, filing a joint return, earns weekly:

   a. $190 and claims 4 exemptions.
   b. $177 and claims 2 exemptions.
   c. $158 and claims 2 exemptions.
   d. $185 and claims 4 exemptions.
   e. $169 and claims 2 exemptions.
   f. $163.50 and claims 2 exemptions.
   g. $172.90 and claims 4 exemptions.
   h. $180.25 and claims 4 exemptions.
   i. $155.75 and claims 2 exemptions.
   j. $191.60 and claims 4 exemptions.

CONSUMER SPENDING

9–7 INSTALLMENT BUYING; ANNUAL PERCENTAGE RATE; DEFERRED PAYMENT PURCHASE PLANS

Installment Buying

Installment buying or *buying on credit* is a purchase plan by which the customer usually pays a certain amount of cash at the time of purchase and then pays in equal installments at regular intervals, usually months, both the balance and the finance charge on this balance. This *finance charge,* also called *carrying charge,* is the interest (see page 484) which usually is applied on the entire balance as being owed for the full time of payment. The carrying charge or finance charge is the difference between the cash price and the total amount paid. However, since in this type of purchase plan part of the original balance owed is being repaid each month, the given percentage rate charged is not the true rate. The true rate charge is much higher since it is based on the lower average balance that is owed during the payment period. Actually the true rate is almost twice the given rate.

Annual Percentage Rate

In order that people who buy on credit may know what this true rate is so that they may compare finance charge rates, the federal government requires each credit purchase contract or loan contract (see section 9–16) to indicate the *annual percentage rate* which is the true finance charge rate for the year. Annual Percentage Rate Tables are generally used to determine these rates. See page 474.

Deferred Payment Purchase Plan

Many stores use *deferred payment* plans. In these plans a specific monthly payment is made to reduce the balance owed. Sometimes these payments are fixed amounts. Other times these payments decrease as the bracket of the balance owed falls. However a finance charge, usually $1\frac{1}{4}\%$ per month or $1\frac{1}{2}\%$ per month, is applied to the previous monthly unpaid balance and is added to this previous balance. The deferred payment plans also must indicate the annual percentage rate. To compute this we multiply the rate per period by the number of periods per year. Thus the finance charge rate of $1\frac{1}{4}\%$ per month is $12 \times 1\frac{1}{4}\%$ or a 15% annual percentage rate. Credit card accounts are charged a late payment finance charge each month on past due amounts. Here also the annual percentage rates must be indicated.

EXERCISES

1. Find the annual percentage rate if the monthly finance charge rate is:

    a. 3%   **b.** 2%   **c.** $1\frac{3}{4}\%$ **d.** $1\frac{1}{3}\%$ **e.** $2\frac{1}{2}\%$

2. Find the annual percentage rate if the monthly late payment finance charge rate on past due amounts is:

    a. 1%   **b.** $\frac{2}{3}$ of 1%   **c.** $\frac{5}{8}$ of 1%   **d.** $\frac{5}{6}$ of 1%   **e.** $\frac{3}{4}$ of 1%

3. Mr. Garcia bought a sewing machine for $19 down and $8.50 a month for 10 months. How much did the sewing machine cost on the installment plan? If the cash price was $92.50, how much more did the sewing machine cost on the installment plan?

Applications of Mathematics

4. Andrew's mother can purchase an air conditioner for the cash price of $240 or $25 down and 12 equal monthly payments of $22.50 each. How much is the carrying charge? How much can be saved by paying cash?

5. Find the carrying charge when each of the following articles is purchased on the installment plan:

	Cash Price	Down Payment	Monthly Payment	Number of Monthly Payments
a. Rug	$89	$15	$7	12
b. Oil Burner	$650	$100	$26.50	24
c. Refrigerator	$249	$49	$12.25	18
d. Camera	$72.50	$13	$6.75	10

6. Find the monthly payment if the amount financed is $385 and the finance charge is $71, both to be paid in 24 equal monthly payments.

7. A washing machine can be purchased for the cash price of $240 or $20 down and both the balance and the finance charge to be paid in 12 equal monthly payments. If 8% is charged on the full balance, how much is paid monthly?

8. Find the monthly payment when each of the following articles is purchased on the installment plan:

	Cash Price	Down Payment	Finance Charge Interest Rate	Number of Monthly Payments
a. Piano	$645	$105	6%	12
b. Sink	$115	$15	8%	12
c. Projector	$96	$24	10%	12
d. TV	$520	$70	9%	12

9. A department store uses this deferred payment schedule:

Monthly Balance	Monthly Payment	Monthly Balance	Monthly Payment
$0 to $10	Full Balance	$161 to $175	$25
$11 to $90	$10	$176 to $210	$30
$91 to $120	$15	$211 to $250	$40
$121 to $160	$20	over $250	$\frac{1}{5}$ of balance

If finance charges are computed at the monthly rate of $1\frac{1}{4}\%$ on the previous month's balance, find the monthly payment due on each of these previous month's balances. First compute and add the finance charge to the previous month's balance.

a. $68   b. $174   c. $145   d. $92.50   e. $234.80

10. Compute the late payment finance charge on each of the following past due amounts at the specified rate:

   a. $65 past due; $1\frac{1}{2}$% monthly rate
   b. $140 past due; $1\frac{1}{4}$% monthly rate
   c. $28.60 past due; 1% monthly rate
   d. $71.42 past due; $\frac{1}{2}$ of 1% monthly rate
   e. $93.80 past due; $\frac{3}{4}$ of 1% monthly rate
   f. $17.49 past due; $\frac{2}{3}$ of 1% monthly rate

11. To find the annual percentage rate when the amount to be financed, the finance charge for this amount, and the number of monthly payments are known,
   (1) Determine the finance charge per $100 of amount financed by multiplying the given finance charge by 100 and dividing this product by the amount financed.
   (2) Then follow down the left column of the table on page 474 to the line indicating the required number of monthly payments, move across this line until you reach the nearest number to the finance charge per $100.
   (3) The rate at the top of this column is the annual percentage rate.

> For 18 monthly payments a finance charge of $12.29 per $100 financed is at the annual percentage rate of 15%.

Find these annual percentage rates (see table on p. 474):

	Total Amount Financed	Finance Charge on Total Amount	Number of Monthly Payments
a.	$100	$9.94	15
b.	$100	$13.88	20
c.	$100	$6.68	9
d.	$200	$16.34	12
e.	$500	$25.10	6
f.	$350	$59.29	24
g.	$840	$96.18	18
h.	$1,000	$265.70	36
i.	$1,250	$344.75	42
j.	$2,000	$498.40	30
k.	$1,500	$211.05	21
l.	$600	$34.80	8
m.	$3,000	$583.50	25
n.	$900	$91.80	16

Applications of Mathematics

The following is a condensed section of the Federal Reserve Board tables for computing annual percentage rates for monthly payment plans:

NUMBER OF PAYMENTS	14.00%	14.25%	14.50%	14.75%	15.00%	15.25%	15.50%	15.75%	16.00%	17.00%	18.00%
\multicolumn{12}{c}{ANNUAL PERCENTAGE RATE}											
\multicolumn{12}{c}{(FINANCE CHARGE PER $100 OF AMOUNT FINANCED)}											
1	1.17	1.19	1.21	1.23	1.25	1.27	1.29	1.31	1.33	1.42	1.50
2	1.75	1.78	1.82	1.85	1.88	1.91	1.94	1.97	2.00	2.13	2.26
3	2.34	2.38	2.43	2.47	2.51	2.55	2.59	2.64	2.68	2.85	3.01
4	2.93	2.99	3.04	3.09	3.14	3.20	3.25	3.30	3.36	3.57	3.78
5	3.53	3.59	3.65	3.72	3.78	3.84	3.91	3.97	4.04	4.29	4.54
6	4.12	4.20	4.27	4.35	4.42	4.49	4.57	4.64	4.72	5.02	5.32
7	4.72	4.81	4.89	4.98	5.06	5.15	5.23	5.32	5.40	5.75	6.09
8	5.32	5.42	5.51	5.61	5.71	5.80	5.90	6.00	6.09	6.48	6.87
9	5.92	6.03	6.14	6.25	6.35	6.46	6.57	6.68	6.78	7.22	7.65
10	6.53	6.65	6.77	6.88	7.00	7.12	7.24	7.36	7.48	7.96	8.43
11	7.14	7.27	7.40	7.53	7.66	7.79	7.92	8.05	8.18	8.70	9.22
12	7.74	7.89	8.03	8.17	8.31	8.45	8.59	8.74	8.88	9.45	10.02
13	8.36	8.51	8.66	8.81	8.97	9.12	9.27	9.43	9.58	10.20	10.81
14	8.97	9.13	9.30	9.46	9.63	9.79	9.96	10.12	10.29	10.95	11.61
15	9.59	9.76	9.94	10.11	10.29	10.47	10.64	10.82	11.00	11.71	12.42
16	10.20	10.39	10.58	10.77	10.95	11.14	11.33	11.52	11.71	12.46	13.22
17	10.82	11.02	11.22	11.42	11.62	11.82	12.02	12.22	12.42	13.23	14.04
18	11.45	11.66	11.87	12.08	12.29	12.50	12.72	12.93	13.14	13.99	14.85
19	12.07	12.30	12.52	12.74	12.97	13.19	13.41	13.64	13.86	14.76	15.67
20	12.70	12.93	13.17	13.41	13.64	13.88	14.11	14.35	14.59	15.54	16.49
21	13.23	13.58	13.82	14.07	14.32	14.57	14.82	15.06	15.31	16.31	17.32
22	13.96	14.22	14.48	14.74	15.00	15.26	15.52	15.78	16.04	17.09	18.15
23	14.59	14.87	15.14	15.41	15.68	15.96	16.23	16.50	16.78	17.88	18.98
24	15.23	15.51	15.80	16.08	16.37	16.65	16.94	17.22	17.51	18.66	19.82
25	15.87	16.17	16.46	16.76	17.06	17.35	17.65	17.95	18.25	19.45	20.66
26	16.51	16.82	17.13	17.44	17.75	18.06	18.37	18.68	18.99	20.24	21.50
27	17.15	17.47	17.80	18.12	18.44	18.76	19.09	19.41	19.74	21.04	22.35
28	17.80	18.13	18.47	18.80	19.14	19.47	19.81	20.15	20.48	21.84	23.20
29	18.45	18.79	19.14	19.49	19.83	20.18	20.53	20.88	21.23	22.64	24.06
30	19.10	19.45	19.81	20.17	20.54	20.90	21.26	21.62	21.99	23.45	24.92
31	19.75	20.12	20.49	20.87	21.24	21.61	21.99	22.37	22.74	24.26	25.78
32	20.40	20.79	21.17	21.56	21.95	22.33	22.72	23.11	23.50	25.07	26.65
33	21.06	21.46	21.85	22.25	22.65	23.06	23.46	23.86	24.26	25.88	27.52
34	21.72	22.13	22.54	22.95	23.37	23.78	24.19	24.61	25.03	26.70	28.39
35	22.38	22.80	23.23	23.65	24.08	24.51	24.94	25.36	25.79	27.52	29.27
36	23.04	23.48	23.92	24.35	24.80	25.24	25.68	26.12	26.57	28.35	30.15
37	23.70	24.16	24.61	25.06	25.51	25.97	26.42	26.88	27.34	29.18	31.03
38	24.37	24.84	25.30	25.77	26.24	26.70	27.17	27.64	28.11	30.01	31.92
39	25.04	25.52	26.00	26.48	26.96	27.44	27.92	28.41	28.89	30.85	32.81
40	25.71	26.20	26.70	27.19	27.69	28.18	28.68	29.18	29.68	31.68	33.71
41	26.39	26.89	27.40	27.91	28.41	28.92	29.44	29.95	30.46	32.52	34.61
42	27.06	27.58	28.10	28.62	29.15	29.67	30.19	30.72	31.25	33.37	35.51
43	27.74	28.27	28.81	29.34	29.88	30.42	30.96	31.50	32.04	34.22	36.42
44	28.42	28.97	29.52	30.07	30.62	31.17	31.72	32.28	32.83	35.07	37.33
45	29.11	29.67	30.23	30.79	31.36	31.92	32.49	33.06	33.63	35.92	38.24

12. The annual percentage rate tables may be used to determine the finance charge and the monthly payment when the annual percentage rate is known.
   (1) Using the column headed by the known annual percentage rate, read down this column until you reach the line corresponding to the given number of monthly payments. This is the finance charge per $100 of the amount to be financed.
   (2) *To determine the finance charge for the given amount financed,* multiply the finance charge per $100 by the amount financed, then divide the product by 100.

(3) *To find the monthly payment,* add the finance charge to the amount financed, then divide this sum by the number of monthly payments.

Using the table on page 474, find the finance charge for each of the following amounts financed at the given annual percentage rates. Also find the monthly payment.

	Amount Financed	Annual Percentage Rate	Number of Monthly Payments
a.	$100	16%	15
b.	$300	15.25%	9
c.	$500	17%	24
d.	$750	15.75%	40
e.	$1,000	14.25%	36
f.	$800	14.50%	6
g.	$1,200	18%	42
h.	$4,000	14.75%	12
i.	$3,600	15.50%	36
j.	$5,500	14.25%	30
k.	$2,300	15%	28
l.	$1,750	14.50%	18
m.	$625	15.25%	10
n.	$1,500	14%	45
o.	$6,000	15.75%	20

9–8 FIRE INSURANCE

Insurance is a plan by which persons share risks so that each person is protected against financial loss. The insured person is charged a sum of money called a *premium.* The insurance company uses the fund created by these premiums to pay the insured persons who suffer losses.

The written contract between the insured person and the insurance company is called the *policy.* The amount of insurance specified in the policy is called the *face* of the policy. The length of time the insurance is in force is called the *term* of the policy.

To find the premium, we multiply the number of $100 of fire insurance by the rate per $100.

The prepaid premium rate on fire insurance for 3 years is 2.7 times the yearly rate. If this premium is paid by yearly installments, then 35% of the total premium is paid each year.

Applications of Mathematics

EXERCISES

1. What is the cost of $5,000 fire insurance for 1 year at the rate of $.20 per $100?
2. Find the premium for each of the following 1-year policies:

Face of policy	$2,000	$8,000	$15,000	$6,500	$30,000	$48,000
Yearly rate per $100	$.22	$.16	$.41	$.35	$.18	$.25

3. What is the premium for a 3-year policy if the yearly rate is $.24 per $100 and the face of the policy is $7,500?
4. Find for each of the following 3-yr. policies (a) the total prepaid premium and (b) the yearly premium if paid by yearly installments:

Face of Policy	$20,000	$35,000	$7,500	$10,000	$12,500	$45,000
Yearly rate per $100	$.14	$.18	$.36	$.25	$.22	$.40

5. Find the total premium for each of the following policies:

Value of building	$10,000	$7,500	$40,000	$30,000	$20,000	$17,500
Per cent of value insured	80%	60%	75%	70%	75%	80%
Yearly rate per $100	$.32	$.14	$.30	$.26	$.35	$.28
Term	1 yr.	3 yr.	3 yr.	1 yr.	3 yr.	3 yr.

6. How much does Mrs. Carroll save by taking out a 3-year policy instead of three 1-year policies if she insures her home for $27,500 at the yearly rate of $.22 per $100?
7. Mr. Johnson insured his house worth $18,400 for 75% of its value. He also insured the contents worth $9,000 for 70% of their value. What was his total premium if he bought a 3-year policy for each? The yearly rate on the building was $.12 per $100 and on the contents was $.24 per $100.
8. In a certain community the rate for a 1-year policy on a frame building is $.42 per $100 and on a brick building is $.14 per $100.

At these rates what is the premium for $15,000 insurance on a frame building? On a brick building?

9. An owner of a building worth $40,000 took out $24,000 fire insurance. Her policy contains the 80% coinsurance clause. How much would the insurance company pay her if she suffered a fire loss of $16,000?

10. During the 5th year of his insurance, Mr. Petranic collected a fire loss claim of $15,000. His house was insured for $38,000 at the annual rate of $.20 per $100 and the contents for $19,000 at the annual rate of $.25 per $100. How much more did he receive from his insurance than he paid in premiums?

9-9 LIFE INSURANCE

Life insurance offers financial protection to the dependents of an insured person in the event of the insured person's death. The person who receives the money when the insured person dies is called the *beneficiary*.

(1) *Term insurance*—Person is insured for a specified period of time. Premiums are paid only during that time. Beneficiary receives face of policy if insured person dies within the specified time.

(2) *Ordinary life insurance*—Person is insured until death. Premiums are paid until death of the insured person. Beneficiary receives face of policy when insured person dies.

(3) *Limited-payment life insurance*—Person is insured until death. Premiums are paid for a specified period of time. Beneficiary receives face of policy when insured person dies.

(4) *Endowment insurance*—Person is insured for a specified period of time. Premiums are paid only during that time. Beneficiary receives face of policy if insured person dies within the specified time. The insured person or policy holder receives face of policy if alive at the end of the specified time.

To find the premium, we select from the table on page 478 the premium for $1,000 insurance, then we multiply it by the number of $1,000 of insurance.

Applications of Mathematics

EXERCISES

1. Average life expectation:

At the age of	Birth	1	5	10	15	20	25
Years male may expect to live	68.9	69.1	65.3	60.5	55.6	51.0	46.5
Years female may expect to live	76.6	76.6	72.7	67.8	62.9	58.1	53.3

At the age of	30	35	40	45	50	55	60
Years male may expect to live	41.8	37.2	32.6	28.1	24.0	20.1	16.5
Years female may expect to live	48.4	43.6	38.9	34.3	29.8	25.5	21.4

   a. To what age may a 20-yr.-old female expect to live?
   b. To what age may a 35 yr.-old-male expect to live?
   c. How may years longer may a 15-yr.-old female expect to live than a 15-yr.-old male?

Annual Life Insurance Premiums per $1,000

Age	5-Year Term Male	5-Year Term Female	Ordinary Life Male	Ordinary Life Female	20-Payment Life Male	20-Payment Life Female	20-Year Endowment Male	20-Year Endowment Female
18	3.77	3,74	13.48	12.95	23.97	23.22	45.17	45.17
20	4.01	4.01	14.20	13.62	24.91	24.11	45.30	45.30
22	4.20	4.20	15.00	14.35	25.89	25.05	45.46	45.46
24	4.40	4.40	15.86	15.15	26.94	26.03	45.64	45.64
25	4.50	4.50	16.34	15.59	27.49	26.55	45.75	45.75
26	4.60	4.60	16.83	16.04	28.06	27.09	45.86	45.85
28	4.83	4.83	17.90	17.02	29.26	28.21	46.14	46.10
30	5.09	5.05	19.09	18.11	30.54	29.42	46.48	46.40
35	6.08	5.86	22.71	21.43	34.24	32.88	47.72	47.49
40	8.03	7.44	27.20	25.56	38.49	36.85	49.45	49.00
45	11.10	10.13	32.79	30.73	43.34	41.40	51.81	51.06
50	16.00	14.45	40.07	37.50	49.28	46.94	55.40	54.18

2. Find the annual premium on each of the following policies:
   a. 20-year endowment for $5,000 issued to a woman at age 24.
   b. Ordinary life for $25,000 issued to a man at age 45.
   c. 20-payment life for $15,000 issued to a man at age 26.

478   CHAPTER 9

    d. 5-year term for $50,000 issued to a woman at age 40.
    e. Ordinary life for $30,000 issued to a woman at age 35.
    f. 20-year endowment for $10,000 issued to a man at age 30.
    g. 20-payment life for $20,000 issued to a woman at age 22.
    h. 5-year term for $40,000 issued to a man at age 28.
    i. 20-year endowment for $12,000 issued to a woman at age 20.
    j. Ordinary life for $75,000 issued to a man at age 25.

3. How much more is the male's life insurance premium per $1,000 than the female's rate on each of the following policies:

    a. 20-payment life, issued at age 25?
    b. Ordinary life, issued at age 30?
    c. 20-year endowment, issued at age 18?
    d. 5-year term, issued at age 50?
    e. Ordinary life, issued at age 22?

4. Mr. Harris bought a 20-year endowment life insurance policy for $10,000 at age 22. If he is alive at the age of 42, how much more will he receive than he paid in?

5. At the age of 24 Mrs. Warren bought a 20-payment life policy for $15,000. What is the total amount of premium she pays for the 20 years? Will the beneficiary receive more or less than Mrs. Warren paid in? How much?

6. Approximately how many times as much ordinary life insurance can be bought for a given premium at age 30 as 20-year endowment insurance?

9–10 SALES TAX

A *sales tax* is a tax on the purchase price of an article.

EXERCISES

1. At the rate of 5% sales tax, what would the tax be on an article that sells for $16.50?

2. Find the sales tax to the nearest cent on each of the following purchases at the given tax rates:

Purchases	$18.00	$26.25	$7.49	$19.20	$68.75	$150.89
Tax Rate	6%	4%	3%	$4\frac{1}{2}$%	$2\frac{1}{2}$%	$1\frac{1}{2}$%

Applications of Mathematics

3. Find the selling price, including tax, of each of the following articles:
   a. A suit selling for $145, tax 7%.
   b. A table selling for $89, tax 2%.
   c. A piano selling for $950, tax 6%.
   d. A blanket selling for $39.95, tax 3%.
   e. A clock selling for $27.49, tax 4%.
   f. A sofa selling for $349.95, tax 8%.
   g. A CB radio selling for $112.50, tax 5%.
   h. A mirror selling for $64.25, tax $3\frac{1}{2}$%.
   i. A dress selling for $39.98, tax $5\frac{1}{2}$%.
   j. A toaster selling for $29.75, tax $4\frac{1}{2}$%.

9–11 EXCISE TAX

An *excise tax* is a federal tax on certain items. The federal excise tax rate on telephone service is 4% and on fares for air travel is 8%.

EXERCISES

1. Find the amount of tax on local telephone service costing:
   a. $6.90   b. $14.70   c. $21.85   d. $9.40   e. $17.25
2. Find the amount of tax on an air travel fare of:
   a. $93   b. $112.50   c. $186.20   d. $321.75   e. $253.40
3. Find the full amount of the bill including federal excise tax and, where given, state tax on telephone service costing:
   a. $7.75
   b. $15.30
   c. $21.25, 4% state tax
   d. $9.50, 5% state tax
   e. $16.84, 6% state tax
   f. $28.95, 3% state tax
4. Find the total cost of airplane fare including excise tax when the fare excluding excise tax is:
   a. $79   b. $165   c. $73.50   d. $240.25   e. $188.60

9–12 PROPERTY TAX

A tax on buildings and land is called a *real property* or *real estate tax*. The value that tax officers place on property for tax purposes is the *assessed value* of the property.

The tax rate on property may be expressed as: cents per $1, mills per $1, dollars and cents per $100, dollars and cents per $1,000, and percent.

There are 10 mills in 1 cent. A mill is one-thousandth of a dollar ($.001) or one-tenth of a cent.

To find the amount of taxes on property, we apply the tax rate to the assessed value of the property.

$.04 per $1 = 40 mills per $1 = $4 per $100
     = $40 per $1,000 = 4%

Find the taxes on a house assessed for $8,500 when the tax rate is $2.50 per $100.

$$\begin{array}{r} 85 \\ \$100 \overline{)\$8,500} \end{array}$$

$$\begin{array}{r} \$2.50 \text{ tax rate per \$100} \\ \times 85 \text{ number of \$100} \\ \hline 1250 \\ 2000 \\ \hline \$212.50 \text{ taxes} \end{array}$$

Answer, $212.50 taxes

EXERCISES

1. Write as a decimal part of a dollar:
   9 mills; 14 mills; 128 mills; 40 mills
2. Write as cents: 30 mills; 23 mills; 7 mills; 115 mills
3. Write as mills: $.02; $.04; $.006; $.131

Write each of the following rates as:

4. Cents per $1: $8 per $100; 64.5 mills per $1; $39 per $1,000; 2.9%
5. Mills per $1: $13 per $100; $47 per $1,000; 1.7%; $.91 per $1
6. Dollars and cents per $100: $.09 per $1; 3.1%; 118 mills per $1; $75 per $1,000
7. Dollars and cents per $1,000: 4.3%; $5.85 per $100; 34.5 mills per $1; $.16 per $1
8. Percent: $3 per $100; $93.60 per $1,000; $.08 per $1; 52 mills per $1
9. A community with an assessed valuation of $2,500,000 requires $50,000 as taxes. What should its tax rate be? Express the rate as dollars per $100.
10. a. If the tax rate is $6.80 per $100, how much must you pay for taxes on a house assessed for $19,900?
    b. How much must you pay for taxes on a building assessed for $75,000 if the tax rate is $5.75 per $100?

Applications of Mathematics

Find the amount of taxes on property having the following assessed valuations and tax rates:

11.

Assessed valuation	$4,300	$9,100	$17,500	$11,000	$22,500
Tax rate per $100	$9.60	$3.40	$1.85	$14.10	$5.80

12.

Assessed valuation	$5,600	$10,000	$18,700	$27,800	$13,000
Tax rate per $1	$.15	$.127	$.04$\frac{1}{2}$%	$.054	$.09

13.

Assessed valuation	$25,300	$22,900	$12,000	$16,800	$39,300
Tax rate per $1	26 mills	74 mills	119 mills	47 mills	61 mills

14.

Assessed valuation	$8,000	$13,500	$7,600	$10,400	$25,000
Tax rate	4%	8.2%	12.9%	9.3%	4.8%

15. A house is assessed for $9,500 and the tax rate is $2.75 per $100. Which will give a greater reduction in taxes, reducing the assessment to $8,200 or reducing the rate to $2.40 per $100? How much greater?

16. A building worth $27,000 is assessed for $\frac{2}{3}$ of its value. How much tax does the owner pay on the property if the tax rate is $7.25 per $100?

17. How much does Lisa's father pay for taxes on his property which is assessed for 40% of its cost of $28,000 if the tax rate is $3.45 per $100?

9–13 BUYING AND RENTING A HOUSE

When a person borrows money to buy a house, the borrower gives the lender a written claim (called a *mortgage*) to the property in case of failure to repay the specified loan or to pay the interest on the loan when due. To find the amount of the mortgage, we subtract the down payment from the purchase price.

To determine the cost of owning a house, we find the sum of the following expenses: interest (see page 484) on mortgage, cost of repairs, depreciation, cost of insurance on the house, property taxes, and interest lost on the principal paid. Although in the computation of the federal income tax we use the property tax and the interest paid on the mortgage as deductions and the interest received on the invested principal as income, we shall disregard these considerations in problems 7 through 10.

EXERCISES

1. What is your rent for a year if your monthly rent is:
   a. $135? **b.** $490? **c.** $92.50? **d.** $268.25? **e.** $387.50?

2. If you should not spend more than 25% of your income for the rent of a house, what is the highest monthly rent you can afford to pay if you earn:
   a. $8,760 per yr.?
   b. $1,000 per mo.?
   c. $217.75 per wk.?
   d. $25,000 per yr.?

3. If you should not buy a house costing more than $2\frac{1}{2}$ times your annual income, what is the highest price you can afford to pay if you earn:
   a. $12,000 per yr.?
   b. $825 per mo.?
   c. $96.25 per wk.?
   d. $335 per wk.?

4. A man bought a house for $25,000. He paid $7,500 in cash and gave a mortage bearing 8% interest for the balance. What is his annual interest? Semi-annual interest?

5. Find the amount of mortgage, annual interest, and semi-annual interest on the mortgage:

Cost of house	$10,800	$26,000	$21,900	$35,000	$40,000
Down payment	$1,800	$2,600	$5,500	$8,000	$8,000
Annual interest rate	8%	9%	$8\frac{1}{2}\%$	$8\frac{1}{4}\%$	$7\frac{3}{4}\%$

6. A mortgage of $10,000 is to be paid off in 20 years. How much must be paid every 6 months if the mortgage is to be repaid in equal semi-annual payments?

7. What are the costs per month of owning a house if the property tax is $920 per yr., semi-annual interest on the mortgage is $220, repairs average $300 per yr., insurance costs $210 for 3 yr., and interest lost on principal paid amounts to $250 per yr.?

8. Find the monthly carrying charges during the first and second years on a new house that costs $24,000. Down payment is

Applications of Mathematics   483

20% of the cost, and 8% interest is charged on the balance. House is insured for 80% of cost at 20¢ per $100. House is assessed for 75% of cost; tax rate is $4.75 per $100. Mortgage is to be paid off in 25 equal annual installments decreasing the interest due on mortgage each year. Principal consisting of the down payment and succeeding annual payments could be invested at 6%.

9. Find the monthly cost of owning a house:

	Cost of House	Down Payment	Interest Rate on Mortgage	Repairs and Depreciation	Insurance	Assessed Value	Tax Rate per $100	Interest Rate Lost on Principal
a.	$28,500	$6,500	8%	$480	$250	$27,000	$4.95	6%
b.	$16,300	$4,300	$8\frac{1}{2}$%	$372	$126	$15,000	$5.10	$5\frac{1}{2}$%
c.	$37,500	$7,500	$7\frac{1}{2}$%	$540	$324	$19,500	$8.70	8%

10. Is it cheaper to buy or rent the house and how much cheaper per year?

	Cost of House	Down Payment	Interest Rate on Mortgage	Repairs and Insurance	Assessed Value	Tax Rate per $100	Interest Rate Lost on Principal	Rent per Month
a.	$17,400	$3,400	8%	$308	$15,200	$3.65	6%	$200
b.	$18,500	$6,000	7%	$250	$11,000	$7.25	$5\frac{1}{2}$%	$250
c.	$32,800	$11,400	$8\frac{1}{2}$%	$516	$16,400	$9.20	8%	$500
d.	$20,000	$4,800	$7\frac{1}{2}$%	$350	$12,000	$6.95	9%	$300

MANAGING MONEY

9-14 SIMPLE INTEREST

Just as people pay rent for the use of a house belonging to someone else, people pay interest for the use of money belonging to others. *Interest* is money paid for the use of money. The money borrowed or invested on which interest is paid is called the *principal*. Interest paid on the principal only is called *simple interest*. The interest charged is generally expressed as a percent of the principal. This percent is called the *rate of interest*. Unless specified otherwise, the rate of interest is the rate per year. The sum of the principal and the interest is called the *amount*.

(1) *To find the interest*, we multiply the principal by the rate of interest per year. Then we multiply this product by the time expressed in years. (2 mo. = $\frac{1}{6}$ yr.; 120 da. = $\frac{1}{3}$ yr.; semi-annual = $\frac{1}{2}$ yr.; quarterly = $\frac{1}{4}$ yr.)

The interest (i) equals the principal (p) times the rate of interest per year (r) times the time expressed in years (t). This rule expressed as a formula is: $i = prt$.

(2) *To find the interest by the formula*, we write the interest formula, substitute the given quantities, then compute as required. The rate is usually expressed as a decimal but a common fraction may be used.

(3) *To find the amount*, we add the interest and the principal.

Find the interest and the amount due on $375 borrowed for 4 years at 5%:

$375 principal
.05 rate
─────
$18.75 interest for 1 year
×4 time
─────
$75.00 interest for 4 years

$375 principal
 75 interest
─────
$450 amount

Answer, $75 interest; $450 amount

Find the interest on $500 for 2 yr. 8 mo. at 6%:

$p = \$500$

$r = 6\% = \frac{6}{100}$

$t = 2 \text{ yr. } 8 \text{ mo. } = 2\frac{2}{3} \text{ yr.}$

$i = ?$

$i = prt$

$i = 500 \times \frac{6}{100} \times 2\frac{2}{3}$

$i = \$80$

Answer, $80 interest

---

(4) *Sixty-day or 6% method.*—To find the interest for 60 days at 6%, we move the decimal point in the principal two places to the left.

(5) *To find the annual rate of interest*, we find what percent the interest for one year is of the principal.

(6) *To find the principal when the annual rate and annual interest are known*, we divide the interest by the rate.

Applications of Mathematics 485

> What is the annual rate of interest if the annual interest on a principal of $150 is $9?
>
> $$\frac{\$9}{\$150} = \$9 \div \$150 = \$150\overline{)\$9.00}^{.06\ =\ 6\%}$$
>
> *Answer,* 6%

EXERCISES

1. Find the interest for 1 year on:
   a. $300 at 7%.
   b. $950 at 6%.
   c. $1,000 at $8\frac{1}{2}$%.
   d. $1,600 at $5\frac{1}{2}$%.

Find the interest on:

2. a. $500 for 2 yr. at 8%.
   b. $625 for 3 yr. at 9%.
   c. $1,400 for 6 yr. at $7\frac{1}{2}$%.
   d. $2,500 for 8 yr. at $3\frac{3}{4}$%.

3. a. $760 for $3\frac{1}{2}$ yr. at 1%.
   b. $420 for $\frac{2}{3}$ yr. at 11%.
   c. $1,200 for $4\frac{3}{4}$ yr. at $5\frac{1}{2}$%.
   d. $2,700 for $\frac{5}{6}$ yr. at $6\frac{1}{2}$%.

4. a. $800 for 4 mo. at 3%.
   b. $5,200 for 10 mo. at 6%.
   c. $624 for 5 mo. at $7\frac{1}{2}$%.
   d. $1,170 for 8 mo. at $2\frac{1}{2}$%.

5. a. $80 for 1 yr. 9 mo. at 9%.
   b. $4,000 for 5 yr. 6 mo. at 2%.
   c. $2,100 for 2 yr. 11 mo. at $8\frac{1}{2}$%.
   d. $600 for 1 yr. 3 mo. at $7\frac{1}{4}$%.

6. Find the exact interest (1 yr. = 365 da.) on:
   a. $580 for 73 da. at 3%.
   b. $1,740 for 15 da. at 6%.
   c. $3,000 for 45 da. at 8%.
   d. $800 for 146 da. at 5%.

7. Find the interest (1 yr. = 360 da.) on:
   a. $900 for 30 da. at 5%.
   b. $1,600 for 60 da. at 4%.
   c. $480 for 72 da. at 6%.
   d. $7,000 for 90 da. at 9%.

8. Find the semi-annual interest on:
   a. $825 at 10%.
   b. $2,350 at 6%.
   c. $3,900 at $5\frac{1}{2}$%.
   d. $6,400 at 7.9%.

9. Find the quarterly interest on:
   a. $700 at 4%.
   b. $1,250 at 8%.
   c. $6,000 at $1\frac{3}{4}$%.
   d. $10,000 at 9.6%.

10. Use the 6%, 60-day method to find the interest on:
    a. $2,850 for 60 da. at 6%.
    b. $349 for 60 da. at 6%.
    c. $1,200 for 30 da. at 6%.
    d. $700 for 30 da. at 3%.

11. What is the amount due when the principal is $360 and the interest is $21.60?

12. What is the interest and amount due on a loan of $800 for 3 yr. at 6%.

13. What is the amount due on a loan of $425 for 8 yr. at $7\frac{1}{2}$%?

14. Find the interest and amounts due on the following loans:
    a. $200 for 1 yr. at 6%.
    b. $500 for 7 yr. at 9%.
    c. $1,750 for 2 yr. at $8\frac{1}{2}$%.

15. What is the annual rate of interest if the principal is $280 and the annual interest is $14?

16. What is the annual rate of interest if the principal is $450 and the interest for 5 yr. is $135?

17. Find the annual rates of interest when the interest for:
    a. 1 yr. on $75 is $3.
    b. 4 yr. on $1,200 is $144.
    c. 9 yr. on $860 is $154.80.

18. What is the principal if the amount is $796 and the interest is $31?

19. On what principal is the annual interest $9 when the annual rate of interest is 6%?

20. Find the principal on which the annual interest is:
    a. $23 when the annual rate is 4%.
    b. $90 when the annual rate is 5%.
    c. $37.50 when the annual rate is 3%.

21. Mr. Thompson owns eight $1,000 bonds, each bearing $3\frac{1}{4}$% interest. How much interest does he receive every 6 months?

22. A woman bought a house for $29,600. She paid 30% down and gave a mortgage bearing 8% interest for the remainder. How much interest does she pay semi-annually?

Applications of Mathematics

23. On which investment does Mrs. Johnson receive a higher rate of interest? She receives $157.50 semi-annually on an investment of $7,500 and $344 annually on an investment of $8,600.
24. How much money must be invested at 8% to earn $10,000 per year?
25. How many years does it take $200 invested at 4% simple interest to double itself?

9–15 BANKING

There are two main kinds of banks: savings banks and commercial banks. A third type of banking institution that has grown greatly in recent years is the federal and state savings and loan association, which is very much like a savings bank. Credit unions also render both savings and loan services, but they are usually offered only to particular groups of people, like city employees, factory workers, teachers, etc.

The life of a bank depends on the use of its depositors' money as loans to business people and others at rates of interest higher than the rates of interest that the bank pays to its depositors. The difference between these rates of interest produces money for the bank to pay its expenses and to earn a profit.

Both commercial banks and savings banks provide similar services. An important service provided by commercial banks is the *checking account*, not offered by many savings banks. To deposit money in a checking account, we use a *deposit slip* (see form at the right). To transfer funds from a checking account to persons or business establishments to whom we owe money, we use a form called a *check* (see form on the next page). This method facilitates the payment of bills without any cash being handled directly. The *check stub* on the left of the check is filled out for our records.

LOCAL TRUST COMPANY
DEPOSIT

YOUR CITY _____ 19__
ACCOUNT NO. _____
NAME _____
ADDRESS _____

	DOLLARS	CENTS
CASH →		
CHECKS BY BANK NUMBER ↓		
1		
2		
3		
4		
5		
6		
7		
TOTAL		

The main service offered by the savings bank is the *savings account,* on which the depositor is paid interest. Savings banks also provide other services such as passbook loans, safe deposit boxes, and money orders. We use a deposit slip to deposit money and a form called a *withdrawal slip* to withdraw money from our accounts.

EXERCISES

Use two deposit slip forms and four check forms if available; otherwise copy them as illustrated to record the following:
1. On June 1, you have a balance of $1,425.83. Write this balance as the balance brought forward on the first check stub.
2. On June 3, you deposit 6 $10 bills, 8 $5 bills, 17 $1 bills, and a check, numbered 1111–3, drawn on the First National Bank for $389.49. Fill in the deposit slip. Write the total amount of the deposit on the first check stub.
3. On June 6, you send a check to the Scott Furniture Co. in the amount of $746.95 for furniture purchased. Fill in the first check stub. Carry the balance to the second check stub. Write out the first check.
4. On June 8, you send a check to the L. & M. Television Co. in the amount of $559.50 for a television set. Fill in the second check stub. Carry the balance to the third check stub. Write out the second check.
5. On June 16, you deposit 3 $20 bills, 2 $10 bills, 4 $5 bills, 11 $1 bills, 7 quarters, 5 dimes, 9 nickels, a check, numbered 2994–3, drawn on the Bankers Trust Co. for $629.84, and a check, numbered 3004–5, drawn on the Second City Trust Co. for $287.39. Fill in the second deposit slip. Write the total amount of the deposit on the third check stub.

Applications of Mathematics

6. On June 17, you send a check to Stephen Jaffe in the amount of $93.25 for merchandise purchased. Fill in the third check stub. Carry the balance to the fourth check stub. Write out the third check.

7. On June 22, you send a check to Peter John Rugs, Inc. in the amount of $379.60 for rugs purchased. Fill in the fourth check stub. Write out the fourth check.

9–16 BORROWING MONEY

Money may be borrowed from other people, from banks, from loan companies, from savings and loan associations, from credit unions, and on your life insurance policy. When money is borrowed, there is a charge (called interest) for the use of this money. See section 9–14.

Loans which are secured by stocks, bonds, property, etc. are called *collateral loans*. If the loan is not repaid at the specified time, then the lending institution can collect the amount due by selling the collateral. The rate of interest charged on collateral loans is less than the rate of interest charged on unsecured loans.

Some banks will lend money to customers, allowing them to keep the principal until a specified time but requiring payment of interest monthly. This type of loan is called a *demand loan*.

A. Discount Loan and Add-on-Interest Loan

The *discount loan* and the *add-on-interest* loan are the two types of loans most generally used. Annual percentage rates are indicated on loan contracts. See page 471.

When a discount loan is made, the lending company immediately takes off the full interest from the principal (amount of note) and the borrower receives only the difference (called net proceeds). The borrower, however, is required to repay the full principal in a specified number of equal monthly payments. See problem on page 491.

When an add-on-interest loan is made, the borrower receives the full principal. The interest is added on to this principal and the borrower is required to repay the full amount (principal plus interest) in a specified number of equal monthly payments.

For example:
   Mr. Harris borrowed $750 at 6% annual interest and is required to repay his loan in 12 equal monthly payments. What is the actual amount of money that he receives and how much is his monthly payment if he takes a discount loan? If he takes an add-on-interest loan?

Discount Loan		Add-on-Interest Loan	
$750	principal (amount of note)	$750	amount received
−45	interest		
$705	amount received		
		$750	principal
		+45	interest
		$795	amount to be repaid
$62.50	monthly payment		
12)$750.00	amount to be repaid		
		$66.25	monthly payment
		12)$795.00	amount to be repaid

Answer, $705, amount received;
        $62.50, monthly payment

Answer, $750, amount received;
        $66.25, monthly payment

Observe in these loans that since the principal is being repaid monthly, the average amount that was borrowed is only about one-half of the original amount of the loan. Therefore the true rate of interest is almost double the stated rate of interest.

Under the terms of many life insurance policies the policyholder is permitted to borrow from the insurance company an amount of money not more than the loan value or the cash surrender value of the policy. The interest charged is simple interest for the period of time the money is borrowed. See section 9–14.

EXERCISES

1. What is the annual percentage rate if the rate of interest charged on a loan:

   a. Per month is: 1%? $\frac{1}{2}$%? 0.8%? 3%? $1\frac{3}{4}$%?

   b. Per week is: 1%? $\frac{3}{4}$%? $\frac{1}{2}$%? $1\frac{1}{4}$%? 0.2%?

   c. Per day is: 1%? $\frac{1}{4}$%? 0.1%? $\frac{3}{4}$%? 0.3%?

Applications of Mathematics

2. What is the annual rate of interest you are paying when you:
   a. Borrow $1 and pay back $2 at the end of 2 months?
   b. Borrow $5 and pay back $5.50 at the end of a week?
   c. Borrow $100 and pay back $125 at the end of a week?
   d. Borrow 25¢ and pay back 30¢ at the end of 1 day?
   e. Borrow $10 and pay back $11 at the end of 1 day?

3. What is the actual amount of money (net proceeds) the borrower receives and how much is the monthly payment in each of the following discount loans?

	Amount of Note	Interest Rate	Period of Loan
a.	$600	6%	12 months
b.	$1,000	7%	18 months
c.	$500	$6\frac{1}{2}$%	24 months
d.	$2,000	$8\frac{1}{4}$%	36 months
e.	$1,800	$7\frac{1}{2}$%	12 months
f.	$3,500	8%	30 months
g.	$5,000	$6\frac{1}{2}$%	42 months
h.	$7,500	6%	48 months
i.	$4,200	$6\frac{3}{4}$%	24 months
j.	$10,000	$7\frac{1}{4}$%	60 months

4. What is the actual amount of money the borrower receives and how much is the monthly payment in each of the following add-on-interest loans?

	Principal	Interest Rate	Period of Loan
a.	$800	6%	12 months
b.	$1,000	8%	30 months
c.	$4,000	$7\frac{1}{2}$%	24 months
d.	$2,500	$6\frac{1}{2}$%	36 months
e.	$1,600	$8\frac{1}{4}$%	18 months
f.	$6,000	7%	24 months
g.	$7,200	$6\frac{3}{4}$%	60 months
h.	$400	9%	42 months
i.	$9,000	$6\frac{1}{4}$%	36 months
j.	$3,600	$6\frac{1}{2}$%	30 months

5. How much interest is due on the following loans on insurance policies?
   a. $900 borrowed for 6 months at 6%.
   b. $1,700 borrowed for 3 months at $6\frac{1}{2}$%.
   c. $2,100 borrowed for 4 months at 5%.
   d. $500 borrowed for 45 days at 6%.
   e. $600 borrowed for 1 week at 6%.

6. How much interest is due each month on each of the following collateral loans?
   a. $2,000 borrowed for 1 year at 8%.
   b. $1,400 borrowed for 6 months at $7\frac{1}{2}$%.
   c. $7,500 borrowed for 18 months at 7%.
   d. $10,000 borrowed for 9 months at $8\frac{1}{4}$%.
   e. $25,000 borrowed for 3 months at $7\frac{3}{4}$%.

B. Financing a Car

Amount to be Borrowed	Amount of Each Payment		
	24 months	30 months	36 months
$1,000	$45.83	$37.50	$31.94
1,200	55.00	45.00	38.33
1,500	68.75	56.25	47.91
1,800	82.50	67.50	57.50
2,000	91.66	75.00	63.88
2,500	114.58	93.75	79.86
3,000	137.50	112.50	95.83
3,500	160.41	131.25	111.80

EXERCISES

1. Find the total payment if $1,200 is borrowed for 24 months. For 30 months. For 36 months. Find the total cost of borrowing $1,200 for 24 months. For 30 months. For 36 months.
2. How much less each month is the payment for a loan of $3,500 for 36 months than it is for 24 months? How many more monthly payments must be made for a 36-month loan than for a 24-month loan? How much more is the total interest charge on the 36-month loan of $3,500 than it is on the 24-month loan?

Applications of Mathematics

3. How much less is the total interest charge when you borrow:
   (1) $1,800 for 24 months instead of 30 months?
   (2) $2,500 for 30 months instead of 36 months?
   (3) $2,000 for 24 months instead of 36 months?

4. If your budget will allow a car financing cost of $50 per month maximum and you require a loan of $1,500 to complete the purchase of the car, for what period of time must you borrow the money?

C. Amortizing a Loan

When a loan is made to purchase a house, each equal monthly payment to pay off the loan consists of (1) a reduction in principal and (2) the interest on the new reduced balance of principal owed each month instead of the entire amount of the loan, since part of each equal monthly payment goes to pay off the principal.

The repayment of a mortgage loan in this way is sometimes described as *amortizing* the loan. This is different from what we found with discount loans and add-on-interest loans where the borrower pays interest on the full amount of the loan for the entire period of the loan instead of on the balance owed on the loan each month.

The following partial schedule indicates the monthly payments necessary to amortize a loan that was made at $8\frac{1}{2}\%$ annual interest:

MONTHLY PAYMENT
Necessary to Amortize a Loan Made at $8\frac{1}{2}\%$ Annual Interest

TERM AMOUNT	10 Years	15 Years	20 Years	25 Years	30 Years	35 Years	40 Years
$ 5,000	62.00	49.24	43.40	40.27	38.45	37.35	36.66
10,000	123.99	98.48	86.79	80.53	76.90	74.69	73.31
15,000	185.98	147.72	130.18	120.79	115.34	112.03	109.97
20,000	247.98	196.95	173.57	161.05	153.79	149.38	146.62
25,000	309.97	246.19	216.96	201.31	192.23	186.72	183.28
30,000	371.96	295.43	260.35	241.57	230.68	224.06	219.93
35,000	433.95	344.66	303.74	281.83	269.12	261.41	256.59
40,000	495.95	393.90	347.13	322.10	307.57	298.75	293.24
45,000	557.94	443.14	390.53	362.36	346.02	336.09	329.90
50,000	619.93	492.37	433.92	402.62	384.46	373.44	366.55

EXERCISES

Use the preceding table of monthly payments in the following problems:

1. If a person borrowed $20,000 at $8\frac{1}{2}\%$ annual interest, what is the monthly payment when the loan is to be amortized in:
   - **a.** 30 yr.?
   - **b.** 20 yr.?
   - **c.** 40 yr.?
   - **d.** 25 yr.?
   - **e.** 15 yr.?
   - **f.** 35 yr.?

2. If a loan at $8\frac{1}{2}\%$ annual interest is to be amortized in 35 years, what is the monthly payment when the amount of this loan is:
   - **a.** $40,000?
   - **b.** $15,000?
   - **c.** $50,000?
   - **d.** $25,000?
   - **e.** $45,000?
   - **f.** $35,000?

3. What is the monthly payment necessary to amortize a loan of $30,000 at $8\frac{1}{2}\%$ annual interest in 25 years? What is the payment per year? What is the total payment for 25 years? How much of this total payment is interest on the loan?

4. Find the total interest that was paid on each of the following loans at $8\frac{1}{2}\%$ annual interest when the loans were amortized in the specified terms:

	Amount	Term
a.	$5,000	10 years
b.	$40,000	20 years
c.	$25,000	15 years
d.	$35,000	30 years
e.	$15,000	40 years
f.	$50,000	25 years

5. Find the total payment over the full term required on a loan of $25,000 at $8\frac{1}{2}\%$ annual interest when it is amortized in 20 years. When it is amortized in 30 years. For which term is the interest less? How much less?

6. Is the monthly payment on a loan of $10,000 at $8\frac{1}{2}\%$ annual interest greater for a term of 15 years or 25 years? On which is the total interest for the full term greater?

Applications of Mathematics

7. Find the monthly payments due on each of the following mortgage loans at $8\frac{1}{2}\%$ annual interest to be amortized in the specified terms:

Amount	Term
a. $15,000	40 years
b. $30,000	20 years
c. $20,000	10 years
d. $50,000	15 years
e. $45,000	25 years
f. $10,000	20 years

8. For each mortgage loan in problem 7, find the total payment for the specified term.

9. For each mortgage loan in problem 7, find the total interest paid on the loan.

10. Mrs. Amato purchased a house for $31,500. She paid $6,500 in cash and obtained a mortgage loan at $8\frac{1}{2}\%$ annual interest for the balance. What is the monthly payment on her loan if it is to be amortized in 20 years?

9–17 COMPOUND INTEREST

Savings and loan associations and some savings banks advertise that savings left with them earn interest compounded semi-annually or quarterly (every 3 months). Some banks even advertise interest compounded daily on passbook savings or on savings certificates.

Compound interest is interest paid on both the principal and the interest earned previously.

To find how much a given principal will amount to compounded at a given rate for a certain period of time by the use of the table, we select from the table on page 497 how much $1 will amount to based on the given rate and the given number of years (periods) when the interest is compounded annually; and on half the given rate and twice as many periods as the given number of years when the interest is compounded semi-annually; and on one-fourth the given rate and four times as many periods as the given number of years when the interest is compounded quarterly. Then we multiply this amount by the given principal.

To find the compound interest, we subtract the principal from the amount.

COMPOUND INTEREST TABLE
Showing How Much $1 Will Amount to at Various Rates

Periods	$\frac{1}{2}\%$	1%	$1\frac{1}{4}\%$	$1\frac{1}{2}\%$	2%	$2\frac{1}{2}\%$
1	1.005000	1.010000	1.012500	1.015000	1.020000	1.025000
2	1.010025	1.020100	1.025156	1.030225	1.040400	1.050625
3	1.015075	1.030301	1.037971	1.045678	1.061208	1.076891
4	1.020151	1.040604	1.050945	1.061364	1.082432	1.103813
5	1.025251	1.051010	1.064082	1.077284	1.104081	1.131408
6	1.030378	1.061520	1.077383	1.093443	1.126162	1.159693
7	1.035529	1.072135	1.090851	1.109845	1.148686	1.188686
8	1.040707	1.082857	1.104486	1.126493	1.171659	1.218403
9	1.045911	1.093685	1.118292	1.143390	1.195093	1.248863
10	1.051140	1.104622	1.132271	1.160541	1.218994	1.280085
11	1.056396	1.115668	1.146424	1.177949	1.243374	1.312087
12	1.061678	1.126825	1.160755	1.195618	1.268242	1.344889
13	1.066986	1.138093	1.175264	1.213552	1.293607	1.378511
14	1.072321	1.149474	1.189955	1.231756	1.319479	1.412974
15	1.077683	1.160969	1.204829	1.250232	1.345868	1.448298
16	1.083071	1.172579	1.219890	1.268986	1.372786	1.484506
17	1.088487	1.184304	1.235138	1.288020	1.400241	1.521618
18	1.093929	1.196147	1.250577	1.307341	1.428246	1.559659
19	1.099399	1.208109	1.266210	1.326951	1.456811	1.598650
20	1.104896	1.220190	1.282037	1.346855	1.485947	1.638616

Periods	3%	4%	$4\frac{1}{2}\%$	5%	$5\frac{1}{2}\%$	6%
1	1.030000	1.040000	1.045000	1.050000	1.055000	1.060000
2	1.060900	1.081600	1.092025	1.102500	1.113025	1.123600
3	1.092727	1.124864	1.141166	1.157625	1.174241	1.191016
4	1.125509	1.169859	1.192519	1.215506	1.238825	1.262477
5	1.159274	1.216653	1.246182	1.276282	1.306960	1.338226
6	1.194052	1.265319	1.302260	1.340096	1.378843	1.418519
7	1.229874	1.315932	1.360862	1.407100	1.454679	1.503630
8	1.266770	1.368569	1.422101	1.477455	1.534687	1.593848
9	1.304773	1.423312	1.486095	1.551328	1.619094	1.689479
10	1.343916	1.480244	1.552969	1.628895	1.708144	1.790848
11	1.384234	1.539454	1.622853	1.710339	1.802092	1.898299
12	1.424561	1.601032	1.695881	1.795856	1.901207	2.012196
13	1.468534	1.665074	1.772196	1.885649	2.005774	2.132928
14	1.512590	1.731676	1.851945	1.979932	2.116091	2.260904
15	1.557967	1.800944	1.935282	2.078928	2.232476	2.396558
16	1.604706	1.872981	2.022370	2.182875	2.355263	2.540352
17	1.652848	1.947901	2.113377	2.292018	2.484802	2.692773
18	1.702433	2.025817	2.208479	2.406619	2.621466	2.854339
19	1.753506	2.106849	2.307860	2.526950	2.765647	3.025600
20	1.806111	2.191123	2.411714	2.653298	2.917757	3.207135

Applications of Mathematics

Amount and compound interest on
$500 compounded at 6% at the end of 4 years.

Annually:	Semi-annually:	Quarterly:
Periods: 4	Periods: 8	Periods: 16
Rate: 6%	Rate: 3%	Rate: $1\frac{1}{2}$%
1.262477	1.266770	1.268986
$500	$500	$500
$631.238500	$633.385000	$634.493000
= $631.24	= $633.39	= $634.49
Amount: $631.24	Amount: $633.39	Amount: $634.49
Compound interest: $131.24	Compound interest: $133.39	Compound interest: $134.49

EXERCISES

1. Find the amount when the interest is compounded annually on:
   - **a.** $200 for 6 years at 3%.
   - **b.** $700 for 11 years at $5\frac{1}{2}$%.
   - **c.** $1,500 for 18 years at $4\frac{1}{2}$%.
   - **d.** $8,000 for 9 years at 6%.
   - **e.** $20,000 for 15 years at 5%.
   - **f.** $35,000 for 20 years at $2\frac{1}{2}$%.

2. Find the amount when the interest is compounded semi-annually on:
   - **a.** $600 for 1 year at 6%.
   - **b.** $450 for 4 years at 5%.
   - **c.** $2,000 for 5 years at 3%.
   - **d.** $4,800 for 7 years at 4%.
   - **e.** $12,000 for 10 years at 6%.
   - **f.** $7,500 for 9 years at $2\frac{1}{2}$%.

3. Find the amount when the interest is compounded quarterly on:
   - **a.** $100 for 3 years at 6%.
   - **b.** $800 for 1 year at 5%.
   - **c.** $2,500 for 5 years at 2%.
   - **d.** $11,000 for 2 years at 5%.
   - **e.** $14,500 for 3 years at 4%.
   - **f.** $8,300 for 4 years at 6%.

4. Find the interest earned when the interest is compounded annually on:

   a. $400 for 6 years at $5\frac{1}{2}\%$.

   b. $950 for 11 years at 6%.

   c. $1,600 for 17 years at 3%.

   d. $2,100 for 13 years at 5%.

   e. $8,000 for 4 years at $4\frac{1}{2}\%$.

   f. $19,000 for 19 years at 6%.

5. Find the interest earned when the interest is compounded semi-annually on:

   a. $700 for 7 years at 5%.

   b. $860 for 5 years at 4%.

   c. $4,000 for 9 years at 6%.

   d. $20,000 for 2 years at 3%.

   e. $35,000 for 8 years at $2\frac{1}{2}\%$.

   f. $8,200 for 3 years at 6%.

6. Find the interest earned when the interest is compounded quarterly on:

   a. $900 for 2 years at 5%.

   b. $3,000 for 5 years at 6%.

   c. $1,800 for 4 years at 4%.

   d. $17,000 for 1 year at 6%.

   e. $9,250 for 3 years at 2%.

   f. $40,000 for 5 years at 5%.

7. A bank advertises a 5% interest rate compounded semi-annually on all deposits.

   a. What would a deposit of $1,000 amount to at the end of 6 months? 1 year? 18 months? 2 years? 30 months? 3 years? 42 months? 4 years? 54 months? 5 years? 6 years? 7 years? 8 years? 9 years? 10 years?

   b. What is the interest earned on a $2,500 deposit at the end of 6 months? 1 year? 2 years? 30 months? 3 years? 4 years? 5 years? 66 months? 6 years? 78 months? 7 years? 8 years? 102 months? 9 years? 10 years?

8. A bank advertises a 5% interest rate compounded quarterly on all deposits.

   a. What would a deposit of $2,000 amount to at the end of 3 months? 6 months? 9 months? 1 year? 15 months? 18 months? 21 months? 2 years? 30 months? 3 years? 42 months? 4 years? 51 months? 54 months? 5 years?

   b. What is the interest earned on a $5,000 deposit at the end of 3 months? 6 months? 9 months? 1 year? 15 months? 18 months? 21 months? 2 years? 27 months? 30 months? 33 months? 3 years? 42 months? 4 years? 5 years?

Applications of Mathematics

9. At the interest rate of 6% compounded monthly (12 times a year):

   a. What will each of the following deposits amount to at the end of one year?
      (1) $600   (2) $1,400   (3) $8,000   (4) $19,000   (5) $50,000
   b. How much interest is earned at the end of 1 year on each of the following deposits?
      (1) $400   (2) $750   (3) $6,000   (4) $12,500   (5) $30,000
   c. What will each of the following deposits amount to?
      (1) A deposit of $300 at the end of 7 months.
      (2) A deposit of $1,100 at the end of 4 months.
      (3) A deposit of $8,500 at the end of 1 month.
   d. How much interest is earned on:
      (1) A deposit of $900 at the end of 8 months?
      (2) A deposit of $2,150 at the end of 5 months?
      (3) A deposit of $15,000 at the end of 10 months?

10. Will $5,000 compounded annually for 5 years at 6% amount to the same as for 6 years at 5%? If not, the amount of which is greater? How much greater?

11. Find the interest earned on a $25,000 deposit at the end of 1 year at each of the following rates:

   a. 6% compounded semi-annually
   b. 6% compounded annually
   c. 6% compounded quarterly
   d. 6% compounded monthly

   Which rate brings the greatest amount of interest? Which rate brings the smallest amount of interest?

9–18 STOCKS; BONDS; MUTUAL FUNDS; SAVINGS CERTIFICATES

Stocks

A person buying a share of stock becomes a part-owner of the business and receives dividends from the profits earned by the business.

Stock quotations are in terms of dollars.

$$28\tfrac{3}{4} \text{ means } \$28.75$$

The P.E. (*price to earnings*) ratio column heading in the stock quotation refers to an approximate ratio comparing the closing price of a share of each specific stock to its annual earnings. For example, if the P.E. ratio of a certain stock is 8, then the price of the stock is approximately 8 times its annual earnings. To find the annual earnings, we divide the closing price by the P.E. ratio number.

Fees are paid to brokers for the service of buying or selling stocks. These fees are subject to change and at present there is no one set of fixed fees for all brokers.

A Typical Fee Schedule

Showing the commissions charged for buying or selling shares of stock when the cost of the stock is:

$100 but under $2,500	1.3% of cost of stock plus $16
$2,500 but under $20,000	0.9% of cost of stock plus $28
$20,000 but under $30,000	0.6% of cost of stock plus $90
Over $30,000	0.4% of cost of stock plus $152

Bonds

A *bond* is a written promise of a private corporation or of a local state, or the national government to pay a given rate of interest at stated times and to repay the face value of the bond at a specified time (*date of maturity*).

A person buying a bond is lending money to the business or government and receives interest on the *face value* of the bond.

The original value of a bond is called *par value* or *face value*. The value at which it sells at any given time is called *market value*. The value is *above par* when the market value is more than the par value and *below par* when the market value is less than the par value.

Bond quotations are in terms of percent.

$$97\tfrac{1}{2} \text{ means } 97\tfrac{1}{2}\%$$

Applications of Mathematics

> In the following bond quotation: TWA 10s 85 $101\frac{1}{4}$
>
> 10s   represents a 10% annual rate of interest.
> 85   is the date of maturity, 1985.
> $101\frac{1}{4}$ is the market price, indicating $101\frac{1}{4}$% of the face value of the bond.

To find the *rate of income* or *yield* of a bond (see column marked Cur. Yld. in the bond quotations on page 505), we find what percent the annual interest is of the market price of the bond.

Varying fees are paid to the brokers for buying and selling bonds.

Mutual Funds

A mutual fund is an investment company. It obtains money for investment by selling shares of stock of the company. Mutual fund quotations are expressed as a *Net Asset Value* (*NAV*) and an *offer price*. When we buy shares, we pay the offer price; when we sell shares, we receive the Net Asset Value (NAV) of each share.

N.L. means *no load* or no commission charged for buying shares. When N.L. is indicated, the offer price is the same as NAV.

Savings Certificates; Certificates of Deposit

Banks and savings and loan associations sell savings certificates and certificates of deposit paying varying rates of interest depending upon the term of the certificate.

Some banks pay $7\frac{3}{4}$% annual interest on savings certificates when the term is 72 mo. to 120 mo., $7\frac{1}{2}$% for 48 mo. to 71 mo., $6\frac{3}{4}$% for 30 mo. to 47 mo., and $6\frac{1}{2}$% for 12 mo. to 29 mo.

Usually banks pay 5% annual interest on certificates of deposit when the term is 30 days to 89 days, $5\frac{1}{2}$% for 90 days to less than 1 year, $5\frac{3}{4}$% for 1 year, and 6% for 2 years.

Government Bonds; Treasury Notes

The United States government borrows money by selling savings bonds and Treasury notes.

The E savings bond is a very popular government bond. It pays 6% interest compounded semi-annually if held to maturity and matures in 5 years.

The Treasury notes (bonds) can be sold at any time. The rates of interest vary.

EXERCISES

STOCK QUOTATIONS

Year High	Year Low	Stocks	Dividends	P.E. Ratio	Sales 100s	High	Low	Close	Net Chg.
$61\frac{1}{4}$	55	CBS	2	10	249	$58\frac{1}{4}$	$57\frac{3}{4}$	$57\frac{7}{8}$	...
$39\frac{5}{8}$	$36\frac{3}{8}$	Cam Sp	1.48	12	53	$39\frac{3}{8}$	$38\frac{3}{4}$	39	$+\frac{3}{8}$
17	$15\frac{3}{4}$	Cdn Pac	.86e	6	113	$16\frac{3}{4}$	$16\frac{5}{8}$	$16\frac{5}{8}$	$+\frac{1}{8}$
$30\frac{1}{8}$	$26\frac{7}{8}$	Cess Air	1.20	7	51	$27\frac{1}{2}$	$27\frac{1}{8}$	$27\frac{1}{2}$	$+\frac{1}{4}$
$33\frac{5}{8}$	$29\frac{5}{8}$	Chase M	2.20	8	473	$30\frac{1}{4}$	$29\frac{3}{4}$	30	$+\frac{1}{4}$
$43\frac{1}{8}$	35	Chessie	2.32	9	195	$41\frac{5}{8}$	$41\frac{1}{4}$	$41\frac{1}{4}$	$-\frac{3}{8}$
22	$16\frac{1}{4}$	Chrysler	.45e	3	1347	$18\frac{5}{8}$	18	$18\frac{1}{4}$	$+\frac{1}{8}$
34	$26\frac{1}{2}$	Citicorp	1.06	8	1581	$27\frac{1}{4}$	27	27	$-\frac{1}{8}$
$12\frac{3}{4}$	$10\frac{1}{8}$	Clorox	.60	9	534	$11\frac{5}{8}$	$11\frac{3}{8}$	$11\frac{3}{8}$	$+\frac{1}{8}$
$27\frac{5}{8}$	$23\frac{3}{8}$	Colg Pal	.88	12	619	$24\frac{3}{4}$	$24\frac{1}{2}$	$24\frac{5}{8}$	...
$29\frac{3}{8}$	$24\frac{1}{4}$	Col Penn	.70	9	97	$28\frac{1}{2}$	$27\frac{1}{2}$	$27\frac{1}{2}$	$-\frac{5}{8}$
$37\frac{1}{4}$	$29\frac{3}{8}$	Comsat	1	10	348	$36\frac{3}{8}$	$35\frac{3}{4}$	$35\frac{7}{8}$	$-\frac{1}{8}$
$23\frac{1}{4}$	$20\frac{3}{4}$	Con Ed	2	5	568	$22\frac{3}{4}$	$22\frac{1}{2}$	$22\frac{3}{4}$	$+\frac{1}{4}$
38	$33\frac{1}{2}$	Cont Oil	1.40	9	855	$36\frac{1}{2}$	$35\frac{7}{8}$	$36\frac{1}{8}$	$+\frac{3}{8}$
$17\frac{5}{8}$	$14\frac{7}{8}$	Cont Tel	1.08	9	607	16	$15\frac{3}{4}$	16	$+\frac{1}{4}$
$26\frac{5}{8}$	20	Cont Dta	.15e	7	397	$21\frac{5}{8}$	$20\frac{1}{2}$	$21\frac{3}{8}$	$+\frac{7}{8}$
$18\frac{7}{8}$	$15\frac{1}{4}$	Curtis Wr	.60	8	66	$17\frac{1}{2}$	$17\frac{1}{8}$	$17\frac{1}{8}$	$-\frac{3}{8}$
$44\frac{1}{8}$	$37\frac{3}{8}$	Cutler H	1.80	9	22	$43\frac{3}{4}$	$42\frac{3}{4}$	$43\frac{3}{4}$	$+1\frac{1}{4}$

1. Find the cost of each of the following purchases, excluding commission, using the closing prices listed in the stock quotations above:

   a. 100 shares of Chase M
   b. 100 shares of Cam Sp
   c. 200 shares of Col Penn
   d. 50 shares of Con Ed
   e. 25 shares of Chessie
   f. 80 shares of Cont Oil

2. Find the cost of each of the following purchases, excluding commission, using the highest prices listed in the stock quotations:

   a. 50 shares of Cont Tel
   b. 100 shares of Curtis Wr
   c. 100 shares of Colg Pal
   d. 60 shares of CBS
   e. 72 shares of Comsat
   f. 400 shares of Clorox

Applications of Mathematics

3. Find the selling price of each of the following sales, excluding commission, using the lowest prices listed in the stock quotations:

   a. 36 shares of Citicorp
   b. 100 shares of Chrysler
   c. 300 shares of Cdn Pac
   d. 50 shares of Cont Dta
   e. 25 shares of Cutler H
   f. 100 shares of Cess Air

4. Using the closing prices listed in the stock quotations and the broker's fee schedule (see page 501), find the cost of each of the following purchases, including commission:

   a. 100 shares of Chrysler
   b. 100 shares of CBS
   c. 200 shares of Citicorp
   d. 25 shares of Colg Pal
   e. 60 shares of Cess Air
   f. 140 shares of Cutler H

5. Using the highest prices listed in the stock quotations and the broker's fee schedule, find the net amount due (selling price of stock less commission) on each of the following sales of stock:

   a. 100 shares of Cont Oil
   b. 100 shares of Comsat
   c. 40 shares of Cdn Pac
   d. 500 shares of Cont Tel
   e. 76 shares of Clorox
   f. 63 shares of Col Penn

6. How many points above the year low and below the year high is each of the following stocks' closing price:

   a. Chrysler?
   b. Cdn Pac?
   c. Cont Dta?
   d. Chessie?
   e. Chase M?
   f. Con Ed?
   g. Cess Air?
   h. Cutler H?

7. Find the annual earnings per share (to nearest cent) for each of the stocks listed in the stock quotations by using its P.E. ratio and closing price.

8. Include the commission in each of the following purchases and sales and find the amount of profit:

   a. 100 shares of NRT purchased at 28, sold for 35.
   b. 100 shares of AC purchased at $14\frac{3}{4}$, sold for $20\frac{1}{2}$.
   c. 50 shares of DZB purchased at $42\frac{5}{8}$, sold for 51.
   d. 85 shares of SOS purchased at $76\frac{1}{2}$, sold for $85\frac{3}{8}$.
   e. 200 shares of MGH purchased at $38\frac{7}{8}$, sold for $62\frac{1}{4}$.

9. Include the commission in each of the following purchases and sales and find the amount of loss:

   a. 100 shares of TEL purchased at 52, sold for 47.

b. 100 shares of EBS purchased at $15\frac{1}{4}$, sold for $11\frac{3}{4}$.

c. 70 shares of ATS purchased at 33, sold for $27\frac{5}{8}$.

d. 35 shares of LRL purchased at $48\frac{3}{8}$, sold for $42\frac{3}{8}$.

e. 300 shares of CMS purchased at $27\frac{3}{4}$, sold for $19\frac{7}{8}$.

10. The dividend listed for each stock in the quotations is for a year. Determine for each of the following stocks the amount of quarterly dividend:

a. Con Ed	d. Cutler H	g. Curtis Wr	j. Cam Sp
b. Comsat	e. Cont Oil	h. Col Penn	k. Cont Tel
c. Cess Air	f. Chessie	i. Colg Pal	l. Citicorp

BOND QUOTATIONS

Bonds	Cur. Yld.	Vol.	High	Low	Close	Net Chg.
ATT 8.80s 05	8.2	135	107	$106\frac{3}{4}$	107	$+\frac{1}{2}$
Beth St 9s 00	8.5	25	$105\frac{3}{8}$	105	$105\frac{3}{8}$	$+\frac{3}{8}$
Chry F 9s 86	9.2	40	$97\frac{3}{4}$	$97\frac{3}{4}$	$97\frac{3}{4}$	$+\frac{3}{8}$
Con Ed $9\frac{1}{8}$ 04	9.1	32	$100\frac{7}{8}$	$100\frac{3}{8}$	$100\frac{3}{4}$	$-\frac{1}{8}$
Duke P $9\frac{1}{2}$ 05	8.8	21	$109\frac{1}{4}$	$107\frac{1}{2}$	$107\frac{1}{2}$	$-1\frac{3}{4}$
GMA $8\frac{1}{8}$ 96	8.1	20	100	100	100	$-\frac{1}{8}$
Ga Pac $7\frac{1}{4}$ 85	7.3	50	$99\frac{5}{8}$	$99\frac{5}{8}$	$99\frac{5}{8}$	....
Grace $6\frac{1}{2}$ 96	cv*	203	$106\frac{1}{2}$	$104\frac{1}{2}$	$105\frac{1}{8}$	$+\frac{1}{8}$
Greyh $9\frac{3}{8}$ 01	8.9	20	$105\frac{1}{4}$	105	105	$+\frac{1}{2}$
Int Hw $4\frac{5}{8}$ 88	6.2	10	74	74	74	$+\frac{7}{8}$
MGIC $8\frac{3}{8}$ 88	8.5	30	$99\frac{3}{8}$	99	99	$+\frac{1}{4}$
Macor $6\frac{1}{2}$ 88	7.6	74	$85\frac{5}{8}$	$85\frac{1}{2}$	$85\frac{1}{2}$	....
MGM 10s 94	10	25	$97\frac{1}{4}$	$96\frac{3}{4}$	$97\frac{1}{4}$	$+1$
St 0 In 5.6s 89	5.6	17	$100\frac{1}{4}$	$100\frac{1}{4}$	$100\frac{1}{4}$	....
Wstg E $8\frac{5}{8}$ 95	8.5	50	102	$101\frac{1}{2}$	102	$+1$
Xerox 6s 95	cv*	49	88	$86\frac{3}{4}$	88	$+\frac{3}{4}$

* The abbreviation cv indicates convertible bonds.

Applications of Mathematics

11. Find the date of maturity and interest rate for each of the following bonds:

   a. Xerox 6s 95
   b. Beth St 9s 00
   c. ATT 8.80s 05
   d. Grace $6\frac{1}{2}$ 96
   e. MGIC $8\frac{3}{8}$ 88
   f. Ga Pac $7\frac{1}{4}$ 85

12. Find the cost of each of the following purchases, using the closing prices in the bond quotations on page 505 (face value of each bond is $1,000):

   a. One Greyh $9\frac{3}{8}$ 01
   b. Five Con Ed $9\frac{1}{8}$ 04
   c. Ten MGIC $8\frac{3}{8}$ 88
   d. Fifty Macor $6\frac{1}{2}$ 88
   e. Thirty MGM 10s 94
   f. Twenty-five Beth St 9s 00

13. Using the highest quoted prices on page 505, find the total cost including commission of:

   a. Twenty Wstg E $8\frac{5}{8}$ 95, face value $1,000 each, commission $7.50 per bond
   b. Five St O In 5.6s 89, face value $1,000 each, commission $10 per bond
   c. Forty MGIC $8\frac{3}{8}$ 88, face value $1,000 each, commission $5 per bond

14. How much interest will the owner of fifty Duke P $9\frac{1}{2}$ 05 receive semi-annually if the face value of each bond is $1,000?

15. Calculate the current yield, using the closing quoted prices for each of the following $1,000 bonds, then check with the current yield given the bond quotations:

   a. GMA $8\frac{1}{8}$ 96
   b. ATT 8.80s 05
   c. Int Hw $4\frac{5}{8}$ 88
   d. Greyh $9\frac{3}{8}$ 01
   e. Macor $6\frac{1}{2}$ 88
   f. Duke P $9\frac{1}{2}$ 05
   g. Beth St 9s 00
   h. Ga Pac $7\frac{1}{4}$ 85
   i. Con Ed $9\frac{1}{8}$ 04

16. Using the mutual funds quotations find the cost of buying each of the following:

   a. 100 shares of Pilgrim Fd
   b. 100 shares of Dreyfus Fd
   c. 75 shares of Windsor
   d. 30 shares of Bullock Fd
   e. 200 shares of Delaware

MUTUAL FUNDS	Net Asset Value (NAV)	Offer Price
Bullock Fd	12.67	13.85
Delaware	11.20	12.24
Dreyfus Fd	11.68	12.78
Fidelity Fd	15.87	17.34
Mass Fd	10.55	11.53
Pilgrim Fd	8.45	8.89
Pioneer Fd	13.92	15.21
Value Line	6.92	7.10
Wellington	11.06	N.L.
Windsor	9.98	N.L.

17. Using the mutual funds quotations, find how much money you should receive when you sell each of the following:
    a. 100 shares of Pioneer Fd
    b. 100 shares of Mass Fd
    c. 20 shares of Wellington
    d. 50 shares of Value Line
    e. 65 shares of Fidelity Fd

18. Using the rates given in the beginning of this section, find the total interest a bank pays a person when the certificate of deposition matures in:
    a. 100 days   b. 73 days   c. 180 days   d. 55 days

19. Mrs. McCormick owns six $5,000 Treasury Notes that pay an annual rate of $7\frac{3}{4}\%$ interest. How much interest does she receive semi-annually?

20. If $25 E bonds cost $18.75 each, $50 E bonds cost $37.50 each, $75 E bonds cost $56.25 each, and $100 E bonds cost $75 each,
    a. What is the total cost of seven $25 E bonds, six $50 E bonds, nine $75 E bonds, and five $100 E bonds?
    b. How much will the owner receive in all at maturity?

APPLICATIONS IN BUSINESS

9–19 DISCOUNT

Discount is the amount that an article is reduced in price. The regular or full price of the article is called the *list price* or *marked price*. The price of the article after the discount is deducted is called the *net price* or *sale price*. The percent taken off is called the *rate of discount*. This rate is sometimes expressed as a common fraction. Manufacturers and wholesale houses sometimes allow trade discounts.

Often when a bill is paid at the time of purchase or within a specified time thereafter, a reduction called *cash discount* is allowed. The terms of sale are usually stated on the bill. The terms 3/10, *n*/30 mean that a 3% discount is allowed if the bill is paid within 10 days and the full amount is due by the 30th day. The *net amount* is the amount of the bill after the cash discount is deducted.

When two or more discounts are given, they are called *successive discounts*.

(1) *To find the discount when the list price and the rate of discount are given,* we multiply the list price by the rate of discount.
(2) *To find the net price,* we subtract the discount from the list price.

Find the discount and net price when the list price is $240 and the rate of discount is 18%:

$240 list price $240.00 list price
×.18 rate   −43.20 discount
─────   ─────────
1920 $196.80 net price
 240
─────
$43.20 discount

Answer, $43.20 discount; $196.80 net price

---

(3) *To find the rate of discount,* we find what percent the discount is of the list price.

What is the rate of discount when the list price is $35 and the discount is $14?

$$\frac{\$14}{\$35} = \frac{2}{5} = 40\%$$

Answer, 40%

---

(4) *To find the list price when the net price and rate of discount are given,* we subtract the given rate from 100%, then divide the answer into the net price.
(5) *To find the net amount,* we subtract the cash discount from the original amount of the bill. When two or more successive discount rates are given, we do not add the rates but use them one at a time.

EXERCISES

1. Find the discount when the list price is $34.75 and the net price is $26.90.
2. Find the net price when the list price is $9.18 and the discount is $.43.
3. Find the discount when the list price is $56.29 and the rate of discount is 6%.
4. Find the discount and net price when the list price is $349.50 and the rate of discount is 27%.
5. Find the net price when the list price is $1,085 and the rate of discount is 4%.
6. Find the sale price when the regular price is $23.69 and the rate of reduction is 12%.
7. Find the discount and net price:

List price	$84	$9.61	$106.75	$28.50	$695
Rate of discount	10%	3%	25%	40%	$33\frac{1}{3}$%

8. Find the rate of discount when the list price is $20 and the discount is $4.
9. Find the rate of discount when the list price is $56 and the net price is $48.
10. Find the rate of discount:

List price	$72	$8.50	$28	$120	$24.75
Net price	$54	$7.65	$23.80	$110.40	$21.78

11. On what list price is the discount $4 when the rate of discount is 5%?
12. Find the list price when the net price is $18 and the rate of discount is $33\frac{1}{3}$%.
13. Find the list price:

Net price	$32	$42.30	$45	$18.75	$27.36
Rate of discount	20%	6%	$37\frac{1}{2}$%	50%	10%

14. What is the trade discount when the catalogue list price is $12.95 and rate of trade discount is 8%?

Applications of Mathematics

15. What is the net price when the catalogue list price is $76.50 and rate of trade discount is 15%?
16. Find the trade discount and net price:

Catalogue list price	$112	$16.75	$86.40	$247	$49.50
Rate of trade discount	35%	10%	$12\frac{1}{2}$%	20%	8%

17. Find the rate of trade discount when the catalogue list price is $150 and net price is $125.
18. What is the cash discount when the amount of a bill is $26 and the rate of cash discount is 3%?
19. What is the net amount when the amount of a bill is $395.82 and the rate of cash discount is 4%?
20. Find the cash discounts and net amounts:

Amount of bill	$55	$142.98	$106.25	$318.40	$569.85
Rate of cash discount	3%	3%	5%	6%	2%

21. Find the cash discount received and net amount paid when the amount of a bill is $48, terms of payment 2/10, *n*/30, and the bill is paid on the third day after receipt.
22. Find the cash discounts received and net amounts paid:

Amount of bill	$16	$239.50	$89.25	$348.20	$169.75
Terms of payment	2/10, *n*/30	5/10, *n*/60	4/15, *n*/60	3/10, *n*/30	4/10, *n*/60
Bill paid on	4th day	15th day	10th day	9th day	12th day

23. Find the cash discounts and net amounts paid:

Amount of bill	$287	$104.90	$643.75	$986.25	$1,000
Terms of payment	3/10, *n*/30	2/10, *n*/30	5/15, *n*/60	1/10, *n*/30	4/15, *n*/60
Date of bill	June 6	April 14	May 6	March 6	July 5
Date of payment	June 12	April 29	May 19	March 28	July 19

24. What is the net price when the list price is $80 and the rates of discount are 20% and 5%?
25. What is the net price when the list price is $400 and the rates of discount are 25%, 10%, and 5%?

26. What is the net amount when the list price is $67, rate of trade discount is 30%, and rate of cash discount is 4%?
27. Find the net prices:

List price	$130	$456.40	$247.50	$625	$870
Rates of discount	15%, 8%	20%, 2%	40%, 10%	10%, 2%	30%, 5%

28. Merriweather and Co. allow all customers 6% discount on cash purchases. Mr. Olsen paid cash when he bought merchandise costing $879. How much discount was he allowed? What was the net price of the merchandise?
29. Ms. Chung bought a dress marked $58 at a discount of 20%. How much did she pay for it?
30. A furniture store advertised a 25% reduction on all articles. What is the sale price of a living room suite marked $375?
31. Find the sale price of each of the following articles:
    a. Hat; regular price, $15; reduced 10%.
    b. Table; marked to sell for $125; reduced 20%.
    c. Freezer; listed at $290; 15% off.
    d. Clock radio; regular price, $49; 25% off.
    e. Mirror; regular price, $67.50; reduced 10%.
    f. Coat; marked to sell for $79.75; 6% off.
    g. Clothes dryer; listed at $199.95; 5% off.
    h. Lamp; marked to sell for $49.89; $\frac{1}{3}$ off.
    i. Camera; listed at $84.25; 20% off.
    j. Living room suite; regular price, $795; reduced 15%.
32. Joe bought a bicycle for $80 less 18% discount. How much did he pay for it?
33. If a dress marked $24 was sold for $19.20, what rate of discount was given?
34. Find the regular price of a washing machine that sold for $202.30 at a 30% reduction sale.
35. What would 40 science workbooks cost at the school discount of 25% if the list price is $2.70 each?
36. A salesperson sold a calculator marked $89 with discounts of 15% and 5%. What was the amount of error if he calculated the net price, using a single discount of 20%?
37. Miss Sullivan purchased 6 storm windows marked $24.50 each at a discount of 10%. How much did she pay for them?
38. What rate of discount was given if a piano marked $1,300 was sold at a special sale for $1,040?

Applications of Mathematics

39. How much trade discount is allowed if the catalogue lists a heating unit at $369 and the discount sheet shows a 14% discount? What is the net price?
40. If blankets listed for $16.50 each are sold for 15% less, how much did Mrs. Dawson save when she bought 3 blankets?
41. On which is the rate of discount greater and how much greater, a rug reduced from $120 to $90 or one reduced from $80 to $56?
42. If shirts cost $48 a dozen less 10% and 5%, what is the cost of one shirt?
43. The net price of a television set is $210.21 when a 22% discount is allowed. What is its list price?
44. What is the net price of a gas range listed at $290 with a trade discount of 15% and an additional cash discount of 3%?
45. Mrs. King saved $9 by buying a digital clock radio at a sale when prices were reduced 18%. What was the regular price of the digital clock radio?

9–20 PROMISSORY NOTES—BANK DISCOUNT

A *promissory note* is a written promise by the borrower to repay the loan. The *date of maturity* is the date when the money is due. The *face of the note* is the amount of money borrowed that is written on the note. When a person borrows money at a bank and gives a note, the bank deducts the interest from the loan in advance. The interest deducted is called *bank discount* and the amount the borrower gets is called *proceeds*. The exact number of days from the date a note is discounted to the date of maturity is called the *term of discount*.

EXERCISES

1. Draw two forms like the following to write notes, dated today, in which you promise to pay:

```
$_____                                    _____19__
_____after date____promise to pay
to the order of_____
_____Dollars
Value received
No.____ Due_____              _____
```

a. A sum of $600 to Lisa R. Lieberman 3 months after date with interest at 4%.
   b. A sum of $1,000 to First National Bank 60 days after date with interest at 6%.
2. Find the date of maturity of each of the following notes:
   a. 30-day note dated April 9.
   b. 90-day note dated June 21.
   c. 4-month note dated March 17.
   d. 60-day note dated November 30.
   e. 45-day note dated August 14.
3. What is the term of discount if the date of discount is January 9 and the date of maturity is February 28?
4. A 30-day note for $800 dated December 11 is discounted at 6% on the same day. Find the bank discount and proceeds.
5. Find the bank discount and proceeds of each of the following notes:

Face of note	$900	$4,000	$5,000	$8,400
Rate of Discount	5%	6%	4%	6%
Term of Discount	4 mo.	90 da.	60 da.	6 mo.

6. A 90-day note for $2,000 dated September 26 with interest at 5% is discounted on November 20 at 6%. Find the bank discount, proceeds, and date of maturity.

9–21 COMMISSION

A salesperson who sells goods for another person is usually paid a sum of money based on the value of the goods sold. This money which is received for the services is called *commission*. When it is expressed as a percent, it is called the *rate of commission*. The amount that remains after the commission is deducted from the total selling price is called *net proceeds*. When a buyer purchases goods for another person, the buyer also is usually paid a commission for the services based on the cost of the goods. Salespeople and buyers are sometimes called *agents*.

   (1) Selling: *To find the commission when the amount of sales and rate of commission are given,* we multiply the amount of sales by the rate of commission. *To find the net proceeds,* we subtract the commission from the amount of sales.

Applications of Mathematics

> Find the commission and net proceeds when the sales are $86.98 and the rate of commission is 7%:
>
> $86.98 sales
> ×.07 rate
> $6.0886
> = $6.09 commission
>
> $86.98 sales
> −6.09 commission
> $80.89 net proceeds
>
> *Answer,* $6.09 commission; $80.89 net proceeds

(2) Buying: *To find the commission when the cost of goods purchased and rate of commission are given,* we multiply the cost of goods purchased by the rate of commission. *To find the total cost,* we add the commission to the cost of goods purchased.

(3) *To find the rate of commission,* we find what percent the commission is of the amount of sales (or cost of goods purchased).

> Find the rate of commission if the commission is $19.20 on sales of $320:
>
> $$\frac{\$19.20}{\$320} = \$320 \overline{)\$19.20}^{.06\ =\ 6\%}$$
>
> $$\underline{19\ 20}$$
>
> *Answer,* 6% commission

(4) *To find the sales when the commission and rate of commission are given,* we divide the commission by the rate.

EXERCISES

1. What is the commission if the sales are $380 and the net proceeds are $323?
2. What are the net proceeds if the sales are $509.82 and the commission is $43.95?

3. What is the commission if the sales are $721.50 and the rate of commission is 8%?

4. Find the commission:

Sales	$640	$96.20	$867.50	$175	$1,300
Rate of commission	10%	25%	9%	8%	5%

5. What are the net proceeds if the sales are $182.75 and the rate of commission is 20%?

6. Find the commission and net proceeds:

Sales	$92	$905.75	$547.20	$345	$2,000
Rate of commission	15%	7%	$16\frac{2}{3}$%	20%	$12\frac{1}{2}$%

7. What is the rate of commission if the commission is $16 on sales amounting to $320?

Find the rate of commission:

8.

Sales	$480	$525	$239.50	$730	$95.50
Commission	$72	$31.50	$47.90	$58.40	$9.55

9.

Sales	$375	$304.24	$681	$480	$750
Net proceeds	$340	$228.18	$619.71	$420	$625

10. What must the sales be for an 8% rate of commission to bring a commission of $60?

11. Find the amount of sales when commission is as follows:

Commission	$46.70	$50.66	$82	$94.50	$101.35
Rate of commission	10%	17%	$12\frac{1}{2}$%	5%	20%

12. What do the sales amount to if the net proceeds are $1,680 at the rate of 4% commission?

13. Find the commission on goods purchased:

Cost of goods	$139	$928	$436.75	$621.90	$240.80
Rate of commission	6%	5%	8%	10%	$12\frac{1}{2}$%

Applications of Mathematics

14. Find the commission and total cost of goods purchased:

Cost of goods	$594	$389	$618.50	$29.40	$439.60
Rate of commission	7%	4%	12%	5%	15%

15. At the rate of 4% commission on sales, how much will a salesclerk receive when she sells $1,358 worth of merchandise?

16. A real-estate agent sold a house for $19,800. If the agent charges 6% commission, how much money does the owner receive?

17. During the week a salesperson sold 20 washing machines at $254.95 each. At the rate of 2% commission on sales, how much money does the salesperson receive?

18. A lawyer collected 70% of a debt of $1,250 for Mr. Gordon. If the lawyer charges 15% for his services, how much will Mr. Gordon receive?

19. If an agent charges $2\frac{1}{2}$% commission for purchasing goods, how much will he receive for buying 750 boxes of oranges at $6.90 a box?

20. An agent received $1,590 commission for selling a house for $26,500. What rate of commission did she charge?

21. If an agent's commission is 8%, how much will he receive for selling 530 bags of potatoes at $6.25 a bag?

22. Marilyn receives $60 per week and 8% commission on sales. If her weekly sales are $1,859, what are her total earnings?

23. Mr. Morgan is paid $75 a week and 4% commission on all weekly sales over $500. How much did he earn during a week when his sales amounted to $5,928?

24. Charlotte received $63.36 commission last week. If she was paid at the rate of 6%, what was the amount of her sales?

25. An auctioneer, charging 7% commission, sold goods amounting to $2,806. How much commission did she receive?

26. How much must a salesperson sell each week to earn $275 weekly if he is paid 5% commission on sales?

27. A merchant, charging 6% commission, sold for a farmer 275 crates of berries at $7.80 a crate. If freight charges were $182.50, how much did the farmer receive per crate?

28. An agent received $63.75 commission for selling 85 bushels of apples at $7.50 each. What rate of commission did she charge?

29. A salesperson sold 148 dozen ties at $20.50 per dozen. At the rate of $4\frac{1}{2}$% commission, how much money will she receive?

9-22 PROFIT AND LOSS

Merchants are engaged in business to earn a profit. To determine the profit, each merchant must consider the cost of goods, the selling price of the same goods, and the operating expenses. The *cost* is the amount the merchant pays for the goods. The *selling price* is the amount the merchant receives for selling the goods. The *operating expenses* or *overhead* are expenses of running the business. They include wages, rent, heat, light, telephone, taxes, insurance, advertising, repairs, supplies, and delivery costs.

The difference between the selling price and the cost is called the *margin*. The terms *gross profit, spread,* and *markup* sometimes are also used. The margin is often erroneously thought of as the profit earned. Actually, the profit earned, called *net profit,* is the amount that remains after both the cost and the operating expenses are deducted from the selling price.

To earn a profit the merchant must sell goods at a price which is greater than the sum of the cost of the goods and the operating expenses. If the selling price is less than the sum of the cost and operating expenses, the goods are sold at a *loss*.

The profit may be expressed as a percent of either the cost or the selling price. This percent is called the *rate of profit*. (The loss may also be expressed in a similar way.)

To determine the selling price of the articles they sell, some merchants use the percent markup (or gross profit) based on the cost while other merchants use the rate based on the selling price. This *percent markup* is an estimated rate large enough to cover the proportionate amount of overhead expenses to be borne by each article and the amount of profit desired. The percent markup (or gross profit) is the percent the margin is of the cost or of the selling price.

To find the selling price: (1) We add the cost, operating expenses, and net profit; or (2) add the cost and margin; or (3) multiply the cost by the percent markup on the cost, then add the product to the cost.

To find the margin or gross profit or markup: (1) We subtract the cost from the selling price; or (2) multiply the cost by the percent markup on cost; or (3) add the operating expenses and net profit.

Applications of Mathematics

To find the net profit: (1) We subtract the operating expenses from the margin; or (2) add the cost and the operating expenses, and subtract this sum from the selling price.

To find the loss: We add the cost and the operating expenses, and subtract the selling price from this sum.

To find the percent markup (or gross profit) on cost, we find what percent the margin is of the cost. When based on selling price, we find what percent the margin is of the selling price.

To find the rate of net profit (or loss) on the selling price, we find what percent the net profit (or loss) is of the selling price. When based on the cost, we find what percent the profit (or loss) is of the cost.

To find the selling price when the cost and percent markup (or gross profit) on the selling price are given, we subtract the percent markup from 100%, then divide the answer into the cost.

If the rate of markup on cost is 34%, what is the selling price of an article that costs $25?

$25 cost
×.34 rate
 100
  75
$8.50 markup

$25.00 cost
+8.50 markup
$33.50 selling price

Answer, $33.50 selling price

Find the rate of markup on cost if an article costs $40 and sells for $55:

$55 selling price
−40 cost
$15 markup

$\dfrac{\$15}{\$40} = \dfrac{3}{8} = 37\tfrac{1}{2}\%$

Answer, $37\tfrac{1}{2}\%$ markup on cost

Find the net profit and rate of net profit on the selling price if an article costs $18.50, operating expenses are $1.50, and it sells for $24:

$18.50 cost
+1.50 operating expenses
$20.00 total cost

$\dfrac{\$4}{\$24} = \dfrac{1}{6} = 16\tfrac{2}{3}\%$

$24 selling price
−20
$4 net profit

Answer, $4 net profit; $16\tfrac{2}{3}\%$ net profit on selling price

Find the selling price of an article that costs $17 and the percent markup on the selling price is 32%:

$\begin{aligned} 100\% &= \text{selling price} \\ -32\% &= \text{markup} \\ \hline 68\% &= \text{cost} \end{aligned}$

$.68 \overline{)\$17.00}$ = $25

$\underline{13\ 6}$
$3\ 40$
$\underline{3\ 40}$ Answer, $25 selling price

EXERCISES

1. What must the selling price of goods be to allow a net profit of $56 if they cost $287 and the operating expenses are $29?
2. Find the margin or gross profit on an article when the operating expenses are $13.25 and the net profit is $68.50.
3. What is the selling price of a chair if it costs $37.50 and the margin is $18.45?
4. A rug costs $171.90 and sells at $298.50. What is the margin or gross profit?
5. What is the margin on a desk which costs $142 and the rate of markup (margin) on the cost is 35%?
6. A cassette tape recorder costs $62.50. If the rate of markup on cost is 40% what is the selling price?
7. Find the selling price:

Cost	$87	$169	$29.40	$145.50	$78.60
Markup on cost	25%	32%	30%	40%	$33\frac{1}{3}\%$

8. Find the net profit if operating expenses are $26.95 and the margin is $48.75?
9. Find the net profit:

Cost	$34	$91.58	$107	$1.98	$367.39
Operating expenses	$7	$8.45	$9.65	$.27	$26.43
Selling price	$50	$129.98	$145.75	$2.75	$430.75

Applications of Mathematics

10. What is the loss if the operating expenses are $21.50 and the margin is $17.35?
11. Find the loss if the cost is $75, operating expenses are $6.95, and the selling price is $78.50.
12. What is the selling price of an article if it costs $60 and the rate of markup on the selling price is 20%?
13. A refrigerator sold for $375. If a 25% markup on the selling price was used, how much did it cost?
14. Find the selling price:

Cost	$140	$54	$22.50	$72.75	$1,200
Markup on selling price	30%	40%	25%	$33\frac{1}{3}$%	15%

15. What is the rate of markup on cost if the margin is $16 and the cost is $48?
16. If the cost is $30, operating expenses are $6.75 and net profit is $5.25, what is the percent markup on cost?
17. Find the rate of markup (margin or gross profit) on cost:

Cost	$24	$6.50	$95	$.10	$137.50
Selling price	$32	$7.80	$125.40	$.12	$200

18. Find the rate of gross profit on selling price:

Cost	$48	$6.30	$190.99	$.18	$100.20
Selling price	$64	$10.50	$269	$.25	$150.30

19. What is the rate of net profit on the selling price if the net profit is $24 and selling price is $120.
20. Find the rate of net profit on the selling price:

Cost	$25	$18.25	$56.94	$.37	$9.69
Operating expenses	$5	$4.25	$8.25	$.05	$1.71
Selling price	$40	$27	$79.50	$.63	$15

21. If the cost is $18, operating expenses are $3.16, and markup is 30% on cost, what is the rate of net profit on the selling price?
22. What is the rate of loss on the selling price if the loss is $9.50 and the selling price is $190?

23. Find the rate of loss on the selling price if the cost is $49.55, operating expenses are $10.45, and selling price is $50.
24. What is the rate of net profit on the cost if the cost is $52 and net profit is $6.50?
25. Find the rate of net profit on the cost:

Cost	$20	$325	$17.80	$.60	$250
Operating expenses	$9	$28.65	$3.93	$.12	$30.50
Selling price	$36	$392.65	$24.40	$.90	$330

26. What is the rate of loss on the cost if the loss is $8.40 and the cost is $140?
27. Find the rate of loss on the cost if the cost is $36, operating expenses are $7, and selling price is $40.
28. Mr. Harrison's sales last year amounted to $80,000. The cost of the goods sold was $48,000. Operating expenses for the year totaled $12,000. What was the net profit? What percent of the selling price was the net profit? Find the rate of net profit on the cost. Find the percent markup on the cost. What percent of the selling price was the gross profit?
29. A kitchen sink costing $130 was sold for $195. If the overhead expenses were 19% of the selling price, what was the net profit?
30. A radio was bought by a merchant for $45 and marked to sell for $69. It was sold at a discount of 15%. Find the selling price and gross profit.
31. Find the selling price of each of the following articles:

    a. Washing machine, costing $220 and 20% markup on cost.
    b. Dress, costing $18.40 and 25% markup on cost.
    c. Pocket calculator, costing $77 and 30% markup on selling price.
    d. Watch, costing $39 and 35% markup on selling price.
    e. Sweater, costing $9.52 and 20% markup on selling price.

32. Find the rate of gross profit on the selling price in each of the following sales:

    a. Gas range which costs $150 and sells for $250.
    b. Television set which costs $215 and sells for $288.
    c. Suit which costs $54 and sells for $75.
    d. Electric heater which costs $18.60 and sells for $30.
    e. Refrigerator which costs $200.60 and sells for $295.

Applications of Mathematics

33. A dealer bought softballs at $21 a dozen and sold them for $3 each. What percent of the selling price was the margin?

34. A merchant purchased 200 lb. of tomatoes, packed in 1-lb. cartons, for $40. She sold 112 lb. at $.49 a lb., 78 lb. at $.39 a lb., and the rest spoiled. What was her gross profit?

35. A dealer purchased 3 dozen footballs for $180. At what price each must he sell them to realize a gross profit of 40% on the cost?

36. A merchant bought 5 dozen pairs of shoes for $720. If she sells them for $18 a pair, what is her gross profit?

37. At what price each must a merchant sell shirts to realize a gross profit of 20% on the selling price if he buys them at $67.20 a dozen?

38. At what price must a dealer sell a rug which costs $208 to realize a profit of 35% on the selling price?

39. A merchant purchased handbags at $95.76 a dozen. What price should each bag be marked so that a 5% discount may be offered and yet a 30% gross profit on the selling price may be realized?

40. A television set was bought by a dealer for $225. She marked it to sell at a margin of 40% of the cost. If it was sold at a discount of 8%, what was the gross profit?

41. A manufacturer sold coats to a retail merchant, making 25% profit on the cost of each coat. If the merchant sold these coats for $45.50 each, making 30% on the cost, how much did each coat cost the manufacturer?

42. A merchant bought 9 dozen ties for $200. He sold 38 ties for $3.50 each, 56 ties for $2.50 each, and the rest for $1.50 each. If his overhead expenses were 20% of the cost what percent of the selling price was his net profit? What percent of the cost was his net profit? Which rate is greater? How much greater? Compare the rate of gross profit on cost with that on selling price.

43. At what price should a dealer sell a hand calculator costing $25.50 to realize a profit of 60% on the cost?

44. What percent of the selling price is the gross profit when a CB radio is purchased for $64 and is sold for $96?

45. At what price should a merchant sell a video recorder costing $480 to realize a profit of 40% on the selling price?

9–23 BALANCE SHEET—PROFIT AND LOSS STATEMENT

A *balance sheet* tells how much a business is worth. This financial statement shows that the *assets* (value of what is owned) minus the *liabilities* (value of what is owed) is equal to the *capital* or *net worth*.

A *profit and loss statement* tells how much profit is earned or loss is suffered by a business during a certain period of time. This financial statement shows that the *sales* minus the *cost of goods sold* is equal to the *gross profit on sales*. The *net profit* is equal to the *gross profit* minus the *operating expenses*.

EXERCISES

1. Copy the form below, and prepare with these facts a balance sheet as of December 31, showing the capital of Mary Santoro. Assets: cash, $4,382.59; accounts receivable, $12,857.16; notes receivable, $875; inventory of merchandise, $25,200; furniture and fixtures, $3,425; delivery trucks, $8,200. Liabilities: accounts payable, $2,195.63; notes payable, $1,450.

MARY SANTORO

BALANCE SHEET, DECEMBER 31, 19___

Assets		Liabilities	
Cash	$	Accounts Payable	$
Accounts Receivable		Notes Payable	
Notes Receivable		Total Liabilities	
Inventory of Merchandise			
Furniture and Fixtures		Mary Santoro, Capital	
Delivery Trucks			
Total Assets	___	Total Liabilities and Capital	___

Applications of Mathematics

2. Copy the form below, then prepare from the following facts a profit and loss statement for the Wilson Company for the year ending December 31. Sales amounted to $193,575. Inventory on January 1 was $13,750 and on December 31 was $29,800. Purchases for the year amounted to $116,500. The operating expenses were: salaries, $50,500; rent, $6,400; delivery expenses, $3,700; advertising, $1,900; insurance, $1,750.

```
                        WILSON COMPANY
                   STATEMENT OF PROFIT AND LOSS
               FOR THE YEAR ENDING DECEMBER 31, 19____

    Sales                                                       $
    Cost of Goods Sold:
        Inventory of Merchandise, January 1     $
        Purchases                               _____
            Total
        Less Inventory of Merchandise, December 31  _____
            Cost of Goods Sold                              $_____
    Gross Profit on Sales                                   $
      Less
    Operating Expenses
        Salaries                                $
        Rent
        Delivery Expenses
        Advertising
        Insurance                               _____
            Total Expenses
                                                            $_____
    Net Profit for the Year                                 $
```

9–24 PAYROLLS

Factories and some businesses use *time cards* for each employee to record the number of hours he or she works. The payroll clerk uses the time cards to prepare the *payroll*. A *currency break-up sheet* is used to determine the number and kinds of bills and coins needed for each pay envelope. The clerk then prepares a *bank cash memorandum* listing the total number of each kind of coin and bill required.

EXERCISES

Make a copy of each of the following forms. Please do not write in this book.

1. For each of the following time cards compute the number of hours (to the nearest half-hour) worked each day and the total time for the week. Then find the amount of wages due each employee.

TIME CARD				
No. 52 NAME Louise Jackson				
WEEK ENDING _____				
DAY	IN	OUT	IN	OUT	HOURS
M.	8:59	12:01	12:59	5:02	
T.	8:56	12:00	12:57	5:00	
W.	9:00	12:02	12:58	4:30	
Th.	8:59	12:00	12:57	5:00	
F.	8:58	12:03	1:00	5:03	
S.	9:00	12:01			

RATE $5.80   TOTAL HOURS _____
WAGES _____

TIME CARD				
No. 128 NAME David Evans				
WEEK ENDING _____				
DAY	IN	OUT	IN	OUT	HOURS
M.	8:00	12:01	12:58	5:00	
T.	7:58	12:00	12:59	5:01	
W.	7:59	12:02	1:00	5:02	
Th.	7:57	12:01	12:59	5:00	
F.	8:00	12:02	12:58	5:01	
S.					

RATE $6.47   TOTAL HOURS _____
WAGES _____

2. Find the weekly take-home pay (see page 465) of each of the following married persons at the given wage rates:

   a. Robert Jones, $242.75 per week; 3 withholding allowances.

   b. Margaret Sullivan, $197.50 per week; 2 withholding allowances.

   c. Todd Elliot, 34 hours at $8.10 per hour; 1 withholding allowance.

   d. Diane Warren, 39 hours at $5.25 per hour; 3 withholding allowances.

   e. Wilbur Ross, $35\frac{1}{2}$ hours at $4.80 per hour; 2 withholding allowances.

Applications of Mathematics

3. Complete the following payroll. Then prepare a currency break-up sheet and bank cash memorandum for payroll as shown below.

PAYROLL FOR WEEK ENDING _____

NO.	NAME	M.	T.	W.	Th.	F.	S.	TOTAL HOURS	HOURLY RATE	WAGES
84	Lopez, Maria	8	7	7	7	7	4		$6.25	
85	Allen, Joseph	8	8	8	8	8	0		4.38	
86	Hooper, Albert	8	7	8	5	8	3		3.51	
87	Bailey, Ralph	8	$7\frac{1}{2}$	$7\frac{1}{2}$	8	7	0		7.46	
88	Mitchell, Jeffrey	$7\frac{1}{2}$	$7\frac{1}{2}$	$7\frac{1}{2}$	7	7	3		5.60	
89	Nelson, Olga	7	8	$7\frac{1}{2}$	8	7	0		7.84	
90	Riley, Irene	7	7	$6\frac{1}{2}$	8	7	3		6.72	
	TOTALS								xxxx	

CURRENCY BREAK-UP SHEET   FOR WEEK ENDING _____

NO.	NAME	NET PAY	NOTES				COINS				
			$20	$10	$5	$1	50¢	25¢	10¢	5¢	1¢
	TOTALS										

YOUR CITY BANK PAY ROLL

_____ 19___

FOR _____

NUMBER		AMOUNT
	$20 Notes	
	$10 "	
	$ 5 "	
	$ 1 "	
	50¢ Coins	
	25¢ "	
	10¢ "	
	5¢ "	
	1¢ "	
TOTAL		

9–25 INVOICES

An *invoice* is a bill that the seller sends to the purchaser listing the terms of purchase and the quantity, description, and unit price of each article purchased. It also includes (1) the extension for each item, the amount obtained by multiplying the quantity of each article by its unit price and written in the column next to the given unit price; and (2) the total amount of the bill, obtained by adding the extensions and written in the last column. A copy of the invoice is usually sent with the order. When the bill is paid, it should be marked paid, dated, and signed by the person receiving the money.

EXERCISES

1. Make a copy of each of the following invoices, then complete them.

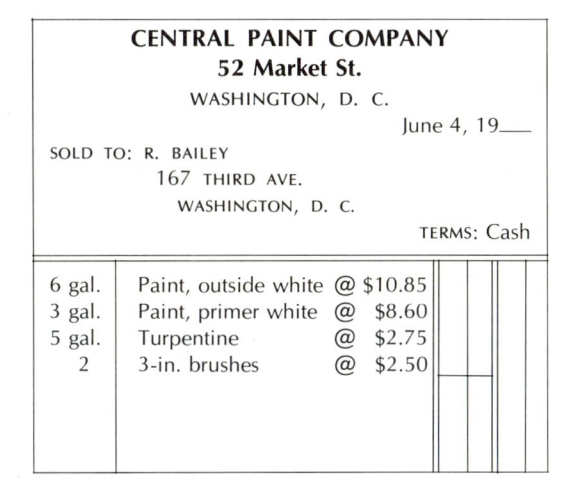

CENTRAL PAINT COMPANY		
52 Market St.		
WASHINGTON, D. C.		
		June 4, 19___
SOLD TO: R. BAILEY		
167 THIRD AVE.		
WASHINGTON, D. C.		
		TERMS: Cash
6 gal.	Paint, outside white	@ $10.85
3 gal.	Paint, primer white	@ $8.60
5 gal.	Turpentine	@ $2.75
2	3-in. brushes	@ $2.50

MODERN CLOTHING COMPANY		
457 Main St.		
NEW YORK, N. Y.		
		December 17, 19___
SOLD TO: MEN'S SPORT SHOP		
46 W. CHESTNUT ST.		
NEW YORK, N. Y.		
		TERMS: n/30
4 doz.	Dress shirts	@ $54.00
2½ doz.	Sport shirts	@ $43.20
5 doz.	Ties	@ $16.40
18 pr.	Slacks	@ $11.75

WILLIAMS LEATHER PRODUCTS		
2917 High St.		
CHICAGO, ILL.		
		September 6, 19___
SOLD TO: JOHNSON LUGGAGE CO.		
863 NINTH ST.		
CHICAGO, ILL.		
		TERMS: 2/10, n/30
2 doz.	Wallets, men's	@ $49.20
3 doz.	Handbags	@ $86.45
1½ doz.	Brief cases	@ $79.00
	Received Payment September 12, 19___	

KITCHEN SUPPLY COMPANY		
917 W. Sixth St.		
LOUISVILLE, KY.		
		Feb. 25, 19___
SOLD TO: WILSON CONSTRUCTION CO.		
471 GIRARD AVE.		
LOUISVILLE, KY.		
		TERMS: 3/10, n/30
4	Heat-Well ranges 40"	@ $179.50
6	Cabinet sinks, 52"	@ $127.50
2	Dependable refrigerators 6 cu. ft.	@ $282.00
	Received Payment March 1, 19___	

2. Prepare invoices for each of the following:

   a. Wholesale Food Company sold to Wilson Grocery Store: 2 cases sardines @ $16.50, 4 cases canned peas @ $7.12, 6 cases canned corn @ $6.75, and 5 cases soap @ $13.25. Terms were 2/10, *n*/30.

Applications of Mathematics

b. Allied Drug Company sold to Walker Pharmacy: 2 gross prescription bottles @ $10.50, 4 doz. eye droppers @ 90¢, 3 doz. tubes shaving cream @ $4.75, 6 doz. tubes toothpaste @ $5.80, and $2\frac{1}{2}$ doz. packages of hair spray @ $9.26. Terms were *n*/30.

```
WHOLESALE FOOD COMPANY
        1648 First St.
         DALLAS, TEXAS
                              19___
SOLD TO:
                      TERMS:
```

```
ALLIED DRUG COMPANY
      21 East Boulevard
     SAN FRANCISCO, CAL.
                              19___
SOLD TO:
                      TERMS:
```

9–26 INVENTORIES AND SALES SUMMARIES

An *inventory* of merchandise is a record of the number of articles of each kind on hand and their value.

A *sales summary* is a record of the goods sold during a specified period of time.

EXERCISES

1. Make a copy of each of the six charts on the next page. Then complete the inventories, sales summaries, and invoices. In the inventory of Feb. 15, determine the quantity of each article that should be on hand.

2. Find the cost of goods sold and the gross profit earned by the school store during the first two weeks of February.

Inventory, Feb. 1	$_____
Purchases	$_____
Total	$_____
Less Inventory Feb. 15	$_____
Cost of goods sold	$_____
Sales	$_____
Less cost of goods sold	$_____
Gross profit	$_____

SCHOOL STORE INVENTORY
FEB. 1, 19__

Article	Quantity	Unit Price	Total Value
Assignment books	163	$.15	
Compasses	38	.19	
Copy books	319	.30	
Erasers	91	.10	
Ball point pens	47	.45	
Loose leaf binders	104	.90	
Loose leaf paper	212	.35	
Pencils	296	.03	
Pencils, automatic	57	.40	
Protractors	83	.25	
Rulers	125	.08	
Tablets	374	.20	
Total	xxxx	xxxx	

SCHOOL STORE SALES SUMMARY
FOR WEEK ENDING FEB. 7, 19__

Article	M.	T.	W.	Th.	F.	Total Sold	Unit Price	Total Sales
Assignment books	26	15	9	11	4		$ .25	
Compasses	8	5	6	3	5		.29	
Copy books	51	28	34	17	24		.49	
Erasers	17	7	8	5	9		.19	
Ball point pens	2	1	3	0	1		.79	
Loose leaf binders	5	12	4	6	3		1.49	
Loose leaf paper	21	18	23	26	35		.69	
Pencils	69	37	31	45	28		.05	
Pencils, automatic	8	4	3	5	9		.59	
Protractors	6	7	4	2	5		.35	
Rulers	39	15	9	12	6		.15	
Tablets	45	34	27	38	29		.35	
Totals							xxxx	

SCHOOL SUPPLY COMPANY
1950 Main St.
OAK LANE, N. Y.
February 5, 19___

SOLD TO: WILSON HIGH SCHOOL
OAK LANE, N. Y.

TERMS: n/30

3 doz.	Compasses	@ $2.28		
15 doz.	Loose leaf paper	@ 4.20		
4 doz.	Erasers	@ 1.20		
18 doz.	Pencils	@ .36		
6 doz.	Rulers	@ .96		

SCHOOL SUPPLY COMPANY
1950 Main St.
OAK LANE, N. Y.
February 11, 19___

SOLD TO: WILSON HIGH SCHOOL
OAK LANE, N. Y.

TERMS: n/30

6 doz.	Assignment books	@ $1.80		
8 doz.	Copy books	@ 3.60		
10 doz.	Tablets	@ 2.40		
9 doz.	Loose leaf paper	@ 4.20		

SCHOOL STORE SALES SUMMARY
FOR WEEK ENDING FEB. 14, 19__

Article	M.	T.	W.	Th.	F.	Total Sold	Unit Price	Total Sales
Assignment books	16	13	8	14	6		$ .25	
Compasses	7	9	5	8	4		.29	
Copy books	29	38	33	46	19		.49	
Erasers	11	6	9	12	5		.19	
Ball point pens	2	0	0	1	1		.79	
Loose leaf binders	5	6	3	4	7		1.49	
Loose leaf paper	34	19	22	27	32		.69	
Pencils	41	28	17	35	21		.05	
Pencils, automatic	3	5	6	4	1		.59	
Protractors	6	3	2	9	5		.35	
Rulers	14	5	6	2	4		.15	
Tablets	27	16	18	24	15		.35	
Totals							xxxx	

SCHOOL STORE INVENTORY
FEB. 15, 19__

Article	Quantity	Unit Price	Total Value
Assignment books		$.15	
Compasses		.19	
Copy books		.30	
Erasers		.10	
Ball point pens		.45	
Loose leaf binders		.90	
Loose leaf paper		.35	
Pencils		.03	
Pencils, automatic		.40	
Protractors		.25	
Rulers		.08	
Tablets		.20	
Total	xxxx	xxxx	

Applications of Mathematics

9–27 PARCEL POST

Parcel post may be insured as follows:

From $.01 to $15 insurance, the fee is $.40; from $15.01 to $50 insurance, the fee is $.60; from $50.01 to $100 insurance, the fee is $.80; from $100.01 to $150 insurance, the fee is $1.00; from $150.01 to $200 insurance, the fee is $1.20.

EXERCISES

In the following examples fractions of pounds are computed as full pounds. Third-class rates are used for weights of 1 pound and under. Use the table on page 531 for the exercises.

1. Find the cost of sending by parcel post a package weighing:

   a. 20 lb. to zone 4.
   b. 11 lb. local.
   c. 16 lb. to zone 7.
   d. 29 lb. to zone 2.
   e. 5 lb. to zone 6.
   f. 3 lb. 8 oz. to zone 3.
   g. 1 lb. 14 oz. to zone 1.
   h. 2 lb. 15 oz. local.
   i. 8 lb. 7 oz. to zone 8.
   j. 24 lb. 12 oz. to zone 5.
   k. 13 lb. local.
   l. 2 lb. 7 oz. to zone 4.
   m. 29 lb. 11 oz. to zone 6.
   n. 19 lb. to zone 8.
   o. 5 lb. 13 oz. to zone 2.

2. Find the cost of sending by parcel post a package weighing:

   a. 26 lb. for 200 miles.
   b. 9 lb. for 1,100 miles.
   c. 17 lb. for 125 miles.
   d. 3 lb. for 750 miles.
   e. 21 lb. for 2,000 miles.
   f. 1 lb. 8 oz. for 100 miles.
   g. 12 lb. 1 oz. for 278 miles.
   h. 23 lb. 14 oz. for 940 miles.
   i. 4 lb. 15 oz. for 1,575 miles.
   j. 9 lb. 6 oz. for 1,300 miles.
   k. 18 lb. for 60 miles.
   l. 19 lb. 10 oz. for 185 miles.

3. Find the cost of sending by parcel post a package weighing:

   a. 16 lb. to zone 5, insured for $50.
   b. 4 lb. to zone 6, insured for $10.
   c. 11 lb. to zone 3, insured for $100.
   d. 3 lb. to zone 4, insured for $5.
   e. 22 lb. to zone 7, insured for $200.
   f. 9 lb. 11 oz. local, insured for $15.
   g. 14 lb. 5 oz. to zone 6, insured for $25.
   h. 12 lb. for 500 miles, insured for $125.
   i. 2 lb. for 40 miles, insured for $5.
   j. 24 lb. for 110 miles, insured for $100.
   k. 5 lb. for 1,350 miles, insured for $10.
   l. 19 lb. for 2,300 miles, insured for $150.

PARCEL POST RATE TABLE

Weight 1 pound and not exceeding (pounds)	Local	Zones 1 and 2 Up to 150 miles	3 150 to 300 miles	4 300 to 600 miles	5 600 to 1,000 miles	6 1,000 to 1,400 miles	7 1,400 to 1,800 miles	8 Over 1,800 miles
2	$0.77	$0.90	$0.93	$1.04	$1.15	$1.28	$1.40	$1.48
3	.82	.97	1.02	1.15	1.29	1.46	1.62	1.74
4	.86	1.04	1.10	1.25	1.42	1.63	1.84	2.00
5	.91	1.11	1.19	1.36	1.56	1.81	2.06	2.26
6	.95	1.18	1.27	1.46	1.69	1.98	2.28	2.52
7	1.00	1.25	1.36	1.57	1.83	2.16	2.50	2.78
8	1.04	1.32	1.44	1.67	1.96	2.33	2.72	3.04
9	1.09	1.39	1.53	1.78	2.10	2.51	2.94	3.30
10	1.13	1.46	1.61	1.88	2.23	2.68	3.16	3.56
11	1.18	1.53	1.70	1.99	2.37	2.86	3.38	3.82
12	1.22	1.60	1.78	2.09	2.50	3.03	3.60	4.08
13	1.27	1.67	1.87	2.20	2.64	3.21	3.82	4.34
14	1.31	1.74	1.95	2.30	2.77	3.38	4.04	4.60
15	1.36	1.81	2.04	2.41	2.91	3.56	4.26	4.86
16	1.40	1.88	2.12	2.51	3.04	3.73	4.48	5.12
17	1.45	1.95	2.21	2.62	3.18	3.91	4.70	5.38
18	1.49	2.02	2.29	2.72	3.31	4.08	4.92	5.64
19	1.54	2.09	2.38	2.83	3.45	4.26	5.14	5.90
20	1.58	2.16	2.46	2.93	3.58	4.43	5.36	6.16
21	1.63	2.23	2.55	3.04	3.72	4.61	5.58	6.42
22	1.67	2.30	2.63	3.14	3.85	4.78	5.80	6.68
23	1.72	2.37	2.72	3.25	3.99	4.96	6.02	6.94
24	1.76	2.44	2.80	3.35	4.12	5.13	6.24	7.20
25	1.81	2.51	2.89	3.46	4.26	5.31	6.46	7.46
26	1.85	2.58	2.97	3.56	4.39	5.48	6.68	7.72
27	1.90	2.65	3.06	3.67	4.53	5.66	6.90	7.98
28	1.94	2.72	3.14	3.77	4.66	5.83	7.12	8.24
29	1.99	2.79	3.23	3.88	4.80	6.01	7.34	8.50
30	2.03	2.86	3.31	3.98	4.93	6.18	7.56	8.76
31	2.08	2.93	3.40	4.09	5.07	6.36	7.78	9.02
32	2.12	3.00	3.48	4.19	5.20	6.53	8.00	9.28
33	2.17	3.07	3.57	4.30	5.34	6.71	8.22	9.54
34	2.21	3.14	3.65	4.40	5.47	6.88	8.44	9.80
35	2.26	3.21	3.74	4.51	5.61	7.06	8.66	10.06

Applications of Mathematics

APPLICATIONS IN INDUSTRY

9-28 LUMBER

The measure of a piece of lumber one foot long, one foot wide, and one inch thick is a unit called a *board foot*. *To find the number of board feet in a piece of lumber,* multiply the length in feet by the width in feet by the thickness in inches.

EXERCISES

1. How many board feet are there in one piece of lumber 2 in. thick, 4 in. wide, and 18 ft. long?
2. How many board feet are there in six pieces of lumber 3 in. thick, 6 in. wide, and 12 ft. long?
3. Find the number of board feet in each of the following:
    a. 1 pc. of pine 1 in. by 4 in. and 16 ft. long.
    b. 4 pc. of fir 1 in. by 10 in. and 12 ft. long.
    c. 9 pc. of spruce 3 in. by 12 in. and 18 ft. long.
    d. 15 pc. of hemlock 2 in. by 8 in. and 10 ft. long.
    e. 20 pc. of pine 4 in. by 6 in. and 14 ft. long.
4. How many 1,000 (M) board feet are in 50 pieces of pine each 3 in. × 4 in. and 12 ft. long?
5. Find the cost of 1 piece of fir 2 in. × 6 in. and 14 ft. long at 35¢ per board foot?
6. What is the cost of 18 pieces of spruce each 4 in. × 12 in. and 16 ft. long at $340 per 1,000 (M) board feet?
7. Find the cost of each of the following:
    a. 10 pc. of fir each 1 in. × 6 in. and 12 ft. long at $385 per M.
    b. 30 pc. of pine each 2 in. × 4 in. and 16 ft. long at $360 per M.
    c. 24 pc. of hemlock each 1 in. × 10 in. and 14 ft. long at $325 per M.
    d. 56 pc. of spruce each 3 in. × 12 in. and 18 ft. long at $330 per M.
    e. 60 pc. of fir each 4 in. × 8 in. and 10 ft. long at $390 per M.
8. How many board feet of flooring 1″ × 3″ cut in 20-ft. lengths are required for a floor 18 ft. by 20 ft.? Allowing 10% for waste, how much will the flooring cost at $360 per M?

9. How many board feet of lumber 1" × 4" in 12-ft. lengths are needed for the four walls of a bin each 12 ft. by 15 ft.? What is the cost of the lumber at $345 per M, allowing 15% for waste?
10. Twelve pieces of lumber 1" × 9" in 3-ft. lengths are required for repairs. Disregarding waste, how many board feet are needed and what is the cost of the lumber at 36¢ per board foot?

9–29 EXCAVATION, BRICKWORK, AND CONCRETE MIXTURES

EXERCISES

1. How many cubic yards of dirt must be removed to make an excavation 40 ft. long, 32 ft. wide, and 15 ft. deep?
2. At $38.88 per cu. yd., what will it cost to dig a well 4 ft. in diameter and 35 ft. deep?
3. How many bricks are needed to construct a wall 1 ft. thick, 18 ft. high, and 48 ft. long if it takes 20 standard bricks to make 1 sq. ft. of a wall with a thickness of 1 ft.? At $110 per M, what is the cost of the brick?
4. A driveway 15 ft. wide and 54 ft. long is to be paved with concrete 4 in. thick. How many cu. yd. of concrete are required? If a mixture of 650 lb. of cement, 1,300 lb. of sand, and 1,700 lb. of gravel make one cu. yd. of concrete, how many pounds of each are needed for the driveway?
5. How many cu. yd. of concrete are required for a sidewalk 9 ft. wide, 24 ft. long, and 6 in. thick? How much will it cost at $39.50 per cu. yd.?

9–30 LATHING, PLASTERING, PAPERHANGING, PAINTING, AND ROOFING

EXERCISES

1. If a bundle of rock laths covers 32 sq. ft., how many bundles of laths are needed to cover the walls and ceiling of a room 16 ft. long, 12 ft. wide, and 10 ft. high, subtracting 80 sq. ft. for

door and window openings? At $2.40 per bundle, how much will the laths cost?

2. At $9.45 per sq. yd., how much will it cost to plaster the walls and ceiling of the room described in Exercise 1?

3. The walls and ceiling of a room 24 ft. long, 14 ft. wide, and 8 ft. high are to be covered with plasterboard. If 112 sq. ft. are deducted for openings, how many sheets of plasterboard, 4 ft. by 8 ft. each, are required? How much will the plasterboards cost at $2.10 per sheet?

4. A single roll of wallpaper covers about 36 sq. ft. How many rolls are needed to cover the walls of a room 16 ft. long, 12 ft. wide, and 9 ft. high if 72 sq. ft. of wall space is deducted for openings and 1 extra roll of paper is allowed for the waste in matching? What is the cost of the wallpaper at $4.80 per roll?

5. The walls and ceiling of a room 18 ft. long, 16 ft. wide, and 12 ft. high are to be papered with double rolls each covering 72 sq. ft. If 10% is to be allowed for matching and 96 sq. ft. is to be deducted for openings, how many rolls are needed for the four walls? What is the cost at $7.20 per roll? How many rolls are needed for the ceiling and what is the cost at $3.60 per roll? Find the total cost.

6. A room is 16 ft. long, 15 ft. wide, and 10 ft. high. Deducting 60 sq. ft. for openings, how many gallons of paint are needed to cover the walls and ceiling with two coats if a gallon will cover 400 sq. ft. with one coat? At $9.60 per gallon, how much will the paint cost?

7. How many liters of varnish are needed to cover a floor 6.3 m by 10 m two coats if a liter covers 7 m² two coats?

8. If the pitch (or slope) of a roof is the ratio of the rise to the span, what is the pitch of a roof whose rise is 1.9 m and span is 7.6 m? Find the rise of a roof whose span is 20 ft. and pitch is $\frac{1}{5}$.

9. If three bundles, each containing 26 shingles, cover a square (100 sq. ft.), how many bundles of shingles are needed to cover a roof of 15 ft. by 40 ft.? What is the total cost of the shingles at $6.95 per bundle?

10. How many rolls of roofing are needed to cover the roof of a house 20 ft. by 35 ft. if each roll covers 100 sq. ft. (or 1 square)? At $9.50 per square, what is the cost of the roofing?

9-31 RIM SPEEDS

The *rim speed* of a revolving object (also called *surface speed* or *cutting speed*) is the number of feet per minute that a point on the circumference of the revolving object travels.

EXERCISES

1. What is the cutting speed of a bandsaw with a diameter of 35 inches when it revolves at 650 r.p.m.?
2. Find the surface speed of a pulley with a diameter of 16 inches turning at 300 r.p.m.
3. What is the surface speed of a grinding wheel 21 inches in diameter when it revolves at 800 r.p.m.?
4. A pulley 15 inches in diameter runs at 700 r.p.m. What is the surface speed?
5. Find the cutting speed of a 28-inch bandsaw when it runs at 900 r.p.m.

9-32 GEARS AND PULLEYS

In industry gears and pulleys are used to change the speed at which power is transmitted.

(1) Two meshed gears are related as follows: the number of teeth (T) of the driving gear times its number of revolutions per minute (R) equals the number of teeth (t) of the driven gear times its number of revolutions per minute (r). Formula: $TR = tr$.

(2) Two belted pulleys (wheels mounted so that they turn on axles) are related as follows: the diameter (D) of the driving pulley times its number of revolutions per minute (R) equals the diameter (d) of the driven pulley times its number of revolutions per minute (r). Formula: $DR = dr$.

EXERCISES

1. A 20-inch pulley turning 175 r.p.m. drives a 5-inch pulley. How many revolutions per minute is the 5-inch pulley turning?

Applications of Mathematics

2. How many teeth are required on a gear if it is to run at 270 r.p.m. when driven by a gear with 42 teeth running at 180 r.p.m.?
3. What size pulley must be used on a motor running at 1,300 r.p.m. in order to drive a 12-inch pulley at 325 r.p.m.?
4. A 30-inch pulley is belted to a 6-inch pulley. If the 30-inch pulley makes 250 r.p.m. how many revolutions per minute does the smaller pulley make?
5. How many revolutions per minute is a 30-tooth gear running when driven by a 45-tooth gear running at 150 r.p.m.? When driven by an 18-tooth gear running at 210 r.p.m.?

9–33 ELECTRICITY

The *ampere* is the unit that measures the *intensity of current* (I); the *volt* is the unit that measures the *electromotive force* (E); the *ohm* is the unit that measures the *resistance* of the conductor (R); the *watt* is the unit that measures *power* or *rate of work* (W) done by the current. These units are related as follows: The number of *volts* equals the number of *amperes* times the number of *ohms*. Formula: $E = IR$. The number of *watts* equals the number of *amperes* times the number of *volts*. Formula: $W = IE$. A *kilowatt* (kW) is 1,000 watts. A *kilowatt-hour* (kW·h) is the unit that measures electrical energy. *To find the amount of energy furnished,* multiply the power in kilowatts by the time in hours.

EXERCISES

1. How many amperes of current are flowing through a 120-volt circuit with a resistance of 40 ohms?
2. How much power in watts is a toaster consuming when it takes 7.2 amperes at 110 volts?
3. Find the power consumed if a motor takes 10 amperes at 125 volts.
4. At 6 cents per kilowatt hour how much will it cost to burn eight 25-watt lamps for 5 hours each evening for 30 days?
5. How much will it cost at 5 cents per kilowatt-hour to operate for 24 hours a motor that uses 15 amperes of current at 120 volts?
6. A 550-watt electric iron is used on a 110-volt circuit. How much current does it take? What is the resistance of the iron?

9–34 LEVERS

A *lever* is a simple machine that is used to move and lift objects. It consists of a board or bar resting on a support called the *fulcrum*. Two forces or weights act on it. One is the lifting force and the other is the weight of the object to be lifted or moved. The distance to the fulcrum from the point where each force or weight is applied is called the lever *arm*.

The common seesaw or teeterboard, on which a heavier person is balanced by a lighter person sitting on the other side of the fulcrum but further away, illustrates the basic property of the lever. It demonstrates that one force or weight (W) times its arm (D) equals the other weight (w) times its arm (d).

As a formula the law of the lever is expressed: $WD = wd$.

EXERCISES

1. Using formula $WD = wd$, find:
   a. W when $D = 8$, $w = 120$, and $d = 6$.
   b. D when $W = 175$, $w = 125$, and $d = 7$.
   c. w when $W = 600$, $D = 4$, and $d = 12$.
   d. d when $W = 1,000$, $D = 2\frac{1}{2}$, and $w = 100$.

2. Two boys sit on the opposite side of the fulcrum of a seesaw. Scott weighs 64 pounds and sits 3 feet from the fulcrum. How far from the fulcrum must Steve, who weighs 48 pounds, sit to balance the seesaw?

3. Marilyn sits 1.5 meters from the fulcrum of a teeterboard and balances Charlotte who sits on the opposite side 1.2 meters from the fulcrum. If Marilyn weighs 40 kilograms, how much does Charlotte weigh?

4. How much force must be applied at a distance 5 feet 6 inches from the fulcrum to balance a downward force of 160 pounds applied 4 feet 8 inches on the other side of the fulcrum?

5. A 1.5-meter crowbar is placed under a 60-kilogram rock. If the fulcrum is 0.25 meter from the rock, how much force must be exerted to raise the rock?

Applications of Mathematics

9-35 READING A MICROMETER

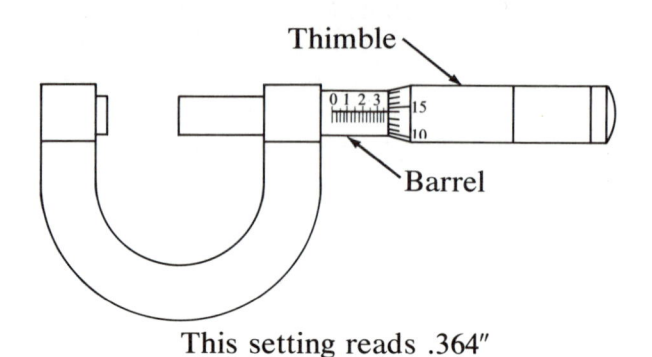

This setting reads .364″

To read a micrometer caliper setting, add the values of the exposed numbers on the barrel (each number = .1″), the uncovered subdivisions on the barrel (each subdivision = .025″), and the number of divisions on the thimble (each division = .001″).

EXERCISES

Read each of the following micrometer settings:

1.

2.

3.

4.

5.

6.

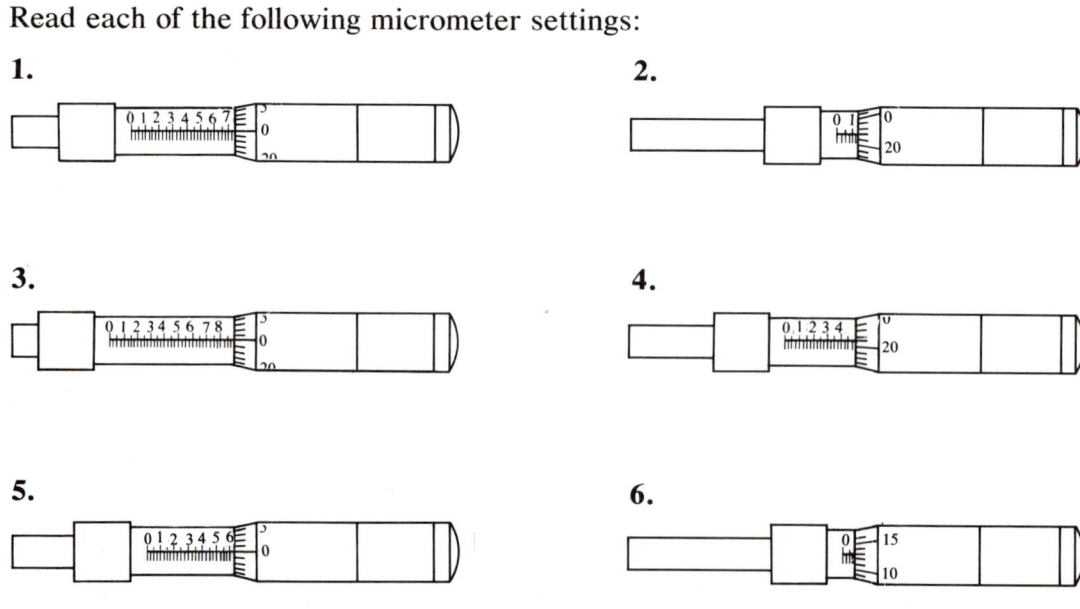

9–36 TOLERANCE

In the manufacture of various parts, a *tolerance* or a difference in the specified size is sometimes allowed and the part is accepted. If a 4-inch part has a tolerance of $\pm \frac{1}{8}''$, it means that any part within the range of $3\frac{7}{8}''$ to $4\frac{1}{8}''$ is acceptable.

EXERCISES

Measure each required part and the corresponding manufactured parts. Which of the manufactured parts are acceptable?

REQUIRED PART	MANUFACTURED PART
1. Tolerance $\pm \frac{1}{8}$ inch	a b c
2. Tolerance $\pm$ 2 mm	a b c
3. Tolerance $\pm \frac{1}{16}$ inch	a b c

Applications of Mathematics

9–37 FINDING MISSING DIMENSIONS

EXERCISES

1. Find the overall length in each of the following figures:

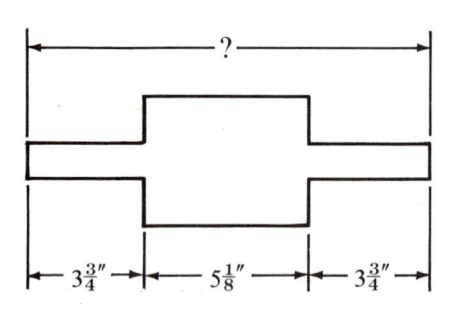

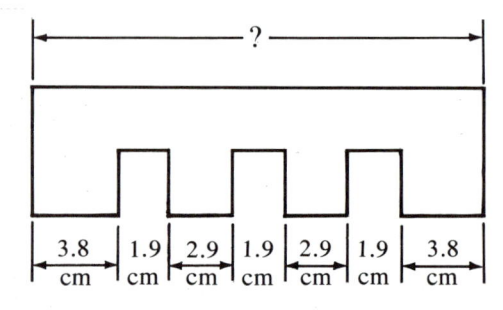

   Left figure: $3\frac{3}{4}''$, $5\frac{1}{8}''$, $3\frac{3}{4}''$; overall = ?

   Right figure: 3.8 cm, 1.9 cm, 2.9 cm, 1.9 cm, 2.9 cm, 1.9 cm, 3.8 cm; overall = ?

2. Find the missing dimension in each of the following figures:

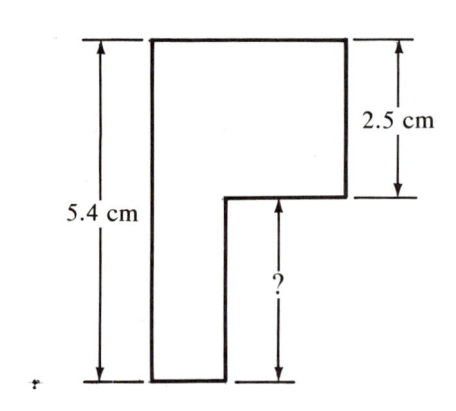

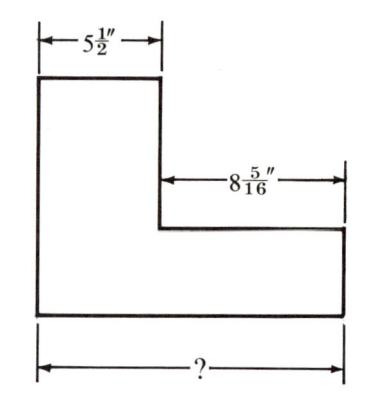

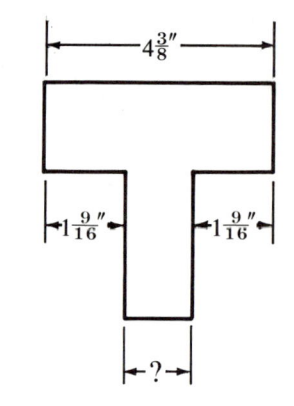

   Left figure: 5.4 cm, 2.5 cm, ?

   Middle figure: $5\frac{1}{2}''$, $8\frac{5}{16}''$, ?

   Right figure: $4\frac{3}{8}''$, $1\frac{9}{16}''$, $1\frac{9}{16}''$, ?

3. Find the missing dimension in each of the following figures:

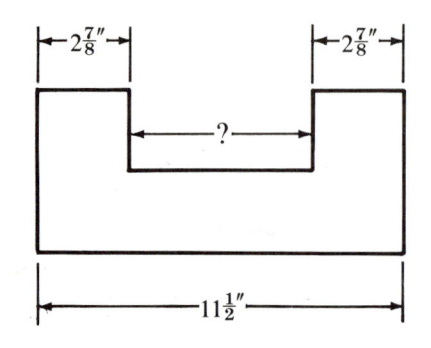

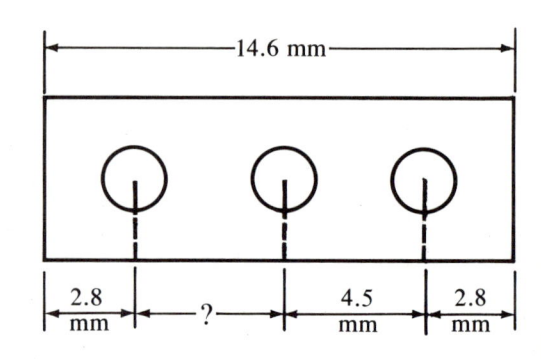

   Left figure: $2\frac{7}{8}''$, $2\frac{7}{8}''$, ?, $11\frac{1}{2}''$

   Right figure: 14.6 mm, 2.8 mm, ?, 4.5 mm, 2.8 mm

540   CHAPTER 9

9–38 AREAS OF IRREGULAR FIGURES

EXERCISES

Find the area of each of the following figures:

1.

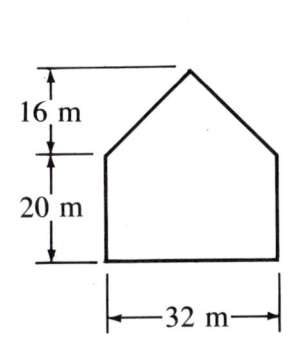

16 m
20 m
32 m

2.

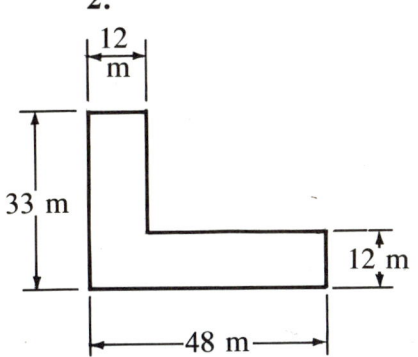

12 m
33 m
12 m
48 m

3.

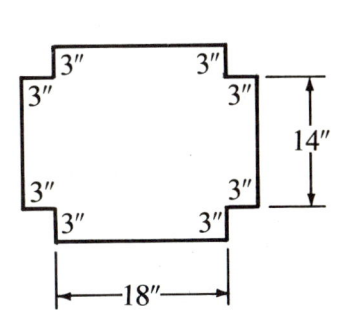

3″, 3″, 3″, 3″, 3″, 3″, 3″, 3″
14″
18″

4.

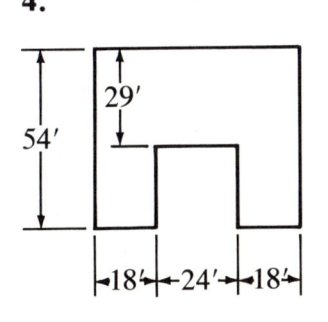

54′
29′
18′ 24′ 18′

5.

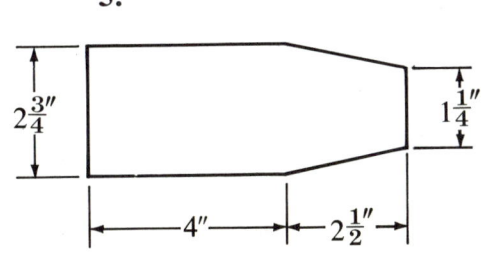

$2\frac{3}{4}″$
$1\frac{1}{4}″$
4″
$2\frac{1}{2}″$

6.

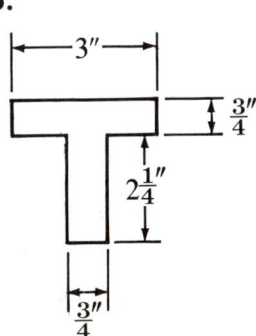

3″
$\frac{3}{4}″$
$2\frac{1}{4}″$
$\frac{3}{4}″$

7.

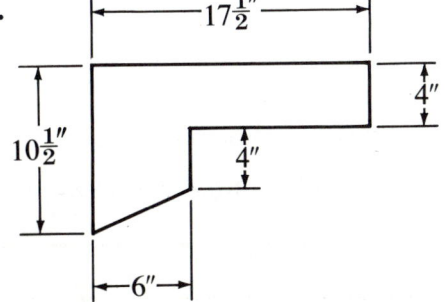

$17\frac{1}{2}″$
4″
$10\frac{1}{2}″$
4″
6″

8.

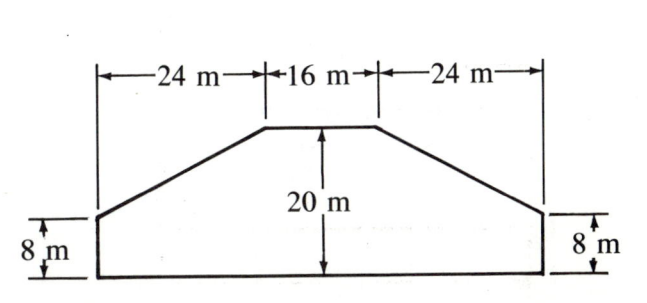

24 m, 16 m, 24 m
20 m
8 m, 8 m

Applications of Mathematics

541

AVIATION

9–39 THE MAGNETIC COMPASS

The direction in which an airplane flies over the earth's surface is called the *course* of the airplane and is expressed as an angle. When it is measured clockwise from the true north (North Pole), it is the *true course*. However, since the compass needle points to the magnetic north, the pilot must correct the true course reading. This correction is called *magnetic variation* or *variation* and the corrected course reading is called the *magnetic course*. The metal parts of the airplane may affect the compass. Thus, the pilot must correct the magnetic course reading. This correction is called *deviation* and the corrected course reading is called the *compass course*.

(1) *To find magnetic course from true course*, we add west variation but subtract east variation. *To find compass course from magnetic course*, we add west deviation but subtract east deviation.

(2) *To find magnetic course from compass course*, we subtract west deviation but add east deviation. *To find true course from magnetic course*, we subtract west variation but add east variation.

EXERCISES

1. Find the magnetic course:

	True Course	Variation
a.	268°	9° W.
b.	95°	4° E.
c.	157°	0°

2. Find the true course:

	Magnetic Course	Variation
a.	67°	2° W.
b.	301°	10° E.
c.	194°	0°

3. What is the variation when the true course is 110° and the magnetic course is 107°?

4. Find the compass course:

	Magnetic Course	Deviation
a.	172°	5° W.
b.	53°	4° E.
c.	298°	0°

5. Find the magnetic course:

	Compass Course	Deviation
a.	42°	6° W.
b.	133°	1° E.
c.	201°	0°

6. What is the deviation when the magnetic course is 89° and the compass course is 93°?

7. Find the compass course:

	True Course	Variation	Deviation
a.	40°	8° W.	1° W.
b.	169°	6° E.	4° E.
c.	324°	11° W.	3° E.
d.	217°	9° E.	6° W.
e.	96°	0°	5° E.

8. Find the true course:

	Compass Course	Deviation	Variation
a.	225°	2° W.	9° W.
b.	61°	3° E.	7° E.
c.	344°	1° W.	5° E.
d.	180°	4° E.	8° W.
e.	15°	0°	3° W.

9. Find the missing numbers:

	True Course	Variation	Magnetic Course	Deviation	Compass Course
a.	28°	5° W.	?	4° E.	?
b.	?	3° E.	?	6° W.	192°
c.	202°	?	199°	?	203°
d.	315°	12° E.	?	5° E.	?
e.	88°	6° E.	?	6° W.	?

10. What compass course should a pilot steer if the true course is 89°, variation is 4°E., and deviation is 7° W.?
11. What is the true course if the compass course is 330°, variation is 5° W., and deviation is 2° W.?
12. The true course between two airports is 127°. If variation is 3° W. and deviation is 5° E., what should the compass course be?
13. A pilot is steering a compass course of 206°. If the variation is 9° E. and the deviation is 5° E., what is the true course?
14. Using the following data, determine a navigator's compass course: true course, 175°; variation, 6° E.; deviation, 4° W.
15. What are the variation and deviation if the true course is 39°, magnetic course is 31°, and the compass course is 36°?

9–40 AVIATION—SOME INTERESTING FACTS

(1) *Lift-drag ratio* is the ratio of the lift of an airplane to its drag.
(2) *Aspect ratio* is the ratio of the length (span) of an airplane wing to its width (chord).
(3) *Tip speed of a propeller* is the number of feet per second that the tip of the propeller travels.

Applications of Mathematics

(4) *Radius of action* is the distance that an airplane can fly on a given amount of fuel in a given direction under known wind conditions and still return to the point where it took off.

(5) *Adiabatic lapse rate*—The temperature of a parcel of air decreases at the rate of $5\frac{1}{2}°$ Fahrenheit for each 1,000 feet it rises following the dry *adiabat*.

(6) A *Mach number* is the ratio of the speed of an object through a gas to the speed of sound through the gas. The air speed of an airplane is Mach 1 if it is equal to the speed of sound, Mach 2 if it is twice the speed of sound, Mach 0.5 if it is half the speed of sound. Speeds faster than the speed of sound are *supersonic* speeds.

EXERCISES

1. The sectional aeronautical maps of the United States are made at a scale of 1 to 500,000. How many kilometers do 6 centimeters on a sectional aeronautical map represent?

2. Using the dry adiabatic lapse rate of $5\frac{1}{2}°$ F per 1,000 ft., find the temperature at an altitude of 8,000 ft. if the temperature at the surface is 58° F.

3. For a certain air density the true altitude is 3% less than the indicated altitude. What is the true altitude if the indicated altitude is 4000 meters?

4. What is the radius of action of an airplane flying at an average ground speed of 820 km/h if it has fuel for 10 hours?

5. The true air speed at a certain altitude and temperature is 8% greater than the indicated air speed. If the indicated air speed is 750 km/h, what is the true air speed?

6. The true course from airport M to airport N is 136°. If the variation is 7° W. and the deviation is 3° E., what compass course should be steered?

7. Find the tip speed of a 7-foot propeller turning at 1,700 revolutions per minute.

8. a. Find the Lift-Drag Ratio if the lift of an airplane is 1,820 lb. and the drag is 140 lb.

   b. Find the Aspect Ratio of an airplane having a span of 54 ft. and a wing area of 324 sq. ft.

9. One minute of arc of latitude is equal to one nautical mile. How many nautical miles apart are Raleigh, S.C., 35°47′ N., 78°30′ W., and Quito, Ecuador, 0°13′ S., 78°30′ W.? How many statute miles apart? If it takes an airplane 12 hours to fly between the two cities, what is the average ground speed of the airplane in knots? in statute miles per hour?
10. Make a drawing at the scale 1 inch = 80 miles showing the position, at the end of one hour, of an airplane flying due east at an air speed of 200 m.p.h. but blown off its course by a 45 m.p.h. north wind. How far was the airplane from its starting point? What course (or track) was it actually flying?
11. If the speed of sound is 740 miles per hour:

   a. What speed in miles per hour is expressed by:
   Mach 1?   Mach 1.6?   Mach 3?   Mach 4.2?   Mach 2.67?

   b. Express each of the following speeds by a Mach number:
   1,480 m.p.h.,   370 m.p.h.,   2,590 m.p.h.,
   1,702 m.p.h.,   2,000 m.p.h.

Applications of Mathematics

TABLE OF SQUARES AND SQUARE ROOTS

No.	Square	Square Root	No.	Square	Square Root	No.	Square	Square Root
1	1	1.000	34	1,156	5.831	67	4,489	8.185
2	4	1.414	35	1,225	5.916	68	4,624	8.246
3	9	1.732	36	1,296	6.000	69	4,761	8.307
4	16	2.000	37	1,369	6.083	70	4,900	8.367
5	25	2.236	38	1,444	6.164	71	5,041	8.426
6	36	2.449	39	1,521	6.245	72	5,184	8.485
7	49	2.646	40	1,600	6.325	73	5,329	8.544
8	64	2.828	41	1,681	6.403	74	5,476	8.602
9	81	3.000	42	1,764	6.481	75	5,625	8.660
10	100	3.162	43	1,849	6.557	76	5,776	8.718
11	121	3.317	44	1,936	6.633	77	5,929	8.775
12	144	3.464	45	2,025	6.708	78	6,084	8.832
13	169	3.606	46	2,116	6.782	79	6,241	8.888
14	196	3.742	47	2,209	6.856	80	6,400	8.944
15	225	3.873	48	2,304	6.928	81	6,561	9.000
16	256	4.000	49	2,401	7.000	82	6,724	9.055
17	289	4.123	50	2,500	7.071	83	6,889	9.110
18	324	4.243	51	2,601	7.141	84	7,056	9.165
19	361	4.359	52	2,704	7.211	85	7,225	9.220
20	400	4.472	53	2,809	7.280	86	7,396	9.274
21	441	4.583	54	2,916	7.348	87	7,569	9.327
22	484	4.690	55	3,025	7.416	88	7,744	9.381
23	529	4.796	56	3,136	7.483	89	7,921	9.434
24	576	4.899	57	3,249	7.550	90	8,100	9.487
25	625	5.000	58	3,364	7.616	91	8,281	9.539
26	676	5.099	59	3,481	7.681	92	8,464	9.592
27	729	5.196	60	3,600	7.746	93	8,649	9.644
28	784	5.292	61	3,721	7.810	94	8,836	9.695
29	841	5.385	62	3,844	7.874	95	9,025	9.747
30	900	5.477	63	3,969	7.937	96	9,216	9.798
31	961	5.568	64	4,096	8.000	97	9,409	9.849
32	1,024	5.657	65	4,225	8.062	98	9,604	9.899
33	1,089	5.745	66	4,356	8.124	99	9,801	9.950

TABLES OF MEASURE

Measure of Length — Metric

10 millimeters (mm) = 1 centimeter (cm)
10 centimeters (cm) = 1 decimeter (dm)
10 decimeters (dm) = 1 meter (m)
10 meters (m) = 1 dekameter (dam)
10 dekameters (dam) = 1 hectometer (hm)
10 hectometers (hm) = 1 kilometer (km)

1 centimeter (cm) = 10 millimeters (mm)
1 meter (m) = 100 centimeters (cm) = 1,000 millimeters (mm)
1 kilometer (km) = 1,000 meters (m)

Tables

Measure of Capacity — Metric

10 milliliters (mL) = 1 centiliter (cL)
10 centiliters (cL) = 1 deciliter (dL)
10 deciliters (dL) = 1 liter (L)
10 liters (L) = 1 dekaliter (daL)
10 dekaliters (daL) = 1 hectoliter (hL)
10 hectoliters (hL) = 1 kiloliter (kL)

Measure of Weight — Metric

10 milligrams (mg) = 1 centigram (cg)
10 centigrams (cg) = 1 decigram (dg)
10 decigrams (dg) = 1 gram (g)
10 grams (g) = 1 dekagram (dag)
10 dekagrams (dag) = 1 hectogram (hg)
10 hectograms (hg) = 1 kilogram (kg)
1,000 kilograms (kg) = 1 metric ton (t)

Metric Measures of Area, Volume, and Miscellaneous Equivalents

See pages 123 and 125.

Measures of Length — Customary

1 foot (ft.) = 12 inches (in.)
1 yard (yd.) = 3 feet (ft.) = 36 inches (in.)
1 rod (rd.) = 16½ feet (ft.) = 5½ yards (yd.)
1 statute mile (stat. mi.) = 5,280 feet (ft.) = 1,760 yards (yd.) = 320 rods (rd.)
1 statute mile (stat. mi.) ≈ 0.87 nautical mile (0.8684 naut. mi.)
1 nautical mile (naut. mi.) ≈ 6,080 feet (6,080.2 ft.)
≈ 1.15 statute miles (1.1515 stat. mi.)
1 fathom (fath.) ≈ 6 feet (ft.)

Measure of Area — Customary

1 square foot (sq. ft.) = 144 square inches (sq. in.)
1 square yard (sq. yd.) = 9 square feet (sq. ft.)
1 square rod (sq. rd.) = 30.25 square yards (sq. yd.)
1 acre = 160 square rods (sq. rd.)
= 4,840 square yards (sq. yd.)
= 43,560 square feet (sq. ft.)
1 square mile (sq. mi.) = 640 acres

Measure of Volume — Customary

1 cubic foot (cu. ft.) = 1,728 cubic inches (cu. in.)
1 cubic yard (cu. yd.) = 27 cubic feet (cu. ft.)

Liquid Measure — Customary

1 pint (pt.) = 16 ounces (oz.)
= 4 gills (gi.)
1 quart (qt.) = 2 pints (pt.)
1 gallon (gal.) = 4 quarts (qt.)

Dry Measure — Customary

1 quart (qt.) = 2 pints (pt.)
1 peck (pk.) = 8 quarts (qt.)
1 bushel (bu.) = 4 pecks (pk.)

Measure of Weight — Avoirdupois — Customary

1 pound (lb.) = 16 ounces (oz.)
1 short ton (sh. tn. or T.) = 2,000 pounds (lb.)
1 long ton (1. ton) = 2,240 pounds (lb.)

Volume, Capacity, and Weight Equivalents — Customary

1 gallon (gal.) ≈ 231 cubic inches (cu. in.)
1 cubic foot (cu. ft.) ≈ 7½ gallons (gal.)
1 bushel (bu.) ≈ 1¼ cubic feet (cu. ft.)
 ≈ 2,150.42 cubic inches (cu. in.)
1 cu. ft. of fresh water weighs 62½ pounds (lb.)
1 cu. ft. of sea water weighs 64 pounds (lb.)

Measure of Time

1 minute (min.) = 60 seconds (sec.)
1 hour (hr.) = 60 minutes (min.)
1 day (da.) = 24 hours (hr.)
1 week (wk.) = 7 days (da.)
1 year (yr.) = 12 months (mo.)
 = 52 weeks (wk.)
 = 365 days (da.)

Angles and Arcs

1 circle = 360 degrees (°)
1 degree (°) = 60 minutes (')
1 minute (') = 60 seconds (")

Measure of Speed

1 knot = 1 nautical m.p.h.

Metric — Customary Equivalents

1 meter — 39.37 inches
 ≈ 3.28 feet (3.2808)
 ≈ 1.09 yards (1.0936)
1 centimeter ≈ .39 or .4 inch (.3937)
1 millimeter ≈ .04 inch (.03937)
1 kilometer ≈ .62 mile (.6214)
1 liter ≈ 1.06 liquid quarts (1.0567)
1 liter ≈ .91 dry quart (.9081)
1 gram ≈ .04 ounce (.0353)
1 kilogram ≈ 2.2 pounds (2.2046)
1 metric ton ≈ 2,204.62 pounds
1 inch = 25.4 millimeters
1 foot ≈ .3 meter (.3048)
1 yard ≈ .91 meter (.9144)
1 mile ≈ 1.61 kilometers (1.6093)
1 liquid quart ≈ .95 liter (.9463)
1 dry quart ≈ 1.1 liters (1.1012)
1 ounce ≈ 28.35 grams (28.3495)
1 pound ≈ .45 kilogram (.4536)
1 short ton ≈ .91 metric ton (.9072)
1 square inch ≈ 6.45 square centimeters (6.4516)
1 cubic inch ≈ 16.39 cubic centimeters (16.3872)

Index

Abscissa, 303
Absolute value, 166–167, 171, 179–182, 186–188, 193, 203–206, 209, 210, 215, 216, 220, 275–276
Accuracy in measurement, 145
Achievement tests, 83, 155, 229, 318, 450
Acute angle, 371
Acute triangle, 373
Addition
 approximate numbers, 149–150
 decimal fractions, 42–44
 degrees, minutes, seconds, 138–139
 denominate numbers, 129–144
 fractions and mixed numbers, 29–31
 integers, 181–184
 monomials, 248–249
 on the number line, 177–180
 polynomials, 249–250
 positive and negative numbers, 177–192
 properties, 99, 184–186, 190–192
 rational numbers, 186–190
 whole numbers, 7–10
Additive identity, 101, 185–186, 192
Additive inverse, 101, 175, 186, 192
Add-on-interest loan, 490–491
Adiabatic lapse rate, 544
Adjacent angle, 406
Air speed, 388
Algebra, 233–322

Algebraic
 expressions, 236, 244
 phrases, 236
 representation, 278–279
Alternate-interior angles, 407
Altitude, 373, 399–400
Amortizing a loan, 494
Amount, 484
Ampere, 536
Angles, 369–372, 385–390
 acute, 371
 adjacent, 406
 alternate-interior, 407
 and arcs, 138–139
 base, 399
 bisecting, 398–400
 central, 375, 401
 complementary, 404–405
 constructing equal angles, 397–398
 copying, 397
 corresponding, 407
 cosine of, 417, 419
 course, 388
 of depression, 417
 drawing, 386–387
 of elevation, 417
 exterior, 406
 kinds of, 371
 meaning of, 369
 measuring, 385–386
 naming, 369–370
 navigation, 388
 obtuse, 371
 opposite, 406
 pairs of, 404–407
 parts of, 369
 reflex, 371
 right, 371
 sine of, 416, 419
 straight, 371

 sum of:
  in polygon, 404
  in quadrilateral, 403–404
  in triangle, 403
 supplementary, 405–406
 tangent of, 416, 418
 vertex, 373
 vertical, 406
Annual percentage rate, 471
Antecedent, 354
Approximate numbers, 149–151
Arc, 374, 401–402
Area
 circle, 433
 definition, 426
 of irregular figures, 541
 lateral, 436
 parallelogram, 429
 rectangle, 426–427
 square, 428
 total, cube, 435
  cylinder, 436
  rectangular solid, 435
 trapezoid, 431
 triangle, 430
 units of measure of, 123–124, 135
Arithmetic computation, 6–18, 29–39, 42–54, 61, 65–71
Arithmetic mean, 328
Aspect ratio, 543
Assessed value, 480
Assets, 523
Associative property
 of addition, 99, 185, 191
 of multiplication, 100, 208, 212–213
Averages, 328–332
Aviation, 388, 542–545
Axioms, 263, 266
Azimuth, 388

Balance sheet, 523
Bank discount, 512
Banking, 488–489
Bar graph, 324
Base, 234
Base (geometric figures), 373, 429
Base angles, 399
Base of numeration system, 5
Base ten, 5
Bearing, 388
Beneficiary, 477
Binary operation, 98
Bisecting
   angle, 398–400
   line segment, 396
Board foot, 532
Bond, 501–502, 505–507
Borrowing money, 490–496
Brickwork problems, 533
Business applications, 507–531
Business forms, 488, 489, 523, 524, 525, 526, 527, 528, 529
Buying, renting a house, 482

Calculator, Check by, 86
Capacity, 121–123, 125–127, 131–134, 136
Capital, 523
Cash discount, 507
Casting out nines, 96–98
Celsius temperature, 140–141
Centered signs, 224
Central angle, 375, 401
Changing units of measure
   to larger, 111
   to smaller, 109
Chapter Reviews, 104, 153, 226, 313, 336, 350, 356, 444
Check, 488, 489
Check by Calculator, 86
Checking by casting out nines, 96–98
Checking equations, 260
Chord, 374, 395, 401
Circle, 374

area, 433
circumference, 375, 423–425
circumscribed, 397
concentric, 375
inscribed, 399
Circle graph, 325
Circumference, 375, 423–426
Circumscribed circle, 397
   triangle, 399
Closure, 100, 185, 192, 197, 201, 208–209, 213–214, 220, 223–224
Collateral loan, 490
Collinear points, 367
Commission, 513–516
Common denominator, 25
Common divisor, 89
Common factor, 89
Common fractions, 19
   expressed as a percent, 63
Common multiple, 92
Commutative property
   of addition, 99, 184, 190–191
   of multiplication, 99, 207–208, 212
Comparing decimals, 41
Comparing fractions, 27
Comparing integers, 170
Compass course, 542
Complementary angles, 404–405
Complementary events, 340
Components, 302
Composite number, 91
Compound interest, 496–500
Computation
   decimals, 38–53
   fractions and mixed numbers, 29–38
   integers, 181–184, 194–196, 201–207, 215–218
   percent, 61–75
   positive and negative numbers, 177–226
   rational numbers, 186–190, 198–200, 209–212, 220–222

using numerals with centered signs, 224–226
   whole numbers, 6–18
Concentric circles, 375
Conclusion, 354
Concrete mixture problems, 533
Concurrent lines, 365
Conditional, 354
Cone, 376
   volume, 443
Congruent triangles, 374, 409
Conjunction, 353
Consequent, 354
Constant, 235
Constructions
   angle equal to a given angle, 397–398
   angle of a given size, 386–387
   bisecting an angle, 398–400
   bisecting a line segment, 396
   copying an angle, 397
   copying a line segment, 379–380
   line parallel to a given line, 400
   perpendicular to a given line at or through a given point on the given line, 394
   perpendicular to a given line from or through a given point not on the given line, 395
   polygons, 401–402
   regular polygons, 401–402
   triangles, 390–393
Contrapositive, 355
Converse, 354
Coordinate, 167, 282, 296
Coordinate axes, 301
Coplanar
   lines, 368
   points, 368
Corresponding angles, 407–408, 409
Corresponding sides, 409
Cosine of an angle, 417, 419
Cost, 517

551

Cost of goods sold, 523, 528
Counting number, 6
Course, 388
   angle, 388
   compass, 542
   magnetic, 542
   true, 542
Cross-products, 27, 287–288
Cross-product test, 27–28
Cube, 376
   total area, 435
   volume, 439–440
Cubic measure, 135, 136, 438
Customary system of measurement, 128
Cylinder, 376
   lateral area, 436
   total area, 436–437
   volume, 441

Date of maturity, 501, 512
Decagon, 373, 402
Decimal fractions, 38
   addition, 42
   comparing, 41
   division, 50–52
   expressed
     as a common fraction, 59
     as a percent, 63
   multiplication, 47
   place value, 5
   rounding, 39
   subtraction, 45
Decimal numeration system, 5
Decimal relationships
   decimal part is known, 73–74
   decimal part of a number, 73–74
   decimal part one number is of another, 73–74
Decimals
   repeating, 39
   terminating, 39
Deferred payment plans, 471
Degree, 371
Demand loan, 490
Denominate numbers, 129

Denominator, 19
   common, 25, 27
Deposit slip, 488
Deviation, 542
Deviation
   from the mean, 334
   mean, 334
   standard, 334
Diagonal, 373
Diameter, 374–375, 423
Digit, 5
   significant, 55, 147–148
Dimensions, finding missing, 540
Direct variation, 308
Directed numbers, 162, 172
Discount, 507–508
   bank, 512
   cash, 507
   loan, 490–491
   rate of, 507
   successive, 507
   trade, 507
Disjunction, 353
Distributive property, 100, 208, 213, 219, 223
Divisibility, tests for, 93–95
Division
   approximate numbers, 150
   by powers of ten, 53
   decimal fractions, 50–53
   degrees, minutes, seconds, 139
   denominate numbers, 129, 131, 132, 133, 134, 138, 139
   fractions and mixed numbers, 36–38
   integers, 215–220
   monomials, 254–255
   polynomials, 255–256
   positive and negative numbers, 215–224
   powers of ten, 53
   properties, 219–220, 222–224
   rational numbers, 220–224
   whole numbers, 15–18
Dodecagon, 373, 402

Drawing: *also see* Construction
   angles, 386–387
   bar graph, 324
   bearing, 388
   circle graph, 325
   course, 388
   frequency polygon, 327
   heading, 388
   histogram, 327
   line graph, 325
   line segments, 378–379
   scale, 380, 384, 388, 391, 392, 393, 415–416, 545
   wind direction, 388
Drawing to scale, 380, 384, 388, 391, 392, 393, 415–416, 545
Dry measure, 121–123, 134

Earnings, 458–460
Edges, 375, 376
Electricity, 536
Endowment insurance, 477
Equality, 263–265
Equations, 239, 260–276, 282–283, 287–289, 311–312
   checking, 260
   equivalent, 261
   graphing, 282
   quadratic, 311–312
   in simplest form, 261
   solving, 73–74, 260, 266–275
   with absolute values, 275
Equiangular triangle, 373
Equilateral triangle, 373
Equivalent equations, 261
Equivalent fractions, 25–26, 27
Equivalent ratios, 285
Error
   greatest possible, 145
   relative, 145
Estimating answers, 85, 151
Euclid, 90, 360
Euclid's algorithm, 90
Euler's formula, 376

Evaluation
 algebraic expressions, 244–245
 formulas, 246–247, 276–277
Even number, 88
Events, 339, 340
 complementary, 340
 independent, 347
 mutually exclusive, 346
 simple, 340
Excavation problems, 533
Excise tax, 480
Exponents, 234, 244, 252, 254, 256–259
 law for division, 254
 law for multiplication, 252
 negative, 256–259
 zero, 256–259
Exterior angle, 406
Extremes, 287

Face
 of note, 512
 of polyhedron, 375–376
Face value, 501
Factor, 89
 common, 89
 greatest common, 89
 prime, 91
Fahrenheit temperature, 140–141
Financing a car, 493
Fire insurance, 475
Flow charts, 352
Formula, 241, 246–247, 308–310
 area, 427–434
 circumference, 423–425
 derivation, 310
 evaluation, 246–247, 276–277
 interest, 485
 lateral area, 436
 numerical trigonometry, 416–417
 perimeter, 421–423
 rule of Pythagoras, 412
 total area, 435–437

 transforming, 310
 translating, 241–243
 volume, 438–444
 writing, 241–243
Fractional relationships
 fractional part is known, 73–74
 fractional part of a number, 73–74
 fractional part one number is of another, 73–74
Fractions
 addition, 29–30
 common, 19
 comparing, 27
 decimal, 38–39
 division, 36–37
 equivalent, 20, 27
 expressing as decimals, 57–58
 expressing as equivalent fractions with a common denominator, 25
 expressing as mixed or whole numbers, 23
 expressing as percents, 63–64
 expressing in higher terms, 21
 expressing in lowest terms, 20
 improper, 19
 meaning, 19
 multiplication, 34
 proper, 19
 subtraction, 32–33
Frequency, 326–327
Frequency distribution, 326–327
Frequency distribution table, 327
Frequency polygon, 327
Fulcrum, 537

Gears, 535
Geometric constructions: see **Constructions**
Geometric figures, 373–376
Geometry, 357–456

Graphing
 equation in one variable on number line, 282
 groups of numbers, 168
 inequality in one variable on number line, 296–300
 number, 167–168
Graphs
 bar, 324
 circle, 325
 frequency polygon, 327
 histogram, 327
 line, 325
Greatest common factor, 89
Greatest possible error, 145
Gross profit, 517, 523
Ground speed, 388

Half-line, 296, 297, 298, 360
Half-plane, 360
Heading, 388
Hexagon, 373, 401
Histogram, 327
Hypotenuse, 373, 412
Hypothesis, 354

Identity element
 for addition, 101, 185–186, 192
 for multiplication, 101, 209, 214
Implication, 354
Improper fraction, 19
Income, 458, 465–466
Income tax, 461–463, 467–468
Independent events, 347
Indirect measurement, 412–420
 numerical trigonometry, 416–418
 rule of Pythagoras, 412
 scale drawing, 415
 similar triangles, 414
Industrial problems, 532–541
Inequality, 239
 graphing in one variable, 296–300

553

solving in one variable, 289-290
Inscribed circle, 399
   triangle, 397
Installment buying, 470
Insurance
   fire, 475
   life, 477
Integer, 161-162
   addition, 181-182
   addition on the number line, 177-178
   comparing, 170
   division, 215-217
   even, 162
   multiplication, 201-206
   negative, 162
   non-negative, 162
   non-positive, 162
   odd, 162
   positive, 162
   properties, 184-186, 197, 207-209, 219-220
   subtraction, 194-195
   subtraction on the number line, 193
Interest, 484
   compound, 496-498
   simple, 484-486
Intersecting lines, 365
Interval, 297, 362
Inventory, 528
Inventory Test, 3-4, 108, 158-160, 358-359
Inverse
   additive, 101, 175, 186, 192
   logic, 354, 355
   multiplicative, 101, 214
   operations, 98, 263-264
   variation, 308
Invoice, 526
Irrational number, 76, 164
Isosceles triangle, 373, 399

Kelvin temperature, 140-141
Kilowatt, 536
Kilowatt hour, 536
Kinds of angles, 371-372
Knot (naut. m.p.h.), 139

Lateral area of a cylinder, 436
Lathing problems, 533
Least common multiple, 92
Length, measure of, 114, 130-131
Levers, 537
Liabilities, 523
Life insurance, 477
Lift-drag ratio, 543
Linear measure, 378
Line graph, 325
Line, number, *see* **Number line**
Lines, 360
   concurrent, 365
   coplanar, 368
   drawing, 378-379
   facts about, 360, 367
   intersecting, 365, 366
   kinds, 364
   measuring, 378-379
   naming, 361-362
   parallel, 365, 366, 400, 407-408
   perpendicular, 365, 366, 394, 395
   position, 365
   skew, 365
Line segment, 298, 361, 378-380
   bisecting, 396
   copying, 379-380
   drawing, 378
   measuring, 378
Liquid measure, 121, 132
List price, 507
Loan, 490-491
   add-on-interest, 490-491
   collateral, 490
   demand, 490
   discount, 490-491
Locating points on a number plane, 303-305
Logic, 353-356
Longitude, 143
Loss, 517
Lumber problems, 532

Mach number, 544

Magnetic
   compass, 542
   course, 542
   variation, 542
Managing money, 484-507
Map
   time zones, 143
Margin, 517
Marked price, 507
Market value (price), 501
Markup, 517
Mathematical
   phrases or expressions, 236, 244
   sentences, 239
Mean, 328
Mean deviation from the mean, 334
Means, 287
Measure, units of, 109
   angles and arcs, 138
   area, 123, 135
   capacity
     dry, 121, 134
     liquid, 121, 132
   changing to larger unit, 111-112
   changing to smaller unit, 109-110
   length, 114, 130
   mass, 118
   metric, 112-127
   nautical mile, 139
   tables of, 547-549
   temperature, 140
   time, 137, 142
     speed, 139
   volume, 125, 135
   weight, 118, 131
Measurement, 107-156, 412-416
   accuracy in, 145
   indirect, 412-420
   precision in, 144-147
Median, 329
Median of triangle, 373, 397
Member
   of an equation, 260
Meridians, 143
Metric system, 112-127

changing to larger unit,
    111–112
changing to smaller unit,
    109–110
comparison to decimal
    system, 113
measure of
    area, 123–124
    capacity, 121–122
    length, 114–118
    mass, 118–120
    volume, 125–127
    weight, 118–120
units of measure, 112–127
Micrometer, 538
Midpoint, 396
Minuend, 10
Minutes, 371
Mixed decimals, 38
Mixed numbers, 20
    addition, 29–30
    definition, 20
    division, 36–37
    expressed as improper
        fraction, 23
    multiplication, 34
    subtraction, 32–33
Mode, 329
Monomials, 247
    addition, 248–249
    division, 254–255
    multiplication, 252–253
    subtraction, 250–251
Mortgage, 482
    amortizing a loan, 494
Multiple, 92
    common, 92
    least common, 92
Multiplication
    approximate numbers, 150,
        151
    by powers of ten, 49
    decimal fractions, 47
    degrees, minutes, seconds,
        139
    denominate numbers, 129,
        131–134, 138, 140
    fractions and mixed
        numbers, 34
    integers, 201–206

monomials, 252–253
polynomials, 253–254
positive and negative
    numbers, 201–214
properties, 99–101,
    207–209, 212–214
rational numbers, 209–212
whole numbers, 12–14
Multiplicative identity, 101,
    209, 214
Multiplicative inverse, 101,
    214
Mutual funds, 502, 506–507
Mutually exclusive events,
    346

Natural number, 6
Nautical mile, 139, 140
Navigation, 388
Negation, 354
Negative exponents, 256–257
Negative number, 161–162,
    163–164
Net amount, 507–508
Net price, 507–508
Net proceeds, 513
Net profit, 517, 523
Net worth, 523
Non-negative integer, 162
Non-positive integer, 162
Notation
    decimal, 38–39
    fraction, 19
    scientific, 55–56
    sigma, 329–330
Note
    face, 512
    promissory, 512
Number,
    approximate, 149–150
    common multiples, 92
    composite, 91
    counting, 6
    denominate, 129
    directed, 162, 172
    even, 88
    graph of, 167–168
    integer, 161–162
    irrational, 76, 164

least common multiple, 92
Mach, 544
mixed, 20
multiples, 92
natural, 6
negative, 161–162, 163
odd, 88
positive, 162, 163
prime, 91
rational, 19, 163–164
real, 163–164
shortened names for large
    numbers, 54
signed, 162
squaring a, 96
whole, 6
Number line, 161–162, 163,
    167–169, 170, 172–175,
    177–179, 193, 203,
    282–283, 297–298,
    299–300
Number pairs, 19, 301–302
Number plane, 301
    locating points, 303–305
    plotting points, 306–307
Number theory, 88–98
Numeral, 2, 5, 6, 19
Numeration, systems of
    decimal, 5
Numerator, 19
Numerical coefficient, 236,
    247
**Numerical phrase or
    expression,** 236

Obtuse angle, 371
Obtuse triangle, 373
Octagon, 373
Odd number, 88
Odds, 341
Ohm, 536
One, properties of, 101–102
Open sentences, 239, 260–301
**Operating (overhead)
    expenses,** 517, 523
Operations, 98
    algebraic, 177–226,
        247–259
    binary, 98

555

inverse, 98, 263–264
properties, 99–102
Opposite angles, 406
Opposite directions, 172
Opposite meanings, 171
Opposites, 162, 175, 195
Order
reverse, 294
same, 290, 292
Ordered number pair,
301–302
Ordinary life insurance, 477
Ordinate, 303
Origin, 305
Outcome, 339
Owning a house, 482

Painting problems, 534
Paperhanging problems, 534
Parallel
lines, 365–366, 400, 407
planes, 368
Parallelogram, 374
area, 429
Parcel post, 530
Parentheses, 235, 244
Par value, 501
Payrolls, 524
Pentagon, 373
Percent, 61–74
expressed as common
fractions, 62
expressed as decimal
fractions, 61
finding a number when a
percent of it is known,
70–72, 73–74
finding a percent of a
number, 65–68, 73–74
finding what percent one
number is of another,
68–70, 73–74
meaning, 61
solving by equations, 73–74
solving by proportions,
72–73
Percentile, 332
Percentile rank, 332
Perfect square, 82

Perimeter
definition, 421
rectangle, 421
square, 422
triangle, 423
Perpendicular bisector,
396–397, 399–400
Perpendicular lines, 365–366,
369, 394–395
Pi (π), 424–425
Pitch of roof, 286, 534
Place value, 5
Plane, 360, 362, 368–369
Plane figures, 373
Plastering problems, 534
Plotting points on the
number plane, 306–307
Point, 360, 361
collinear, 367
coplanar, 368
locating on number plane,
303
plotting on number plane,
306
Points on a number plane,
310–313
Polygons, 373–374, 401–402
inscribed, 401
regular, 373, 401
sum of angles, 403–404
Polyhedron, 375–376
Polynomial, 247
addition, 249–250
division, 255–256
multiplication, 253–254
subtraction, 251
Positive number, 161–162,
163–164
Powers, 234
Powers of ten
dividing by, 53
multiplying by, 49
Precision in measurement,
144–147
Price
list, 507
marked, 507
market, 501
net, 507–508
sale, 507

selling, 517
Prime
factor, 91
number, 91
relatively, 91
Primes, twin, 91
Principal, 484, 496
Principal square root, 82
Probability, 337–349
Problems, 280
Proceeds, 512
net, 513
Product, 89
Profit
gross, 517
and loss, 517
net, 517
rate, 517
Profit and Loss statement,
523
Promissory note, 512
Proper fraction, 19
Properties, 99–102, 184–186,
190–192, 197, 200–201,
207–209, 212–214,
219–220, 222–224
of equality, 263–264
number and properties,
99–102
one, 101–102
zero, 101–102
Property tax, 480
Proportion, 72, 287–289
Protractor, 385–387
Pulleys, 535–536
Pyramid, 376
volume, 443
Pythagoras, 412–413

Quadrant, 304
Quadratic equation, 311–312
Quadrilateral, 374
Quartile, 332–333

Radius, 375, 423–425
Radius of action, 544
Range
statistics, 334

Rank, 332
  percentile, 332
Rate
  adiabatic lapse rate, 544
  commission, 513
  discount, 507
  interest, 484-486
  loss, 517
  profit, 517
  ratio, 284-285
Rates of speed, 139-140
Ratio, 284-285
  aspect, 543
  cosine, 419
  lift-drag, 543
  sine, 419
  tangent, 418
Rational number, 19,
    163-164, 186-192,
    198-201, 209-214,
    220-224
  addition, 186-190
  division, 220-222
  multiplication, 209-212
  properties, 190-192,
    200-201, 212-214,
    222-224
  subtraction, 198-200
Ray, 296, 362
Real number, 164
Real number line, see
  Number line
Real number plane, see
  Number plane
Reciprocal, 101
Rectangle, 374
  area, 426-427
  perimeter, 421
Rectangular solid, 376
  total area, 435
  volume, 438
Reflex angle, 371
Regrouping, 10
Regular polygon, 373, 401
Relative error, 145
Relatively prime, 91
Renting a house, 482
Representative fraction, 380
Reviews: *also see* Tests
  Chapter, 104, 153, 226,
    313, 336, 350, 356, 444
Right angle, 371, 373
Right circular cone, volume
    443
Right circular cylinder,
    volume, 441
Right triangle, 373, 412-413,
    416
Rim speed, 535
Roofing problems, 534
Root
  of equation, 260
  square, 76, 79-82, 235
Rule of Pythagoras, 412-413

Sale price, 507
Sales summary, 528
Sales tax, 479
Sample space, 339
Savings bond, 502
Savings certificate, 502
Scale, 380
Scale drawing, 380, 384,
    388, 390, 391, 392,
    393, 415-416, 545
Scalene triangle, 373
Scientific notation, 55-56,
    259
Seconds, 371
Sector, 325, 433
Segment, line, 298, 361,
    378-380
Selling price, 517
Semi-circle, 374
Sentences, 239-240
  graphing, 282, 296-300
  mathematical, 239
  open, 239, 260-301
Shortened whole number
    names, 54
Sigma notation, 329-330
Signed numbers, 162
  addition, 181-182
  division, 215-217
  multiplication, 201-206
  subtraction, 194-195
Significant digits, 55,
    147-148
Signs of operation, 234

Similar triangles, 374, 410,
    414
Simple event, 340
Simple interest, 484-486
Sine of angle, 416, 419
Skew lines, 365
Social security, 464
Solid, rectangular, 375-376,
    435, 438
Solids, geometric, 375-376
Solution, 260
Solving equations in one
    variable, 260, 266-276
Solving inequalities in one
    variable, 289-296
Space, 360, 438-444
Speeds
  air, 388
  cutting, 535
  ground, 388
  rates, 139-140
  rim, 535
  supersonic, 544
  tip, 543
Sphere, 376
  volume, 442
Spread, 517
Square, 374
  area, 428
  perimeter, 422
Square measure: *see*
  Measure, Area
Square of a number, 76-78
Square root, 76, 79-82, 235
  alternate method, 80
  by average, 79
  by division, 79
  by estimation, 79
  by table, 81-82
  principal, 82
  table, 547
Standard deviation, 334
Standard time, 143
Statement, 239, 353
Statistics, 324-334
Stocks, 500-501
Straight angle, 371
Subtraction
  approximate numbers, 150
  decimal fractions, 45

557

degrees, minutes, seconds, 139
denominate numbers, 129, 131–134, 138–139
  fractions and mixed numbers, 32–33
  integers, 194–195
  monomials, 250–251
  on the number line, 193
  polynomials, 251
  positive and negative numbers, 194–195, 198–200
  properties, 197, 200–201
  rational numbers, 198–200
  whole numbers, 10–12
Subtrahend, 10
Successive discounts, 507
Supersonic speed, 544
Supplementary angles, 405–406
Symbols, 234–235, 236–237
  absolute value, 166–167

Tables
  amortizing a mortgage, 494
  annual percentage rate, 474
  compound interest, 497
  income tax, 468
  life expectancy, 478
  life insurance premium, 478
  measure, 547–549
  parcel post rates, 531
  squares and square roots, 547
  trigonometric values, 420
  withholding tax, 462
Take-home pay, 465–466, 467, 525
Tangent of angle, 416, 418
Tangent to a circle, 375, 394
Tax
  excise, 480
  income, 467
  property, 480
  sales, 479
  social security, 464
  wage, 464
  withholding, 461

Temperature, 140
Terminating decimal, 39
Term insurance, 477
Terms of a fraction, 19
Tests: *also see* **Reviews**
  Achievement, 83, 155, 229, 318, 450
  for divisibility, 93–95
  Inventory, 3, 108, 158, 358
Time
  measure of, 137
  24-hour clock, 142
Time cards, 524
Time zones, 143
Tip speed, 543
Tolerance, 146, 539
Total area
  cube, 435
  cylinder, 436–437
  rectangular solid, 435
Track (of airplane), 545
Trade discount, 507
Transforming formulas, 310
Transversal, 407
Trapezoid, 374
  area, 431–432
Triangle, 373
  acute, 373
  area, 430
  circumscribed, 399
  congruent, 374, 409
  construction, 390–393
  equiangular, 373
  equilateral, 373
  inscribed, 397
  isosceles, 373
  obtuse, 373
  perimeter, 423
  right, 373, 412, 416–420
  scalene, 373
  similar, 374, 410, 414
  sum of angles, 403
Trigonometry, numerical, 416–420
True course, 542
Twenty-four hour clock, 142
Twin primes, 91

Units of measure, 109

angles and arcs, 138
area, 123, 135
capacity
  dry, 121, 134
  liquid, 121, 132
changing to a larger unit, 111
changing to a smaller unit, 109
customary, 128
length, 114, 130
mass, 118
metric, 112–127
speed, 139
tables, 547–549
time, 137, 142
volume, 125, 135
weight, 118, 131

Variable, 235, 260, 266, 273, 308
Variance, 334
Variation
  direct, inverse, etc., 308
  magnetic, 542
Vector, 172–175, 388
Vertex, 373, 375, 376
Vertex angle, 373
Vertical angle, 406
Volt, 536
Volume
  cube, 439–440
  definition, 438
  measure of, 438–444
  pyramid, 443
  rectangular solid, 438
  right circular cone, 443
  right circular cylinder, 441
  sphere, 442

Wage tax, 464
Watt, 536
Weight, 112, 118, 131
Whole numbers, 6
  addition, 7–10
  division, 15–18
  multiplication, 12–14
  rounding, 6

shortened names, 54
subtraction, 10–12
Wind direction, 388, 389
Wind velocity, 388, 390
Withholding tax, 461

X-axis, 301
x-coordinate, 303

Y-axis, 301
y-coordinate, 303

Zero, 6, 92, 101–102
properties of, 101–102
Zero exponents, 256–259